Society of Earth Scientists Series

Series Editor

Satish C. Tripathi, Lucknow, India

The Society of Earth Scientists Series aims to publish selected conference proceedings, monographs, edited topical books/text books by leading scientists and experts in the field of geophysics, geology, atmospheric and environmental science, meteorology and oceanography as Special Publications of The Society of Earth Scientists. The objective is to highlight recent multidisciplinary scientific research and to strengthen the scientific literature related to Earth Sciences. Quality scientific contributions from all across the Globe are invited for publication under this series. Series Editor: Dr. Satish C. Tripathi

Binita Phartiyal • Rahul Mohan •
Supriyo Chakraborty • Venkatesh Dutta •
Anil Kumar Gupta
Editors

Climate Change and Environmental Impacts: Past, Present and Future Perspective

The Society of Earth Scientists
2008

Springer

Editors
Binita Phartiyal
Birbal Sahni Institute of Palaeosciences
Lucknow, Uttar Pradesh, India

Rahul Mohan
National Centre for Polar and Ocean Research
Vasco-da-Gama, Goa, India

Supriyo Chakraborty
Indian Institute of Tropical Meteorology
Pune, India

Venkatesh Dutta
School of Earth and Environmental Sciences
Babasaheb Bhimrao Ambedkar University
Lucknow, India

Anil Kumar Gupta
National Institute of Disaster Management,
Ministry of Home Affairs, Govt of India
New Delhi, India

ISSN 2194-9204 ISSN 2194-9212 (electronic)
Society of Earth Scientists Series
ISBN 978-3-031-13118-9 ISBN 978-3-031-13119-6 (eBook)
https://doi.org/10.1007/978-3-031-13119-6

This Springer imprint is published by the registered company Springer Nature Switzerland AG
The registered company address is: Gewerbestrasse 11, 6330 Cham, Switzerland

Series Editor's Foreword

Earth's climate varies even without human influence, but the acceleration in the changing pattern with cause and effect by/to the civilisation is a matter of concern to scientists. These patterns are lessons to understand future trends and ways and means for mitigation. The extreme weather events in almost every region of the globe involving excessive loss of human life and property are causing anxiety in society and posing challenges before scientists and planners. Cyclical variations in the Earth's climate occur at multiple timescales, from years to decades, centuries, and millennia. Cycles at each scale are caused by a variety of physical mechanisms. In the last 65 Ma only, there have been several cycles of glacial advances and retreat, with the abrupt end of the last ice age about 11,700 years ago marking the beginning of the modern climate era and human civilization. A multidisciplinary approach in studying the Earth's changing climate will provide a holistic view and guide us in future planning and programming.

In order to discuss these issues, an Expert Talk and Group Discussion followed by multi-domain international virtual conference was organised (13 October and15–17 October, 2020), wherein recent researches on palaeo-climatic changes and disasters, Quaternary climate variations and climate cycles, extreme weather events and meteorological studies, natural atmospheric climate forcing, ocean warming, coastal ecosystem, and disaster mitigation planning and management will be presented and the scope of mutual research cooperation in future will be discussed. COVID-19's impact on the environment and future strategies to retain a positive impact on the environment by sustainable development were presented. The outcome of this brainstorming is presented in this volume.

Our sincere acknowledgement to Prof. Dame Jane Francis, Director, British Antarctic Survey; Prof. Kim Holmén, International Director, Norwegian Polar Institute; Prof. A. Singhvi, Physical Research Laboratory; Sri Surya Sethi, Former Principal Adviser (Power & Energy), Plg. Com. & UNFCCC Negotiator; Sri Mukul Sanwal, Retd. IAS, Formerly with UNEP & UNFCCC Negotiator; Prof. R. Srikanth, NIAS, Bangalore; Dr. J R Bhat, Adviser, MoEF & CC; Dr. Akhilesh Gupta, Adviser, DST; Dr. R. Krishnan, IITM, Pune; Dr. Ajay Mathur, Director General, TERI;

Dr. Satish Shenoy, Former Director, INCOIS, Hyderabad; Dr. M. Ravichandran, Director, NCPOR, Goa; Prof. Ravi Shankar Nanjundiah, Director, IITM, Pune; Dr. M V Ramana Murthy, Director, NCCR; Prof. Anil Kulkarni, IISc, Bangalore; Prof. Chandra Venkataraman, IIT Bombay, for taking part in Pre-conference Expert Talks and International Group Discussions. Special thanks to Dr. Vandana Prasad, Director, BSIP, Lucknow, for all-round support in the virtual organisation of the conference. I also sincerely thank all the editors and contributors of this volume for bringing out this valuable scientific information and research.

The Society of Earth Scientists
Lucknow, India

Satish C. Tripathi

Contents

About the Editors

Binita Phartiyal has been a scientist at the Birbal Sahni Institute of Palaeosciences, Lucknow, India, since 2001. She has expertise in geomorphology, Quaternary climate, and neotectonics of the cold arid deserts. She received her early education from Nainital and a Ph.D. in 2000 from Kumaun University and was a DAAD fellow at the University of Tuebingen, Germany (1999–2000). Being a geologist by training, field work is her passion, and she has participated and led several expeditions to the Tethyan and Trans-Himalayan region regularly in the last two decades. She is the first Indian woman to have taken part in the Indian Scientific Expeditions to both the Poles (Antarctica and the Arctic). Dr. Phartiyal has made significant contribution towards quantitative reconstruction of Quaternary climate using sedimentary archives from various parts of Himalaya (especially Third Pole region (Ladakh)), Gujarat, and Polar Regions. She also helps in popularising Quaternary science and bringing the Quaternary scientist community in the Indian subcontinent over one platform through the Association of Quaternary Researchers.

Rahul Mohan is currently Scientist F and Group Director at the National Centre for Polar and Ocean Research (erstwhile NCAOR), Ministry of Earth Sciences, Goa, India. His research interests include oceanic micropaleontology and environmental geosciences with a special emphasis on paleoclimatology and paleoceanography with special interest in Southern Ocean, Antarctica lakes, and the Arctic as areas of current interest. With more than 27 years of research experience in oceanic micropaleontology and more than 100 published papers currently, he is part of the INSA-National Committee's of Scientific Committee on Antarctica Research (SCAR) and INSA-IUGS National Committee. Dr. Rahul is also the National Coordinator from India for the Asian Forum for Polar Sciences and a member of the SCAR CBET.

Supriyo Chakraborty is a senior scientist and is the Head of the Mass Spectrometry Group and the MetFlux India Project at the Indian Institute of Tropical Meteorology, Pune. He received his Ph.D. from the M.S. University of Baroda. He undertook postdoctoral work at the University of California at Santa Barbara and then at San Diego. His work is focused on monsoon reconstruction using the natural archives, isotope hydrology, moisture dynamical processes, ecosystem GHGs fluxes, and energy transfer processes at various natural ecosystems. He has about 30 years of research experience. He has guided students for Ph.D. and M.Sc. dissertations. He has about 80 peer-reviewed research papers to his credit. He is a recipient of the Certificate of Merit award by the Ministry of Earth Sciences, Government of India. He is a coordinating lead author and an editor of the book *Assessment of Climate Change over the Indian Region. A Report of the Ministry of Earth Sciences (MoES), Government of India* published in 2020 by Springer, Singapore.

Venkatesh Dutta is currently Professor at School of Earth and Environmental Sciences at Babasaheb Bhimrao Ambedkar Central University, Lucknow, India. His research interests include environmental management with a special emphasis on water policy, catchment planning, river restoration, and eco-hydrology. He has about twenty years of professional experience in areas of water quality management, land-use planning, and environmental impact assessment. Dr. Dutta has contributed to significant changes in legislation and policy for sustainable water resource management in India. As part of his public policy work, he has served as a member of the Expert Group for the formulation of the State Water Policy of Government of Uttar Pradesh in 2020. He is a Fulbright-Nehru Fellow and a British Chevening Scholar.

Anil Kumar Gupta is full Professor of India's National Institute of Disaster Management at New Delhi, and is a sustainability and risk management strategist working in the area of disaster management, environment, and climate resilience. He steered international and national projects of policy research - planning and capacity building and established new approaches, viz. ecoDRR, CCA-DRR integrated mainstreaming, Loss and Damage focused Climate Risk Management Framework, industrial, urban, business continuity resilience, and tools like DIA (with EIA/SEA, Auditing, and LCA), mitigation analysis, PDNA, etc. Dr. Gupta is credited with the preparation of national and sectoral disaster management plans, national human resource capacity development plans, national action plan for chemical disaster management, national drought management manual, etc. He has to his credit over 180 publications of international and national level. He was recipient of National Science Day Young Scientist Award in Forestry & Environment in 1996 and Excellence Award by the Society of Environmental and Occupational Health and bestowed with IDRC Canada's Thank Tank Initiative Senior Fellowship 2011 for policy research.

Floral Diversity and Climate Change in the Siwalik Succession

Harshita Bhatia, Gaurav Srivastava, R. C. Mehrotra, and Khum N. Paudayal

Abstract Siwalik sediments were deposited during the Himalayan orogeny in the Himalayan Foreland Basin. They were deposited in a coarsening upward succession to form the Siwalik Group which is further classified into Lower, Middle and Upper Siwalik. These sediments archive abundant plant fossils in the form of wood, leaves, flowers, fruits, seeds and pollen throughout the succession. These fossils have been used to understand the depositional environment of the Siwalik basin. In the Siwalik succession, the vegetation reconstruction indicates an increasing trend of deciduous forest taxa over the evergreen ones due to increased seasonality in rainfall and temperature. The quantitative estimation of climate based on plant megafossils indicates a monsoonal climate, particularly South Asia Monsoon, since the Lower Siwalik.

Keywords Plant fossils · Climate · Miocene · Neogene

1 Introduction

Siwalik has been fascinating for all the geoscientists because its sediments were derived from the Himalayan orogeny. The uplift of the Himalaya had a direct and indirect impact on the regional as well as global climate. The shaping of modern monsoon system and vegetation shift in south Asia are directly linked with the uplift history of the Himalaya (Ding et al. 2017; Srivastava et al. 2018c; Bhatia et al. 2021). Globally, the monsoonal climate is confined to low latitudes, except for a few places, and is a planetary phenomenon which occurs due to seasonal migration of the Intertropical Convergence Zone (ITCZ). The movement of ITCZ depends on the seasonal migration of insolation and is mainly modified by the land-ocean

H. Bhatia · G. Srivastava (✉) · R. C. Mehrotra
Birbal Sahni Institute of Palaeosciences, Lucknow, India

K. N. Paudayal
Central Department of Geology, Tribhuvan University, Kirtipur, Kathmandu, Nepal

B. Phartiyal et al. (eds.), *Climate Change and Environmental Impacts: Past, Present and Future Perspective*, Society of Earth Scientists Series,
https://doi.org/10.1007/978-3-031-13119-6_1

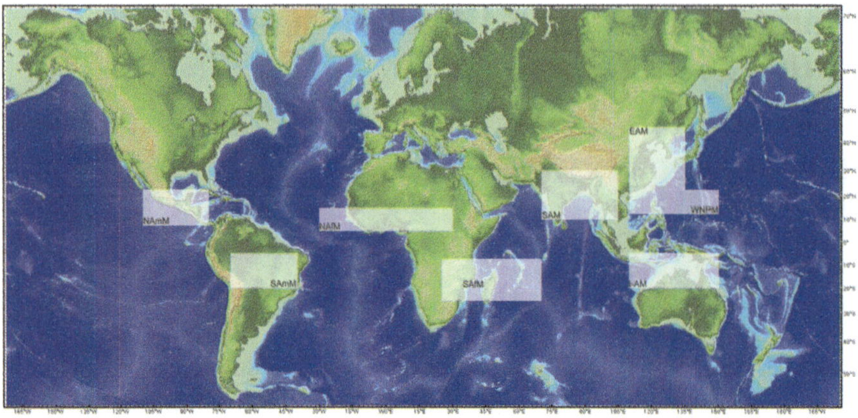

Fig. 1 Worldmap showing different monsoon domains namely North America Monsoon (NAmM), South America Monsoon (SAmM), North Africa Monsoon (NAfM), South Africa Monsoon (SAfM), South Asia Monsoon (SAM), East Asia Monsoon (EAM), Western North Pacific Monsoon (WNPM) and Indonesia-Australia Monsoon (I-AM) (after Wang et al. 2017)

configuration and topography of the specific region (Wang et al. 2017). Based on the regional characteristic of ITCZ, the monsoonal areas are classified into eight domains, namely South Asia Monsoon (SAM), Western North Pacific Monsoon (WNPM), East Asian Monsoon (EAM), Indonesian-Australian Monsoon (I-Am), North American Monsoon (NAmM), South American Monsoon (SAmM), North African Monsoon (NAfM) and South African Monsoon (SAfM) (Yim et al. 2014; Wang et al. 2017) (Fig. 1). The monsoon is characterised by seasonal reversal of surface wind, often associated with rainy summer and dry winter seasons (Wang et al. 2017). The Asian monsoon system (AMS) collectively consisting of three domains, namely SAM, EAM and WNPM, is the largest and strongest monsoon system on the earth. In AMS, SAM and EAM are continental monsoons, while WNPM is oceanic in nature (Wang et al. 2017). The circulation pattern of SAM is characterised by annual reversal of both zonal and cross equatorial wind, while EAM is delineated by annual reversal of meridional wind (Wang et al. 2017). The high orography in Asia such as Tibetan Plateau (TP) and Himalaya has a direct impact on the modern AMS (Molnar et al. 2010; Borah et al. 2020). The Himalaya acts as a mechanical barrier to insulate the warm moist air from extra tropical cool dry air, while the TP acts as an elevated "heat pump" in building the low pressure in warm season (Molnar et al. 2010). Recent climate modelling study suggests that non-elevated topography of northern India is important in generating characteristic SAM circulation (Molnar et al. 2010; Boos and Kuang 2013; Acosta and Huber 2020).

The biotic and abiotic proxy records from both continental and marine sediments indicate that the intensification of SAM is linked to the uplift of the Himalaya at ~8 Ma (Quade et al. 1989; Prell et al. 1992; Dettman et al. 2001; Zhisheng et al. 2001; Barry et al. 2002; Guo et al. 2002; Clift et al. 2008; Betzler et al. 2016;

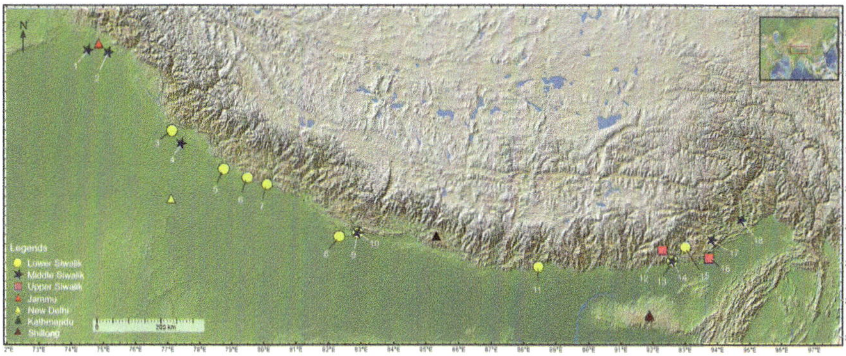

Fig. 2 Physiographic map of northern India showing fossil localities of Siwalik; (1) Bameal, Jammu and Kashmir; (2) Gandla, Jammu and Kashmir; (3) Ambala, Haryana; (4) Saharanpur, Uttar Pradesh; (5) Kalagarh, Uttar Pradesh; (6) Kathgodam, Uttarakhand; (7) Tanakpur, Uttarakhand; (8) Koilabas, Nepal; (9, 10) Surai Khola, Nepal; (11) Darjeeling, West Bengal; (12–15) Bhalukpong, Arunachal Pradesh; (16) Nirjuli, Arunachal Pradesh; (17) Kimin, Arunachal Pradesh; (18) Likabali, Arunachal Pradesh

Srivastava et al. 2018c; Bhatia et al. 2021) and tectonic setting of Indian Ocean gateway to the Mediterranean Sea (Bialik et al. 2019). The carbon and oxygen isotope data unearthed from the Siwalik sediments of western Himalaya reveal that vegetation shift from C_3 to C_4 occurred due to intensification of SAM which is linked to the uplift of the Himalaya during the late Miocene (Sanyal et al. 2004, 2010; Singh et al. 2011). However, recent isotopic and plant fossil records derived from the Siwalik reveal that winter rainfall, brought by the westerlies, also plays an important role in vegetation shift during the late Miocene (Vögeli et al. 2017; Srivastava et al. 2018c). The Himalayan Foreland Basin (HFB) was formed due to the uplift of the Himalaya which caused the deposition of fluvial muds, silts, sands, and gravels between the Lesser Himalaya in the north and the Gangetic Plains in the south, achieving a thickness of ~6 km. These sediments were deposited in the HFB in a coarsening upward succession since the middle Miocene to form the Siwalik Group extending all along the Himalaya from Sindh of Pakistan in the west to Arunachal Pradesh in the east (Fig. 2). Based on the sediment size, the Siwalik Group is further sub-divided into three sub-groups: mudstone dominated Lower, sandstone dominated Middle and conglomerate in the Upper Siwalik (Pilgrim 1910). The magnetostratigraphy of the Siwalik basin from different regions constrains the boundaries of the Lower, Middle and Upper Siwalik succession to be 18.3, 11, 5.3, and 0.22 Ma, respectively (Johnson et al. 1982, 1983; Tandon et al. 1984; Ranga Rao et al. 1988; Appel et al. 1991; Ranga Rao 1993; Gautam and Appel 1994; Sangode et al. 1996; Gautam and Fujiwara 1999; Chirouze et al. 2012).

In this communication, we synthesize the megafossil data excavated from the different sectors of the Siwalik (Fig. 2) to understand the relationship between climate change and vegetation shift caused due to the uplift of the Himalaya.

2 Siwalik Flora

The plant megafossils are archived all along the Siwalik sediments in the form of seeds, fruits, leaves, and wood. These fossils have been reported from the Lower, Middle and Upper Siwalik of India, Nepal and Bhutan (Fig. 2). In India, these fossils have been reported from Jammu and Kashmir, Himachal Pradesh, Uttarakhand, Uttar Pradesh, West Bengal and Arunachal Pradesh. The plant megaremains reported from the Siwalik sediments are as follows:

2.1 Lower Siwalik Flora

The Lower Siwalik flora is mainly reported from Himachal Pradesh, Uttarakhand, Uttar Pradesh, West Bengal and Arunachal Pradesh of India and Surai Khola of Nepal (Fig. 2). The most important nearest living relatives of the plant megafossils are: *Marantochloa* Brongn. ex Gris of the Marantaceae, *Cyclosorus* Link of the Thelypteridaceae, *Caryota* L. of the Arecaceae, *Bambusa* Schreb. of the Poaceae, *Smilax* L. of the Smilacaceae, *Gynocardia* R. Br. and *Hydnocarpus* Gaertn. of the Achariaceae, *Bouea* Meisn., *Dracontomelon* Bl., *Mangifera* L., *Nothopegia* Bl., *Swintonia* Griff. and *Tapiria* of the Anacardiaceae, *Cananga* Hook. f. & Thomson, *Fissistigma* Griff., *Goniothalamus* Hook. f. & Thomson, *Miliusa* Lesch. ex A. DC., *Mitrephora* Hook. f. & Thomson, *Polyalthia* Bl., and *Uvaria* L. of the Annonaceae, *Calophyllum* L. and *Kayea* Wall. of the Calophyllaceae, *Garcinia* L., and *Mesua* L. of the Clusiaceae, *Combretum* Loefl., *Getonia* Roxb. and *Terminalia* L. of the Combretaceae, *Mastixia* Bl. of the Cornaceae, *Dipterocarpus* Gaertn., *Hopea* Roxb., and *Shorea* Roxb. ex Gaertn. of the Dipterocarpaceae, *Dillenia* L. of the Dilleniaceae, *Diospyros* L. of the Ebenaceae, *Glochidion* Forst. & Forst., *Homonoia* Lour., and *Mallotus* Lour, of the Euphorbiaceae, *Acacia* Mill., *Albizzia* Durazz., *Bauhinia* Plum. ex L., *Caesalpinia* Plum. ex L., *Cassia* L., *Cynometra* L., *Dalbergia* L. f., *Derris* Lour., *Dialium* L., *Entada* Adans, *Ormosia* Jacks., *Pongamia* Adans., *Millettia* Wight & Arn., *Mucuna* Adans., and *Samanea* Merrs. of the Fabaceae, *Actinodaphne* Nees, *Cinnamomum* Schaeff., *Litsea* Lam., *Machilus* Nees, and *Persea* Mill. of the Lauraceae, *Lagerstroemia* L. and *Woodfordia* Salisb. of the Lythraceae, *Grewia* L., *Pterospermum* Schreb., and *Sterculia* L. of the Malvaceae, *Memecylon* L. of the Melastomataceae, *Chukrasia* A. Juss., *Dysoxylum* Bl., *Toona* Roem. and *Trichilia* Browne of the Meliaceae, *Artocarpus* Forst. & Forst. and *Ficus* Tourn. ex L. of the Moraceae, *Syzygium* Gaertn. of the Myrtaceae, *Bridelia* and *Phyllanthus* L. of the Phyllanthaceae, *Gardenia* Ellis and *Morinda* L. of the Rubiaceae, *Berchemia* Neck. ex DC. and *Ziziphus* Mill. of the Rhamnaceae, *Canthium* Lam., *Morinda* L., *Nauclea* L., and *Randia* L. of the Rubiaceae, *Evodia* Gaertn. and *Murraya* Koenig ex L. of the Rutaceae, *Filicium* Thwaites ex Bength. & Hk., *Nephelium* L., *Paranephelium* Miq. and *Xerospermum* Bl. of the Sapindaceae, and *Sarcosperma* Hk. f. of the Sapotaceae (Prasad and Prakash 1984; Prasad 1990a, b, 1994c, d, e, 2006; Prasad et al. 1997, 1999, 2004; Tripathi et al. 2002; Prasad and Dwivedi 2008) (Fig. 3).

Fig. 3 Representative fossil taxa from the Lower Siwalik sediments. (1) *Cananga tertiara* Prasad (2) *Combretum sahnii* Antal & Awasthi (3) *Polyalthium palaeosimiarum* Awasthi & Prasad (4) *Cleistanthus suraikholaensis* Prasad & Awasthi (5) *Nephelium palaeoglabrum* Prasad (6) *Shorea neoassamica* Prasad (7) *Gaertnera siwalica* Prasad (8) *Nothopegia eutravancoria* Antal & Awasthi (9) *Filicium koilabasensis* Prasad et al. (10) *Cynometra siwalika* Awasthi & Prasad (11) *Millettia chriensis* Prasad & Awasthi (12) *Terminalia koilabasensis* Prasad (13) *Albizzia microfolia* Prasad & Awasthi (All scale bars are of 1 cm, unless specified)

2.2 Middle Siwalik Flora

The Middle Siwalik flora is known from Jammu and Kashmir, Himachal Pradesh, Uttarakhand, West Bengal and Arunachal Pradesh of India, and Surai Khola of Nepal and Bhutan (Fig. 2). The most important and common nearest living relatives of the plant megaremains are: *Gynocardia* R. Br., *Hydnocarpus* Gaertn. of the Achariaceae, *Buchanania* Spreng., *Mangifera* L., *Sorindeia* Thouars of the Anacardiaceae, *Artabotrys* R. Br., *Fissistigma* Griff., *Meiogyne* Miq., *Pseuduvaria* Miq., and *Uvaria* L. of the Annonaceae, *Cerbera* L., *Chonemorpha* Don of the Apocynaceae, *Bombax* L. of the Bombaceae, *Calophyllum* L. of the Calophyllaceae, *Kokoona* Thwaites and *Lophopetalum* Wight ex Arn. of the Celastraceae, *Garcinia* L. of the Clusiaceae, *Terminalia* L. of the Combretaceae, *Dipterocarpus* Gaertn., *Hopea* Roxb., *Shorea* Roxb. ex Gaertn, and *Vatica* L. of the Dipterocarpaceae, *Diospyros* L. of the Ebenaceae, *Homonoya*, and *Mallotus* Lour. of the Euphorbiaceae, *Afzelia* Sm., *Albizia* Durazz., *Bauhinia* Plum. ex L., *Callerya* Endl., *Cassia* L., *Cynometra* L., *Dalbergia* L. f., *Intsia* Thouars, *Millettia* Wight & Arn., *Ormosia* Jacks., *Pongamia* Adans, *Sindora* Miq., and *Spatholobus* Hassk. of the Fabaceae, *Premna* L. of the Lamiaceae, *Beilschmiedia* Nees, *Lindera* Adans., *Litsea* Lam., *Machilus* Nees and *Persea* Mill. of the Lauraceae, *Duabanga* Buch. -Ham. and *Lagerstroemia* L. of the Lythraceae, *Sterculia* L. of the Malvaceae, *Chukrasia* Juss., *Dysoxylum* Bl. of the Meliaceae, *Artocarpus* Forst. & Forst., *Ficus* Tourn. ex L. of the Moraceae, *Syzygium* Gaertn. of the Myrtaceae, *Chionanthus* L. of the Oleaceae, *Bischofia* Bl. and *Glochidion* Forst. & Forst. and *Phyllanthus* L. of the Phyllanthaceae, *Myrsine* L. of the Primulaceae, *Rhamnus* L., *Ventilago* Gaertn. and *Ziziphus* Mill. of the Rhamnaceae, *Gardenia* Ellis, *Mitragyna* Korth., *Neolamarckia* Bosser and *Randia* L. of the Rubiaceae, *Geijera* Schott, *Murraya* Koenig ex L. and *Zanthoxylum* L. of the Rutaceae, *Sabia* Colebr. of the Sabiaceae, *Cupania* L., *Filicium* Thwaites and *Paranephelium* Miq. of the Sapindaceae, and *Vitis* L. of the Vitaceae (Prasad 1994a, b, 2010; Mehrotra et al. 1999, 2004; Khan and Bera 2012, 2014a, b, 2017; Prasad et al. 2013, 2015, 2017; Khan et al. 2019; Srivastava et al. 2018c) (Fig. 4).

2.3 Upper Siwalik Flora

The plant megafossils have been reported mainly from Arunachal Pradesh. Their NLRs are *Cyathea* Smith of the Cyatheaceae, *Bambusa* Schreb. of the Poaceae, *Gynocardia* R.Br. of the Achariaceae, *Polyalthia* Bl. of the Annonaceae, *Mangifera* L. of the Anacardiaceae, *Chonemorpha* Don of the Apocynaceae, *Canarium* L. of the Burseraceae, *Calophyllum* L. and *Kayea* Wall. of the Calophyllaceae, *Combretum* Loefl. and *Terminalia* L. of the Combretaceae, *Mastixia* Bl. of the Cornaceae, *Dipterocarpus* Gaertn. and *Shorea* Roxb. ex Gaertn. of the Dipterocarpaceae, *Elaeocarpus* Burm. ex L. of the Elaeocarpaceae, *Croton* L. and

Fig. 4 Representative fossil taxa from the Middle Siwalik sediments. (1) *Vitis siwalicus* Prasad et al. (2) *Buchanania palaeosessifolia* Prasad et al. (3) *Lophopetalumoxylon indicum* Mehrotra et al. (4) *Swintonia palaeoschwenckii* Prasad & Awasthi (5) *Hopea siwalika* Antal & Awasthi (6) *Bambusium arunachalense* Srivastava & Mehrotra (7) *Entada palaeoscandens* Awasthi &

Macaranga Thouars of the Euphorbiaceae, *Cynometra* L., *Dalbergia* L. f., *Millettia* Wight & Arn. and *Pongamia* Adans of the Fabaceae, *Quercus* L. of the Fagaceae, *Actinodaphne* Nees, *Lindera* Thunb. and *Litsea* Lam. of the Lauraceae, *Lagerstroemia* L. of the Lythraceae, *Dysoxylum* Bl. of the Meliaceae, *Knema* Lour of the Myristicaceae, and *Berchemia* Neck. ex DC. of the Rhamnaceae (Joshi et al. 2003a, b; Bera et al. 2004, 2014; Joshi and Mehrotra 2007; Bera and Khan 2009; Khan et al. 2011, 2014a, b, 2015, 2016, 2017a, b; Khan and Bera 2014b, 2016; Srivastava et al. 2018a, b) (Fig. 5).

3 Siwalik Climate

The Siwalik climate has been reconstructed using plant megaremains reported from different sectors (Fig. 2). The qualitative palaeoclimate reconstruction is based on the floristic assemblages, while the quantitative data is based on the methodologies such as Coexistence Approach (CA) (Utescher et al. 2014) and CLAMP (climate leaf analysis multivariate program) Analysis (Spicer et al. 2020).

The CA is based on the Nearest Living Relative (NLR) approach and can be applied to any fossil assemblage which includes leaves, fruits, wood, seeds and pollen (Utescher et al. 2014). The CA relies on the fact that the plant fossils have close affinity with their modern analogs and use the same climatic requirements of their NLRs to reconstruct the past climate. The CA is robust in the reconstruction of Neogene and Quaternary climates where in majority of the cases no significant change in climatic requirements of any taxon is expected (MacGinitie 1941; Hickey 1977; Chaloner and Creber 1990; Mosbrugger 1999; Utescher et al. 2014). The CA can robustly reconstruct seven climate variables such as MAT (mean annual temperature), CMMT (cold month mean temperature), WMMT (warm month mean temperature), MAP (mean annual precipitation), MPwet (mean precipitation during the wettest season), MPdry (mean precipitation during the driest season) and MPwarm (mean precipitation during the warm season). In contrast to the CA, the CLAMP analysis trusts on the relationship between leaf morphological traits and their prevailing climatic conditions. The CLAMP utilises morphological/physiognomy traits of the woody dicot leaves for paleoclimate reconstruction, which are ecologically specialised to adapt in all the seasons to perform maximum photosynthesis with minimum loss of water in the from transpiration (Givnish 1984). The CLAMP can reconstruct climate up to 100 Ma and their results have been validated by the oxygen isotopes (Spicer et al. 2003, 2020). It can robustly reconstruct MAT, CMMT, WMT, LGS (length of the growing season), GSP (growing season precipitation), MMGSP (mean monthly growing season precipitation), 3-WET

Fig. 4 (continued) Prasad (8) *Pongamia kathgodamensis* Prasad (9) *Sabia eopaniculata* Prasad (All scale bars are of 1 cm, unless specified)

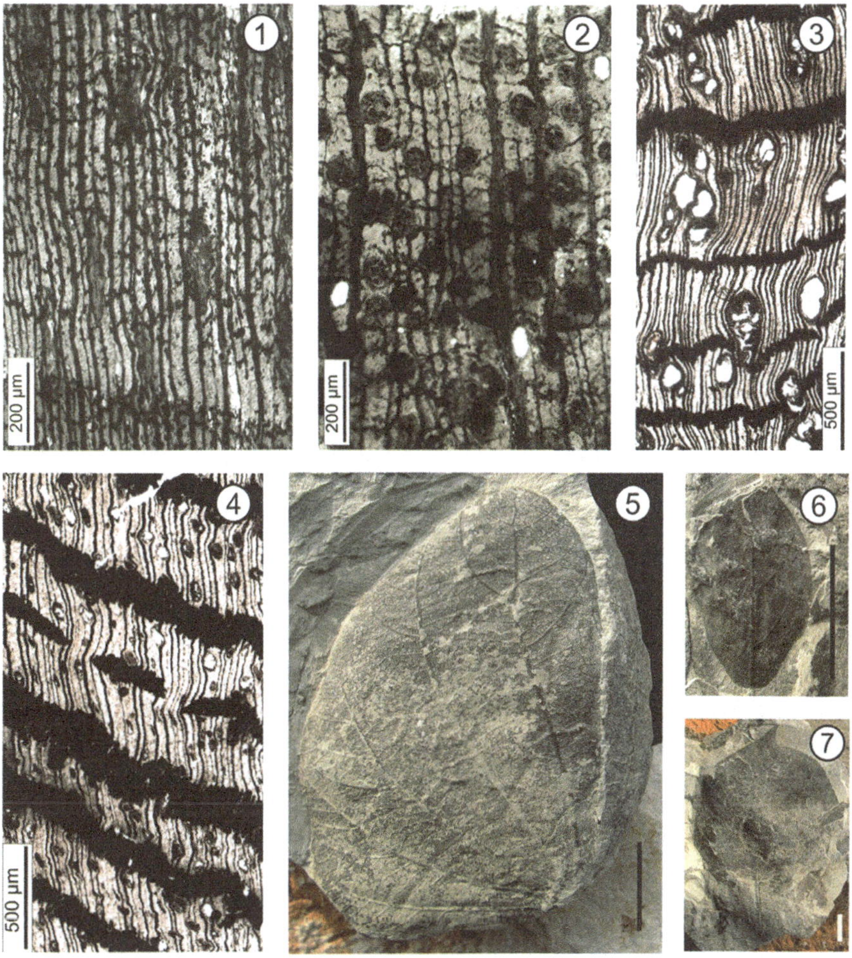

Fig. 5 Representative fossil taxa from the Upper Siwalik sediments. (1) *Ebenoxylon siwalicus* Prakash (2) *Polyalthioxylon arunachalensis* Srivastava, Mehrotra & Srikarni (3) *Calophyllum cuddalorense* Lakhanpal & Awasthi (4) *Kayeoxylon assamicum* Chowdhary & Tandon (5) *Bauhinia nepalensis* Awasthi & Prasad (6) *Breynia prerhamnoides* Awasthi & Prasad (7) *Diospyros miokaki* Hu & Chang (All scale bars are of 1 cm, unless specified)

(precipitation during the three wettest months), 3-DRY (precipitation during the three driest months), RH (relative humidity), SH (specific humidity), Enthalpy, minTempW_1 (minimum temperature of the warmest month), maxTempC_1 (maximum temperature of the coldest month), VPD.ann (mean annual vapour pressure deficit), VPD.Sum (mean summer vapour pressure deficit), VPD.Win (mean winter vapour pressure deficit), VPD.Spr (mean spring vapour pressure deficit), VPD.Aut (mean autumn vapour pressure deficit), annual PET (potential evapotranspiration), PETWarm_1 (mean month PET of the warmest quarter), PETCold_1 (mean monthly

PET of the coldest quarter), growing D0_div1000 (growing degree days above 0 °C) and growing D5 × 10^{-3} (growing degree days above 5 °C) (Spicer et al. 2020; Bhatia et al. 2021).

The CA has been used for the reconstruction of Lower and Middle Siwalik climate of Surai Khola, Nepal (Srivastava et al. 2018c) and Upper Siwalik climate of Arunachal Pradesh, India (Srivastava et al. 2021), while CLAMP has been used for the reconstruction of Lower and Middle Siwalik of Darjeeling and Arunachal Pradesh of India, and Surai Khola of Nepal and Bhutan (Khan et al. 2014a, b, 2019; Bhatia et al. 2021). The Lower and Middle Siwalik climate reconstruction of Surai Khola using CA indicates a MAT of 23.2 ± 2 °C and 26.5 ± 0.5 °C, CMMT of 22.45 ± 1.85 °C and 24.5 ± 0.2 °C, WMMT of 27.8 ± 0.3 °C and 27.25 ± 0.55 °C, MAP 2308.5 ± 560.5 mm and 2871.5 ± 279.5 mm, MPwet of 314 ± 0.5 mm and 277 ± 52 mm, MPdry of 90.5 ± 44.5 mm and 33.5 ± 25.5 mm, MPwarm of 174.5 ± 46.5 mm and 213.5 ± 7.5 mm, respectively (Table 1) (Srivastava et al. 2018c). The CLAMP analysis on the Lower and Middle Siwalik sediments of Surai Khola section indicates a MAT of 22.2 ± 2.3 °C and 24.7 ± 2.3 °C, CMMT of 14.7 ± 3.5 °C and 19 ± 3.5 °C, WMMT of 28.3 ± 3 °C and 28.5 ± 3 °C, GSP of 2130 ± 640 mm and 2320 ± 640 mm, 3-WET of 1050 ± 400 mm and 1210 ± 400 mm, 3-DRY of 160 ± 98 mm and 170 ± 98 mm, SH of 12.3 ± 1.7 g/kg and 15 ± 1.7 g/kg, and Enthalpy of 3430 ± 80 kJ/kg and 3550 ± 80 kJ/kg, respectively (Table 1) (Bhatia et al. 2021).

The CLAMP analysis has been used to reconstruct the climate of Lower, Middle and Upper Siwalik of Arunachal Pradesh (Khan et al. 2014a, b). The Lower, Middle and Upper Siwalik climate indicates a MAT of 25.3 ± 2.8 °C, 23.6 ± 2.8 °C and 25.4 ± 2.8 °C, CMMT of 21.3 ± 4 °C, 16.9 ± 4 °C and 20.8 ± 4 °C, WMMT of 27.8 ± 3.4 °C, 28.1 ± 3.4 °C and 28 ± 3.4 °C, MAP of 1741.3 ± 916.2 mm, 1981.2 ± 916.2 mm and 1898.6 ± 916.2 mm, MPwet of 961.5 ± 528 mm, 994.1 ± 528 mm and 1016.4 ± 528 mm, MPdry of 73.4 ± 115 mm, 137.8 ± 115 mm and 89.7 ± 115 mm, SH of 14.9 ± 2.1 g/kg, 14 ± 2.1 g/kg and 14.9 ± 2.1 g/kg, and Enthalpy of 355.8 ± 10.3 kJ/kg, 351.3 ± 10.3 kJ/kg and 356.1 ± 10.3 kJ/kg, respectively (Table 1). Moreover, the CA supports the above data and indicates a warm climate with plenty of rainfall during the deposition of the Upper Siwalik sediments of Arunachal Pradesh (Srivastava et al. 2021) (Table 1).

4 Discussion

4.1 Changing Pattern in the Siwalik Flora

The NLRs of the plant megafossils from the Siwalik sediments indicate the existence of tropical forest all along the HFB. The majority of taxa in these forests were evergreen, such as *Aglaia*, *Albizia*, *Anisoptera*, *Bauhinia*, *Calophyllum*, *Cassia*, *Cinnamomum*, *Cynometra*, *Dialium*, *Diospyros*, *Dipterocarpus*, *Dracontomelum*, *Duabanga*, *Dysoxylum*, *Gluta*, *Hopea*, *Koompassia*, *Litsea*, *Mallotus*, *Mangifera*,

Table 1 Showing comparison of the reconstructed quantitative climate data of the Lower, Middle and Upper Siwalik

Fossil assemblages	MAT (°C)	CMMT (°C)	WMMT (°C)	MAP/GSP (mm)	3-WET (mm)	3-DRY (mm)	SH (g/kg)	Enthalpy (kJ/kg)	MPwet (mm)	MPdry (mm)	MPwarm (mm)
Upper Siwalik (AP)[a]	25.4 2.8	20.8 ± 4	28 ± 3.4	1898.6 ± 916.2	1016.4 ± 528	89.7 ± 115	14.9 ± 2.1	356.1 ± 10.3	–	–	–
Upper Siwalik (AP)[b]	23.5 ± 0.5	22.8 ± 0.8	26.6 ± 1.5	2354 ± 41	–	–	–	–	531.3 ± 15.8	52.8 ± 7.8	168.5 ± 16.5
Middle Siwalik (AP)[a]	23.6 ± 2.8	16.9 ± 4	28.1 ± 3.4	1981.2 ± 916.2	994.1 ± 528	137.8 ± 115	14 ± 2.1	351.3 ± 10.3	–	–	–
Lower Siwalik (AP)[a]	25.3 ± 2.8	21.3 ± 4	27.8 ± 3.4	1741.3 ± 916.2	961.5 ± 528	73.4 ± 115	14.9 ± 2.1	355.8 ± 10.3	–	–	–
Middle Siwalik (SK)[c]	24.7 ± 2.3	19 ± 3.5	28.5 ± 3	2320 ± 640	1210 ± 400	170 ± 98	15 ± 1.7	3550 ± 80	–	–	–
Lower Siwalik (SK)[c]	22.2 ± 2.3	14.7 ± 3.5	28.3 ± 3	2130 ± 640	1050 ± 400	160 ± 98	12.3 ± 1.7	3430 ± 80	–	–	–
Middle Siwalik (SK)[d]	26.5 ± 0.5	24.5 ± 0.2	27.25 ± 0.55	2871.5 ± 279.5	–	–	–	–	277 ± 52	33.5 ± 25.5	213.5 ± 7.5
Lower Siwalik (SK)[d]	23.25 ± 2.15	22.45 ± 1.85	27.8 ± 0.3	2308.5 ± 560.5	–	–	–	–	314 ± 0.5	90.5 ± 44.5	174.5 ± 46.5

[a]CLAMP Analysis on Arunachal Pradesh (AP) assemblages (Khan et al. 2014a)
[b]Coexistence Approach on Arunachal Pradesh (AP) assemblage (Srivastava et al. 2021)
[c]CLAMP Analysis on Surai Khola (SK) assemblages (Bhatia et al. 2021)
[d]Coexistence Approach on Surai Khola (SK) assemblages (Srivastava et al. 2018c)

Ormosia, *Phoebe*, *Polyalthia*, *Pongamia*, *Shorea*, *Sindora*, *Sterculia*, *Swintonia* and *Syzygium*, etc. Some of these evergreen taxa such as *Anisoptera*, *Dipterocarpus*, *Gluta*, *Hopea*, *Koompassia*, *Sindora*, and *Swintonia* are important constituent of evergreen forests of Southeast Asia. It is interesting to note that megafloral assemblage from the Siwalik bears predominantly angiosperms followed by pteridophytes. There is only one gymnosperm genus *Pinus* with poorly preserved winged seed from the Lower Siwalik of Arunachal Pradesh (Khan and Bera 2017). However, palynological assemblages represent a sizable number of gymnosperms such as *Abies* Mill., *Cedrus* Trew, *Picea* Mill., *Pinus* L., *Podocarpus* L'Hér ex Pers., and *Tsuga*, along with several temperate angiosperms like *Alnus* Mill., *Betula* L., *Carya* Nutt. and *Juglans* L., etc. (Banerjee 1968; Lukose 1968; Nandi 1972, 1975; Mathur 1974; Saxena and Singh 1980, 1982a, b; Singh and Saxena 1980, 1981; Saxena et al. 1984; Singh and Sarkar 1984; Saxena and Bhattacharyya 1987) indicating that since the middle Miocene Himalaya attained a height which supported the growth of sub-tropical to temperate taxa.

A complete and uninterrupted sequence of Lower, Middle and basal part of the Upper Siwalik is exposed in Surai Khola of Nepal. The plant megafossils reported from there indicates a gradual change in the floristic composition that is linked to changing climatic conditions (Awasthi and Prasad 1990; Srivastava et al. 2018c). The Lower Siwalik (~13–11 Ma) megafossil assemblages dominantly represent wet evergreen taxa over the moist deciduous ones, along with the littoral and swampy taxa (Awasthi and Prasad 1990; Srivastava et al. 2018c). However, the Middle Siwalik (9.5–6.8 Ma) floristic assemblage shows a significant increase in moist deciduous taxa over the evergreen ones, with the complete absence of littoral and swampy taxa. The NLRs of Surai Khola assemblage suggest that the flora of Middle Siwalik was more diverse than that of the Lower Siwalik (Fig. 3) (Awasthi and Prasad 1990; Srivastava et al. 2018c).

4.2 Changing Pattern in the Siwalik Climate and the Establishment of Modern Monsoon

The CA and CLAMP have been used for the quantitative reconstruction of the palaeoclimate of Lower, Middle and Upper Siwalik of eastern and central Himalayan regions (Khan et al. 2014a; Srivastava et al. 2018c, 2021; Bhatia et al. 2021). The reconstructed climate data of both the regions indicates that the Siwalik sediments were deposited in a warm (tropical to sub-tropical) climate having plenty of rainfall (Table 1). The seasonality in temperature was less pronounced (Table 1). Further observations on the reconstructed data reveals that there was a decreasing trend in CMMT in the eastern Himalaya in contrast to the central Himalaya where an increasing trend existed (Table 1). Moreover, the MAP in both the regions showed an increasing trend from the Lower to Upper Siwalik. As far as the rainfall seasonality is concerned, it was present throughout the Siwalik in both eastern and central

Himalaya (Table 1). Both CA and CLAMP analysis indicates that a prominent wet summer and dry winter was present there.

In general, the monsoon can be defined as the seasonal reversal of surface wind, often associated with a prominent rainy summer and dry winter season (Webster 1987; Wang et al. 2017). A ratio of summer monsoon season (June–September/JJAS) rainfall and winter dry season (December–February/DJF) rainfall greater than 5:1 suggests a monsoon type of climate (Lau and Yang 1997; Zhang and Wang 2008) and has often been used in the estimation of deep time monsoon (Srivastava et al. 2012, 2018a, b; Khan et al. 2014a; Shukla et al. 2014; Ding et al. 2017). The reconstructed climate data from both the methodologies reveals that this ratio is greater than five in both eastern and central Siwalik regions (Table 1). This suggests that the Siwalik vegetation had enjoyed a monsoonal type of climate having a characteristic wet summer and dry winter season.

The prediction of monsoonal climate based on the ratio of wet summer and dry winter seasons does not always provide the exact picture about the monsoonal climate (Bhatia et al. 2021) as the areas having equable rainfall throughout the year can give low ratio of wet summer and dry winter seasons, despite having monsoonal type of climate (Khan et al. 2014a). To avoid such ambiguity in the estimation of monsoonal climate leaf physiognomy plays an important role in decoding the deep time monsoonal climate (Yang et al. 2015; Spicer et al. 2016, 2017). The leaves of woody dicot plants are directly exposed to their prevailing climate and are thus strategically tuned to perform maximum photosynthesis with minimum loss from the transpiration, respiration and structural investments. Therefore, these dicot leaves have structural adaptation in monsoonal and non-monsoonal climates. The CLAMP has ability to decode these adaptations in modern and fossil dicot leaves (Yang et al. 2015; Spicer et al. 2016, 2017). Recently, Bhatia et al. (2021) have used CLAMP methodology to decode the monsoonal behaviour of Lower and Middle Siwalik leaves unearthed from Surai Khola of Nepal. They have suggested that fossil leaves recovered from the Bankas (Lower Siwalik) and Chor Khola (Middle Siwalik) formations of Surai Khola have similar adaptations as seen in the modern vegetation sites mainly influenced by the SAM (Fig. 6). This suggests that modern SAM or something very similar to that, was already well established during the late middle Miocene (~13–11 Ma). The climate modelling studies have suggested that high topography of Himalaya and Tibetan Plateau had major influence in shaping the modern (Boos and Kuang 2010; Zhang et al. 2018) and ancient (Farnsworth et al. 2019) AMS. Acosta and Huber (2020) have also suggested the role of Asian orography in shaping the modern AMS. In Asia, Spicer et al. (2016, 2017) on the basis of fossil leaf morphology have suggested the prevalence of I–Am during the Paleogene in the absence or low elevation of Asian orography. The aforesaid discussion indicates that characteristic modern AMS was mainly shaped by the high orography in Asia.

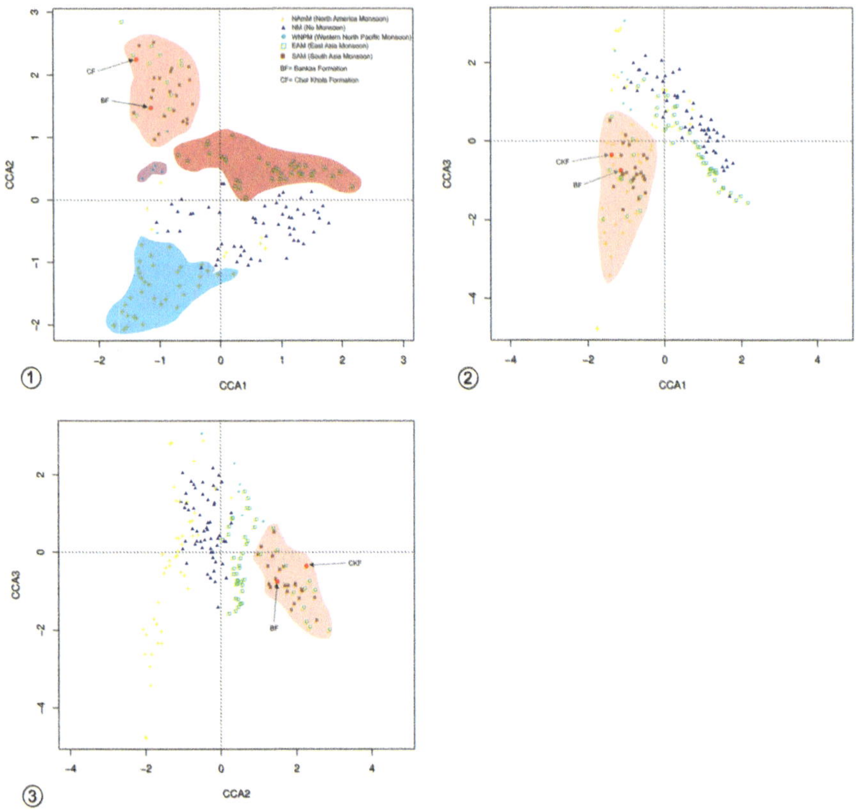

Fig. 6 (1–3) Canonical correspondence analysis plots showing the relationship between leaf physiognomy displayed by modern and fossil forests in the physiognomic space according to their corresponding climates. Study sites are clumped into Asian monsoons (light pink, dark pink and purple shading indicating Asian monsoons and blue shading representing North American monsoon (after Bhatia et al. 2021)

Acknowledgments GS, HB and RCM are grateful to the Director of the Birbal Sahni Institute of Palaeosciences, Lucknow for providing necessary facilities and encouragement to carry out this research work. This work was supported by the Science and Engineering Research Board, New Delhi (grant number CRG/2019/002461) sanctioned to GS.

References

Acosta RP, Huber M (2020) Competing topographic mechanisms for the summer Indo-Asian monsoon. Geophys Res Lett 47:e2019GL085112

Appel EW, Rosler W, Corvinus G (1991) Magnetostratigraphy of the Miocene Pleistocene Surai Khola Siwaliks in West Nepal. Int Geophys J 105:191–198

Awasthi N, Prasad M (1990) Siwalik plant fossils from Suria Khola area, western Nepal. Palaeobotanist 38:298–318

Banerjee D (1968) Siwalik microflora from Punjab, India. Rev Palaeobot Palynol 6:171–178

Barry JC, Morgan MLE, Flynn LJ, Pilbeam D, Behrensmeyer AK, Raza SM, Khan IA, Badgley C, Hicks J, Kelley J (2002) Faunal and environmental change in the late Miocene Siwaliks of northern Pakistan. Paleobiology 28:1–71

Bera S, Khan M (2009) Record of fruit and leaflet cf. *Pongamia pinnata* (L.) Pierre from the upper Siwalik sediments (Kimin formation) of Arunachal Pradesh. In: Mondal S, Bhattacharya S (eds) Advances in plant biology, D Bhattacharya birth centenary commemorative volume. Springer, Berlin, pp 432–441

Bera S, De A, De B (2004) First record of *Elaeocarpus* Linn. fruits from the upper Siwalik sediments (Kimin formation) of Arunachal Pradesh, India. J Geol Soc India 64:350–352

Bera S, Gupta S, Khan MA, De A, Mukhopadhyay R (2014) First megafossil evidence of Cyatheaceous tree fern from the Indian Cenozoic. J Earth Syst Sci 123(6):1433–1438

Betzler C, Eberli GP, Kroon D, Wright JD, Swart PK, Nath BN, Alvarez-Zarikian CA, Alonso-García M, Bialik OM, Blättler CL, Guo JA, Haffen S, Horozal S, Inoue M, Jovane LL, Luca L, Juan C, Mee ALH, Lüdmann T, Nakakuni M, Niino K, Petruny LM, Pratiwi SD, Reijmer JJG, Reolid J, Slagle AL, Sloss CR, Su X, Yao Z, Young JR (2016) The abrupt onset of the modern south Asian monsoon winds. Sci Rep 6:29838

Bhatia H, Srivastava G, Spicer RA, Farnsworth A, Spicer TEV, Mehrotra RC, Paudayal K, Valdes PJ (2021) Leaf physiognomy records the miocene intensification of the South Asia monsoon. Glob Planet Chang 196:103365

Bialik OM, Frank M, Betzler C, Zammit R, Waldamann ND (2019) Two-step closure of the Miocene Indian Ocean gateway to the Mediterranean. Sci Rep 9:8842

Boos WR, Kuang Z (2010) Dominant control of the south Asian monsoon by orographic insulation versus plateau heating. Nature 463:218–222

Boos WR, Kuang Z (2013) Sensitivity of the South Asian monsoon to elevated and non-elevated heating. Sci Rep 3:1192

Borah PJ, Venugopa V, Sukhatme J, Muddebihal P, Goswami BN (2020) Indian monsoon derailed by a North Atlantic wavetrain. Science 370:1335–1338

Chaloner WG, Creber GT (1990) Do fossil plants give a climatic signal? J Geol Soc Lond 147:343–350

Chirouze F, Dupont-Nivet G, Huyghe P, van der Beek P, Chakraborti T, Bernet M, Erens V (2012) Magnetostratigraphy of the Neogene Siwalik Group in the far eastern Himalaya: Kameng section, Arunachal Pradesh, India. J Asian Earth Sci 44:117–135

Clift PD, Hodges KV, Heslop D, Hannigan R, van Long H, Calves G (2008) Correlation of Himalayan exhumation rates and Asian monsoon intensity. Nat Geosci 1:875–880

Dettman DL, Kohn MJ, Quade J, Ryerson FJ, Ojha TP, Hamidullah S (2001) Seasonal stable isotope evidence for a strong Asian monsoon throughout the past 10.7 My. Geology 29:31–34

Ding L, Spicer RA, Yang J, Xu Q, Cai F, Li S, Lai Q, Wang H, Spicer TEV, Yue Y, Shukla A, Srivastava G, Khan MA, Bera S, Mehrotra RC (2017) Quantifying the rise of the Himalaya orogen and implications for the south Asian monsoon. Geology 45:215–218

Farnsworth A, Lunt DJ, Robinson SA, Valdes PJ, Roberts WHG, Clift PD, Markwick P, Tao S, Wrobel N, Bragg F, Kelland S-J, Pancost RD (2019) Past east Asian monsoon evolution controlled by paleogeography, not CO_2. Sci Adv 5:eaax1697

Gautam P, Appel E (1994) Magnetic-polarity stratigraphy of Siwalik Group sediments of Tinaukhola section in west-central Nepal. Int Geophys J 117:223–234

Gautam P, Fujiwara Y (1999) Magnetic polarity stratigraphy of Siwalik Group sediments from the Karnali River section in western Nepal. In: Sobel E, Appel E, Strecher M, Ratschbacher L, Blisniuk P (eds) Terra Nostra Koln. Kloster Ettal, Bavaria, Germany, pp 53–54

Givnish TJ (1984) Leaf and canopy adaptations in tropical forest. In: Medina E, Mooney HA, Vázquez-Yáñez C (eds) Physiological ecology of plants of the wet tropics. Junk, The Hague, pp 51–84

Guo ZT, Ruddiman WF, Hao QZ, Wu HB, Qian YS, Zhu RX, Peng SZ, Wei JJ, Yuan BY, Liu TS (2002) Onset of Asian desertification by 22 Ma ago inferred from loess deposits in China. Nature 416:159–163

Hickey LJ (1977) Stratigraphy and paleobotany of the Golden Valley Formation (early tertiary) of western North Dakota. Geol Soc Am Memoir 150:1–183

Johnson NM, Opdyke ND, Johnson GD, Lindsay EH, Tahirkheli RAK (1982) Magnetic-polarity stratigraphy and ages of Siwalik Group rocks of the Potwar Plateau, Pakistan. Palaeogeogr Palaeoclimatol Palaeoecol 37:17–42

Johnson GD, Opdyke NM, Tandon SK, Nanda AC (1983) The magnetic-Polaritystratigraphy of the Siwalik Groups at Haritalyangar, District Bilaspur. H.P. Himal Geol 12:118–144

Joshi A, Mehrotra RC (2007) Megaremains from the Siwalik sediments of West and East Kameng Districts, Arunachal Pradesh. J Geol Soc India 69:1256–1266

Joshi A, Mehrotra RC, De A (2003a) A fossil wood from the Upper Siwalik sediments of West Kameng District, Arunachal Pradesh, India. In: Proceedings of Fourth South Asia Geological Congress (GEOSAS – IV), The Director General Geological Survey of India, Kolkata, pp 312–315

Joshi A, Tewari R, Mehrotra RC, Chakraborty PP, De A (2003b) Plant remains from the Upper Siwalik sediments of West Kameng District, Arunachal Pradesh, India. J Geol Soc India 61(3): 319–324

Khan M, Bera S (2012) *Glochidion palaeogamblei* sp. nov.—a new fossil leaf of Euphorbiaceae from the Pliocene sediments of Arunachal Pradesh, Eastern India and its palaeoclimatic significance. In: Panda S, Ghosh C (eds) Diversity and conservation of plants and traditional knowledge. Bishen Singh Mahendra Pal Singh, Dehradun, pp 149–154

Khan M, Bera S (2014a) New lauraceous species from the Siwalik forest of Arunachal Pradesh, eastern Himalaya, and their palaeoclimatic and palaeogeographic implications. Turk J Bot 38: 453–464

Khan M, Bera S (2014b) On some Fabaceous fruits from the Siwalik sediments (Middle Miocene-Lower Pleistocene) of Eastern Himalaya, India. J Geol Soc India 83:165–174

Khan MA, Bera S (2016) First fossil evidence of Connaraceae R. Br. from Indian Cenozoic and its phytogeographical significance. J Earth Syst Sci 125:1079–1087

Khan MA, Bera S (2017) First discovery of fossil winged seeds of *Pinus* L. (family Pinaceae) from the Indian Cenozoic and its palaeobiogeographic significance. J Earth Syst Sci 126:63p

Khan MA, Ghosh R, Bera S, Spicer RA, Spicer TEV (2011) Floral diversity during Plio-Pleistocene Siwalik sedimentation (Kimin Formation) in Arunachal Pradesh, India, and its palaeoclimatic significance. Palaeobiodivers Palaeoenviron 91:237–255

Khan MA, Spicer RA, Bera S, Ghosh R, Yang J, Spicer TEV, Guo SX, Tao S, Frédéric J, Grote PJ (2014a) Miocene to Pleistocene floras and climate of the Eastern Himalayan Siwaliks, and new palaeoelevation estimates for the Namling–Oiyug Basin, Tibet. Glob Planet Change 113:1–10

Khan MA, Spicer TEV, Spicer RA, Bera S (2014b) Occurrence of *Gynocardia odorata* Robert Brown (Achariaceae, formerly Flacourtiaceae) from the Plio-Pleistocene sediments of Arunachal Pradesh, northeast India and its palaeoclimatic and phytogeographic significance. Rev Palaeobot Palynol 211:1–9

Khan MA, Bera S, Ghosh R, Spicer RA, Spicer TEV (2015) Leaf cuticular morphology of some angiosperm taxa from the Siwalik sediments (middle Miocene to lower Pleistocene) of Arunachal Pradesh, eastern Himalaya: systematic and palaeoclimatic implications. Rev Palaeobot Palynol 214:9–26

Khan MA, Spicer RA, Spicer TEV, Bera S (2016) Occurrence of *Shorea* Roxburgh ex C. F. Gaertner (Dipterocarpaceae) in the Neogene Siwalik forests of eastern Himalaya and its biogeography during the Cenozoic of Southeast Asia. Rev Palaeobot Palynol 233:236–254

Khan MA, Bera M, Spicer RA, Spicer TEV, Bera S (2017a) First occurrence of mastixioid (Cornaceae) fossil in India and its biogeographic implications. Rev Palaeobot Palynol 247: 83–96

Khan MA, Spicer RA, Spicer TEV, Bera S (2017b) Evidence for diversification of *Calophyllum* L. (Calophyllaceae) in the Neogene Siwalik forests of eastern Himalaya. Plants Syst Evol 303: 371–386
Khan MA, Bera M, Spicer RA, Spicer TEV, Bera S (2019) Floral diversity and environment during the middle Siwalik sedimentation (Pliocene) in the Arunachal sub-Himalaya. Palaeobiodivers Palaeoenviron 99(3):401–424
Lau KM, Yang S (1997) Climatology and interannual variability of the southeast Asian summer monsoon. Adv Atmos Sc 14:141–162
Lukose NG (1968) Microfossils from the Middle Siwalik of Bihar, India. J Palynol 4(2):107–112
MacGinitie HD (1941) A middle Eocene flora from the central Sierra Nevada, vol 534. Carnegie Institution of Washington Publication, Washington
Mathur AK (1974) A new seed (Boraginaceae) from the Siwalik Group. Bull Indian Geol Assoc 7(1):43–49
Mehrotra RC, Awasthi N, Dutta SK (1999) Study of fossil wood from the upper tertiary sediments (Siwalik) of Arunachal Pradesh, India and its implication in palaeoecological and phytogeographical interpretations. Rev Palaeobot Palynol 10:223–247
Mehrotra RC, Pande N, Ralimongla. (2004) Two fossil woods from Miocene sediments of Changki, Mokokchung district, Nagaland. Geophytology 32(1–2):79–82
Molnar P, Boos WR, Battisti DS (2010) Orographic controls on climate and paleoclimate of Asia: thermal and mechanical roles for the Tibetan Plateau. Annu Rev Earth Planet Sci 38:77–102
Mosbrugger V (1999) The nearest living relative method. In: Jones TP, Rowe NP (eds) Fossil plants and spores modern techniques. Geological Society of London, London, pp 261–265
Nandi B (1972) Some observations on the microflora of Middle Siwalik sediments of Mohand (East) Field, Himachal Pradesh. In: Proceedings of the Seminar on Palaeopalynology, Indian Stratigraphy, 1971. Kolkata, Department of Botany, University of Calcutta, pp 375–383
Nandi B (1975) Palynostratigraphy of the Siwalik Group of Punjab. Himal Geol 5:411–423
Pilgrim GE (1910) Preliminary note on a revised classification of the Tertiary fresh-water deposits of India. Records Geol Surv India 40:185–205
Prasad M (1990a) Fossil flora from the Siwalik sediments of Koilabas, Nepal. Geophytology 19: 79–105
Prasad M (1990b) Some more leaf impressions from the Lower Siwalik beds of Koilabas, Nepal. Palaeobotanist 37:299–315
Prasad M (1994a) Angiospermous leaf remains from the Siwalik sediments of Hardwar, Uttar Pradesh, and their bearing on palaeoclimate and phytogeography. Himal Geol 15:83–94
Prasad M (1994b) Siwalik (Middle-Miocene) woods from the Kalagarh area in the Himalayan foot hills and their bearing on palaeoclimate and phytogeography. Rev Palaeobot Palynol 76:49–82
Prasad M (1994c) Siwalik (Middle-Miocene) leaf impressions from the foot hills of the Himalaya, India. Tertiary Res 15:53–90
Prasad M (1994d) Morphotaxonomical study on angiospermous plant remains from the foot hills of Kathgodam, north India. Phytomorphology 44:115–126
Prasad M (1994e) Plant megafossils from the Siwalik sediments of Koilabas, central Himalaya, Nepal and their impact on palaeoenvironment. Palaeobotanist 42:126–156
Prasad M (2006) Plant fossils from Siwalik sediments of Himachal Pradesh and their palaeoclimatic significance. Phytomorphology 56:9–22
Prasad M (2010) Carbonised fossil woods from the Siwalik Group of Himachal Pradesh, India and their significance. J Palaeontol Soc India 55:23–28
Prasad M, Dwivedi HD (2008) Studies on plant megafossils from the Sub-himalayan zone (Siwalik) of western Nepal and their palaeoclimatic implications. J Palaeontol Soc India 53:51–64
Prasad M, Prakash U (1984) Leaf impressions from the lower Siwalik beds of Koilabas, Nepal. In: Proceeding of Indian Geophytological Conference, Lucknow, 1983. Special Publication, Lucknow, pp 246–256
Prasad M, Antal JS, Tiwari VD (1997) Investigation on plant fossils from Seria Naka in the Himalayan foot hills of Uttar Pradesh, India. Palaeobotanist 46:13–30

Prasad M, Antal JS, Tripathi PP, Pandey VK (1999) Further contribution to the Siwalik flora from the Koilabas area, western Nepal. Palaeobotanist 48:49–95
Prasad M, Ghosh R, Tripathi A (2004) Floristics and climate during the Siwalik (Middle Miocene) near Kathgodam in the Himalayan foothills of Uttaranchal, India. J Palaeontol Soc India 49:35–93
Prasad M, Mohan L, Singh SK (2013) First record of fossil leaves from Siwalik (Upper Miocene) sediments of Mandi District, Himachal Pradesh, India: palaeoclimatic and phytogeographical implications. Palaeobotanist 62:165–180
Prasad M, Kannaujia AK, Alok S, Singh SK (2015) Plant megaflora from the Siwalik (Upper Miocene) of Darjeeling District, West Bengal, India and its palaeoclimatic and phytogeographic significance. Palaeobotanist 64:13–94
Prasad M, Kumar S, Singh SK, Pandey SM, Alok (2017) Fossil woods from the Middle Siwalik (Upper Miocene) of Tanakpur area in the Himalayan foot hills of Uttarakhand and their palaeoclimatic significance. Geophytology 47:161–191
Prell WL, Murray WD, Clemens SC, Anderson DM (1992) Evolution and variability of the Indian Ocean summer monsoon: evidence from the western Arabian Sea drilling program. In: Duncan RA, Rea DK, Kidd RB, von Rad U, Weissel JK (eds) Synthesis of results from scientific drilling in the Indian Ocean, vol 70. Geophysical Monograph, American Geophysical Union, Washington, DC, pp 447–469
Quade J, Cerling TE, Bowman JR (1989) Development of Asian monsoon revealed by marked ecological shift during the latest Miocene in northern Pakistan. Nature 342:163–166
Ranga Rao A (1993) Magnetic-polarity stratigraphy of Upper Siwaliks of Northwestern Himalayan foothills. Curr Sci 64:863–872
Ranga Rao A, Agarwal RP, Sharma UN, Bhalla MS, Nanda AC (1988) Magnetic polarity stratigraphy and vertebrate palaeontology of the Upper Siwalik subgroup of Jammu Hills, India. J Geol Soc India 31:361–385
Sangode SJ, Kumar R, Ghosh SV (1996) Magnetic-polarity stratigraphy of the Siwalik sequence of Haripur area (H.P), NW Himalaya. J Geol Soc India 47:683–705
Sanyal P, Bhattacharya SK, Kumar R, Ghosh SK, Sangode SJ (2004) Mio-Pliocene monsoonal record from Himalayan foreland basin (Indian Siwalik) and its relation to vegetational change. Palaeogeogr Palaeoclimatol Palaeoecol 205:23–41
Sanyal P, Sarkar A, Bhattacharya SK, Kumar R, Ghosh SK, Agrawal S (2010) Intensification of monsoon, microclimate and asynchronous C4 appearance: isotopic evidence from the Indian Siwalik sediments. Palaeogeogr Palaeoclimatol Palaeoecol 296:165–173
Saxena RK, Bhattacharyya AP (1987) Palynology of the Siwalik Sediments of Kala Amb-Nahan area in Sirmaur District, Himachal Pradesh. Palaeobotanist 35:187–195
Saxena RK, Singh HP (1980) Occurrence of palynofossils from the Pinjor Formation (Upper Siwalik) exposed near Chandigarh. Curr Sci 49(12):479–480
Saxena RK, Singh HP (1982a) Palynology of the Pinjor Formation (Upper Siwalik) exposed near Chandigarh, India. Palaeobotanist 30:325–339
Saxena RK, Singh HP (1982b) Palynological investigation of the Upper Siwalik sediments exposed along Hoshiarpur-Una Road Section in Punjab and Himachal Pradesh. Geophytology 12:287–306
Saxena RK, Sarkar S, Singh HP (1984) Palynological investigation of Siwalik sediments of Bhakra Nangal area Himachal Pradesh. Geophytology 14:178–198
Shukla A, Mehrotra RC, Spicer RA, Spicer TEV, Kumar M (2014) Cool equatorial terrestrial temperatures and the South Asian monsoon in the Early Eocene: evidence from the Gurha Mine, Rajasthan, India. Palaeogeogr Palaeoclimatol Palaeoecol 412:187–198
Singh HP, Sarkar S (1984) A Kasauli palynoflora from Banelhi area of Himachal Pradesh, India. Geophytology 14:40–51
Singh HP, Saxena RK (1980) Upper Siwalik palynoflora from Gagrel-Bharwain Road section, Himachal Pradesh. Geophytology 10:278–229

Singh HP, Saxena RK (1981) Palynology of the Upper Siwalik sediments in Una District, Himachal Pradesh. Geophytology 11:173–181

Singh S, Parkash B, Awasthi AK, Kumar S (2011) Late Miocene record of palaeo-vegetation from the Siwalik palaeosols of the Ramnagar sub-basin. Curr Sci 100:213–222

Spicer RA, Harris NBW, Widdowson M, Herman AB, Guo S-X, Valdes PJ, Wolfe JA, Kelley SP (2003) Constant elevation of southern Tibet over the past 15 million years. Nature 421:622–624

Spicer RA, Yang J, Herman AB, Kodrul T, Maslova N, Spicer TEV, Aleksandrova GN, Jin J (2016) Asian Eocene monsoons as revealed by leaf architectural signatures. Earth Planet Sci Lett 449: 61–68

Spicer RA, Yang J, Herman AB, Kodrul T, Aleksandrova G et al (2017) Paleogene monsoons across India and South China: drivers of biotic change. Gondwana Res 49:350–363

Spicer RA, Yang J, Spicer TEV, Farnsworth A (2020) Woody dicot leaf traits as a palaeoclimate proxy: 100 years of development and application. Palaeogeogr Palaeoclimatol Palaeoecol 757. https://doi.org/10.1016/j.palaeo.2020.110138

Srivastava G, Spicer RA, Spicer TEV, Yang J, Kumar M, Mehrotra RC, Mehrotra NC (2012) Megaflora and palaeoclimate of a late Oligocene tropical delta, Makum Coalfield, Assam: evidence for the early development of the South Asia Monsoon. Palaeogeogr Palaeoclimatol Palaeoecol 342–343:130–142

Srivastava G, Mehrotra RC, Srikarni C (2018a) Fossil wood flora from the Siwalik Group of Arunachal Pradesh, India and its climatic and phytogeographic significance. J Earth Syst Sci 127:6

Srivastava G, Mehrotra RC, Srikarni C (2018b) *Lagerstroemia* L. wood from the Kimin Formation (Upper Siwalik) of Arunachal Pradesh and its climatic and phytogeographic significance. J Geol Soc India 91:695–699

Srivastava G, Paudayal KN, Utescher T, Mehrotra RC (2018c) Miocene vegetation shift and climate change: evidence from the Siwalik of Nepal. Glob Planet Change 161:108–120

Srivastava G, Farnsworth A, Bhatia H, Mehrotra RC, Shekhar M, Tao S, Utescher T, Valdes PJ (2021) Climate and vegetation change during the Upper Siwalik: a study based on the palaeobotanical record of the eastern Himalaya. Palaeodivers Palaeoenviron. https://doi.org/10.1007/s12549-020-00457-w

Tandon SK, Kumar R, Koyama M, Niitsuma N (1984) Magnetic-polarity Stratigraphy of the Upper Siwalik east of Chandigarh, Panjab sub-Himalaya. J Geol Soc India 25:45–55

Tripathi PP, Pandey SM, Prasad M (2002) Angiospermous leaf impressions from Siwalik sediments of the himalayan foot hills near Jarva, U.P. and their bearing on palaeoclimate. Biol Memoirs 28: 79–90

Utescher T, Bruch AA, Erdei B, François L, Ivanov D, Jacques FMB, Kern AK, Liu Y-S, Mosbrugger V, Spicer RA (2014) The coexistence approach—theoretical background and practical considerations of using plant fossils for climate quantification. Palaeogeogr Palaeoclimatol Palaeoecol 410:58–73

Vögeli N, Najman Y, van der Beek P, Huyghe P, Wynn PM, Govin G, van der Veen I, Sachse D (2017) Lateral variations in vegetation in the Himalaya since the Miocene and implications for climate evolution. Earth Planet Sci Lett 471:1–9

Wang PX, Wang B, Cheng H, Fasullo J, Guo Z-T, Liu Z-Y, Kiefer T (2017) The global monsoon across time scales: mechanisms and outstanding issues. Earth Sci Rev 174:84–121

Webster PJ (1987) The variable and interactive monsoon. In: Fein JS, Stephens P (eds) Monsoon. Wiley, New York, pp 268–330

Yang J, Spicer RA, Spicer TEV, Arens NC, Jacques FMB, Su T, Kennedy EM, Herman AB, Steart DC, Srivastava G, Mehrotra RC, Valdes PJ, Mehrotra NC, Zhou Z, Lai J (2015) Leaf form–climate relationships on the global stage: an ensemble of characters. Glob Ecol Biogeogr 24: 1113–1125

Yim S-Y, Wang B, Liu J, Wu Z (2014) A comparison of regional monsoon variability using monsoon indices. Clim Dyn 43:1423–1437
Zhang S, Wang B (2008) Global summer monsoon rainy seasons. Int J Climatol 28:1563–1578
Zhang R, Jiang D, Ramstein G, Zhang Z, Lippert PC, Yu E (2018) Changes in Tibetan Plateau latitude as an important factor for understanding East Asian climate since the Eocene. A modelling study. Earth Planet Sci Lett 484:295–308
Zhisheng A, Kutzbach JE, Prell WL, Porter SC (2001) Evolution of Asian monsoons and phased uplift of the Himalaya–Tibetan plateau since Late Miocene times. Nature 411:62–66

Early Paleogene Megaflora of the Palaeoequatorial Climate: A Case Study from the Gurha Lignite Mine of Rajasthan, Western India

Kajal Chandra, Anumeha Shukla, and R. C. Mehrotra

Abstract The early Paleogene was a crucial time for the diversification of the Indian flora, as palaeogeographically the Indian subcontinent was situated near the equator (~10°N), and was moving northwards towards Eurasia. In this chapter we mainly focus on the Gurha lignite mine as it is very rich in plant remains. It is situated ~47 km WSW of Bikaner in western Rajasthan. The plant fossils of this mine are late Paleocene-early Eocene in age and show similarities with the evergreen taxa largely constituents of tropical wet evergreen forests of the Western Ghats (south India) and experience a high annual precipitation. Further, the presence of legumes and a few drier elements is indicative of monsoon driven seasonality in and around Bikaner. In contrast, at present the fossil locality falls in the Thar Desert and experiences a harsh arid climate.

Keywords Plant megafossils · Climate change · Thar Desert · Bikaner · Late Paleocene-early Eocene

1 Introduction

The longest voyage of the Indian subcontinent started when it separated from the Gondwanaland around 140 Ma and finally from Madagascar ~90 Ma ago (Chatterjee and Scotese 1999). Coinciding with its changing latitudinal position, the Indian subcontinent had gone through several climatic fluctuations that affected its biota since the Cretaceous (Morley 2003). The Indian subcontinent entered the tropical climate zone in the Late Cretaceous that changed its core vital elements and finally paved the way for the establishment of existing vegetation. Nowadays the Indian subcontinent consists of diverse biomes existing in the different climatic scenarios. In the Paleogene, the earth witnessed the warm greenhouse climate, marked by the close interaction between global climate and biological evolution. During this time

K. Chandra · A. Shukla (✉) · R. C. Mehrotra
Birbal Sahni Institute of Paleosciences, Lucknow, India

B. Phartiyal et al. (eds.), *Climate Change and Environmental Impacts: Past, Present and Future Perspective*, Society of Earth Scientists Series,
https://doi.org/10.1007/978-3-031-13119-6_2

frame, the Indian subcontinent was positioned near the equator, experiencing the warm and humid climate, which supported the luxurious growth of tropical elements eventually establishing a lush evergreen forest. Any spatial and temporal change in climatic parameters is tend to affect the distribution, structure and ecology of any forest and if the climate change is severe or abrupt, the plant species will be unable to adapt itself and will become locally extinct (Parmesan and Yohe 2003; Sykes 2009). A picture-perfect model for the effects of climate change on the plant communities can be seen by studying the plant fossils assemblage of Cenozoic age. Cenozoic plant relicts that were once widespread across the entire Indian subcontinent, have migrated and are now restricted to some parts of the Western Ghats as refugia having suitable conditions for their growth. Fossil floral analysis and correlations between leaf physiognomic traits and climate are widely used to estimate palaeoclimate and serve as effective proxies to understand the response of plants to climatic change provided the requirements of their Nearest Living Relatives (NLRs) are known. Our attempt in this study provides a useful concept on regional climatic influence on plants in view of climatic forcing prior to their deposition and burial in sediments of Gurha lignite mine.

The early Paleogene was a crucial time for the diversification of the Indian flora, as palaeogeographically the Indian subcontinent was situated near the equator (~10°N), and was moving northwards towards Eurasia (Molnar and Stock 2009). The Paleogene was also a time of profound reorganization in the biosphere, following the important Cretaceous-Paleocene (K-P) catastrophic extinction characterised by global warmth (Zachos et al. 2001, 2003; Eberle and Greenwood 2012) preceded by the Paleocene–Eocene Thermal Maximum or PETM. A more persistent early Eocene 'climatic optimum', that lasted for several million years until around 52 Ma (Zachos et al. 2001, 2008; Lowenstein and Demicco 2006; Kroeger et al. 2011) with a significant impact on the environmental evolution of terrestrial taxa, is followed by a gradual cooling. The paleobotanical investigations dealing with paleoclimatic and paleobiogeographic implications of the plant based proxies i.e. plant megafossils (leaves, woods, seeds etc.) have been carried out and reviewed by many researchers (Shukla and Mehrotra 2013; Mehrotra et al. 2014; Srivastava et al. 2014; Shukla and Mehrotra 2018).

With this background, here in this chapter we mainly focus on:

What is the importance plant megafossils recovered from early of Paleogene Gurha lignite mine in context with paleoclimate the Indian subcontinent?

2 Study Site

Authors have been making a huge collection of fossil leaves, flowers and fruits from the Gurha lignite mine of Rajasthan for the past many years. It is an open cast mine located (27°52′26″ N; 72°52′20″ E; Fig. 1) in western Rajasthan, ~47 km WSW of Bikaner. The lignite of this mine belongs to the Palana Formation (Paleogene) which

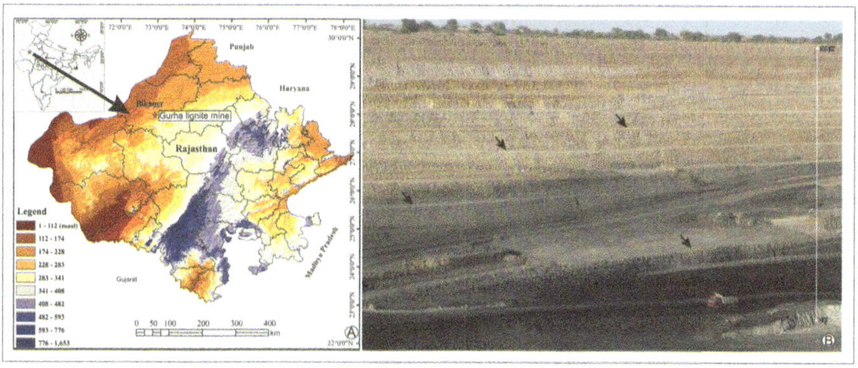

Fig. 1 (**a**) Shuttle Radar Topographic Mission (SRTM) digital elevation model (DEM) of Rajasthan (red star showing the location of Gurha lignite mine (fossil locality)). (**b**) Field photograph showing exposed sedimentary section from where plant fossils have been collected (indicated by arrows)

is exposed near Kolayat and Nagaur and according to Khosla (1971) and Pareek (1984) this formation is of the Paleocene age. Based on palynomorph assemblage and fish fossils, the Paleocene age for this formation has also been advised by Kulshreshtha et al. (1989) and Kumar et al. (2005). Although an early Eocene age was proposed for the Gurha lignite mine based on its palynoassemblage and its comparison with that of Vastan lignite mine (Shukla et al. 2014), yet the Paleocene age seems more suitable for this formation because of the overlying Marh/Kolayat Formation which is considered to be late Paleocene-early Eocene on the basis of planktonic foraminifers (Singh 1971Kalia and Sharma 1985) and lower vertebrates which are quite similar to those from the early-middle Eocene (~47 My BP) Subathu Formation of northwest Outer Himalaya (Jolly and Loyal 1985; Kumar and Loyal 1987). Detailed palynological studies being carried out on various lignite mines from Bikaner and Barmer areas also support an older age (Prasad et al. 2020). Sedimentologically, this mine is characterized by the association of grey clay, silty clay, sand, lignite and volcanic ash. The sedimentary succession has been studied and described in detail by Shukla et al. (2014) (Fig. 2). The plant fossils were collected from light to dark, banded clays overlying the lignite beds. Fossil leaves, flowers and legume fruits were found in abundance at ~70–72 m level from bottom to top (Shukla et al. 2014). Most of the exposed section of this mine represents the deposition under lacustrine system including small fluvial channels, ponds, fluvial plains, swamps and abandoned channels. The fossil leaf flora was preserved in horizontally laminated/ poorly laminated fine grained clays/silts (Fig. 2). The sedimentation of this mine begins with a significant influx of volcanic ash that is now altered to clay. The lignite deposits are more or less uniform containing abundant amber and charcoal particles which are dispersed throughout (Shukla et al. 2014).

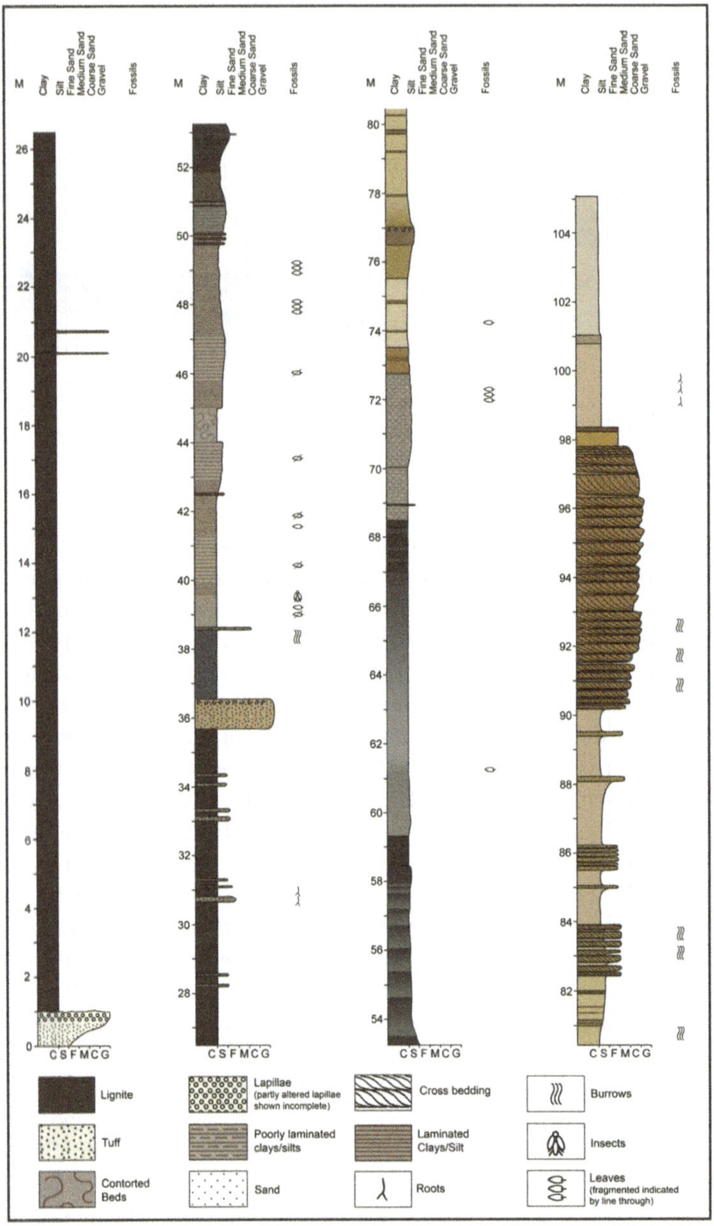

Fig. 2 Generalised litholog of the Gurha lignite mine (after Shukla et al. 2014) showing different stratigraphic levels

3 Early Paleogene Biodiversity of the Gurha Mine

The Paleocene-Eocene position of the Indian subcontinent favoured luxuriant growth of the tropical evergreen elements. The Gurha lignite mine is known for its remarkable plant fossil assemblage, including both mega and micro fossils (Shukla and Mehrotra 2018). From this mine, Kumar et al. (2016) have described taxonomically highly diverse angiosperm pollen signifying a rich near-coastal tropical environment under a strong seasonal precipitation regime (Shukla et al. 2014). The abundant and diverse terrestrial palynoflora from this mine includes fresh water algae (*Botryococcus braunii, Debarya* and *Spirogyra,* etc.), pteridophytes and angiosperms belonging to Arecaceae, Gunneraceae, Bombacaceae, Thymelaeaceae, Calophyllaceae, Proteaceae, and Acanthaceae (Kumar et al. 2016).

Plant megafossil assemblage from the Gurha lignite mine of western India includes leaves and fruits, and has been summarised in Table 1 and shown in Figs. 3 and 4. Fossil leaves resembling *Aporosa acuminata* Thw. (Phyllanthaceae) (Shukla et al. 2016), *Uvaria zeylanica* L. (Annonaceae) (Shukla and Mehrotra 2014), *Syzygium* sp., (Myrtaceae), *Gardenia* sp. (Rubiaceae), *Pterygota alata* (Roxb.) R. Br. (Malvaceae) and *Holigarna grahamii* (Anacardiaceae) (Shukla et al. 2018) and a monocot leaf resembling *Dioscorea pubera* Blume (Dioscoreaceae) (Shukla and Mehrotra 2014) are described from this mine. Apart from these fossil leaves, a few fossil fruits i.e. *Cajanus crassus* (King) Maesen, *Saraca asoca* (Roxb.) Wilde (Fabaceae) (Shukla and Mehrotra 2016) and *Mallotus* (Shukla and Mehrotra 2019) have also been reported from this mine.

4 Paleoclimate Versus Modern Climate of the Study Site

The biogeographical distribution and plant species richness are supposed to be correlated with the climatic conditions of the region. Due to the tectonic movement from the south of equator to the north and its collision with Eurasia, the Indian plate experienced a drastic change in the climatic conditions. The plant fossils provide direct evidence of the past life. The studies have shown that the evolution and diversification of tropical rainforests took place somewhere during the Late Cretaceous and/or early Paleogene time and were significantly influenced by the plate tectonics (Givnish and Renner 2004). The conventional approach to Cenozoic paleobotany and analysis of the climatic implications of floristic change is based on an assumption of relative stability of the taxon-climate relationship. For the reconstruction of paleoclimate, the nearest living relative (NLR) approach is being used since 1953 (MacGinitie 1953), and in this approach the fossil flora is matched to the climatic conditions of an area containing a large number of NLRs as the climatic tolerances of the NLRs is assumed to mirror their relics in the earlier geological time. Most of the fossils mentioned in Table 1 show the similarities with the evergreen taxa largely constituents of tropical wet evergreen forests of the

Table 1 List of megafossils recovered from the Gurha lignite mine, Bikaner, Rajasthan

Sr. no.	Name of the fossil taxa	Nearest living relatives (NLRs)	Habitat	Biogeography of NLRs
1.	*Uvaria palaeozeylanica* Shukla and Mehrotra (2014) (leaf)	*Uvaria zeylanica* L. (Annonaceae)	Tropical rainforest	Endemic to South India and Sri Lanka
2.	*Aporosa ecocenicus* Shukla et al. (2016) (leaf)	*Aporosa acuminata* Thwaites (Phyllanthaceae)	Tropical evergreen forest	Endemic to the Western Ghats
3.	*Leguminocarpon cajanoides* Shukla and Mehrotra (2016) (fruit)	*Cajanus crassus* (King) Maesen (Fabaceae)	Tropical semi-arid forest	Northeast India, Central India
4.	*Leguminocarpon saracoides* Shukla and Mehrotra (2016) (fruit)	*Saraca asoca* (Roxb.) Wilde (Fabaceae)	Tropical rainforest	Native to India
5.	*Gardenia eocenicus* Shukla et al. (2018) (leaf)	*Gardenia gummifera* L.f. (Rubiaceae)	Deciduous forest	Throughout India
6.	*Holigarna palaeograhamii* Shukla et al. (2018) (leaf)	*Holigarna grahamii* (Wight) Kurz (Anacardiaceae)	Wet evergreen forest	Endemic to the Western Ghats
7.	*Pterygota eocenica* Shukla et al. (2018) (leaf)	*Pterygota alata* (Roxb.) R.Br. (Malvaceae)	Evergreen and semi-evergreen forests	Northeast India (particularly Assam) and the Western Ghats
8.	*Syzygium gurhaensis* Shukla et al. (2018) (leaf)	*Syzygium cumini* L. (Myrtaceae)	Evergreen forest	Throughout India
9.	*Dioscorea eocenicus* Mehrotra and Shukla (2019) (leaf)	*Dioscorea pubera* Blume (Dioscoreaceae)	Evergreen and moist deciduous forests	India, Myanmar, Thailand and China
10.	*Mallotocarpon gurhaensis* Shukla and Mehrotra (2019) (fruit)	*Mallotus mollissimus* (Geiseler) airy Shaw (Euphorbiaceae)	Deciduous and evergreen forests	Sumatra to East Australia and West Pacific (Solomon islands).

Western Ghats (south India) and other parts of the Indian subcontinent and experience a high annual precipitation (Fig. 5). The climatic preferences of nearest living relatives (NLRs) witnessed increased atmospheric temperatures during the late Paleocene-early Eocene. During the early Paleogene, the Indian subcontinent was located near the equator ~10°N (Chatterjee and Scotese 1999) which provided the conducive environment for the luxuriant growth of tropical evergreen elements. As shown in Fig. 5, most of the NLRs of the recovered fossils are biogeographically located in the humid climatic zone and experience high precipitation regime. In contrast, at present the fossil locality falls in the Thar Desert (Singh and Kumar

Fig. 3 Showing fossils of the Gurha lignite mine. (**a**) *Dioscoreaeocenicus* Mehrotra and Shukla. (**b**) *Legumino carponsaracoides* Shukla and Mehrotra. (**c**) *Legumino carponcajanoides* Shukla and Mehrotra. (**d**) *Malloto carpongurhaensis* Shukla and Mehrotra. (**e**) *Uvaria palaeozeylanica* Shukla and Mehrotra. (**f**) *Syzygium gurhaensis* Shukla et al. (**g**) *Gardenia eocenicus* Shukla et al.

Fig. 4 Showing fossils of the Gurha lignite mine. (**a**) *Holigarna palaeograhamii* Shukla et al. (**b**) *Pterygota eocenica* Shukla et al. (**c**) *Aporosa ecocenicus* Shukla et al.

2015) and experiences a harsh arid climate with a mean annual rainfall of ~310 mm. This shows that the fossil locality has witnessed a drastic change in climate since the Cenozoic. Among the described fossils, a few NLRs i.e., *Aporosa acuminata, Gardenia gummifera, Holigarna grahamii,* and *Syzygium cumini* are found in evergreen forests of the Western Ghats (Fig. 5) and require high precipitation and humid conditions for flourishing. This indicates similar climatic scenario in Rajasthan during the early Paleogene was similar to western Ghats.

Further, the presence of legumes and a few drier elements is indicative of monsoon driven seasonality in and around the area. Plant fossil based palaeo–temperature reconstructions by Shukla et al. (2014) estimated mean annual temperatures (MATs) of 24.7 and 23.9 ± 2.82 °C, for two different horizons in the Gurha mine. The warm month mean temperatures (WMMTs) are 28.2 and 27.9 ± 3.39 °C, and the cold month mean temperatures (CMMTs) are 19 and 18.2 ± 4 °C. Shukla et al. (2014) have also found that the reconstructed palaeo–temperatures are unexpectedly cooler for an equatorial region. Similar kinds of unforeseen results were observed from a higher palaeo–latitude (30°N) locality (Keating-Bitonti et al. 2011).

The collision of the Indian plate with Eurasia around ~55 Ma had a great impact on the evolution of South Asian Monsoon (SAM) (Spicer 2017) that played an important role in the development of Asian biodiversity. It is assumed that the SAM originated at ~25–22 Ma (Myers et al. 2000; Licht et al. 2014; Spicer et al. 2017), however, the record of an earlier monsoon (e.g. Eocene; 55–34 Ma ago) that has a great significance for understanding the past global changes is still controversial, debated and wanting. It is evident that while the Indian subcontinent traversed the equator, it was exposed to ITCZ-driven seasonality with no sign of orographically

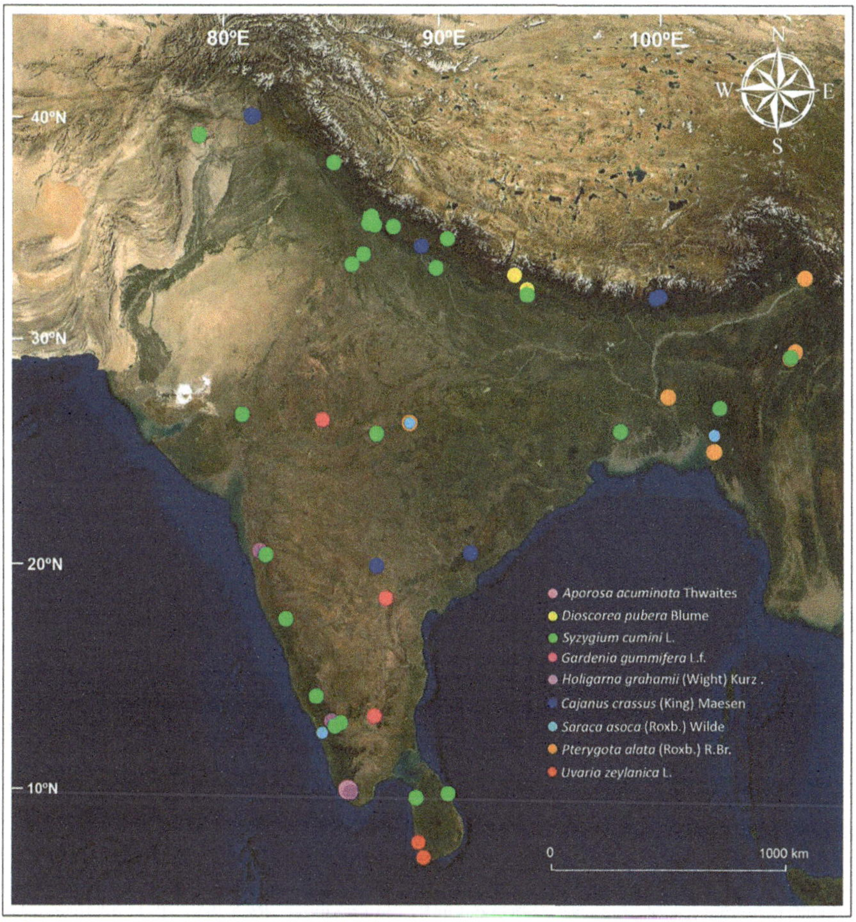

Fig. 5 Shuttle Radar Topographic Mission (SRTM) image showing present biogeographic distribution of NLRs of described fossils from the Gurha lignite mine, Rajasthan

modified monsoon circulation. The palaeoenvironmental reconstructions based on the Gurha fossil assemblage do not show any evidence of an ever-wet climatic condition (Shukla et al. 2014) and suggest the existence of tropical evergreen vegetation with marked seasonality. The Gurha lignite mine of western India shows stratigraphic record of plant macro- and microfossils which firmly correlate to the global late Paleocene-early Eocene warming. Plant fossil data from this region may have certain limits in correlating the global PETM events. So, there is a need for more studies regarding the evolution and intensification of monsoon and development of vegetation in the early Paleogene for a better picture of SAM during the period.

Acknowledgments Authors are thankful to the Director, Birbal Sahni Institute of Palaeosciences, Lucknow for the permission to publish this work. They are grateful to the anonymous reviewer for his constructive suggestions.

References

Chatterjee HG, Scotese CR (1999) The breakup of Gondwana and the evolution and biogeography of the Indian plate. Proceedings of the Indian National Science Academy 65A:397–425

Eberle JJ, Greenwood DR (2012) Life at the top of the greenhouse Eocene world—a review of the Eocene flora and vertebrate fauna from Canada's high Arctic. Bulletin 124(1–2):3–23

Givnish TJ, Renner SR (2004) Tropical intercontinental disjunctions: Gondwana breakup, immigration from the boreotropics, and transoceanic dispersal. Int J Plant Sci 165:S1–S6

Jolly A, Loyal RS (1985) Record of microvertebrates from the Middle Eocene Marh Stage of Sri Kolayat-Ji, Rajasthan. Bull Geol Mining Metall Soc India 52:374–384

Kalia P, Sharma R (1985) Planktonic foraminiferal biostratigraphy of the Kolayat Formation (lower Middle Eocene), Bikaner District, Rajasthan. Bull Geol Mining Metall Soc India 52:20–36

Keating-Bitonti CR, Ivany LC, Affek HP, Douglas P, Samson SD (2011) Warm, not super-hot, temperatures in the early Eocene subtropics. Geology 39:771–774

Khosla SC (1971) Classification of the lower tertiary beds of Rajasthan. Bull Indian Geol Assoc 4: 54–60

Kroeger KF, di Primio R, Horsfield B (2011) Atmospheric methane from organic carbon mobilization in sedimentary basins—the sleeping giant? Earth Sci Rev 107:423–442

Kulshreshtha SK, Singh RY, Sobeh AY (1989) Stratigraphy of the lower tertiary sediments in Bikaner, western Rajasthan. In: Kalia P (ed) Micro-palaeontology of the shelf sequences of India. Proceedings of XII Indian Colloquium on Micropalaeontology and Stratigraphy. Papyrus Publishing House, New Delhi, pp 193–202

Kumar K, Loyal RS (1987) Eocene ichthyofauna from the Subathu formation, northwestern Himalaya, India. J Palaeontol Soc India 32:60–84

Kumar K, Rana RS, Paliwal BS (2005) Osteoglossid and lepisosteid fish remains from the Palaeocene Palana formation, Rajasthan, India. Palaeontology 48(6):1187–1209

Kumar M, Spicer RA, Spicer TEV, Shukla A, Mehrotra RC, Monga P (2016) Palynostratigraphy and palynofacies of the early Eocene Gurha lignite mine, Rajasthan, India. Palaeogeogr Palaeoclimatol Palaeoecol 461:98–108

Licht A, van Cappelle M, Abels HA, Ladant JB, Trabucho-Alexandre J, France-Lanord C, Donnadieu Y, Vandenberghe J, Rigaudier T, Lecuyer C, Terry D, Adriaens R, Boura A, Guo Z, Soe AN, Quade J, Dupont-Nivet G, Jaeger JJ (2014) Asian monsoons in a late Eocene greenhouse world. Nature 513:501–506

Lowenstein TK, Demicco RV (2006) Elevated Eocene atmospheric CO_2 and its subsequent decline. Science 313:1928

MacGinitie HD (1953) Fossil plants of the florissant beds, vol 599. Carnegie Institution of Washington Publication, Washington, pp 1–198

Mehrotra RC, Shukla A (2019) First record of *Dioscorea* from the early Eocene of northwestern India: its evolutionary and palaeoecological importance. Rev Palaeobot Palynol 261:11–17

Mehrotra RC, Shukla A, Srivastava G, Tiwari RP (2014) Miocene megaflora of peninsular India: present status and future prospect, vol 5. Special Paper Palaeontological Society of India, Lucknow, pp 283–290

Molnar P, Stock JM (2009) Slowing of India's convergence with Eurasia since 20 Ma and its implications for Tibetan mantle dynamics. Tectonics 28:TC3001

Morley RJ (2003) Interplate dispersal paths for megathermal angiosperms. Perspectives in plant ecology. Evol Syst 6(1–2):5–20

Myers N, Mittermeir CG, da Fonseca GAB, Kent J (2000) Biodiversity hotspots for conservation priorities. Nature 403:218–222
Pareek HS (1984) Pre-quaternary geology and mineral resources of northwestern Rajasthan. Memoirs Geol Surv India 115:1–99
Parmesan C, Yohe G (2003) A globally coherent fingerprint of climate change impacts across natural systems. Nature 421:37–42
Prasad V, Uddandam PR, Agarwal S, Bajpai S, Singh I, Mishra AK, Sharma A, Kumar M, Verma P (2020) Biostratigraphy, palaeoenvironment and sea level changes during pre-collisional (Palaeocene) phase of the Indian plate: palynological evidence from Akli formation in Giral Lignite Mine, Barmer Basin, Rajasthan, Western India. Episodes 43(1):476–488
Shukla A, Mehrotra RC (2013) Cenozoic flora of western India and its significance in palaeoclimatic and palaeophytogeographic interpretation, India. Chin Sci Bull (Chinese Version) 58(Suppl I):134–141
Shukla A, Mehrotra RC (2014) Paleoequatorial rain forest of western India during the EECO: evidence from *Uvaria* L. fossil and its geological distribution pattern. Hist Biol 26(6):693–698
Shukla A, Mehrotra RC (2016) Early Eocene (~50 my) legume fruits from Rajasthan. Curr Sci 111(3):465
Shukla A, Mehrotra RC (2018) Early Eocene plant megafossil assemblage of western India: Paleoclimatic and paleobiogeographic implications. Rev Palaeobot Palynol 258:123–132
Shukla A, Mehrotra RC (2019) First fossil fruit of *Mallotus* Lour. (Euphorbiaceae) from the early Eocene lignite mine of Rajasthan, India. J Geol Soc India 94(2):206–210
Shukla A, Mehrotra R, Spicer RA, Kumar M (2014) Cool equatorial terrestrial temperatures and the South Asian monsoon in the early Eocene: evidence from the Gurha Mine, Rajasthan, India. Palaeogeogr Palaeoclimatol Palaeoecol 412:187–198
Shukla A, Mehrotra RC, Spicer RA, Spicer TEV (2016) *Aporosa* Blume from the paleoequatorial rainforest of Bikaner, India: its evolution and diversification in deep time. Rev Palaeobot Palynol 232:14–21
Shukla A, Mehrotra RC, Ali SN (2018) Early Eocene leaves of northwestern India and their response to climate change. J Asian Earth Sci 166:152–161
Singh SN (1971) Planktonic foraminifera in the Eocene stratigraphy of Rajasthan, India. In: Proceedings of the 2nd international conference on planktonic microfossils, Rome, vol 2. Springer, New York, pp 1169–1181
Singh RB, Kumar A (2015) Climate variability and water resource scarcity in dry lands of Rajasthan, India. Geoenviron Disasters 2:7
Spicer RA (2017) Tibet, the Himalaya, Asian monsoons and biodiversity – in what ways are they related? Plant Divers 39:233–244
Spicer RA, Yang J, Herman A, Kodrul T, Aleksandrova G, Maslova N, Spicer TEV, Ding L, Xu Q, Shukla A, Srivastava G, Mehrotra RC, Jin JH (2017) Paleogene monsoons across India and South China: drivers of biotic change. Gondwana Res 49:350–363
Srivastava G, Mehrotra RC, Shukla A, Tiwari RP (2014) Miocene vegetation and climate in extra peninsular India: megafossil evidences. J Palaeontol Soc India 5:283–290
Sykes MT (2009) Climate change impacts: vegetation. Wiley, Chichester
Zachos J, Pagani M, Sloan L, Thomas E, Billups K (2001) Trend, rhythms, and aberrations in global climate 65 Ma to present. Science 292:686–693
Zachos JC, Wara MW, Bohaty S, Delaney ML, Petrizzo MR, Brill A, Bralower TJ, Premoli Silva I (2003) A transient rise in tropical sea surface temperature during the Paleocene–Eocene thermal maximum. Science 302:1551–1554
Zachos JC, Dickens GR, Zeebe RE (2008) An early Cenozoic perspective on greenhouse warming and carbon–cycle dynamics. Nature 451:279–283

Development of Cenogram Technique Over the Past Six Decades with Some Insights into the Varied Habitats Occupied by Diverse Mammalian Communities Across Spain, China, and India Transiting the Middle Miocene Climatic Optimum

Vivesh V. Kapur, Blanca A. García Yelo, and M. G. Thakkar

Abstract The climatic evolution of the Neogene, with long-term cooling disrupted by the Middle Miocene Climatic Optimum (MMCO; ~17–14.75 Ma), arises as a suitable baseline to analyze the effects of these transcendent climatic changes on the mammalian community structures. The present investigation is an attempt to examine the palaeohabitat of a Neogene (Middle Miocene: ~15–11.5 Ma) geographically distant (i.e., from Spain, China, and India) extinct mammalian communities utilizing the cenogram approach (in both qualitative and quantitative framework). The detailed statistical analyses (presented herein) incorporating a total of eight mammalian communities allows us to infer predominance of Tropical Deciduous Forest environments between ~15 and ~11.5 Ma interval, with several pulses of distinctive aridity experienced by some communities thriving within the Iberian region. On the contrary, stable forested conditions were witnessed by the middle Miocene communities of Asia [i.e., the ~11.5 million-year-old mammalian community of Laogou (China), and the ~13.5 million-year-old mammalian community of Ramnagar (north India)]. Our present investigation also infers that additional mammalian remains (particularly of body mass of <35 kg) are warranted to decipher the habitat (based on cenogram approach) of the Middle Miocene (~13 Ma) mammalian community of Kalagarh (Himalayan Foreland Basin, north India) and the Middle Miocene (~14 Ma) mammalian community of Palasava (Kutch Basin, western India).

V. V. Kapur (✉)
Birbal Sahni Institute of Palaeosciences, Lucknow, India
e-mail: viveshvir_kapur@bsip.res.in

B. A. García Yelo (✉)
Dpto. Didáctica de las Ciencias Experimentales, Sociales y Matemáticas - Facultad de Educación (UCM), Madrid, Spain
e-mail: bgyelo@ucm.es

M. G. Thakkar
Krantiguru Shyamji Krishna Verma Kachchh University, Bhuj, Gujarat, India

B. Phartiyal et al. (eds.), *Climate Change and Environmental Impacts: Past, Present and Future Perspective*, Society of Earth Scientists Series,
https://doi.org/10.1007/978-3-031-13119-6_3

Nonetheless, the Cenogram technique (being continuously developed over the past six decades) may become an important tool to decipher any habitat change(s) of western India's mammalian communities considering renewed palaeontological efforts within the Neogene of the region.

Keywords Cenogram analysis · Neogene · Mammals · Palaeoenvironment · Palaeohabitat

1 Introduction

The extended Cenozoic cooling was interrupted by warm Middle Miocene Climatic Optimum (MMCO) ~17 to 14.75 Ma (Zachos et al. 2001, 2008; Methner et al. 2020 and references therein). Consequently, the planet witnessed global scale ecological changes in the geographic expansion of grassland ecosystems across many continents during the Miocene Epoch. Further, the fossil mammal data suggest domination (over browsers) of grazing mammals dwelling in grassland ecosystems due to adaption to the C3 diet during the late Miocene (Domingo et al. 2013). However, still unclear are the causal links between climate and ecosystem changes (Flower and Kennett 1994; Zachos et al. 2001, 2008; You et al. 2009; Domingo et al. 2013; Holbourn et al. 2015; Harris et al. 2020). Habitat reconstructions employing changes in the community structure of geographically distant deep-time mammalian communities due to changes in the environment provide useful palaeontological inferences. The cenogram technique was introduced in the late half of the twentieth century (Valverde 1964) and subsequently developed to include both qualitative and quantitative analysis to potentially characterize the habitat of deep time mammalian communities (Valverde 1967; Legendre 1986, 1987, 1989; Gingerich 1989; García Yelo et al. 2014 and references therein; also refer to Sect. 2). Published literature records numerous Neogene (Miocene) mammalian communities within India, particularly from the Himalayan Foreland Basin (north India) and the Kutch Basin (western India) (Sahni and Mishra 1975; Sehgal and Patnaik 2012; Sehgal 2013; Kapur et al. 2019 and references therein). However, the cenogram technique (in both a qualitative and a quantitative framework) has not been utilized to characterize the habitat of the previously cited Neogene (Miocene) mammalian communities of India. Fortunately, well-characterized are the habitats of the Neogene (Miocene) mammalian communities of the Iberian region (Spain) and Laogou (Linxia Basin, China) based on the cenogram approach (García Yelo et al. 2014; Kapur et al. 2020). Thus, it provides some basis to characterize extinct mammalian communities' habitats across distant geographic areas (i.e., Spain, China and India) spanning the Middle Miocene Climatic Optimum (MMCO) that we attempt in the present study, utilizing the cenogram approach.

2 A Brief Account of the Development of Cenogram Technique Over the Past Six Decades

Biotic (fossil flora and fauna) and abiotic proxies (e.g. geochemical) are quite well-known to provide pieces of evidence for past life and developing an understanding of the surrounding habitat. In a similar context, cenogram technique has also been utilized and developed over the past six decades to reconstruct deep-time terrestrial habitats. In general, the application of cenogram technique entails three pre-requisites: (a) it is exclusively utilized for terrestrial extant or extinct mammalian communities, (b) considers mammalian palaeocommunities restricted in time-frame, and (c) estimation of mammalian body weight for an individual taxon to be utilized in the cenogram analysis. However, varied methods have been used to estimate extinct terrestrial mammals' body weights. These include the dental area of the first lower molar (Gingerich et al. 1982; Legendre 1986); mid-shaft cross-section dimension(s) (Anderson et al. 1985), distal and proximal limb bone dimensions (Gingerich 1990; Andersson 2004; Christiansen 2004), craniodental measurements (Myers 2001; Mendoza et al. 2006), and anklebone dimensions (Dagosto and Terranova 1992; Martinez and Sudre 1995). In general, (a) owing to excellent preservation, (b) potential for recovery, (c) being taxonomically diagnostic in many mammalian groups, and (d) showing a strong correlation between occlusal surface dimensions and body weights, the first lower molar (m1) in mammals is favoured in estimating body weights (Kay 1975; Creighton 1980; Gingerich et al. 1982; Legendre 1989; Legendre and Roth 1988; Damuth and MacFadden 1990; Bown et al. 1994; Egi et al. 2004; Millien and Bovy 2010; Kapur et al. 2020).

Ecologist Jose Valverde proposed the term 'Cenogram' for the line obtained when species in a mammalian community are arranged by size (Valverde 1964). Thus, a cenogram plot was initially a graphical display of the body-size distribution of mammalian communities, produced by plotting rank-ordered taxa versus head-body length to analyze the trophic interactions between predators and prey (Valverde 1964, 1967). Subsequently, cenograms were modified by plotting rank-ordered taxa versus bodyweight to analyze modern and extinct (late Eocene and Oligocene) terrestrial mammalian assemblages of Africa and Europe by establishing relationships between cenogram structure and the characteristics of the environment (Legendre 1986, 1987, 1989). Although cenograms at this stage were visualized qualitatively; however, the initial methodology formed the basis for the future development of cenograms, mainly considering the gap between the small/medium-sized and large-sized mammals within a community. Qualitative visualization also suggested that a large number of small-sized mammals in a community hinted at the prevalence of a closed habitat (Legendre 1986, 1987, 1989). Later, Gingerich (1989) utilized cenogram technique to study the palaeoecology of the earliest Wasatchian mammalian fauna from northwest Wyoming and established a relationship between the canopy of the environment and the difference (offset) between the regression lines for the medium to large-sized and small-sized mammal species. Salient features of Gingerich's observations on the quantitative visualization of the cenogram plots include (a) each species differs from the next larger or smaller species by a nearly constant proportion, (b) any gap in the uniform distribution, when they occur, are always at about 500 g "*Kay's threshold*" and at ~250 kg, and (c) extant mammals are naturally divided into three broad size categories, i.e., *Small* (<500 g), *Medium* (>500 g and <250 kg) and *Large* (>250 kg). Further, Gingerich (1989) supported that present-day body size scaling patterns in ecology were found to be operating in deep-time considering that Kay's Threshold (500 g break/gap) can be traced, in primates, from the Eocene to the present (refer Fleagle 1978). Since the past decade or more, studies incorporating cenogram technique have widely utilized 12 cenogram variables to describe the terrestrial mammalian community structure and have additionally considered 100 modern mammalian communities across the globe for biome characterization (Gómez Cano et al. 2006; García Yelo et al. 2014; Kapur et al. 2020; this study). Interestingly, cenogram methodology is shown to be relevant even for the assemblages with a random species loss of >60–70% (Gómez Cano et al. 2006). Although few studies have argued against the correlation of cenogram patterns to palaeoclimate and/or temperature (Rodríguez 1999; Croft 2001; Nieto and Rodríguez 2003), over the years cenogram technique has proved useful for identifying the preferred biomes of mammalian communities with the general consensus on the inferences on palaeoenvironments particularly within the tropical realms (Alroy 2000; Nieto and Rodríguez 2003; Travouillon and Legendre 2009; Travouillon et al. 2009; García Yelo et al. 2014; Kapur et al. 2020). For instance, utilizing the cenogram technique Gingerich (1989) suggested that the early Eocene (Wasatchian: Wa0) mammal fauna of Polecat Bench, North America, sustained in a moist forested environment. Cenogram analysis on the middle Eocene Pondaung mammals of Myanmar indicated that the terrestrial community inhabited a forest/woodland with a prevalence of humid/sub-humid conditions (Tsubamoto et al. 2005). Costeur (2005) argued favouring tropical to subtropical humid forest-savanna-like mosaic habitat for the late Miocene Rudabánya community from Hungary. The predominance of semiarid environments with pulses of higher aridity was inferred based on cenogram technique on the middle Miocene assemblages from the Madrid Basin (García Yelo et al. 2014). Very recently, Kapur et al. (2020) utilized the cenogram analysis on the early Eocene Cambay Shale mammalian fauna of India known to dwell in humid tropical forests. Kapur et al. (2020) explained the unanticipated (i.e., tropical desert habitat) inference based on the cenogram approach due to the hidden diversity [i.e., lack of mammals within the logarithmic body weight (g) ranges 3–3.99 to 6–6.99 (1000–10,000,000 g)] in the early Eocene Cambay Shale mammals. It should also be noted that no cenogram analysis has involved fossil Neogene mammals from India. Thus, as a first attempt, we herein utilize the cenogram technique to examine the palaeohabitat of taxonomically diverse extinct mammal communities that thrived in distant geographic areas (i.e., north and western India) during the Miocene.

3 Methodology

Initially, the individual mammalian species' body weight was estimated using the occlusal dimensions (mean size of all specimens in a single species) of the m1, utilizing Legendre's regression parameters (Legendre 1989). Besides, we also incorporated the bodyweight estimations following some previous studies (Flynn et al. 1995; Bernor and Fessaha 2000; Barry et al. 2002; Liu 2003; DeSilva et al. 2010; Göhlich 2010; Sehgal 2013; Aiglsstorfer et al. 2014; Parmar et al. 2018; Larramendi 2016; Grabowski and Jungers 2017; Gilbert et al. 2017, 2020; Guzmán 2018; Becker and Tissier 2019). All the cenogram structures were plotted by depicting the natural log (ln) of body weight (g) on the y-axis and the rank-ordered taxa (in decreasing size) on the abscissa (x-axis). The ecological, faunal data from 100 modern localities uniformly distributed all around the world, and of known biome (Hernández Fernández et al. 2006 and references therein; Fig. 1) was established as a comparative framework for palaeoenvironmental inference based on multivariate discriminant analysis of the dataset containing both modern and fossil mammals. Additionally, in a separate analysis, the different geographic distributions of the continents were taken into account, considering the extinct communities' latitudinal position during Miocene. The mammalian community structure of both modern and fossil fauna was described through 12 cenogram variables (Table 2 in Kapur et al. 2020). The detailed methodology followed herein is also provided in García Yelo et al. (2014) and Kapur et al. (2020).

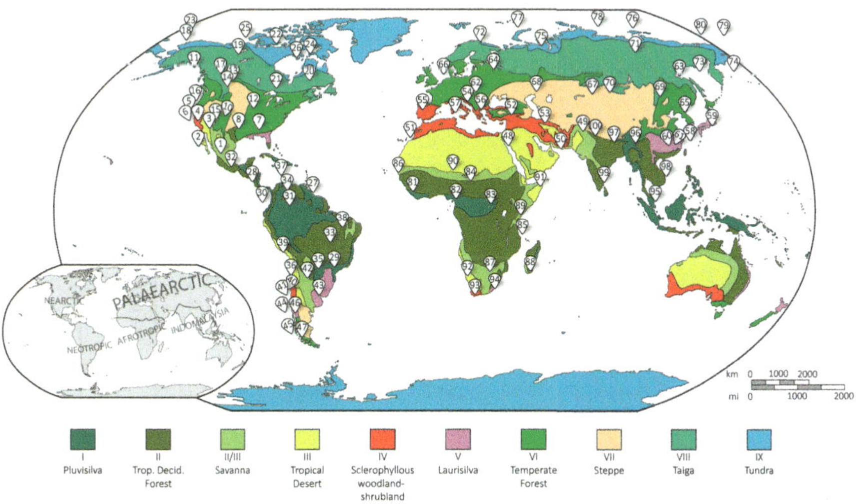

Fig. 1 Geographic distribution of one hundred modern mammalian communities utilized in the biogeographic analysis

4 Results (Cenogram Analysis)

Cenogram analysis of a total of eight Neogene (middle Miocene: ~15 to 11.5 Ma; Fig. 2) extinct mammalian communities from Spain, China and India are resumed as Fig. 3. These include six extinct middle-Miocene communities of the Madrid Basin, Spain [Estación Imperial (EI); Paseo de las Acacias (PA); Arroyo del Olivar-Puente de Vallecas (AO-PV); Somo-saguas (Som.); Paracuellos 5 (P5); Paracuellos 3 (P3)], one from the Linxia Basin, China [Laogou (Lao.)] and one from Himalayan Foreland Basin, India [Ramnagar (Ram.)] (Figs. 2 and 3). The recently recorded middle Miocene (~14 Ma) mammalian community of Palasava, Kutch Basin, India (Kapur et al. 2019) and the well-known middle Miocene (~13.5 Ma) extinct mammalian community of Kalagarh (Pauri Garhwal, Uttarakhand, India) (Sehgal 2013) were excluded from the cenogram analysis as the majority of the faunas from these two localities were formed by large or mega prey (i.e., body masses > 35 kg) and low (null) representation of small-sized mammals in the community. Consequently, several of the cenogram variables were absent (Table 1), preventing a cenogram analysis. Further, discriminant analyses of the above-mentioned eight mammalian communities show a low biome discriminant capability at the global scale (56.7% of correctly classified localities, respectively). However, the results were significantly improved when the fossil communities' geographical situation was considered

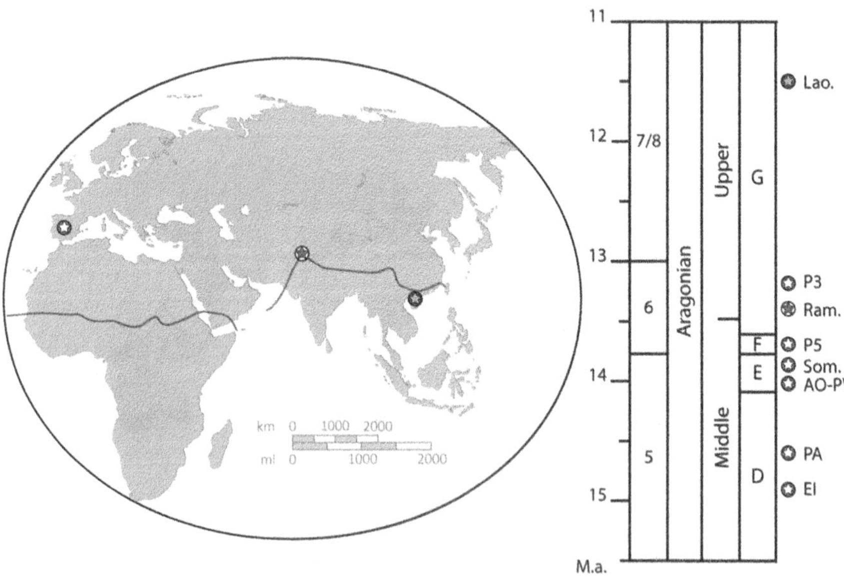

Fig. 2 Map of the globe showing the geographic locations of the Neogene (Middle Miocene) fossil mammalian communities of Spain, China, and India utilized in the cenogram analysis presented herein and their chronostratigraphic position within the Miocene Epoch. Note: For further geographic details on the Spanish Miocene localities refer to Fig. 1 in Kapur et al. (2020)

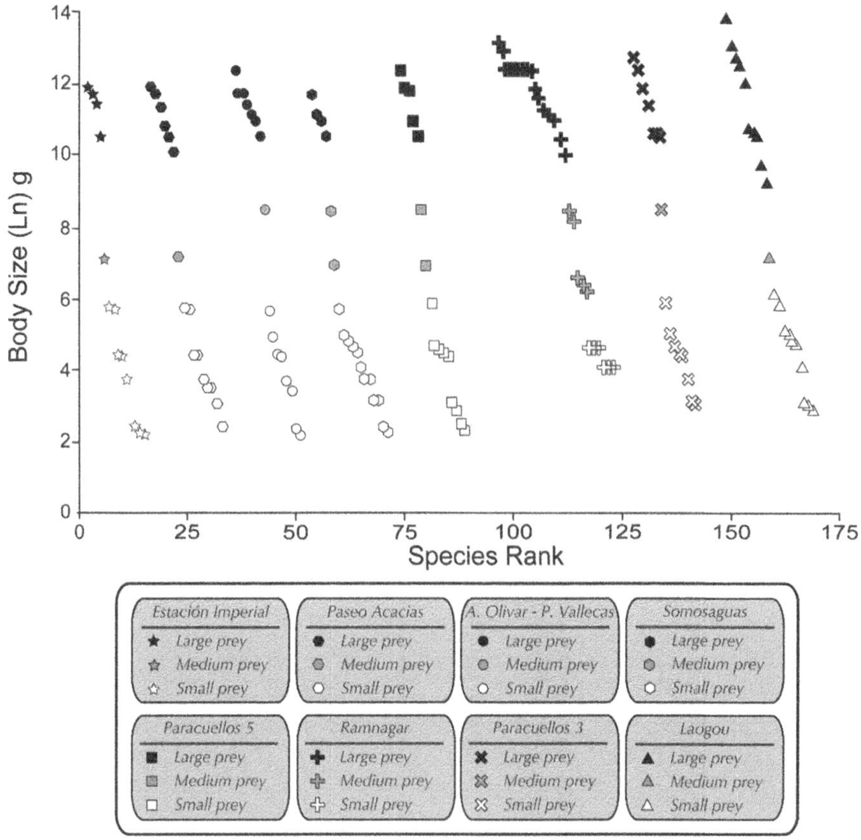

Fig. 3 Cenograms of the eight Neogene (Middle Miocene) fossil mammalian communities utilized in the present analysis

(Table 2). According to the discriminant analyses, all predicted biomes detected are arid and warm (Tropical Desert and Tropical Deciduous Forest) (Table 2).

5 Discussion

Again, it is noteworthy that the present investigation, utilizing the cenogram approach (in both qualitative and quantitative framework), is a first attempt to reconstruct the palaeohabitat of an extinct Neogene (Miocene) community of India. Body size community structures of the Aragonian mammalian communities present in the Madrid Basin allow inferring a predominance of Tropical Deciduous Forest environments between 15 and 10 Ma, with several pulses of distinctive aridity throughout the whole sequence. In contrast, stable forested conditions are detected

Table 1 Results for the cenogram variables for the Neogene (Middle Miocene) sites of Palasava (Kutch Basin, western India) and Kalagarh (Himalayan Foreland Basin, north India) that were excluded from the analysis presented herein. Note: Null data (no representation) for the discriminant analysis (considering biogeography) for both these two locales

		Cenogram variables											
Mammalian Communities excluded from the original analysis.		W1	W2	W5	S1	S5	S5-S1	G	mG	MG	WG	WmG	WMG
Kutch District, Gujarat State (India)	Palasava (West India)	–	–	10.82	–	–	–	–	–	–	–	–	–
Pauri Garhwal District, Uttarakhand State (India)	Kalagarh (North India)	–	8.41	10.94	–	−0.57	–	–	2.05	2.05	–	9.44	9.44

Table 2 Results for the twelve cenogram variables for the eight Neogene (Middle Miocene) sites from Spain, China and India included in the present analysis

Mammalian communities included in the present analysis		Cenogram variables												Discriminant analysis	
		W1	W2	W5	S1	S5	S5-S1	G	mG	MG	WG	WmG	WMG	Global (56.7%)	Paleotropic (100%)
Madrid Basin (Spain)	Estación Imperial	3.80	7.15	10.55	−0.49	−1.07	−0.58	1.44	3.38	3.38	6.44	8.84	8.84	IV	III
	Paseo de las Acacias	4.04	7.15	10.51	−0.40	−0.66	−0.26	1.44	2.93	2.93	6.44	8.62	8.62	IV	II
	Arroyo del Olivar-Puente de Vallecas	3.91	8.48	11.04	−0.49	−0.43	0.06	2.77	2.05	2.77	7.09	9.50	7.09	VIII	II
	Somosaguas	3.91	7.70	9.95	−0.30	−0.92	−0.63	1.22	2.05	2.05	6.31	9.50	9.50	VI	III
	Paracuellos 5	3.84	7.70	10.42	−0.42	−0.87	−0.45	1.06	2.05	2.05	6.39	9.50	9.50	VII	III
	Paracuellos 3	4.45	8.48	11.08	−0.38	−0.52	−0.14	2.62	2.05	2.62	7.17	9.50	7.17	VIII	II
Linxia Basin (China)	Laogou	4.34	7.15	10.71	−0.14	−0.33	−0.19	1.61	0.17	0.17	5.41	6.30	6.30	I	II
Himalayan Foreland Basin (India)	Ramnagar	4.34	7.15	10.71	−0.14	−0.33	−0.19	1.61	0.17	0.17	5.41	6.30	6.30	III	II

for the Asian localities. According to previous studies (García Yelo et al. 2014), in the Madrid Basin, arid conditions dominated the landscapes at the beginning of the sequence. A change towards a more humid and forested condition was developed in the middle of the sequence. Finally, arid conditions and grass-dominated habitats returned to central Iberia, associated with the Global Cooling Event of the middle Miocene (coinciding with the final stage of the sequence). Finally, at the end of the sequence (Paracuellos 3), the landscape returned to a more covered and humid scenario. Our results (considering biogeography) also reiterate tropically forested (tropical deciduous forest) habitat for the middle Miocene mammalian community of Laogou (Linxia Basin, China) (also refer to Kapur et al. 2020). The cenogram analysis and additional discriminant analysis (considering the biogeography) of the Middle Miocene Ramnagar mammalian community also suggest a prevalence of tropical deciduous forested habitat. Thus, taking into account the time-slices covered in the present investigation (i.e., from ~15 to 11.5 Ma), the Asian contemporaneous communities of Laogou (Linxia Basin, China) and Ramnagar (Himalayan Foreland Basin, north India) predicting a Tropical Deciduous Forest, reflect more humid conditions at the end of the sequence. Our results based on the Cenogram approach for the Indian mammalian community (i.e., Ramnagar mammals thriving in tropical deciduous forests) support previous studies that argue in favour of the establishment of seasonality (in terms of the development of Monsoon System) during the early Neogene (Retallack et al. 2018 and references therein). A similar habitat (tropical deciduous forest) was most likely sustained by the middle Miocene (~14 Ma) extinct mammalian community of Palasava (Kutch Basin, western India) considering the overall assemblage of mammals and associated floral evidence (Kapur et al. 2019 and references therein); however, cenogram analysis (in both qualitative and quantitative framework) is not possible at this stage for the Palasava mammalian community due to lack of small-sized mammals (i.e., lack of <35 kg mammalian body mass estimations based on m1).

Interestingly, the Neogene (Miocene: ~18 to 8 Ma) interval of the Kutch and its adjoining regions has gained renewed attention in recent times owing to the record of mammalian communities from numerous locales, namely Pipar and Fetehgarh (~18 Ma); Palasava, Junagia, Samda, and Jangadia (~16 to 14 Ma); Tapar and Pasuda (~11 Ma) and Piram Islands within the Gulf of Cambay (~8 Ma) (Prasad 1974; Sahni and Mishra 1975; Patnaik et al. 2014; Ferreira et al. 2018; Bhandari et al. 2018; Kapur et al. 2019; Singh et al. 2020). Thus, providing researchers with an opportunity to study any change(s) in these mammalian communities' habitat across the Neogene interval of India's western part linked to any climatic fluctuation (s) such as Monsoon variability. The cenogram technique can prove to be a useful tool in this context; however, we should await recovery of additional mammalian remains from the Neogene of Kachchh region, western India. Nonetheless, our present analysis provides baseline data and valuable insights into the varied habitats occupied by diverse mammalian communities across distant geographic areas (Spain, China and India) that transited the MMCO interval.

6 Conclusions

(a) The present investigation is a first attempt to examine the palaeohabitat of a Neogene (Miocene) mammalian community from India utilizing the cenogram approach (in both qualitatively and quantitatively framework).
(b) The present study suggests that additional mammalian remains (particularly of body mass of <35 kg) are warranted to decipher the habitat (based on cenogram approach) of the Middle Miocene (~13 Ma) mammalian community of Kalagarh (Himalayan Foreland Basin, north India) and the Middle Miocene (~14 Ma) mammalian community of Palasava (Kutch Basin, western India).
(c) The detailed statistical analysis (presented herein) incorporating a total of eight geographically distant Neogene mammalian communities allow to infer predominance of Tropical Deciduous Forest environments between ~15 and ~11.5 Ma, with several pulses of distinctive aridity experienced by some communities thriving within the Iberian region. In contrast, stable forested conditions were witnessed by the Middle Miocene (~11.5 Ma), Laogou (China) and Middle Miocene (~13.5 Ma), Ramnagar (north India) mammalian communities.
(d) Moreover, the Cenogram technique (being continuously developed over the past six decades) can become instrumental in deciphering any habitat change(s) of the mammalian communities of western India considering renewed palaeontological efforts in the Neogene of the region.

Acknowledgments At the onset, VVK acknowledges the use of the infrastructural facilities at Birbal Sahni Institute of Palaeosciences (BSIP), Lucknow, India and is thankful to Dr Vandana Prasad (Director, BSIP) for constant encouragement and necessary permissions (BSIP/RDCC/ Publication no. 93/2020-2022). VVK also acknowledges funding support from BSIP in the form of in-house project 4.2 (2019-2021) & project 3 (2021-2025). BAGY acknowledges funding support from Ministerio de Educación, Ciencia e Investigación in the form of supporting projects (PGC2018-094122-B-I00; PGC2018-094955-A-I00). The authors would like to thank the anonymous reviewers and the editor for critical and constructive commentaries that improved the manuscript.

Disclosure Statement We have no potential conflict of interest.

References

Aiglsstorfer M, Bocherens H, Böhme M (2014) Large mammal ecology in the late middle Miocene Gratkorn locality (Austria). Palaeodivers Palaeoenviron 94:189–213. https://doi.org/10.1007/s12549-013-0145-5

Alroy J (2000) New methods for quantifying macroevolutionary patterns and processes. Paleobiology 26:707–733

Anderson JF, Hall-Martin A, Russell DA (1985) Long-bone circumference and weight in mammals, birds and dinosaurs. J Zool 207:53–61

Andersson K (2004) Predicting carnivoran body mass from a weight-bearing joint. J Zool 262:161–172

Barry JC, Morgan ME, Flynn LJ, Pilbeam D, Behrensmeyer AK, Raza SM, Khan IA, Badgley C, Hicks J, Kelley J (2002) Faunal and environmental change in the Late Miocene Siwaliks of northern Pakistan. Paleobiology 28(S2):1–71

Becker D, Tissier J (2019) Rhinocerotidae from the early middle Miocene locality Gračanica (Bugojno Basin, Bosnia-Herzegovina). Palaeobiodivers Palaeoenviron. https://doi.org/10.1007/s12549-018-0352-1

Bernor RL, Fessaha N (2000) Evolution of late Miocene Hungarian Suinae (Artiodactyla, Suidae). Carolina 58:83–92

Bhandari A, Kay RF, Williams BA, Tiwari BN, Bajpai S, Hieronymus T (2018) First record of the Miocene hominoid *Sivapithecus* from Kutch, Gujarat State, western India. PLoS One 13: e0206314. https://doi.org/10.1371/journal.pone.0206314

Bown TM, Holroyd PA, Rose KD (1994) Mammal extinctions, body size, and paleotemperature. Proc Natl Acad Sci USA 91:10403–10406

Christiansen P (2004) Body size in proboscideans, with notes on elephant metabolism. Zool J Linnean Soc 140:523–549

Costeur L (2005) Cenogram analysis of the Rudabánya mammalian community: palaeoenvironmental interpretations. Palaeontogr Ital 90:303–307

Creighton GK (1980) Static allometry of mammalian teeth and the correlation of tooth size and body size in contemporary mammals. J Zool (Lond) 191:435–443

Croft DA (2001) Cenozoic environmental change in South America as indicated by mammalian body size distributions (cenograms). Divers Distrib 7:271–287. https://doi.org/10.1046/j.1366-9516.2001.00117.x

Dagosto M, Terranova CJ (1992) Estimating the body size of Eocene primates: a comparison of results from dental and postcranial variables. Int J Primatol 13(3):307–343

Damuth J, MacFadden BJ (1990) Introduction: body size and its estimation. In: Damuth J, MacFadden BJ (eds) Body size in mammalian paleobiology: estimation and biological implications. Cambridge University Press, Cambridge, UK, pp 1–10

DeSilva JM, Morgan ME, Barry JC, Pilbeam D (2010) A hominoid distal tibia from the Miocene of Pakistan. J Hum Evol 58:147–154

Domingo L, Koch PL, Hernández Fernández M, Fox DL, Domingo MS, Alberdi MT (2013) Late Neogene and early quaternary paleoenvironmental and paleoclimatic conditions in southwestern Europe: isotopic analyses on mammalian taxa. PLoS One 8(5):e63739. https://doi.org/10.1371/journal.pone.0063739

Egi N, Takai M, Shigehara N, Tsubamoto T (2004) Body mass estimates for Eocene eosimiid and amphipithecid primates using prosimians and anthropoid scaling models. Int J Primatol 25:211–236

Ferreira GS, Bandyopadhyay S, Joyce WG (2018) A taxonomic reassessment of *Piramys auffenbergi*, a neglected turtle from the late Miocene of Piram Island, Gujarat, India. PeerJ. https://doi.org/10.7717/peerj.5938

Fleagle JG (1978) Size distributions of living and fossil primate faunas. Paleobiology 4:67–76

Flower BP, Kennett JP (1994) The middle Miocene climatic transition: east Antarctic ice sheet development, deep ocean circulation and global carbon cycling. Palaeogeogr Palaeoclimatol Palaeoecol 108:537–555

Flynn LJ, Barry JC, Morgan ME, Pilbeam D, Jacobs LL, Lindsay EH (1995) Neogene Siwalik mammalian lineages: species longevities, rates of change, and modes of speciation. In: Badgley C, Behrensmeyer AK (eds) Long records of continental ecosystems. Palaeogeography, palaeoclimatology, palaeoecology, vol 115. Cambridge University Press, Cambridge, pp 249–264

García Yelo BA, Gómez Cano AR, Cantalapiedra JL, Alcalde GM, Sanisidro O, Oliver A, Hernández-Ballarín V, López-Guerrero P, Fraile S, Hernández-Fernández M (2014) Palaeoenvironmental analysis of the Aragonian (middle Miocene) mammalian faunas from the Madrid Basin based on body-size structure. J Iber Geol 40(1):129–140

Gilbert CC, Patel BA, Singh NP, Campisano CJ, Fleagle JG, Rust KL, Patnaik R (2017) New sivaladapid primate from lower Siwalik deposits surrounding Ramnagar (Jammu and Kashmir State), India. J Hum Evol 102:21–41. https://doi.org/10.1016/j.jhevol.2016.10.001
Gilbert CC, Ortiz A, Pugh KD, Campisano CJ, Patel BA, Singh NP, Fleagle JG, Patnaik R (2020) New middle Miocene ape (primates: hylobatidae) from Ramnagar, India fills major gaps in the hominoid fossil record. Proc R Soc B 287. https://doi.org/10.1098/rspb.2020.1655
Gingerich PD (1989) New earliest Wasatchian mammalian fauna from the Eocene of northwestern Wyoming: composition and diversity in a rarely sampled high-floodplain assemblage, vol 28. University of Michigan, Ann Arbor, pp 1–97
Gingerich PD (1990) Prediction of body mass in mammalian species from long bone lengths and diameters. Contrib Mus Paleontol Univ Mich 28(4):79–92
Gingerich PD, Smith BH, Rosenberg K (1982) Allometric scaling in the dentition of primates and prediction of body weight from tooth size in fossils. Am J Phys Anthropol 58:81–100
Göhlich UB (2010) The Proboscidea (Mammalia) from the Miocene of Sandelzhaunsen (southern Germany). Paläontol Z 84:163–204. https://doi.org/10.1007/s12542-010-0053-1
Gómez Cano AR, García Yelo BA, Hernández Fernández M (2006) Cenogramas, análisis bioclimático y muestreo en faunas de mamíferos: implicacion espara la aplicación de métodos de análisis paleoecológico. Estud Geol 62:135–144
Grabowski M, Jungers WL (2017) Evidence of a chimpanzee-sized ancestor of humans but a gibbon-sized ancestor of apes. Nat Commun. https://doi.org/10.1038/s41467-017-00997-4
Guzmán JA (2018) Palaeobiology of tragulids (Mammalia: Artiodactyla: Ruminantia). Dissertation zur Erlangung des Doktorgrades an der Fakultät für Geowissenschaften der Ludwig-Maximilians-Universität München, pp 1–230
Harris EB, Kohn MJ, Strömberg CAE (2020) Stable isotope compositions of herbivore teeth indicate climatic stability leading into the mid-Miocene climatic optimum, in Idaho, U.S.A. Palaeogeogr Palaeoclimatol Palaeoecol 546:109610. https://doi.org/10.1016/j.palaeo.2020.109610
Hernández Fernández M, Alberdi MT, Azanza B, Montoya P, Morales J, Nieto M, Peláez-Campomanes P (2006) Identification problems of arid environments in the neogene–quaternary mammal record of Spain. J Arid Environ 66:585–608
Holbourn A, Kuhnt W, Kochhann KGD, Andersen N, Sebastian Meier KJ (2015) Global perturbation of the carbon cycle at the onset of the Miocene climatic optimum. Geology 43:123–126. https://doi.org/10.1130/G36317.1
Kapur VV, Pickford M, Chauhan G, Thakkar MG (2019) A middle Miocene (~14 Ma) vertebrate assemblage from Palasava, Rapar Taluka, Kutch (Kachchh) district, Gujarat State, western India. Hist Biol. https://doi.org/10.1080/08912963.2019.1648451
Kapur VV, GarcíaYelo BA, Morthekai P (2020) Cenogram analyses as habitat indicators for the paleogene-neogene mammalian communities across the globe, with an emphasis on the early Eocene Cambay Shale mammalian community from India. J Iber Geol 46(3):291–310. https://doi.org/10.1007/s41513-020-00131-2
Kay RF (1975) The functional adaptations of primate molar teeth. Am J Phys Anthropol 43:195–216
Larramendi A (2016) Shoulder height, body mass, and shape of proboscideans. Acta Palaeontol Pol 61(3):537–574
Legendre S (1986) Analysis of mammalian communities from the late Eocene and Oligocene of southern France. Palaeovertebrata 16:191–212
Legendre S (1987) Les communautés de mammifères d'Europe occidentale de 'Eocene supérieur et Oligocène: structures et milieux. Münchner Geowissenschaft liche Abhandlungen A10:301–312
Legendre S (1989) Les communautés de mammifères du Paléogène (Eocène supérieur et Oligocène) d'Europe occidentatructures, milieu et évolution. Münchner Geowissenschaft liche Abhandlungen A16:1–110

Legendre S, Roth C (1988) Correlation of carnassial tooth size and body weight in recent carnivores (mammalia). Hist Biol 1:85–98

Liu L-P (2003) The Chinese fossil Suoidea: systematics, evolution, and paleoecology. Yliopistopaino, Helsinki, pp 1–41

Martinez JN, Sudre J (1995) The astragalus of paleogene artiodactyls: comparative morphology, variability and prediction of body mass. Lethaia 28:197–209

Mendoza M, Janis CM, Palmqvist P (2006) Estimating the body mass of extinct ungulates: a study on the use of multiple regression. J Zool 270:90–101

Methner K, Campani M, Fiebig J, Löffler N, Kempf O, Mulch A (2020) Middle Miocene long-term continental temperature change in and out of pace with marine climate records. Sci Rep 10: 7989. https://doi.org/10.1038/s41598-020-64743-5

Millien V, Bovy H (2010) When teeth and bones disagree: body mass estimation of a giant extinct rodent. J Mammal 91(1):11–18

Myers TJ (2001) Prediction of marsupial body mass. Aust J Zool 49:99–118

Nieto M, Rodríguez J (2003) Inferencia paleoecológica en mamíferos cenozoicos: limitaciones metodológicas. Coloquios de Paleontología 1:459–474

Parmar V, Prasad GVR, Norboo R (2018) Middle Miocene small mammals from the Siwalik Group of Northwestern India. J Asian Earth Sci 162:84–92. https://doi.org/10.1016/j.jseaes.2017.11.023

Patnaik R, Sharma KM, Mohan L, Williams BA, Kay RF, Chatrath P (2014) Additional vertebrate remains from the early Miocene of Kutch, Gujarat. Spec Publ Paleontol Soc India 5:335–351

Prasad KN (1974) The vertebrate fauna from Piram Island, Gujarat, India. Mem Geol Surv India 1974:1–22

Retallack GJ, Bajpai S, Liu X, Kapur VV, Pandey SK (2018) Advent of strong south Asian monsoon by 20 million years ago. J Geol 126:1–24

Rodríguez J (1999) Use of cenograms in mammalian palaeocology - a critical review. Lethaia 32: 331–347

Sahni A, Mishra VP (1975) Lower tertiary vertebrates from western India. Monogr Palaeontol Soc India 3:1–48

Sehgal RK (2013) Revised mammalian biostratigraphy of the lower Siwalik sediments of Ramnagar (J. & K.), India and its faunal correlation. J Palaeontol Soc India 58(1):87–92

Sehgal RK, Patnaik R (2012) New muroid rodent and *Sivapithecus* dental remains from the lower Siwalik deposits of Ramnagar (J & K, India): age implication. Quat Int 269:69–73

Singh NP, Jukar AD, Patnaik R, Sharma MK, Singh NA, Singh YP (2020) The first specimen of *Deinotherium indicum* (Mammalia, Proboscidea, Deinotheriidae) from the late Miocene of Kutch, India. J Paleontol. https://doi.org/10.1017/jpa.2020.3

Travouillon KJ, Legendre S (2009) Using cenograms to investigate gaps in mammalian body mass distributions in Australian mammals. Palaeogeogr Palaeoclimatol Palaeoecol 272:69–84

Travouillon KJ, Legendre S, Archer M, Hand SA (2009) Palaeoecological analyses of Riversleigh's oligo-Miocene sites: implications for oligo-Miocene climate change in Australia. Palaeogeogr Palaeoclimatol Palaeoecol 276:24–37

Tsubamoto T, Egi N, Takai M, Sein C, Maung M (2005) Middle Eocene ungulate mammals from Myanmar: a review with description of new specimens. Acta Palaeontol Pol 50(1):117–138

Valverde JA (1964) Remarquessur la structure et l'évolution des communautés de vertebras terrestres. 1. Structure d'une communauté 2, Rapport entre prédateurs et proies. La Terre et la Vie 111:121–154

Valverde JA (1967) Estructura de unacommunidad de vertebra dos terrestres. Monografías de la Estación Biológica de Doñana 1:1–129

You Y, Huber M, Müller RD, Poulsen CJ, Ribbe J (2009) Simulation of the middle Miocene climate optimum. Geophys Res Lett 36. https://doi.org/10.1029/2008GL036571

Zachos J, Pagani M, Sloan L, Thomas E, Billups K (2001) Trends, rhythms, and aberrations in global climate 65 Ma to present. Science 292:686–693. https://doi.org/10.1126/science.1059412

Zachos JC, Dickens GR, Zeebe RE (2008) An early Cenozoic perspective on greenhouse warming and carbon-cycle dynamics. Nature 451:279–283. https://doi.org/10.1038/nature06588

Palynofloral Diversity During Mid-Miocene Warming in Kerala Basin, South-Western India: Palaeoclimatic Implications

Yogesh Pal Singh, Poonam Verma, and Abha Singh

Abstract The sedimentary successions of Kerala Basin are one of the rare onshore opportunity to study Neogene palaeoclimate of southwest India. Among the climate events of the Neogene, MMCO (Mid-Miocene Climate Optimum) that occurred around ~17–15 Ma is considered to be the most recent extreme global warming episode. The global warming during MMCO, created annual surface temperature ~3–4 °C higher than the present, is equivalent to the warming predicted for the next century. Later to this event, the earth's climate showed a gradual cooling with the increasing global ice volumes at the poles, giving rise to modern world. Assessment of such warming events particularly in the lower latitudes is perplexing. Palynological study is one of the most useful technique to reconstruct marginal marine and terrestrial climate change as climate directly governs the vegetation types in any given area. In the Kerala Basin Neogene sequence consists of Quilon and Warkalli Formation within Malabar Super Group. The present palynological study has been done on sediment samples from Quilon Formation exposed at Padappakara section near Ashtamudi Kayal in Kollam district, Kerala. The assemblage consists of diverse and well-preserved terrestrial as well as marine palynological components. The preliminary data of dinoflagellate cysts along with mangrove and back-mangrove taxa mainly of Arecaceae family suggest that deposition took place in mangrove-dominated, fluctuating lagoon in marginal marine environment. Dominance of pollen belonging to typical rain forest families such as Ctenolophonaceae (*Retistephanocolpites* and *Ctenolophonidites*), Dipterocarpaceae (Dipterocarpuspollenites), Bombacaceae (*Lakiapollis*, *Dermatobrevicolporites* and *Tricolporopollis*) and Oleaceae (*Retitrescolpites*) indicate warm and humid climate and receiving heavy rainfall throughout the year, unlike the present climate which is

Y. P. Singh (✉)
Birbal Sahni Institute of Palaeosciences, Lucknow, India

Department of Geology, University of Lucknow, Lucknow, India
e-mail: yogesh.pal@bsip.res.in

P. Verma · A. Singh
Birbal Sahni Institute of Palaeosciences, Lucknow, India
e-mail: poonam_verma@bsip.res.in

B. Phartiyal et al. (eds.), *Climate Change and Environmental Impacts: Past, Present and Future Perspective*, Society of Earth Scientists Series,
https://doi.org/10.1007/978-3-031-13119-6_4

characterized by heavy precipitation in summer months and 3–7 dry months in winter season. Interestingly, a megathermal family, Ctenolophonaceae now extinct in India was present in high diversity during extreme humid Miocene period. Overall palynological data from studied succession has provided evidence of dense tropical megathermal rain forest vegetation flourished under warmer and wetter climate with low seasonality around Middle Miocene warming in south-western tropical region of India.

Keywords Palynology · Mid-Miocene warming · Palaeoclimate · Quilon formation · Kerala · India

1 Introduction

The extant endemic flora in South-Western Ghats, Kerala is known as 'biodiversity hotspot' in tropical region (Myers et al. 2000). The region exhibits rich floral diversity with more than 8000 taxa of flowering plants (Nayar et al. 2014). The distribution of flora, species richness and endemicity pattern in the region is distinctly controlled by temperature and rainfall gradients prevailing across south-north and west-east directions of the Ghats (Pascal 1982; Gimaret-Carpentier et al. 2003). Pollen based studies explains that it is biological legacy of Cenozoic flora that survived as 'refugia' through the cool Quaternary glacial periods (Prasad et al. 2009; Farooqui et al. 2010). Several studies focusing on palynofloral based palaeovegetation of the region reveal the tropical vegetation of Paleogene-Neogene times was different from the composition of modern vegetation type (Prasad et al. 2009). However, due to unavailability of precise dating of majority of studied sedimentary successions cannot be discussed in term of vegetational and palaeoclimatic evolution.

The Neogene sedimentary successions of Kerala Basin provide such opportunity to study palaeovegetational succession during Miocene period. Among these, the sedimentary successions of Quilon Formation exposed at type section in Kollam district has been considered to be of late Early Miocene age on the basis of fossil content and biostratigraphic correlation (Dey 1961; Verma 1977; Reuter et al. 2011). The Quilon Formation corresponds to a global warming event known as Mid Miocene Climate Optimum (MMCO) at 17–15 Ma (Zachos et al. 2001). It was the time when earth experienced average increase of ~3–4 °C in global annual surface temperature as compare to present. Interestingly, it is equivalent to the rise in temperature predicted for next century (You et al. 2009; You 2010). In the present study we present late Early Miocene pollen flora from the mixed siliciclastic-carbonate Quilon Formation. Since the general geographical position and bathymetry is similar when compare to late Early Miocene, the palynoflora from the studied succession can be considered as direct proxy for comprehension of palaeovegetation and palaeoclimate prevailing at the onset of the Middle Miocene Climate Optimum (MMCO).

2 Geological Setting and Stratigraphy

Kerala Basin is the southern sub-basin of the pericratonic Konkan-Kerala Basin on the western Indian passive continental margin. It is separated from the northerly Konkan Basin by the Tellicherry Arch basement high and bordered by steep escarpments of up to 2695 m high Western Ghats in the east. To the West, the basin continues into deep-waters to the Chagos-Laccadive Ridge. The Cenozoic sedimentary succession in the onshore part of the Kerala Basin is dominated by siliciclastic sediments with interbedded lignite seams. Elevation and denudation of the Western Ghats was the source for siliciclastics (Campanile et al. 2008) in the basin.

Firstly, King (1882) and Foote (1883) divided the Cenozoic rocks of the Quilon-Varkala area into Quilon beds, consisting of limestone and calcareous clays, Warkalli beds comprising sandstone and clays with lignites. Bore hole data of CGWB (Soman 2013) suggests three units: Warkalli Formation—consists of variegated clays, carbonaceous clays, sands and seams of peaty lignite; Quilon Formation—characterized by fossiliferous limestone and intercalations of calcareous clays (Poulose and Narayanaswamy 1968); subsurface Vaikom Formation—consist of gravel, coarse sand and carbonaceous clays with thin seams of lignite (Raghav Rao 1976), resting un-conformably over the Archean crystalline complex (Figs. 1 and 2). The Lower-Middle Miocene mixed siliciclastic-carbonate Quilon Formation extends between Edava in the south and Alappuzha in the north (Narayanan et al. 2007). However due to intense weathering and erosion the exposures are very patchy.

3 Material and Methods

For the present study systematic sampling was done from around 5 m thick sedimentary succession exposed at the type section of Quilon Formation (N 8°58′36.6″: E 76°38′08.4″) along the banks of Ashtamudi Kayal at Padappakara village, district Kollam (Fig. 1). Total nine samples were collected from the basal 1.5 m succession consisting of fossiliferous limestone and carbonaceous shale with well-preserved foraminifera test and bivalve shells. Starting from the base, five samples were taken from 60 cm thick sandy shale and two samples each from overlying 40 cm thick carbonaceous shale and 50 cm thick limestone (Fig. 1).

For the recovery of palynomorphs, the samples were processed as per the standard procedure of maceration techniques (Traverse 2007). Twenty-five grams of each sample was crushed and was initially treated with 10% HCl and 40% HF for removal of carbonate and silicate minerals respectively. Thereafter, the samples were rinsed thoroughly with distilled water and treated with 40% HNO_3 for oxidation and recovery of spores-pollen. The residue was then treated with 5% KOH and washed repeatedly in distilled water to completely remove the alkali and was finally sieved through a 15 μm mesh sieve. The final residue, separated through gravity separation

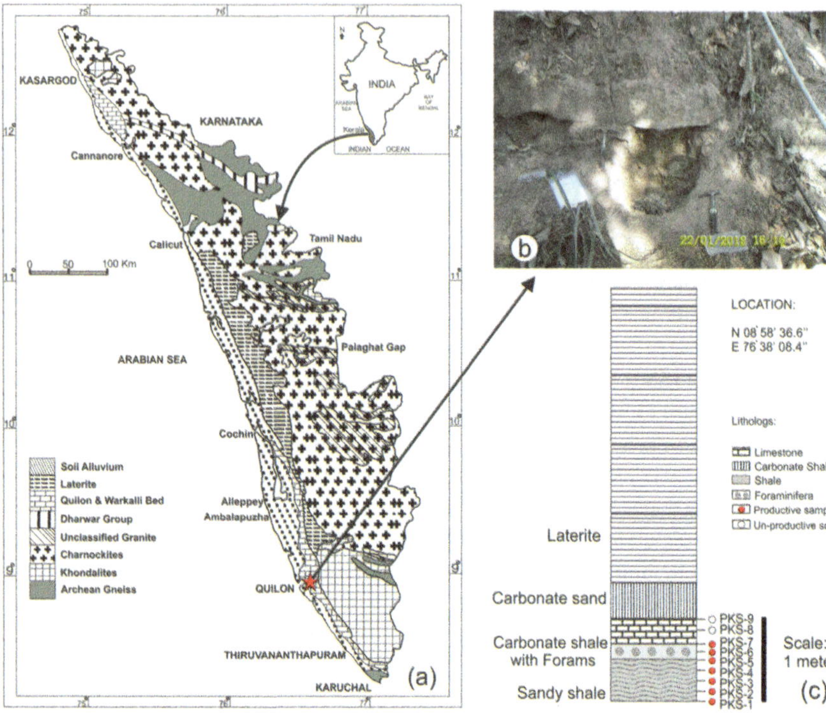

Fig. 1 (**a**) Geological map of Kerala (after Soman 2013); (**b**) Field photograph, (**c**) Lithocolumn of sampled type section of Quilon Formation, Kollam district, Kerala, India

method, was mixed with polyvinyl alcohol, spread evenly on the coverslips and dried. Permanent slides were prepared by using Canada balsam as mounting media. Photomicrography of palynomorphs was carried out using a Nikon Eclipse 90i microscope with a Canon digital camera under normal transmitted light and DIC. Out of nine samples, seven (PKS-1 to 7) were found palynologically productive. The significant specimens are illustrated in Fig. 3. The England Finder coordinates (EF) of each palynomorph on the respective slides are given in the figure caption. All the slides are deposited in the repository of the Birbal Sahni Institute of Palaeosciences, Lucknow, with statement number-1555 (BSIP Slide number-16742 to 16758).

4 Results and Interpretations

Palynoflora recovered from seven productive samples (PKS-1 to 7) collected from Type section of Quilon Formation exposed at Padappakara near Ashtamudi kayal in Kollam district have yielded variety of pteridophytic spores, angiosperm pollen, dinoflagellate cysts, algal and fungal remains. Highly diverse pteridophytic spores

	Formation	Lithology
MALABAR SUPERGROUP	Recent to sub-Recent	Soils and alluvium Beach sand deposits Lime shell deposits of backwaters Old and red Teri sands of subrecent marine and lacustrine formations Peat beds with semi-carbonised woods Calcareous clays with shell etc. Laterite
	~~Unconformity~~~	
	Warkalli Formation (Late Miocene to Pliocene)	Current-bedded friable variegated sandstone interbedded with plastic clay and variegated clays Carboncaeous and alum clays with (Mio-Pliocene) lignite seams Gravel and pebble beds. Base marked by gibbsitic sedimentary clays
	Quilon Formation (Middle Miocene)	Fossiliferous shell limestone alternating with thick beds of sandy clays Calcareous clays and sandstones. Base Unknown
	~~Unconformity~~~	
		Archean Crystalline Complex

Fig. 2 Lithostratigraphic classification of Kerala Basin, India (After Poulose and Narayanaswamy 1968)

and angiosperm pollen assemblage consists of around 43 genera and 55 species. Qualitative and quantitative estimates reveal that the assemblage is dominated by angiosperm pollen followed by pteridophytic spores. The palynomorphs have affinities with 25 extant families and subfamilies: Polypodiaceae, Dicksoniaceae, Matoniaceae, Schizaeaceae, Lycopodiaceae, Cyatheaceae, Gleicheniaceae, Ophioglossaceae, Anacardiaceae, Annonaceae, Arecaceae, Bombacaceae, Ctenolophonaceae, Dipterocarpaceae, Euphorbiaceae (Crotonoideae), Liliaceae, Lythraceae (Sonneratioideae), Meliaceae, Myricaceae, Onagraceae, Oleaceae, Fabaceae (Cesalpinoideae and Ceridoideae), Proteaceae, Sapotaceae, Thymelaeaceae. The botanical affinities of the palynotaxa and comparable families are given in Table 1. Of these, 7 families are restricted to tropical climate, 9 families pertain to tropical-subtropical, rest families are cosmopolitan (Table 1).

The palynological assemblage is dominated by angiospermous pollen grains (65%) followed by pteridophytic spores (25%), dinoflagellate cysts (8%) and fungal and algal remains (2%). Amongst the pteridophytes, the spores of the families Polypodiaceae (*Polypodiisporites* spp.), Schizaeaceae (*Lygodiumsporites* spp.) and Osmundaceae (*Todisporites* sp.) are the most dominant families, prefer moist and shady habitat with perennial water in close vicinity. Ferns belonging to the families Cyatheaceae (*Cyathidites*) and Lycopodiaceae (*Lycopodiumsporites*) generally make understory of modern thick tropical rain forests (Rao 1995). Among the dinoflagellate cysts several species of *Cleistosphaeridium*, *Hystrichokolpoma*, *Polysphaeridium*, *Spiniferites*, *Homotryblium*, *Pentadinium* have been recorded.

Fig. 3 Significant palynomorphs recovered from the samples of Quilon Formation, Kerala Basin, India. The morphotype names are followed by BSIP Slide number and the England Finder (EF) Coordinates (scale on all photographs is of 10 μm). (1) *Biretisporites* sp., BSIP Slide no. 16742, EF E40. (2) *Crassoretitriletes vanraadshooveni*, BSIP Slide no. 16743, EF K47. (3) *Monolites* sp., BSIP Slide no. 16744, EF F49/2. (4) *Polypodiisporites ratnamii,* BSIP Slide no. 16745, EF E18/1. (5) *Cheilanthoidospora monoleta*, BSIP Slide no. 16746, EF H45/2. (6) *Clavaperiporites jacobii,* BSIP Slide no. 16746, EF H40/3. (7) *Spinizonocolpites* sp., BSIP Slide no. 16747, EF O18/2. (8) *Spinizonocolpites echinatus*, BSIP Slide no. 16748, EF D28/1. (9) *Quilonipollenites* sp., BSIP Slide no. 16746, EF M45. (10) *Quilonipollenites sahnii*, BSIP Slide no. 16749, EF H36/2. (11) *Proteacidites triangulus*, BSIP Slide no. 16750, EF V44.

Amongst the angiosperms, morphospecies having modern affinity with the family Bombacaceae (*Lakiapollis*, *Tricolporopollis*, *Tricolporocollumellites*, *Dermatobrevicolporites*), Ctenolophonaceae (*Ctenolophonidites* and *Retistephanocolporites*) and Dipterocarpaceae (*Dipterocarpuspollenites*) and subfamily Caesalpinioideae (*Margocolporites*) are most diverse and dominant. Pollen grains showing affinity with the family Arecaceae represented by several morphogenera of *Spinizonocolpites*, *Monocolpopollenites* and *Quilonipollenites*, *Neocouperipollis* are present throughout the succession.

Abundantly recorded *Lakiapollis ovatus* and *Dermatobrevicolporites dermatus* show close affinities with the extant *Durio*-type pollen particularly *Durioconicus Beccari* and to *Durio graveolens* Wyath-Smith, respectively of the subfamily Bombacaceae (Thanikaimoni et al. 1984; Venkatachala et al. 1989; Mandal 2005). *Tricolporopollis* spp. are considered to be extinct members of the same tribe Durioneae (Mandal 2005). Highly diversified Bombacaceous pollen grains are recorded in present deposit. However, the tribe Duroneae is presently restricted to Southeast Asian islands with only two species of the genus *Cullenia* present in India (Mabberley 1997; Bose et al. 1998).

The species of *Spinizonocolpites* are nearest living relative of modern brackish water Palm Nypa (Arecaceae) (Muller 1968). The genus *Nypa* is presently represented by one living species (*N. fruticans*) which is restricted to a narrow tropical area of Southeast Asia, including Indian mangroves of Sundarbans and Andaman Islands. The diverse palm pollen in the present assemblage indicates presence of dense tropical rainforest under warm and humid climate (Greenwood and Wing 1995; Morley 2000, 2003) fringed by mangroves in the region.

The several morphospecies of family Ctenolophonaceae described under the generic name *Ctenolophonidites* and *Retistephanocolpites* are abundantly recovered from Quilon sediments. However, the family is now extinct from Indian subcontinent. Presently, it occurs as mono-generic family restricted in riverine forest of West African tropics mainly in Nigeria, Congo, Angola and Zaire and lowland primary forest and peat swamps in Malesian islands (Kubitzki 2014). Another morphogenus *Margocolporites* belonging to subfamily Caesalpinioideae are represented by

Fig. 3 (continued) (12) *Albertpollenites retibaculatus*, BSIP Slide no. 16749, EF E27/2. (13) *Margocolporites tsukadae*, BSIP Slide no. 16751, EF F32. (14) *Margocolporites sitholeyi*, BSIP Slide no. 16750, EF 41. (15) *Retitrescolpites megareticulatus*, BSIP Slide no. 16752, EF K39. (16, 19) *Dermatobrevitricolporites dermatus*, BSIP Slide no. 16753, EF M27/3, BSIP Slide no. 16755, EF N26. (17) *Trisyncolpites ramanujamii*, BSIP Slide no. 16754, EF N27/3. (18) *Sapotaceoidaepollenites keralaensis*, BSIP Slide no. 16751, EF L35/1. (20) *Dermatobrevicolporites exaltus*, BSIP Slide no. 16756, EF W48/4. (21) *Dipterocarpuspollenites retipilatus*, BSIP Slide no. 16757, EF O44/4. (22) *Lakiapollis ovatus*, BSIP Slide no. 16745, EF M22. (23) *Tricolporopollis matanomadhensis*, BSIP Slide no. 16750, EF Q40/2. (24) *Tribrevicolporites duttae*, BSIP Slide no. 16758, EF N13/3. (25) *Ctenolophonidites keralensis*, BSIP Slide no. 16745, EF K48/1. (26) *Ctenolophonidites saadii*, BSIP Slide no. 16752, EF U44/2. (27) *Ctenolophonidites costatus*, BSIP Slide no. 16750, EF V46/1. (28) *Retistephanocolpites williamsii*, BSIP Slide no. 16750, EF H39/4

Table 1 List of palynotaxa recovered from sediments of Quilon Formation, Kerala Basin, India, and extant ecological distribution of families

Palynotaxa	Family	Ecological distribution
Cyathidites australis (Couper 1953)	Cyatheaceae	Tropical–sub tropical
Lygodiumsporites lakiensis (Sah and Kar 1969)	Schizaeaceae	Tropical–sub tropical
Lygodiumsporites padappakkarensis (Rao and Ramanujam 1978)		
Schizaeoisporites multistriatus (Rao and Ramanujam 1978)		
Crassoretitriletes vanraadshoovenii (Germeraad 1968)		
Dictyophyllidites granulatus (Saxena 1978)	Dicksoniaceae	
Biretisporites convexus (Sah and Kar 1969)	Matoniaceae	
Verrucosisporites dakshinensis (Rao and Ramanujam 1978)	Lycopodiaceae	Tropical–sub tropical
Gleicheniidites senonicus (Ross 1949)	Gleicheniaceae	
Foveosporites miocenicus (Ramanujam 1972)	Ophioglossaceae	
Todisporites sp.	Osmundaceae	Tropical to temperate, cosmopolitan
Cingulatisporites sinuatus (Rao and Ramanujam 1978)	Unidentified	
Laevigatosporites ovatus (Wilson and Webster 1946)	Polypodiaceae	Tropical–sub tropical
Polypodiisporites miocenicus (Rao and Ramanujam 1976)		
Polypodiisporonites ornatus (Sah) (Saxena and Trivedi 2009)		
Polypodiisporites perverrucatus (Couper) (Khan and Martin 1971)		
Polypodiisporites ratnamii (Ramanujam) (Rao and Ramanujam 1978)		
Polypodiisporites usmensis (Germeraad) (Khan and Martin 1971)		
Polypodiisporites verrucosus (Sah and Kar) (Singh et al. 1985)		
Monolites sp.		
Quilonipollenites sahnii (Rao and Ramanujam 1978) *Quilonipollenites* sp.	Arecaceae	Tropical–sub tropical
Monocolpopollenites kutchensis (Venkatachala and Kar)(Saxena and Trivedi 2009)		
Spinizonocolpites echinatus (Muller 1968)		
Margocolporites sitholeyi (Ramanujam 1966)	Fabaceae	Tropical–sub tropical

(continued)

Table 1 (continued)

Palynotaxa	Family	Ecological distribution
Margocolporites tsukadae (Ramanujam 1966), *Margocolporites sahnii* (Ramanujam 1966)		
Lakiapollis ovatus (Venkatachala and Kar 1969)	Bombacaceae	Tropical–sub tropical
Dermatobrevicolporites dermatus (Sah and Kar) (Kar 1985)		
Dermatobrevicolporites alleppeyensis (Rao 1996)		
Dermatobrevicolporites exaltus (Kar 1985)		
Tricolporocolumellites pilatus (Kar 1985)		
Tricolporopollis matanomadhensis (Venkatachala and Kar) (Tripathi and Singh 1992)		
Tricolporopollis decorus (Dutta and Sah 1970)		
Tribrevicolporites duttae (Rao and Rajendran 1996)		
Tribrevicolporites sarkarii (Rao and Rajendran 1996)		
Crotonoidaepollenites euphorbioides (Rao and Ramanujam 1983)	Euphorbiaceae (Crotonoideae)	Tropical to temperate, cosmopolitan
Crotonisulcites grandis (Rao and Ramanujam 1978)		
Verrutriporites annulatus (Kar and Jain 1981)	Lythraceae (Sonneratioideae)	
Dipterocarpuspollenites retipilatus (Kar and Jain 1981) (Kar 1992)	Dipterocarpaceae	Tropical–sub tropical
Albertipollenites retibaculatus (Saxena) (Mandal and Rao 2001)		
Albertipollenites robustus (Sah and Kar) (Mandal and Rao 2001)		
Ctenolophonidites costatus (von Hoeken Klinkenberg 1966) *Ctenolophonidites keralensis* (Ramanujam and Rao 1971) *Ctenolophonidites saadi* (Ramanujam and Rao 1971)	Ctenolophonaceae	Tropical
Retistephanocolpites williamsi (Germeraad 1968)		
Meliapollis quadrangularis (Ramanujam) (Sah and Kar 1969)	Meliaceae	Tropical
Triorites quilonensis (Rao and Nair 1998)	Onagraceae	
Myricipites singhii (Rao 1995)	Myricaceae	Tropical to Temperate, Cosmopolitan

(continued)

Table 1 (continued)

Palynotaxa	Family	Ecological distribution
Retitrescolpites megareticulatus (Mathur) (Mandal and Rao 2001)	Oleaceae	Tropical to temperate, cosmopolitan
Trisyncolpites ramanujamii (Kar 1979)	Fabaceae (Cesalpinoideae)	Tropical–sub tropical
Striacolporites ovatus (Sah and Kar 1969)	Fabaceae (Ceridoideae)	
Proteacidites triangulus (Kar and Jain 1981)	Proteaceae	Tropical
Liliacidites sp.	Liliaceae	Tropical
Matanomadhiasulcites kutchensis (Saxena) (Kar 1985)	Annonaceae	Tropical–sub tropical
Sapotaceoidaepollenites keralaensis (Rao and Ramanujam 1983)	Sapotaceae	Tropical
Clavaperiporites jacobii (Ramanujam 1966)	Thymelaeaceae	Tropical to temperate, cosmopolitan
Foveotricolporites sp.	Unidentified	
Cheilanthoidspora monoleta (Sah and Kar 1972)		
Cheilanthoidspora mioceneca (Kar and Jain 1981)		

several species (Table.1). *M. sahnii* show close resemblance with extant pollen of *Agrostistachys meeboldii* that is endemic plant of Western Ghats, generally occur in low elevation of wet evergreen forest (Prasad et al. 2009).

The palynological assemblage contained the major families that are characteristic of today's rain forest of the region, however, the dominance of components of megathermal families such as Bombacaeae, Ctenolophonaceae, Dipterocarpacae and Arecaceae are noteworthy. It reveals that tropical rainforest of Kerala in Early Miocene times were dominated by many megathermal plant families that indicates intense warmer and wetter climate condition in comparison to today's climate of Kerala Basin.

5 Discussion

The Miocene Period is considered to be period of "making of the modern world" (Potter and Szatmari 2009). The major uplift of modern mountain chains, the initiation of bipolar glaciations, the origin of modern ocean currents, the aridification of the continental interiors, the reduction in atmospheric CO_2 levels and the overall cooling trend of the global climate occurred in Miocene time (Zachos et al. 2008; Potter and Szatmari 2009; Beerling and Royer 2011). The late Miocene is also important from the perspective of environmental changes, such as the expansion of open landscapes and the spread of C4 grasses (Quade et al. 1989), the aridification of the continental interior and the intensification of the Asian monsoon climate (Quade et al. 1989; Molnar et al. 1993; Xing et al. 2012). The palaeovegetational changes under Asian monsoon system during the late Miocene has been reconstructed mainly from Siwaliks (Quade et al. 1989; Sanyal et al. 2004, 2005a, b, 2010), Northeast India (Khan et al. 2014) and China (Wang 1994; Xu et al. 2008; Xia et al. 2009; Jacques et al. 2011a, b; Xing et al. 2012) by using micro- and mega-floras and stable isotopes. However, other part of Indian subcontinent remains poorly explored. Present study have been done to analyse palynological data from the Kerala basin which have provided insights into palaeofloral changes occurred around Middle Miocene warming in tropical region.

The Quilon beds have an important position in Indian stratigraphy. The succession represents the evidence of marine transgression along the south-west coast of India during Miocene, similar to the marine deposits along northwestern and western parts of the Indian subcontinent (Verma 1977). Ostracods and palynological studies suggested depositional environment of these sediments ranges from marginal marine, brackish lagoon as well as brackish and freshwater swamps (Rao and Ramanujam 1975; Rao 1995). The open marine shelf shallower than 20 m depth with coral reef had also been inferred during deposition of the Quilon Formation (Menon 1967a, b; Raha and Sinha-Roy 1982; Narayanan et al. 2007). Recently, facies and faunal assemblage from the same formation suggested sea grass environments developed during marine transgression into marginal marine lagoons and swamps in the region (Reuter et al. 2011).

Palynological studies of the Cenozoic sediments of the Kerala basin had been carried by many workers such as Rao and Vimal (1953), Potonie and Sah (1958), Ramanujam (1977, 1987), Rao and Ramanujam (1975, 1978, 1982), Kar and Jain (1981), Raha et al. (1987), Rajendran et al. (1989), Rao (1990, 1995, 1996), Rao and Rajendran (1996) and Rao and Nair (1998). These studies also provided floristically diverse palynological assemblages consist of fungal remains, pteridophytic spores and angiospermous pollen from Quilon and Warkalli successions. The pollen of Schizaeaceae, Parkeriaceae, Osmundaceae, Adiantaceae, Polypodiaceae, Caesalpiniaceae, Clusiaceae, Ctenolophonaceae, Oleaceae and Rubiaceae families together with the tropical rain forest elements (Ctenolophonaceae, Oleaceae, Proteaceae and Moraceae) indicate high rainfall. It was also suggested that since Miocene time the climate pattern of Kerala region have not much changed (Ramanujam 1987). The palynoflora assemblage from present study suggests humid and tropical climate with plenty of rainfall throughout the year. The dominance of megathermal families in palynological assemblage indicates enhanced warming and precipitation in region around mid-Miocene times.

6 Conclusions

- A rich and diverse palynological assemblage consisting of algal and fungal remains, dinoflagellate cysts, pteridophytic spores, angiosperm pollen recovered from the sediment samples Quilon Formation, Kerala.
- The recovered assemblage belongs to 25 extant families of pteridophyte and angiosperm and is dominated by megathermal families such as Bombacaceae, Ctenolophonaceae, Dipterocarpaceae and Oleaceae along with mangrove and back-mangrove taxa mainly of Arecaceae family.
- Overall, the palynological assemblage of Quilon Formation indicates a warmer and more humid climate in Kerala during late Early Miocene (Burdigalian) time period.
- The deposition took place in mangrove-dominated, fluctuating lagoon in marginal marine environment surrounded by megathermal tropical rainforest.

Acknowledgments The authors would like to thank the Director, Birbal Sahni Institute of Palaeosciences (BSIP) for providing infrastructural facilities, and Research Development Co-ordination Cell (RDCC/78/2020-21), BSIP, Lucknow for necessary permission. The authors also acknowledge financial support from Department of Science and Technology (DST-SERB), New Delhi (Project no. EMR/2016/005983).

References

Beerling DJ, Royer DL (2011) Convergent cenozoic CO_2 history. Nat Geosci 4:418–420

Bose TK, Das P, Maiti GG (1998) Tress of the world. Bhubaneshwar, Regional Plant Resource Centre

Campanile D, Nambiar CG, Bishop P, Widdowson M, Brown R (2008) Sedimentation record in the Konkan-Kerala Basin: implications for the evolution of the Western Ghats and the Western Indian passive margin. Basin Res 20:3–22

Dey AK (1961) The Miocene Mollusca from Quilon, Kerala (India). Mem Geol Surv India 36:1–129

Farooqui A, Ray JG, Farooqui SA, Tiwari RK, Khan ZA (2010) Tropical rainforest vegetation, climate and sea level during the Pleistocene in Kerala, India. Quat Int 213(1–2):2–11

Foote RB (1883) On the geology of South Travancore. Records Geol Surv India 16:20–35

Gimaret-Carpentier C, Dray S, Pascal JP (2003) Broad-scale biodiversity pattern of the endemic tree flora of the Western Ghats (India) using canonical correlation analysis of herbarium records. Ecography 26:429–444

Greenwood DR, Wing SL (1995) Eocene continental climates and latitudinal temperature gradients. Geology 23(11):1044–1048

Jacques FMB, Su T, Spicer RA, Xing YW, Huang YJ, Wang WM, Zhou ZK (2011a) Leaf physiognomy and climates: are monsoon systems different? Glob Planet Chang 76:56–62

Jacques FMB, Guo SX, Su T, Xing YW, Huang YJ, Liu YS, Ferguson DK, Zhou ZK (2011b) Quantitative reconstruction of the late Miocene monsoon climates of Southwest China: a case study of the Lincang flora from Yunnan Province. Palaeogeogr Palaeoclimatol Palaeoecol 304(3–4):318–327

Kar RK, Jain KP (1981) Palynology of the Neogene sediments around Quilon and Varkala, Kerala coast, South India-2. Spores and pollen grains. Palaeobotanist 27(2):113–131
Khan MA, Spicer RA, Bera S, Ghosh R, Yang J, Spicer TEV, Guo S, Su T, Jacques FMB, Grote PJ (2014) Miocene to Pleistocene floras and climate of the eastern Himalayan Siwaliks, and new palaeoelevation estimates for the Namling-Oiyug Basin, Tibet. Global Planet Chang 113:1–10
King W (1882) General sketch of the geology of Travancore state. The Warkalli beds and associated deposits at Quilon in Travancore. Records Geol Surv India 15:93–102
Kubitzki K (ed) (2014) Flowering plants. Eudicots. The families and genera of vascular plants 11. Springer, Berlin
Mabberley DJ (1997) The plant-book. Cambridge University Press, Cambridge, p 680
Mandal J (2005) Bombacaceae pollen from the Indian tertiary sediments and its bearing on evolution and migration. Rev Palaeobot Palynol 133(3–4):277–293
Menon KK (1967a) The lithology and sequence of the Quilon beds. Proc Indian Acad Sci 65:20–25
Menon KK (1967b) Origin of diagenetic pyrite in the Quilon limestone, Kerala, India. Nature 213: 1219–1220
Molnar P, England P, Martinod J (1993) Mantle dynamics, uplift of the Tibetan plateau, and the Indian monsoon. Rev Geophys 31:357–396
Morley RJ (2000) Origin and evolution of tropical rain forests. Wiley, London
Morley RJ (2003) Interplate dispersal routes for megathermal angiosperms. Perspect Plant Ecol Evol Syst 6:5–20
Muller J (1968) Palynology of the Pedawan and plateau sandstone formations (cretaceous-Eocene) in Sarawak, Malaysia. Micropaleontology 14:1–37
Myers N, Mittermeier RA, Mittermeier CG, da Fonseca GAB, Kent J (2000) Biodiversity hotspots for conservation priorities. Nature 403:853–858
Narayanan V, Anirudhan S, Grottoli AG (2007) Oxygen and carbon isotope analysis of the Miocene limestone of Kerala and its implications to palaeoclimate and its depositional setting. Curr Sci 93:1155–1158
Nayar TS, Beegam AR, Sibi M (2014) Flowering plants of the Western Ghats, India. Tropical Botanic Garden and Research Institute, Thiruvananthapuram, Kerala, India
Pascal JP (1982) Bioclomates of the Western Ghats at 1/250,000 (2 sheets). Institut francais de Pondichéry, trav sec sci tech Hors série 17
Potonie R, Sah SCD (1958) Sporae dispersae of the lignites from Cannanore beach on the Malabar coast of India. Palaeobotanist 7:121–135
Potter PE, Szatmari P (2009) Global Miocene tectonics and the modern world. Earth Sci Rev 96: 279–295
Poulose KV, Narayanaswamy S (1968) The tertiaries of Kerala coast. Mem Geol Soc India 2:300–308
Prasad V, Farooqui A, Tripathi SKM, Garg R, Thakur B (2009) Evidence of late Palaeocene-early Eocene equatorial rain forest refugia in southern Western Ghats, India. J Biosci 34(5):777–797
Quade J, Cerling TE, Bowmen JR (1989) Development of Asian monsoon revealed by marked ecological shift during the latest Miocene in northern Pakistan. Nature 342:163–166
Raghav Rao KV (1976) Ground water exploration, development and long term aquifer management in Kerala. In: Proceedings symposium mineral resources of Kerala and their utilization, Trivandrum, October 1975, pp 30–36
Raha PK, Sinha-Roy S (1982) Glauconite in the Quilon limestone (early Miocene) of coastal Kerala Sedimentary Basin. Curr Sci 51:608–609
Raha PK, Rajendran CP, Kar RK (1987) Record of early tertiary deposits in Kerala, India and its palaeogeographic significance. Geophytology 17:209–218
Rajendran CP, Raha PK, Kar RK (1989) Palynological assemblages from Neogene outcrops of Kerala coast, India. Indian Miner 43:39–46
Ramanujam CGK (1977) A palynological approach to the study of Warkalli deposits of Kerala in South India. Geophytology 7(2):160–164

Ramanujam CGK (1987) Palynology of the Neogene Warkalli beds of Kerala state in South India. J Palaeontol Soc India 32:26–46
Rao MR (1990) Palynological investigation of Arthungal borehole, Alleppey District, Kerala. In: Jain KP, Tiwari RS (eds) Proceedings of Symposium in Vistas in Indian Palaeobotany The Palaeobotanist, vol 38, pp 243–255
Rao MR (1995) Palynostratigraphic zonation and correlation of the Eocene-early Miocene sequences in Alleppey district, Kerala, India. Rev Palaeobot Palynol 86:325–348
Rao MR (1996) An early Miocene palynofloral assemblage from Turavur borehole, Alleppey District, Kerala–its palaeoecological and stratigraphical significance. Geophytology 25:155–163
Rao MR, Nair KK (1998) Palynological investigation of Miocene sediments exposed in Kannanellur–Kundara area, Quilon District, Kerala. Geophytology 27:49–59
Rao MR, Rajendran CP (1996) Palynological investigations of tertiary lignite and associated sediments from Cannanore, Kerala Basin, India. Palaeobotanist 43:63–82
Rao KP, Ramanujam CGK (1975) A palynological approach to the study of Quilon beds of Kerala state in South India. Curr Sci 44:730–732
Rao KP, Ramanujam CGK (1978) Palynology of the Neogene Quilon beds of Kerala state in South India-I. Spores of pteridophytes and pollen of monocotyledons. Palaeobotanist 25:397–427
Rao KP, Ramanujam CGK (1982) Palynology of the Quilon beds of Kerala state in South India-II. Pollen of dicotyledons and discussion. Palaeobotanist 30:68–100
Rao AR, Vimal KP (1953) Preliminary observation on the microfossil content of some lignites from Warkalli in Travancore. Curr Sci 21:210–215
Reuter M, Piller WE, Harzhauser M, Kroh A, Rogl F, Coric S (2011) The Quilon limestone, Kerala Basin, India: an archive for Miocene Indo-Pacific seagrass beds. Lethaia 44:76–86
Sanyal P, Bhattacharya SK, Kumar R, Ghosh SK, Sangode SJ (2004) Mio-Pliocene monsoonal record from Himalayan foreland basin (Indian Siwalik) and its relation to the vegetation change. Palaeogeogr Palaeoclimatol Palaeoecol 205:23–41
Sanyal P, Bhattacharya SK, Kumar R, Ghosh SK, Sangode SJ (2005a) Palaeovegetational reconstruction in late Miocene: a case study based on early diagenetic carbonate cement from the Indian Siwalik. Palaeogeogr Palaeoclimatol Palaeoecol 228:245–259
Sanyal P, Bhattacharya SK, Prasad M (2005b) Chemical diagenesis of Siwalik sandstone: isotopic and mineralogical proxies from Surai Khola section, Nepal. Sediment Geol 180:57–74
Sanyal P, Sarkar A, Bhattacharya SK, Kumar R, Ghosh SK, Agrawal S (2010) Phased intensification of monsoon, microclimate and asynchronous C4 appearance: isotopic evidence from the Indian Siwalik sediments. Palaeogeogr Palaeoclimatol Palaeoecol 296:165–173
Soman K (2013) Geology of Kerala. Geological Society of India, Bangalore
Thanikaimoni G, Caratani C, Venkatachala BS, Ramanujam CGK, Kar RK (1984) Selected angiosperm pollens from India and their relationship with African tertiary pollens. Trav Sec Sci Tech 19:1–12
Traverse A (2007) Paleopalynology. Unwin Hyman, Boston, MA, p 600
Venkatachala BS, Caratini C, Tissot C, Kar RK (1989) Paleocene-Eocene marker pollen from India and tropical Africa. Palaeobotanist 37(1):1–25
Verma KK (1977) Cancroid crabs from the Quilon beds (lower Miocene) of Kerala, India. J Palaeontol Soc India 20:305–313
Wang WM (1994) Paleofloristic and paleoclimatic implications of Neogene palynofloras in China. Rev Palaeobot Palynol 82:239–250
Xia K, Su T, Liu YS, Xing YW, Jacques FMB, Zhou ZK (2009) Quantitative climate reconstructions of the late Miocene Xiaolongtan megaflora from Yunnan, Southwest China. Palaeogeogr Palaeoclimatol Palaeoecol 276:80–86
Xing YW, Utescher T, Jacques FMB, Su T, Liu YSC, Huang YJ, Zhou Z (2012) Paleoclimatic estimation reveals a weak winter monsoon in southwestern China during the late Miocene: evidence from plant macrofossils. Palaeogeogr Palaeoclimatol Palaeoecol 358–360:19–26

Xu JK, Ferguson DK, Li CS, Wang YF (2008) Late Miocene vegetation and climate of the Lühe region in Yunnan, southwestern China. Rev Palaeobot Palynol 148:36–59

You Y (2010) Climate model evaluation of the contribution of sea–surface temperature and carbon dioxide to the middle Miocene climate optimum as a possible analogue of future climate change. Aust J Earth Sci 57:207–219

You Y, Huber M, Müller RD, Poulsen CJ, Ribbe J (2009) Simulation of the middle Miocene climate optimum. Geophys Res Lett 36:L04702

Zachos JC, Shackleton NJ, Revenaugh JS, Palike H, Flower BP (2001) Climate response to orbital forcing across the Oligocene-Miocene boundary. Science 292:274–277

Zachos JC, Dickens GR, Zeebe RE (2008) An early Cenozoic perspective on greenhouse warming and carbon cycle dynamics. Nature 451:279–283

Non-Pollen Palynomorphs from the Late-Holocene Sediments of Majuli Island, Assam (Indo-Burma Region): Implications to Palaeoenvironmental Studies

Arya Pandey, Swati Tripathi, and Sadhan Kumar Basumatary

Abstract A total 31 sediment samples from a 150 cm deep sedimentary core was examined for the Non-Pollen Palynomorphs (NPPs) analysis from the Sakali wetland in order to provide an overview of palaeoenvironment in Majuli Island (world largest river island), Assam for the late Holocene. About 25 varieties of non-pollen palynomorphs were reported, out of which fungal spores were at high abundance along with scanty occurrence of zoological remains reflecting the past climate vegetation and faunal interactions in the region. The dominance of coprophilous fungi like *Sordaria, Podospora, Ascodesmis, Coniochaeta* (almost 40%) indicates the past occurrence of vast open-land areas with grazing activities of herbivorous animals. Some non-coprophilous fungi like *Tetraploa, Dictyosporium, Cookeina* indicates the rich floral diversity around the study site. Other fungal remains like *Valsaria, Alternaria, Geastrum* and *Diploidia* along with the presence of zoological remains like *Neorhabdocoela* are indicative of the freshwater ecosystem with diversified rich flora indicating warm and humid climate conditions in the region. The presence of *Entophlyctis lobata* at the bottom of the sedimentary core indicates the relatively dry climatic conditions in the island because this fungal spore is specific of the temperate region. The frequent soil erosional activities could be evident through the dominance of branched and solitary *Glomus*, attributable to the high flood-prone region resulting in the mixing of local vegetation with the outlandish vegetation. The scanty occurrence of *Botryococcus*, supports the high energy levels in wetland water, attributed to frequent flood activities. All these NPP varieties of fungal, algal and zoological affinities collectively display the past forest cover, palaeo-depositional environment, past climatic conditions, anthropogenic response and grazing activities in Majuli Island of Assam.

A. Pandey · S. Tripathi (✉) · S. K. Basumatary
Quaternary Palynology Laboratory, Birbal Sahni Institute of Palaeosciences, Lucknow, Uttar Pradesh, India

B. Phartiyal et al. (eds.), *Climate Change and Environmental Impacts: Past, Present and Future Perspective*, Society of Earth Scientists Series,
https://doi.org/10.1007/978-3-031-13119-6_5

Keywords Late-Holocene · Non-pollen palynomorphs · Palaeoenvironment · Ecology, fungal spores · Majuli Island · Assam

1 Introduction

The Quaternary was a period of major environmental changes that were possibly greater than any other time in the last 60 million years (Prell and Kutzbach 1987; Clement et al. 1999) and the Holocene is a major epoch that had been the witness of some major rapid climatic events, like Holocene climatic optimum (HCO), Medieval Warm Period (MWP) and Little Ice Age (LIA) (Alley et al. 1997; Mayewski et al. 2004). The climatic changes that occurred during the Quaternary can be elaborated by biotic and abiotic proxies and records. Pollen and spores analysis may lead to strong inferences about the past climate and have a great role to reflect the vegetation due to their extreme resistant sporoderm and high production with wide distribution (Bradley 1985). However, owing to some taphonomic issues, such as bioturbation, translocation, sorting destruction (Birks and Birks 1980; Moore and Webb 1978), as well as differences in pollen production and dispersal between taxa, which depend on the plant species themselves and climatic conditions (Hicks 2001; Spieksma et al. 2003), the past ecological inferences are not possible to reconstruct accurately. Therefore, there is a dire need to look after an auxiliary proxy tool which could complement the pollen-inferred past climatic and depositional signals.

The utility of NPPs as paleoecological indicators has grown rapidly during the last decase or so, because of their frequent occurrence in palynological slides and potential to improve the reconstruction of past communities and environments, local grazing pressure and fire and human pressure (van Geel et al. 1994; van Geel and Aptroot 2006). Their fast germination and reproductive ability maintain their presence in the high amount in the sediments, even for a thousand years. Some of the fungal and algal remain often found in a mycorrhizal form, pathogenic form, or in a symbiotic relationship and they apprise about the presence of those co-occurred specific taxa around that area. Though many NPPs are still not properly identified but some of them can be used as palaeoecological indicators. In the present investigation, 25 different and distinguished NPP types, grouped comprehensively under three classifications as fungal, algal and zoological remains, have been considered, archived and represented from the Sakali wetland of Majuli Island, alongside their palaeoecological implications.

The Majuli is the largest river island in the world which is located on the Brahmaputra River in the state of Assam, northeast India (falls under the Indo-Burma biodiversity hot spot). The geographical and physiographic conditions of Majuli Island makes it a sensitive region for any natural calamities such as flood, soil erosion, cyclone, etc. Therefore, to overcome these natural calamities posing threats of physiographic disturbances and ecological imbalances, which would have serious consequences in Assam state, the present study could provide an insight to identify and demonstrate the Non-Pollen Paynomorphs (NPPs) especially fungal and zoological remains in tracing the palaeoenvironment in the Majuli Island.

The palyno-investigations provide valuable insights concerning changes in the vegetation and contemporaneous climatic oscillation on the Indian subcontinent since the Quaternary epoch. The few published records on vegetation succession and past climatic oscillations in Assam include studies of reserve forest of lower Brahmaputra valley and upper Assam region (Bhattacharya and Chanda 1992; Bera and Dixit 2011; Dixit and Bera 2012; Tripathi et al. 2020). However, despite the frequent occurrence in pollen slides, the NPP studies have never been carried out on the modern surface soil and Quaternary sediments from the Assam, except a solitary publication from the Rhino dung samples of Kaziranga National Park of Assam dealing with the coprophilous fungi (Basumatary and McDonald 2017). Nevertheless, the fungal spores have also been studied on the Holocene sediments of Mizoram, northeast India, for the palaeoclimatic interpretation (Mandaokar et al. 2008). The present study thus assesses the behaviour of non-pollen palynomorphs with the respective depositional and ecological setting around the Sakali wetland which may further strengthen and substantiate the pollen-based palaeoclimatic reconstruction in the Majuli Island and the surrounding region of northeast India.

2 Regional Setting

Majuli is the largest tropical river island in the Brahmaputra River located in the Indian state of Assam. It is allocated between latitude 93°30′–94°35′ E and longitude 26°50′–27°10′N at an elevation ranging from 60 to 85 m above mean sea level (Basumatary et al. 2018). Majuli had a total area of 1250 square kilometres (483 sq. mi), but having lost significantly to erosion it has an area of only 421.65 square kilometres (163 sq. mi) in 2001. With a population of 1.6 Lakhs, the majority being tribal have a very rich heritage and has been the abode of Assamese Vashnavite culture with a tremendous option for spiritual and Eco-tourism. The island comprises of many small endangered wetlands like Sakali, Duboi, Johai, Belguri, etc. The sedimentary core for the present study has been retrieved from the Sakali wetland. The Sakali wetland (26°52′N, 94°12′E) in the Majuli Island is located in close vicinity to the Brahmaputra River (Fig. 1). The wetland is bordered by open area and cultivated land. The wetland comprises approximately 0.70 km^2 but is much larger during heavy summer rains. During heavy rainfall, the infiltration of more rainwater from the Brahmaputra River results in flooding of the surrounding area. During winter it becomes smaller due to low rainfall (Basumatary et al. 2018). Majority of the research on Majuli has focused on bank erosion, rainfall pattern, drainage discharge of the Brahmaputra river, geomorphic changes in the river basin and vegetation (Das 2015). This island is one of the most dynamic landforms and highly sensitive towards the natural calamities, like earthquake and floods.

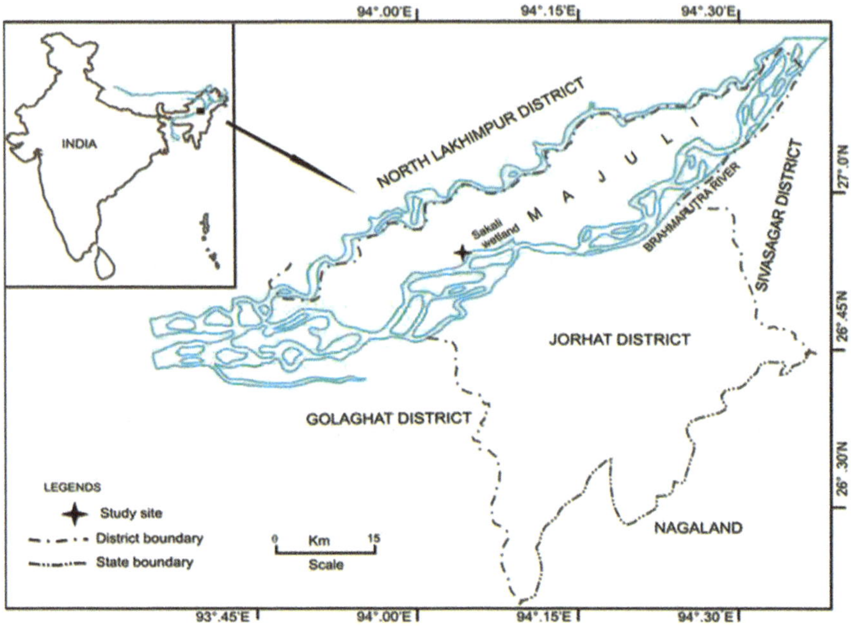

Fig. 1 Location map showing the study area (after Basumatary et al. 2018)

3 Climate and Soil Type

Majuli Island has a subtropical monsoon climate, as is found in the other parts of Assam, in general, the study area enjoys a warm and humid climate. The climate of the region is controlled by the southwest and northeast monsoons since area possesses high rainfall i.e. the average annual rainfall in the area is around 215 cm. The relative humidity is very high and ranges from 75% to 86%. During summers it is hot and humid with the maximum temperature rise up to 37 °C and cold and dry during the winters. All the major festivals in the island are held in the winter season when it is cool and pleasant. The average temperature hovers from 7 to 18 °C.

The soil in the Sakali wetland and adjoining areas is largely characterized by the moderately deep to very deep deposition of alluvium, with variation in colour from grey to mottled grey. It is mostly composed of sandy to silty loams and slightly acidic in nature. It is less acidic on the riverbanks and sometimes slightly alkaline. The soil lacks in the development of intact profile and is deficient in phosphoric acid, nitrogen and humus (Basumatary et al. 2018).

4 Material and Methods

A 1.5 m deep sedimentary core has been procured from the centre of the Sakali wetland through PVC pipe after the several trials coring (Fig. 2). The sedimentary core was sub-sectioned at about 5 cm interval initially for the palynological assessment. Fortunately, after the chemical processing and preparation of permanent slides for the pollen analysis, we realized that the samples were rich in the diversity of non-pollen palynomorphs and with this we started our studies on NPPs as an auxiliary tool for the palaeoenvironmental reconstruction in Majuli Island. The soil sediment of this region is composed of varied proportions of organic clay, silt and sand (Fig. 2). The different types of Non-pollen palynomorphs with their affinity and environmental implications are provided in Table 1. For the chemical processing, acetolysis technique was utilized for the extraction of pollen, spores and NPPs (Erdtman 1943). The step-wise chemical processing of soil samples are enumerated below:

1. 10 g of soil sample was treated with 10% potassium hydroxide (KOH) and boiled for up to 10–15 min until the effervesces comes.
2. After getting cool sieve the samples through 150-μm mesh size sieve for the deflocculating of the organic matter.
3. Then pour the samples into a centrifuge tube and centrifuge until a speed of at least 4500 rpm.
4. Wash the material with water and centrifuge until the supernatant is clear.
5. Dehydrate with 96% acetic acid and centrifuge.
6. Prepare an acetolysis mixture of acetic anhydride and sulphuric acid in (9:1) ratio.
7. Acetolyse the material by heating the sample in the acetolysing mixture to 100 °C for at least 10 min.
8. Cool the sample tubes at room temperature and then again centrifuge.
9. Wash the material with distilled water and centrifuge twice.
10. Pass the sample through 60-μm mesh size (0.56-μm pore size) sieve and the residue (>10 μm) was collected and stored for NPP studies.
11. The counting with photo-documentation of NPPs was done using an Olympus BX-50 Microscope with attached DP-26 Olympus camera (Plates 1 and 2).

5 Results

Due to the high diversification and complex nature of Non-pollen palynomorphs in palynological slides, most researchers often encounter difficulties in identification. It is possible to observe NPPs with variable views and their counting tells us about their abundance in a specific area. The identification and grouping of NPP types are necessary to reflect the vegetation and ecological dynamics rather than assuming a species entity based on their morphology alone (Adojoh et al. 2019). Description of all the NPP types are strictly based on morphological features (size, shape, number,

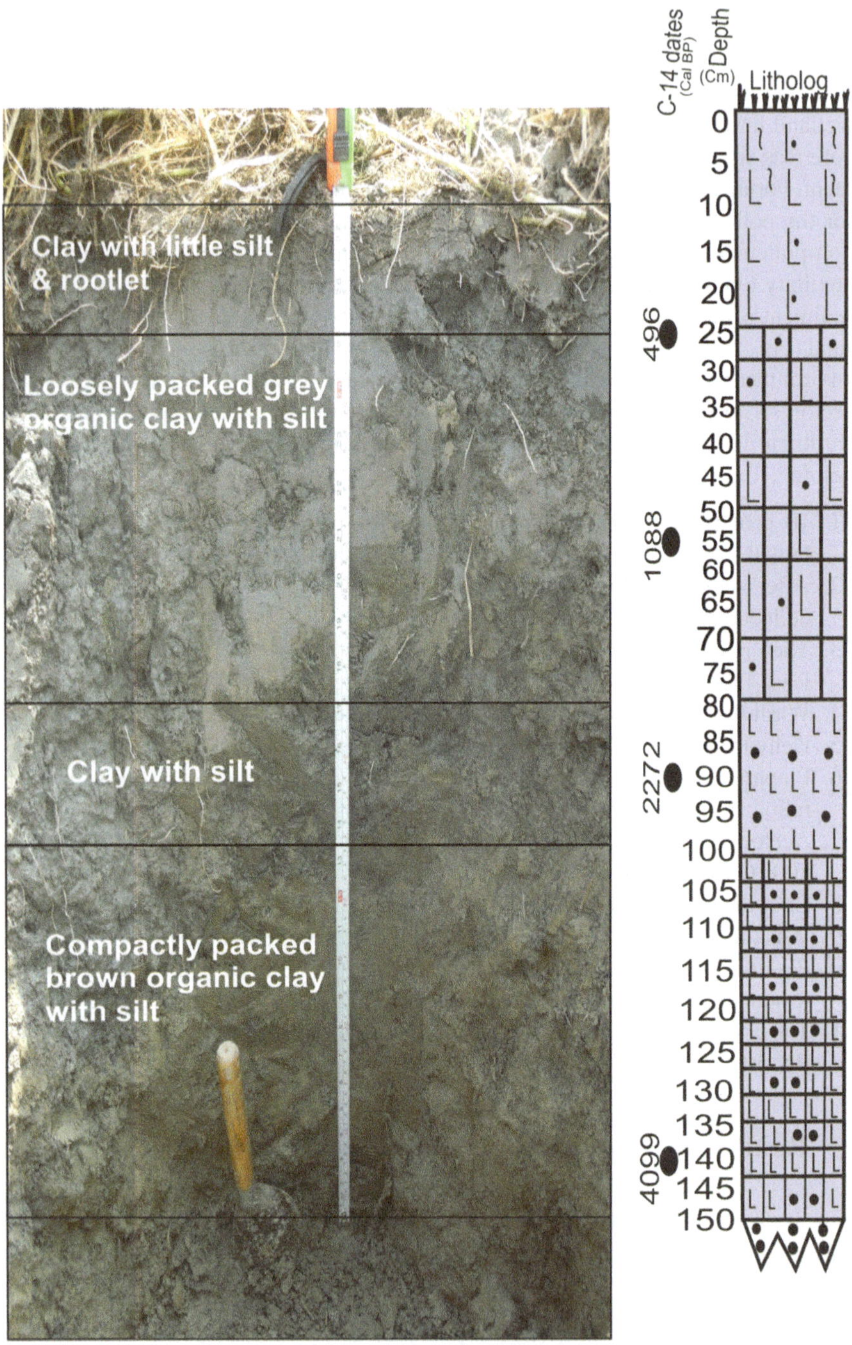

Fig. 2 A litholog of 150 cm deep sedimentary core procured from the Sakali wetland of Majuli Island, Assam

Table 1 The characteristic non-pollen palynomorphs and their affinities along with an environmental implications

S. no	Genus	Affinity/family	Environmental implication
1	*Helminthosporium*	Fungal spore (Pleosporaceae)	Best indicator of cultivated land and pathogen to members of Poaceae family
2	*Gelasinospora*	Non-coprophilous fungi (Sordariaceae)	Found in highly decomposed peat, indicate the human settlement around the study site
3	*Nigrospora*	Coprophilous fungi (Trichosphaeriaceae)	Frequently found at the cropland and open-land. Pathogenic to grasses and some other angiosperms like tea, palms and banana.
4	*Ascodesmis*	Coprophilous fungi (Ascodesmidaceae)	Found on dung of both wild and domesticated animals, indicates the grazing activity around the study site
5	*Podospora*	Coprophilous fungi (Lasiosphaeriaceae)	Found on dung, indicates the animal diversity and grazing activity
6	*Sordaria*	Coprophilous fungi (Sordariaceae)	Found on dung, also occur on moulds indicates grazing activity
7	*Telitia*	Pathogenic fungus (Ustilaginaceae)	Plant pathogen, effects the members of Poaceae family
8	*Delitschia*	Coprophilous fungi	Dung loving fungi, indicates grazing activity
9	*Entophlyctis lobata*	Non-coprophilous fungi (Chytriomycetaceae)	Specific of temperate region. Cold climate loving fungi.
10	*Dictyosporium*	Non-coprophilous fungi (Dictyosporiaceae)	Indicate warm and humid climate
11	*Diporotheca*	Parasitic fungi (Diporothecaceae)	Parasites on the roots of *solanum* species; wet zone highly trampled by livestock.
12	*Coniochaeta*	Coprophilous fungi (Coniochaetaceae)	Widely found on dung of both wild and domesticated animals; also grow on decaying woods
13	*Diploidia*	Non-coprophilous fungi (Botryosphaeriaceae)	Shows cosmopolitan distribution, and no specific temperature is needed for their growth
14	*Glomus*	Mycorrhizal fungi (Glomeraceae)	Frequently found near the wetland and open land area that indicate soil erosion.
15	*Arnium* type	Coprophilous fungi (Lasiosphaeriaceae)	Mostly occur on dung, but some species are also abundant on rotting herbaceous stems and wood
16	*Alternaria*	Non-coprophilous fungi (Pleosporeaceae)	Shows cosmopolitan distribution, and no specific temperature is needed for their growth
17	*Valsaria*	Non-coprophilous fungi (Valsariaceae)	Warm loving, occur on decayed bark or woods

(continued)

Table 1 (continued)

S. no	Genus	Affinity/family	Environmental implication
18	*Geastrum*	Saprophytic fungi (Geastraceae)	Cosmopolitan fungi, grow in tropical climate mainly found to grow scattered or clustered in leaf litter or humus
19	*Tetraploa*	Non-coprophilous fungi (Tetraplosphaeriaceae)	Indicate comparatively warm and humid climatic condition but favourable habitat is near water.
20	*Ustilago*	Smut fungi (Ustilaginaceae)	Indicate warm and humid climatic condition
21	*Cookeina*	Saprobic or parasitic fungi (Pezizaceae)	Commonly found attached with woody substrate may be living or dead. Indicate the presence of woody plants.
22	*Botryococcus*	Algal remain (Dictyosphaeriaceae)	Typically found in tropical fresh water wetlands
23	Biostructured phytoclast	Organic remains	It shows the presence of woody plants
24	*Neorhabdocoela*	Zoological remain	Found in fresh water ecosystem
25	Flatworm residue	Zoological remain	Mostly found near the wetland area that represents the freshwater ecosystem near around.
26	Fungal hyphae (septate)	Fungal body part	Bundle of hyphae ultimately results in formation of the mycelium. Mycelium is highly branched and plays an important role in vegetative reproduction in fungi. Warm and humid climatic conditions are favourable for their growth.

and characteristics of apertures and appendices, wall colour, surface structure) provided by van Geel (1972, 1986, 2001) and others (e.g. Bakker and van Smeerdijk 1982; Vander Wiel 1982; Kuhry 1985; Haas 1996; Guy-Ohlson 1998) have systematically developed the practice of examining all fossils from a wide variety of sediment types. High fungal abundance can be seen in the studied sediment samples of Sakali wetland, a few algal, plant and animal remains were also retrieved. All the identified NPPs that were spotted in the palynological slides have been elaborated below.

5.1 Fungal Remains

Most of the fossil fungi recovered from the sedimentary core of Sakali wetland appeared to be ascospores, conidia, and chlamydospores produced by 21 different fungi belonging to 16 families like Sordariaceae, Pleosporaceae, Glomeraceae, Ascodesmidaceae, Chytriomycetaceae, Dictyosporiaceae, Lasiosphaeriaceae, Valsariaceae, etc.). Many of the recorded fungal spores were found in peat deposits, especially in peat layers which were formed under relatively dry conditions. In lake

Plate 1 Non-pollen palynomorphs retrieved from the sedimentary core of Sakali wetland of Majuli Island, Assam. 1. *Botryococcus*, 2. *Nigrospora*, 3. *Ascodesmis*, 4–5. *Geastrum*, 6. *Diporotheca*, 7. *Coniochaeta*, 8–9. *Glomus*, 10. *Entophlyctis lobata*, 11. *Arnium type*, 12. Unidentified, 13. *Telitia*, 14–15. *Valsaria*, 16. *Ustilago*, 17. Ascospore of *Cookeina*, 18. *Delitschia*, 19. Unidentified, 20. *Diploidia*, 21. *Dictyosporium*, 22. Segmented mycelium, 23. *Alternaria*

Plate 2 Non-pollen palynomorphs retrieved from the sedimentary core of Sakali wetland of Majuli Island, Assam. 24–26. *Sordaria*, 27–28. *Podospora*, 29. *Helminthosporium*, 30. *Tetraploa*, 31. *Gelasinospora*, 32. Unidentified, 33. Unidentified, 34. Unidentified, 35.Unidentified, 36. Biostructured phytoclast, 37. Flatworm residue, 38. *Neorhabdocoela*

deposits, however, fungal remains normally are of rare occurrence (in open water there is no strictly local production of fungal spores which do preserve as fossils). Another factor that influences the fossil record is the fact that only relatively big (heavy) fungal spores with thick walls are normally preserved. Most of the thin-walled spores, which disperse better and which are known from the records of spores in the present atmosphere, obviously do not fossilize. From the various studies of fossil fungal spores, it became clear that the recorded spores in most cases were of strictly local occurrence. Some of the reported fungal remains are listed below with details.

***Nigrospora* [Plate 1, (2)]**
It is a dark hyaline round-shaped spore, the diameter is about 25–32 µm. *Nigrospora* is a ubiquitous, cosmopolitan fungus, especially abundant in a warm climate. It is mostly found on decaying plant material and soil. It belongs to phylum Ascomycota and class Trichosphaeriaceae. It is a common pathogen of grasses.

***Ascodesmis* [Plate 1, (3)]**
It is a coprophilous fungus with a complex fruiting body. Its diameter is about 12–15 µm, round in shape, yellow colour body with several spins like outgrowth. Widely found on the dung of both wild and domesticated animals, indicates the grazing activity.

***Geastrum* [Plate 1, (4, 5)]**
Its diameter is about 25 to 30 µm, spherical with an irregular outer surface having some ridges all around the surface wall. The outer layer of the tissue splits like a star (7–10 pointed rays) that eventually bends back to point downwards. It is a saprophytic fungus which grows mainly in the tropical climate, cosmopolitan in nature with proliferating growth in leaf litter or humus in scattered or clustered form. It belongs to the earth star family of fungi.

***Diporotheca* [Plate 1, (6)]**
These ascospores are one-celled, ellipsoidal, 32–49 × 18–25 µm in size, both ends truncate with a wide germ pore of ca. 1.5 µm in diameter. Parasites on the roots of *Solanum* species, wet zone highly trampled by livestock (Cugny et al. 2010).

***Coniochaeta* [Plate 1, (7)]**
It is a coprophilous fungus with domb shape likes structure placed one above the other, spherical shaped and very small in size (12 × 10 µm), and the measurement fluctuates because of its complex structure. The edge of the spore is dark in colour, rarely visible. The occurrence of *Coniochaeta* indicates the presence of dung and peat. The species of *Coniochaeta* and their anamorphs occurs on dung, wood, or bark of trees, soil, leaves, and leaf litter and rarely in non-woody host plants like Poaceae.

***Glomus* [Plate 1, (8, 9)]**
It is an Arbuscular mycorrhizal fungus, that dependent on their mycorrhizal association with plant root to complete their life cycle. They are dark yellowish to light brown in colour, about 10 µm in size and have a tail-like structure. They mostly occur in soil in solitary form, rarely seen in the branched form. *Glomus* species are found in nearly all terrestrial habitats, including arboreal land, deserts, grassland, tropical forest, and tundras. The endomycorrhizal fungus *Glomus* occurs on a variety of host plants, and chlamydospores are of a regular occurrence on pollen slides (Cook et al. 2011).

Entophlyctis lobata **[Plate 1, (10)]**
It is an endobiotic fungus with multiple morphological variations among their species (van Geel 1978): sporangia crenulate, 20–38 × 18–33 μm, about 6 μm high, walls about 0.5 μm thick (Kuhry 1997): Crenulate sporangia, with a central pore. Mostly found in the temperate region, reflecting the cold temperature.

Arnium **[Plate 1, (11)]**
These ascospores are oval-shaped with 38–45 × 20–25 μm in size, with some ornamentation could be seen on the surface of the spore. These are coprophilous fungi but some species are also abundant in rotten herbaceous wood and stem.

Telitia **[Plate 1, (13)]**
Fungal spore is globose, 11.8–18 μm in diameter, bi-layered with the reticulate body surface, ca. 0.5 μm thick superficial sculpturing. Spines like outgrowth, with 5.6–10.4 μm long, ca. 0.3 μm in diameter. It is a plant pathogenic fungus that generally affects the grasses, warm and humid climatic condition is favourable for their growth.

Valsaria **[Plate 1, (14, 15)]**
Ascospores uniseptate, septum very pronounced and slightly protruding, bi-layered thick verrucose wall, measuring ca. 32 × 22 μm. Warmer climatic condition is favourable for its growth. An indication in this direction may be the observation that species of *Valsaria* have become more common in warmer parts. In times of global warming, it seems justified to expect that these fungi will enlarge their distribution area (Jaklitsch et al. 2015).

Ustilago **[Plate 1, (16)]**
It is a round-shaped smut fungus, globose 16–24 μm in diameter (van Geel et al. 2011). It shows pathogenic characters, especially in maize and wheat.

Ascospore of *Cookeina* [Plate 1, (17)]
Spore is bi-celled with elliptical shaped body, measuring 30–35 × 5–20 μm, ends are slightly pointed, surface provided with longitudinal striations; septum lanceolate and distinct, (Mandaokar et al. 2008). Non-coprophilous fungus, mostly saprophytic and parasitic taxon commonly associated with woody substrates like a living, dead or decaying wood and leaves or litters. It indicates the dense forest under humid and rainy climate.

Delitschia **[Plate 1, (18)]**
The spore is divided by a thick septum, ends are slightly conical, measuring 32–35 × 13–15 μm, the thickness of septa is around 0.41 μm; spore wall is neither differentiated nor visible. It is a coprophilous fungus.

Diploidia **[Plate 1, (20)]**
The ascospore is an ellipsoidal, dark brown in colour. It indicates a warm and humid climate.

***Dictyosporium* [Plate 1, (21)]**
Pale brown conidia of 36–41 × 11–14 μm, consisting of one truncate cell with 3 vertical straight or slightly curved cylindrical, septate arms of more or less similar length, arranged close to each other. Mostly found in deadwood.

***Alternaria* [Plate 1, (23)]**
The tri-septate spore is curved at the axis at one side and slightly pointed on the other, around 5–6 longitudinal septa are present that divide the body of the spore into chambers. It is a cosmopolitan fungus found generally in areas with a warm and humid climate.

***Sordaria* [Plate 2, (24, 25, 26)]**
Spores ellipsoidal, one-celled, brown, 18–20 × 10–11 μm with a protruding apical pore of about 1.5 μm diameter at one end with a slight annulus (van Geel 1978), the spore wall is thick and pigmented with high survival capacity. It is also a coprophilous fungus which indicates grazing activity and sometimes grows on moulds as well.

***Podospora* [Plate 2, (27, 28)]**
A dark-pigmented upper cell and a hyaline lower cell (pedicel); they often have a gelatinous sheath and gelatinous caudal, brown-to-black coloured upper cell. Spore length is 17–34 μm, inflatable-type species (van Geel and Aptroot 2006). It is a very common coprophilous fungus, representing a warm and humid climate with an indication of grazing activities.

***Helminthosporium* [Plate 2, (29)]**
Hyphae are septate. Multicellular conidia (six or more-celled), are brown to dark brown, erect, parallel-walled, and ceasing to elongate when the terminal conidium is formed. Large 40 × 9 μm, solitary, club-shaped, and pale to dark brown. *Helminthosporium* is an indicator of cultivated land.

***Tetraploa* [Plate 2, (30)]**
Conidia ellipsoid to rectangular, 3–4 columns with 4 cells in each column, yellowish-brown, 33–41 × 18–23 μm, verruculose, thick-walled; terminating in septate appendages, 12–80 μm long (frequently broken), 5–8 μm wide. Distributed mainly in tropical and sub-tropical regions. Found as saprophytes, on decaying wood, leaf litters, grasses and some monocotyledons as well (Gelorini et al. 2011).

***Gelasinospora* [Plate 2, (31)]**
Ascospores ellipsoidal, 21–32 × 15–20 μm. Spore surface almost black, evenly ornamented with ca. 1 μm wide round hyaline pits (van Geel 1978).

5.2 Algal Remains

Algal palynomorphs, observed in the pollen slides besides pollen grains and spores of terrestrial and aquatic plants and fungal spores, are mainly represented by

Botryococcus, Pseudoschizia, Pediastrum, Spirogyra and *Zygnema*. However, in the sediments of Majuli Island, we only noticed the trace occurrence of *Botryococcus*.

***Botryococcus* [Plate 1, (1)]**
It is green algae, found in colonies of cells with irregular shape and size. The cell body is oval-shaped folded in mucilage and about 8–10 μm long and 4–6 μm wide. Modern *Botryococcus* is widely dispersed in temperate and tropical regions and is known to tolerate seasonally cold climates.

5.3 Zoological Remains

***Neorhabdocoela* [Plate 2, (38)]**
Oocyte without operculum, yellow, funnel-shaped or oval, 123–147 × 119–150 μm, with a smooth surface, stalk typical but often only partly preserved or not preserved, with articulation just beneath the body (Gelorini et al. 2011). Found in pollen preparations from Holocene freshwater lake sediments (Haas 1996).

Flatworm Residue [Plate 2, (37)]
Flatworm plays a major role in cultivations; hence their remains reflect the farming activities in the respective field. The digestive cavity of the Flatworms has only one opening for both ingestion (intake of nutrients) and egestion (removal of undigested wastes); as a result, the food cannot be processed continuously (Zaborski 2002).

5.4 Organic Matter

Bio-structured Phytoclast [Plate 2, (36)]
Indicates the remaining part of the plant like cortex, epidermis, vascular bundles, etc.

6 Discussion

6.1 General View of NPPs Recorded from the Majuli Sediments

Non-pollen palynomorphs play a very vital role in the prediction of past palynostratigraphical and palaeoclimatic condition. Because of its wide variety and distinguish characters they provide valuable information regarding the palaeoenvironmental reconstruction. The NPPs recovered from the 31 sediment samples of the 150 cm deep sedimentary core procured from the Sakali wetland of Majuli Island have dissimilar origin and nature and include fungal spores, algal and zoological remains of different environment. During our pollen-based late Holocene climatic studies in the aforesaid sediment core, it seems that the vegetation was not fully reflected by the available palynoassemblage, owing to the recurring floods and massive soil erosion in the Majuli Island (Barman et al. 2013). Thus, there is a dire need to look after the NPPs in the palynological slides, in order to provide an auxiliary tool that corroborates the pollen data for inferring the past vegetation and climate of this largest river island.

6.2 Major Inferences Drawn from the Retrieval of Fungal Spore from the Late-Holocene Sediment Samples

Majority of the research on Majuli has focused on climate change, disturbance in vegetation, bank erosion, rainfall pattern, drainage discharge of the Brahmaputra river, geomorphic changes in the river basin and the impact of the 1950 earthquake on settlements and fluvial pattern of the river. The studied sedimentary core gives variant forms of non-pollen palynomorphs especially the coprophilous and non-coprophilous fungal remains that provide very valuable insight into the palaeoenvironmental conditions. A total of 21 different fungal spore types were reported in the present investigation from the late Holocene sediments of Sakali wetland of Majuli Island, Assam. The coprophilous fungi have been broadly used to unravel the presence of herbivores (van Geel and Aptroot 2006; Cugny et al. 2010; Gelorini et al. 2011; Ghosh et al. 2017; Loughlin Nicholas et al. 2018). Their presence could be used as indicators of the presence of the livestock, particularly bovines. The *Sordaria* sp. is particularly found on cattle dung (Ejarque et al. 2011; van Geel and Aptroot 2006) and is commonly used for indicating the grazing activities (Ejarque et al. 2011). Moreover, the *Podospora* sp. and *Delitschia* sp. are strictly coprophilous, whereas *Coniochaeta* sp. and ascospores of Sordariaceae are saprophytic in nature, growing on dead plant parts and other decaying substrates (Ghosh et al. 2017). Coprophilous fungi in surface and sedimentary soil profile helps to document or infer the former presence and subsequent decline of, herbivorous animals in a region. Many coprophilous fungi have rather very narrow host range so; they can be indicator for a specific region which reflects the climate and vegetation around the area. Apart from the coprophilous fungi, the other non-coprophilous fungi like *Alternaria*, *Dictyosporium*, *Telitia* and *Valsaria* were also reported consistently in the sample that shows the proliferation of dense vegetation under warm and humid climate in the region (Ellis and Ellis 1985).

The *Tetraploa* and *Glomus* indicate rich plant diversity and availability of water in the region. These fungi are commonly found as mycorrhizal fungi on roots, leaves and stems of Poaceae and Cyperaceae (Tanaka et al. 2009). *Glomus* is mostly found during the early phase of late Holocene, often used as an indicator of soil erosion, and dry climatic conditions, since this fungus primarily lives underground (Anderson et al. 1984; van Geel et al. 2003). The presence of *Entophlyctis lobata* at the bottom of the sediment core reflects the cold and dry temperature since it is specific of temperate regions. The fungal spores of *Arnium* and *Gelasinospora* mostly occur in highly decomposed peat; formed under relatively dry, oligotrophic conditions, also indicate layers with charred material.

Further, the other fungal spores like *Nigrospora* sp., and *Tetraploa* sp., could be used as indicators of open-land vegetation which could be related to the human pressure (Gelorini et al. 2011). The *Helminthosporium* (a key indicator of cultivated land), and *Telitia* are grass pathogen and especially associated with the Poaceae member's example, *Oryza sativa* (rice), and *Bambusa vulgaris* (Bamboo) and these taxa are abundant in the region that reflects the anthropogenic activities around the area, whereas *Diporotheca* shows the symbiotic relationship with *Solanum* sp.

The corroboration of these NPPs helps us to reconstruct the past climate vegetation interaction. The NPPs collectively display the holistic information of mixed settlement of domestic living, anthropogenic capacities and environmental elements prevailing during the late Holocene in the Majuli Island.

6.3 Overall Inferences from the Algal and Zoological Remains

The occurrence of *Neorhabdocoela*, a freshwater zoological remain demonstrates the existence of cropland along with the presence of perennial water system around it. We have also encountered the scanty amount of algal remain like *Botryococcus*, where the flow of water would be in stagnant condition (Kumar et al. 2017). The complete absence of algal remains and rare occurrence of *Botryococcus* in sediments of Sakali wetland, supports the high energy levels in wetland water, attributed to frequent flood activities, owing to the close vicinity of Brahmaputra river. The overall NPP assemblage shows that this island possesses the grazing activities, cultivation process and domestic settlement during the recent past.

Besides the aforesaid NPPs, there are large numbers of NPPs, which are still unidentified and their ecological, as well as environmental values, are yet to be established. Additionally, the unidentified NPPs particularly fungal spores and conidia, which are in predominance, could be prescribing the probability to improve the typification and knowing of new bio-markers for their future examinations in palaeoecology, palaeoenvironment and palaeoclimate.

7 Conclusions

From the overall reviewing of the inferences discussed in the present paper, it can be suggested that the generated database of Non-pollen palynomorphs could provide baseline information for the precise reconstruction of past vegetation and climate in this high flood and erosion-prone region of Majuli Island. Twenty-five different and well-identified NPP types, classified broadly under three categories as fungal, algal and zoological remains, have been studied, documented and illustrated from the Sakali wetland of Majuli Island, along with their palaeoecological implications. Based on the recovered NPPs we concluded that the study area is providing a blend scenario of scattered forest near the wetland, and the presence of some non-coprophilous fungi indicates the practice of cultivation around with adequate signature of human settlement. The coprophilous fungi especially *Sordaria* and *Podospora* could be used to decode the palaeoherbivory in the Majuli Island. In future, the multiproxy analysis including both biotic and abiotic proxies would definitely unfold the past hidden ambience of the region.

Moreover, the vigorous dataset on NPPs aspect from sediment samples was needed with a view to strengthen the palaeoclimatic and palaeoecological interpretation.

Acknowledgments We thank, Dr. Vandana Prasad, Director, BSIP, Lucknow, India for the encouragement and permission to publish the paper. We acknowledge the Forest Department, Govt., of Assam for permission and necessary help during the field study. Here, we like to thank Dr. Biswajeet Thakur for his kind support during the investigation. This research was supported by the Science and Engineering Research Board (SERB), New Delhi, India under Women Excellence Award (grant number SB/WEA-06/2019).

References

Adojoh O, Fabienne M, Duller R, Osterloff P (2019) Taxonomy and phytoecology of palynomorphs and non-pollen palynomorphs: a refined compendium from the West Africa margin. Biodivers Int J 3(5):188–200

Alley RB, Mayewski PA, Sowers T, Stuiver M, Taylor KC, Clark PU (1997) Holocene climatic instability: a prominent widespread event 8200 years ago. Geology 25:483–486

Anderson RS, Homola RA, Davis RB, Jacobson GL Jr (1984) Fossil remains of the mycorrhizal fungal Glomus fasciculatum complex in post glacial lake sediments from marine. Can J Bot 62: 2325–2328

Bakker M, van Smeerdijk DG (1982) A palaeoecological study of a late Holocene section from "Het Ilperveld" The Netherlands. Rev Palaeobot Palynol 36:95–163

Barman S, Aggarwal SP, Dutta MK (2013) Soil erosion due to vegetated area in the Majuli Island of Assam. Int J Adv Remote Sens GIS Geogr I(1):9–18

Basumatary SK, McDonald HG (2017) Coprophilous fungi from dung of the greater one-horned rhino in Kaziranga National Park, India and its implication to Paleoherbivory and Paleocology. Quat Res 88:14–22

Basumatary SK, Nautiyal CM, Ghosh R, Tripathi S (2018) Modern pollen deposition in wetlands of Majuli Island and its implication to decipher palaeoflood episodes in Northeast India. Grana 57(4):273–283

Bera SK, Dixit S (2011) Pollen analysis of late Holocene lacustrine sediment from Jeypore Reseve Forest, Dibrugarh, Assam. In: Geological processes and climate change. Macmillan, New York, pp 85–94

Bhattacharya K, Chanda S (1992) Late-quaternary vegetational history of upper Assam, India. Rev Palaeobot Palynol 72:325–333

Birks HJB, Birks HH (1980) Quaternary palaeoecology. Blackburn Press, London

Bradley RS (1985) Quaternary palaeoclimatology: methods of palaeoclimatic reconstruction. Allen & Unwin, Boston and London

Clement AC, Seager R, Cane MA (1999) Orbital controls on the El Niño/southern oscillation and tropical climate. Paleoceanography 14:441–456

Cook EJ, van Geel B, van der Kaars S, van Arkel J (2011) A review of the use of non-pollen palynomorphs in palaeoecology with examples from Australia. Palynology 35(2):155–178

Cugny C, Mazier F, Galop D (2010) Modern and fossil non-pollen palynomorphs from the Basque mountains (western Pyrenees, France): the use of coprophilous fungi to reconstruct pastoral activity. Veg Hist Archaeobotany 19:391–408

Das D (2015) Bandung changing climate and its impacts on Assam, Northeast India. J Glob South 2:26

Dixit S, Bera SK (2012) Holocene climatic fluctuations from lower Brahmaputra flood plain of Assam, Northeast India. J Earth Syst Sci 121(1):135–147

Ejarque A, Miras Y, Riera S (2011) Pollen and non-pollen palynomorph indicators of vegetation and highland grazing activities obtained from modern surface and dung datasets in the eastern Pyrenees. Rev Palaeobot Palynol 167(1–2):123–139

Ellis MB, Ellis JP (1985) Microfungi on land plants. Croom Helm, Kent

Erdtman G (1943) An introduction to pollen analysis. Chronica Botanica Co, Waltham, MA

Gelorini V, Verbeken A, van Geel B, Cocquyt C, Verschuren D (2011) Modern non-pollen palynomorphs from east African lake sediments. Rev Palaeobot Palynol 164:143–173

Ghosh R, Paruya DK, Acharya K, Ghorai N, Bera S (2017) How reliable are non-pollen palynomorphs in tracing vegetation changes and grazing activities? Study from the Darjeeling Himalaya, India. Palaeogeogr Palaeoclimatol Palaeoecol 475:23–40

Guy-Ohlson D (1998) The use of the microalga *Botryococcus* in the interpretation of lacustrine environments at the Jurassic–Cretaceous transition in Sweden. Palaeogeogr Palaeoclimatol Palaeoecol 140(1–4):347–356

Haas JN (1996) Neorhabdocoela oocytes – palaeoecological indicators found in pollen preparations from freshwater sediments. Rev Palaeobot Palynol 91:371–382

Hicks S (2001) The use of annual arboreal pollen deposition values for delimiting tree-lines in the landscape and exploring models of pollen dispersal. Rev Palaeobot Palynol 117:1–29

Jaklitsch WM, Fournier J, Dai DQ, Hyde KD, Voglmayr H (2015) *Valsaria* and the Valsariales. Fungal Divers. https://doi.org/10.1007/s13225-015-0330-0

Kuhry P (1985) Transition of a raised bog across a coversand ridge originally covered with an oak-lime forest. Palaeoecological study of a middle Holocene local vegetational succession in the Amstven (Northwest Germany). Rev Palaeobot Palynol 44:303–353

Kuhry P (1997) The Palaeoecology of a treed bog in Western boreal Canada: a study based on microfossils, macrofossils and physico-chemical properties. Rev Palaeobot Palynol 96:183–224

Kumar M, Monga P, Shukla A, Mehrotra RC (2017) *Botryococcus* from the early Eocene lignite mines of western India: inferences on morphology, taphonomy and palaeoenvironment. Palynology 41:462–471

Loughlin Nicholas JD, Gosling William D, Montoya E (2018) Identifying environmental drivers of fungal non-pollen palynomorphs in the montane forest of the eastern Andean flank, Ecuador. Quat Res 89:119–113

Mandaokar BD, Chauhan MS, Chatterjee S (2008) Fungal remains from late Holocene Lake deposits of Demagiri, Mizoram, India and their palaeoclimatic implications. J Palaeontol Soc India 53(2):197–205

Mayewski PA, Rohling EE, Stager JC, Karlen W, Maasch K, Meeker LD, Meyerson EA, Gasse F, van Kreveld S, Holmgren K, Lee-Thorp J, Rosqvist G, Rack F, Staubwasser M, Schneider RR, Steig EJ (2004) Holocene climate variability. Quat Res 62:243–255

Moore P, Webb J (1978) An illustrated guide to pollen analysis. Hodder & Stoughton, London

Prell WL, Kutzbach JE (1987) Monsoon variability over the past 150,000 years. J Geophys Res 92: 8411–8425

Spieksma FTM, Cordon JM, Detandt M, Millinngton WM, Nikkels H, Nolard N, Schoemakers CHH, Wachter R, de Weger LA, Willems R, Emberlin J (2003) Quantitative trends in annual totals of five common airborne pollen type (*Betula*, *Quercus*, Pinaceae, *Urtica* and *Artemisia*), of five pollen monitoring stations in western Europe. Aerobiologia 19:171–184

Tanaka K, Hirayama K, Yonezawa H, Hatakeyama S, Harada Y, Sano T, Shirouzu T, Hosoya T (2009) Molecular taxonomy of bambusicolous fungi: Tetraplosphaeriaceae, a new pleosporalean family with Tetraploalike anamorphs. Stud Mycol 64:175–209

Tripathi S, Thakur B, Nautiyal CM, Bera SK (2020) Floristic and climatic reconstruction in Indo-Burma region for the last 13000 cal. yr: a palynological interpretation from endangered wetlands of Assam, Northeast India. Holocene 30(2):315–331

van Geel B (1972) Palynology of a section from the raised peat bog "Wietmarscher Moor", with special reference to fungal remains. Acta Bot Neerland 21:261–284

van Geel B (1978) A palaeoecological study of Holocene peat bog sections in Germany and the Netherlands, based on the analysis of pollen, spores and macro-and microscopic remains of fungi, algae, cormophytes and animals. Rev Palaeobot Palynol 25:1–120

van Geel B (1986) Application of fungal and algal remains and microfossils in palynological analyses. In: Berglund BE (ed) Handbook of holocene palaeoecology and palaeohydrology. Wiley, Chichester, pp 497–505

van Geel B (2001) Non-pollen palynomorphs. In: Smol JP, Birks HJP, Last WM (eds) Tracking environmental change using lake sediments, Terrestrial, algal and siliceous indicators, vol 3. Kluwer Academic Press, Dordrecht, pp 99–119

van Geel B, Aptroot A (2006) Fossil ascomycetes in quaternary deposits. Nova Hedwigia 82:313–329

van Geel B, Mur LR, Ralska-Jasiewiczowa M, Groslar T (1994) Fossil akinetes of Aphanizomenon and anabaena as indicators of medieval phosphate-eutrophication of Lake Gosciaz (Central Poland). Rev Palaeobot Palynol 83:97–105

van Geel B, Buurman J, Brinkkemper O, Schelvis J, Aptroot A, van Reenen G, Hakbijl T (2003) Environmental reconstruction of a Roman period settlement site in Uitgeest (The Netherlands), with special reference to coprophilous fungi. J Archaeol Sci 30:873–883

van Geel B, Gelorini V, Lyaruu A, Aptroot A, Rucina S, Marchant R, Sinninghe D, Jaap S, Verschuren D (2011) Diversity and ecology of tropical African fungal spores froma 25,000-year palaeoenvironmental recorded in southeastern Kenya. Rev Palaeobot Palynol 164:174–190

Vander Wiel AM (1982) A palaecological study of a section from the foot of the Hazendonk (Zuid-Holland), based on the analysis of pollen, spores and macroscopic remains. Rev Palaeobot Palynol 38:35–90

Zaborski ER (2002) Observations on feeding behavior by the terrestrial flatworm Bipalium adventitium (Platyhelminthes: Tricladida: Terricola) from Illinois. Am Midl Nat 48:401–408

Climate Variability and Its Causal Mechanisms Over the Northeastern Indian Himalaya

Prachita Arora, Priyanka Singh, S. Nawaz Ali, P. Morthekai, Mayank Shekhar, and Ruby Ghosh

Abstract The Indian Summer Monsoon (ISM) is a part of Asian monsoon system and forms a fascinating component of the global climate dynamics. Its long-term evolution and variability at different time scales (centennial to millennial) has been attributed to changes in large-scale forcing or boundary conditions like orbital parameters, mountain-plateau orography (tectonic), glacial surface boundary conditions and atmospheric CO_2 concentrations. The changes in ISM have large scale impacts on the diverse ecosystems and life of billions of people living in south Asia. Despite its importance, the reliable future ISM predictions (model simulations) remain a challenge. The proxy palaeoclimatic reconstruction data serves as an analogue against which the future models are tested and hence their importance. Towards understanding the ISM variability, the eastern Himalayan region is important as it is dominantly influenced by the Bay of Bengal branch of the ISM. Its tropical character and high biodiversity is—in large—a consequence of topography-climate interactions and the influence of the ISM precipitation. Till now, in-depth and quantitative (semi) characterizations of ISM variability and precipitation patterns prior to the instrumental period are rare. As far as the paleoclimatic/ paleoenvironmental proxy records of the tropical Himalaya are concerned, these data requirements are only rudimentarily fulfilled, especially in relation to the sheer size of this sector. The present work provides a semi-quantative synthesis of all the available (11) palaeoclimatic records from this sector that are qualitative in nature. Towards this, the available information on moisture or precipitation associated with ISM (published data) has been evaluated and given a weightage (1–6; driest to wettest) following a consistent criteria, to arrive at meaningful synthesis. In the present study, modern precipitation and temperature (categorized into summer and winter) data have been analyzed to understand the driving forces that potentially

P. Arora (✉) · S. Nawaz Ali · P. Morthekai
Birbal Sahni Institute of Palaeosciences, Lucknow, India

Academy of Scientific & Innovative Research (AcSIR), Ghaziabad, India

P. Singh · M. Shekhar · R. Ghosh
Birbal Sahni Institute of Palaeosciences, Lucknow, India

B. Phartiyal et al. (eds.), *Climate Change and Environmental Impacts: Past, Present and Future Perspective*, Society of Earth Scientists Series,
https://doi.org/10.1007/978-3-031-13119-6_6

influence the variability in the climate. To establish the links between various climate driving indices and the climate parameters, redundancy and Granger-causality analysis have been employed.

Keywords Indian summer monsoon variability · Semi-quantative synthesis · Ordination analysis · Granger-causality test

1 Introduction

Understanding of the past climate variability as well as its present trend, is fundamental for any agrarian society, as it paves the way for predicting/forecasting future scenarios. With ongoing changes in the global climate, our community is eager to know and understand climate variations, their possible impacts and potential effects on the future generations (Dixit 2020). The Indian Summer Monsoon (ISM) is a dominant source of precipitation in the three most important river basins (Indus, Ganga, and Brahmaputra) and plays a significant role in south Asian and Indian hydrology (Hasson et al. 2016). The Himalaya, also acknowledged as the water tower of Asia, has the largest concentration of glacier outside the poles. The contributions however, to the river systems, suggest dominance of rainfall-runoff (~66%) with respect to glacier melt water in case of both Ganga and Brahmaputra (Shrestha et al. 2015). Therefore, it is crucial to understand the dynamics and variability of ISM as ~830 million people involved in agriculture, forestry, fisheries, and livestock directly depend on it and are vulnerable to any water stress due to the ongoing climate change (Webster et al. 1998; Shrestha et al. 2012a, b; Gadgil and Kumar 2006; Dixit 2020). Further, the droughts and floods associated with the ISM rainfall have devastating effects on the people, agriculture and economy of the Indian subcontinent (Xavier et al. 2007; Kaushal et al. 2018).

The ISM is a component of the Asian summer monsoon system and originates as a result of the differential heating of the land and ocean surfaces during boreal spring and is considered as the most pronounced monsoon system in the world (Tao 1987; Webster 1987; Wu 2017). The ISM rainfall has shown both spatial as well as seasonal to multi-decadal variability (Parthasarathy and Yang 1995; Koul et al. 2018). The variability is largely controlled by insolation driven coupled ocean-atmosphere processes such as El Nino Southern Oscillations and Indian Ocean Dipole (Sikka 1980; Pant and Parthasarathy 1981; Rasmusson and Carpenter 1983; Ashok et al. 2019; Webster and Yang 1992; Kumar et al. 1995). Towards understanding the past (Late Quaternary) ISM variability, reasonable research has been done on the Ocean sediments (Gupta et al. 2003, 2005, 2013; Kessarkar et al. 2013; Govil and Naidu 2011; Rashid et al. 2011; Schulz et al. 1998; Overpeck et al. 1996; Sirocko et al. 1993). However, such reconstructions are more reliable for wind speed and monsoon strength and less valid for moisture content (Sinha et al. 2005; Sarkar et al. 2000; Ruddiman 1997). Hence, to understand the ISM variability, terrestrial (fluvial, lacustrine and speleothem) records of climate variability with greater spatial and temporal coverage are needed to improve the predictive capabilities of the future ISM trends (Ali et al. 2018; Chen et al. 2015; Sinha et al. 2005). However, such records are scanty and often suffer poor age control (Ali et al. 2018; Bhushan et al. 2018; Ghosh et al. 2015; Mishra et al. 2015; Prasad et al. 2014; Sinha et al. 2005).

The Himalayan region is variably influenced by two main weather systems viz. the ISM and the mid-latitude westerlies (Yang et al. 2008; Finkel et al. 2003) and the southern and eastern parts are dominated by the ISM, while the northwestern parts are dominated by winter westerlies (Ali et al. 2013; Benn and Owen 1998). These two weather systems have also influenced the inter-regional variability in the pattern and extent of past glacier advances and present dynamics (Bookhagen and Burbank 2010; Finkel et al. 2003). In a recent study, Ménégoz et al. (2013) suggested a differential concentration and limited role of the ISM in the western and some parts of eastern Himalaya. Hence it is important to understand the past ISM variability from different sectors of the Himalayan orogen.

The eastern Himalaya is dominated by the ISM precipitation and comprises of some spectacular landscapes and is a biodiversity hot spot, where the moist winds from the low lands of India and Bangladesh rise and interact with the cold dry winds of Tibetan highland (O'Neill et al. 2017). The eastern part of the Himalayan cordillera is an important sector to understand the past ISM variability, as the contributions from the winter westerlies is very less (Ali et al. 2019; Bookhagen and Burbank 2006). The dominance of ISM and its controls on the landscape evolution, ecosystems and steep climatic and ecological gradient makes the region special in terms of important implications for understanding past environmental/ climate changes (Meyer et al. 2009, 2017; O'Neill et al. 2017; Bookhagen and Burbank 2006; Grujic et al. 2006). Along with this, the tele-connections of ISM with other modes of climate variables such as El Nino Southern Oscillation (ENSO), ENSO-Modoki, Total Solar Irradiance (TSI), Pacific Decadal Oscillation (PDO), Southern Oscillation Index (SOI), Atlantic Multidecadal Oscillation (AMO), Indian Ocean Dipole (IOD), and the Arctic Oscillation (AO) at multiple spatio-temporal scales have been a topic of discussion within the scientific community. The nature and extent of these forcing factors is not quantified and is debatable. Despite the importance of the north eastern Himalayan region in understanding the dynamics of ISM, the palaeoenvironmental and palaeoclimatic trajectory and its underlying mechanisms is—till today—largely under-researched (Ali et al. 2018).

Here we propose an innovative research agenda that will provide a semi-quantative synthesis of palaeoclimatic records during the Late Quaternary in the Eastern Himalaya. The study further aims at understanding the climate drivers and establishing possible links between past climate variability and associated drivers.

2 Regional Setting (Study Area)

The northeastern Indian Himalaya (NEIH) comprising of Sikkim and Arunachal Pradesh and is bordered by Tibet (China), Nepal and Bhutan. The region hosts some of the highest peaks like the Kanchenjunga (~8475 m asl), the third highest mountain peak in the world and a number of glaciers and glacial lakes (Fig. 1). The altitudinal gradient of the area and its direct trajectory to ISM (Bay of Bengal Branch) results in direct confrontation of monsoon winds and results into heavy precipitation characterized by gradual decrease towards the north, as a consequence of orography (Chakravarty et al. 2012; Ali et al. 2019). The diverse landscapes and steep altitudinal gradient of the region also entails diverse microclimatic and climatic conditions, ranging from tropical humid at altitudes <1500 m asl to alpine-humid between ~1500 and 3500 m asl to progressively tundra-type at altitudes above ~3500 m asl (e.g. Ali et al. 2018). Owing to these unique setting, the NEIH shelters richest biological environment and vegetation types (rain forest, alpine meadows and cold desert taxa) that accounts for over one-third of the country's total biodiversity (Arrawatia and Tambe 2011; Chakravarty et al. 2012; Venkataraman and Sivaperuman 2018).

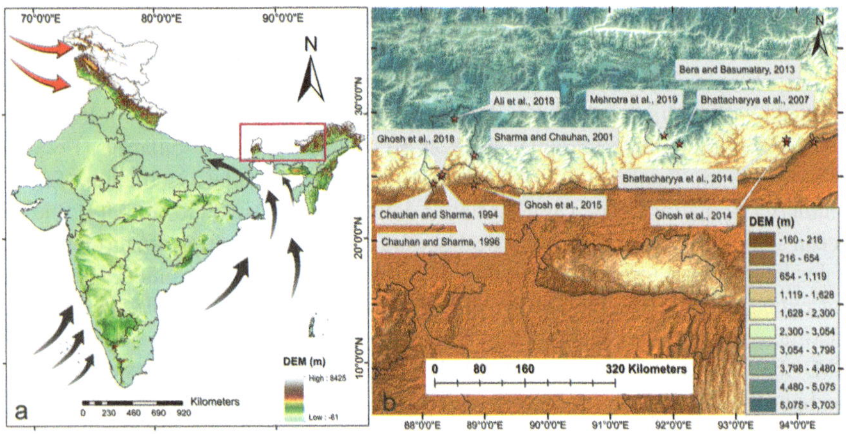

Fig. 1 (**a**) Shuttle Radar Topographic Mission (SRTM) digital elevation model (DEM) showing the regional setup of the area along with the trajectories of the two major weather systems—the Indian summer monsoon (black arrows) and the mid-latitude westerlies (red arrows), (**b**) Shows the close up of the study area along with the location of the available palaeoclimatic records

3 Climate

The two main meteorological parameters that help in understanding the climate of an area are temperature and precipitation. Any variation or extremity in one or both of these parameters can have substantial impacts on the ecology as well as on the socioeconomic gradient of any region. In the Himalayan context, the complexity in terms of topography is accountable for extensive variations in these meteorological parameters over comparatively smaller spatial gradients (Bhutiyani et al. 2010). However, due to the unavailability of instrumental meteorological data, the variations in temperature and precipitation over the NEIH including the foothills have been extracted from the Climatic Research Unit Time series gridded data sets (CRU TS.4.04; 0.5 latitude × 0.5 longitude; 1901–2018 CE; Harris et al. 2020). A seasonal time series has been compiled from the monthly datasets to get annual, summer (June–July–August–September; JJAS) and winter averages (December–January–February; DJF). The climate observations from the CRU dataset show a rise in both summer and winter temperatures over the last 100 years. The mean minimum and maximum winter temperatures recorded in the area are −3 °C and 1.6 °C, respectively. During the summer months the average temperature often reaches to 9 °C. Being in the trajectory of the ISM, the area receives rainfall from June to September. The average annual precipitation recorded from the area is ~1585 mm with minimum and maximum values of ~1192 and ~1991 mm. Out of the total precipitation, ~80% is contributed by the Indian Summer Monsoon during JJAS with contributions of ~2%, .3%, 5% from winter, pre- and post-monsoon months respectively.

4 Methodology

To understand the palaeoclimatic history of the NEIH, 11 previously published studies from the region that infer about the moisture conditions during the last ~46 ka have been used. Basic information about the records including the location, dating technique and time coverage is summarized in Table 1. The description of the moisture (precipitation) conditions in each of the study is qualitative and the nomenclature is quite generalized. Thus, to quantify the information on precipitation associated with ISM, inferences drawn from the 11 palaeoclimatic records have been used for the calculation of a palaeoclimatic index (weightage) and each climatic event from all the records has been assigned a weightage (1–6; driest to wettest; Table 2) following a consistent criteria. It has to be mentioned that the general values for pre- and post-LGM cannot be exactly compared with each other, since only two reconstructions date back to pre LGM time, while only a handful of reconstructions cover the time frame since LGM.

Table 1 Table showing the locations, chronological method used, inferred chronological zones, climatic inferences and assigned weightage of 11 palaeoclimatic records from the northeastern Indian Himalaya

Study site	Dating	Chr. zones	Climatic inference	Wtg.	References
Mirik lake (Darjeeling)	Radio carbon	20,000–18,000	Cold temperate	2	Chauhan and Sharma (1996)
		18,000–12,000	Warm temperate	4	
		12,000–11,000	Cool oscillation	1	
		10,000–4000	Optimum climate	4	
		4000–2000	Barren	4	
		2000–500	Warm temperate with gradual deterioration	3	
		500–0	Barren	3	
Jore-Pokhari (Darjeeling)	Radio carbon	2500–1600	Warm-temperate and humid	5	Chauhan and Sharma (1996)
		1600–1000	Cool oscillation	1	
		1000–300	Warm and humid	5	
		300–0	Warm-temperate and humid	5	
Sixth mile, Darjeeling	Radio carbon	2313–1819	Warm and moist	5	Ghosh et al. (2018)
		1819–1326	Warm and weak monsoon	3	
		1326–832	Warm and strong monsoon	6	
		832–148	Cooler with monsoon intensification	2	
		148–0	Warm with no direct inference on monsoon	3	
Chopta valley	AMS 14C	13,000–12,250	Intensified ISM	6	Ali et al. (2018)
		12,250–10,600	Cool and dry	3	
		10,600–8000	Enhanced ism	6	
		8000–6500	Moderate to low	4	
		6500–3000	Weak	3	
		3000–1500	Moderate	4	
		1500–0	Moderate to high	4	

(continued)

Table 1 (continued)

Study site	Dating	Chr. zones	Climatic inference	Wtg.	References
Darjeeling foothills	AMS 14C	46,400–41,200	Monsoon intensification	6	Ghosh et al. (2015)
		41,200–31,000	Weak monsoon	3	
		31,000–22,300	Monsoon intensification	6	
		22,300–18,300	Low temperature and precipitation	1	
		18,300–15,600	Enhanced monsoon and temperature restoration	6	
		5400–4300	Monsoon intensification	6	
		4300–3500	Rainfall minima (weak monsoon)	3	
Kupup Lake (Sikkim)	Radio carbon	2000–1800	Warm and moist	5	Sharma and Chauhan (2001)
		1800–1450	Cold climate	1	
		1450–450	Warm and moist (MWP)	5	
		400–200	Colder and drier	1	
		200–0	Less drier and colder	1	
PT Tso Lake	Radio carbon	8010–6579	Cool humid	2	Mehrotra et al. (2019)
		6198–4625	Cool humid	2	
		4195–2527	Dry temperate (cool)	1	
		2336–1240	Cool dry	1	
		906–0	Cool wet	2	
Paradise Lake	Radio carbon	1780–684	Warm and moist	5	Bhattacharyya et al. (2007)
		684–0	Cooler and less moist	2	
Ziro lake	Radio carbon/OSL	22,000–19,500	Humid climate	5	Ghosh et al. (2014)
		19,500–10,200	Cooler and drier	1	
		10,200–3800	Monsoon intensification	6	
		3800–0	Warmer and drier	3	
Ziro valley	Radio carbon	~66,000	Cool dry	1	Bhattacharyya et al. (2014)
		44,000–34,000	Increased SW monsoon	6	

(continued)

Table 1 (continued)

Study site	Dating	Chr. zones	Climatic inference	Wtg.	References
		36,000–34,145	Intensified monsoon	6	
		~34,000–30,000	Increased monsoon	6	
		30,000–20,000	Cool moist	2	
Subansari river	Radio carbon	12,500–7960	Cold and dry	1	Bera and Basumatary (2013)
		7960–6421	On set of warm and humid	5	
		6421–4270	Warm and humid	5	
		4270–0	Warm and relatively dry	3	

4.1 Calculation of Palaeoclimate Index

The above mentioned weightage or palaeoclimate indices (pCI) has been assigned to quantify the palaeo–monsoon dynamics over the available time bracket and plotted at an age interval of 100 years. This quantification is based on the intensity of monsoon discussed in different studies (published data). A rank (number) is given to the climate events in an ascending order such that the lower most rank will represent dry conditions while the higher most rank will be the indicative of enhanced monsoon (Ali et al. 2019). The weighted averaging palaeoclimate index (WApCI) is calculated per time slice of 100 years and weight (wi) will either be 1 if the particular time slice has published data or 0 for otherwise (Eq. 2). The probability density function of WApCI is calculated by normalizing WApCI over the time (Eq. 2).

$$WApCI = \frac{\sum_j w_j \times pCI_j}{\sum_j w_j}; w_j = \begin{cases} w_j = 1 \, if \, published \, data \, available \\ w_j = 0 \, otherwise \end{cases} \tag{1}$$

$$p(WApCI) = \frac{WApCI_t}{\sum_t WApCI_t} \tag{2}$$

where pCI = palaeoclimate index assigned for each time slice of 100 years; (dry to intensified monsoon); w is the weight, and the subscripts j and t represent the number of publications and the number of time slices respectively.

Table 2 Climatic classes and their weightage used in the present study

S. no.	Class	Category	Weight moisture
1	Intensified monsoon Enhanced monsoon Increased SW monsoon	Intensified monsoon	6
2	Moist Humid Humid climate Temperate and humid	Enhanced monsoon	5
3	Temperate Temperate with gradual deterioration Optimum climate	Moderate monsoon	4
4	Weak monsoon Rainfall minima	Weak monsoon	3
5	Moist Less moist Cold temperate Drier Relatively dry	Moist	2
6	Low precipitation Cool oscillation Cold climate Cooler and drier Cool and dry Drier and colder Cold dry Colder and drier	Dry/arid	1

S. no.	Class	Category	Weight temperature
1	Intensified monsoon Enhanced monsoon and temperature Increased SW monsoon	Warmer	4
2	Warm moist Warm humid Humid climate Warm temperate and humid Warm temperate Warm temperate with gradual deterioration Optimum climate Weak monsoon Rainfall minima Warmer and drier Warm but relatively dry	Warm	3
3	Cool Cool oscillation Low temperature Cooler	Cool	2
4	Cold Colder	Cold	1

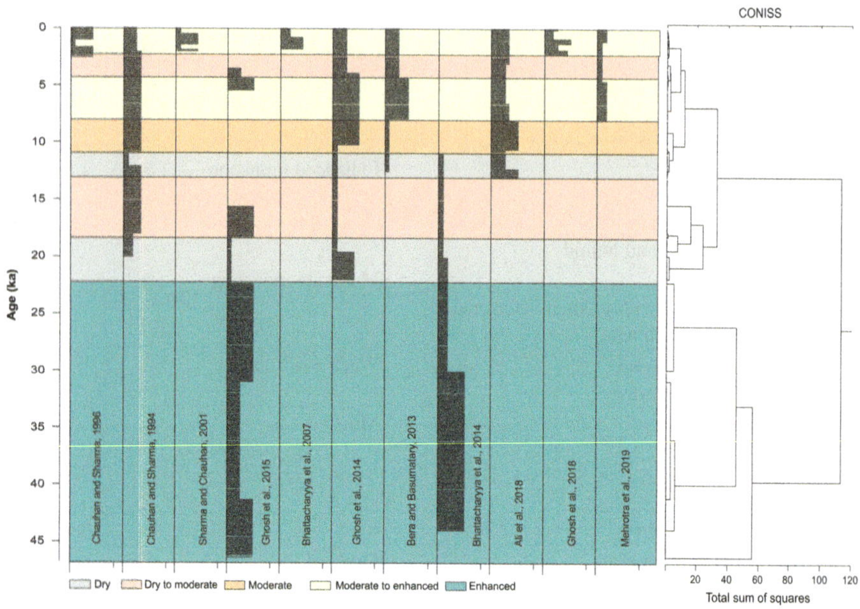

Fig. 2 Synthesis of climate variability data using weighted climatic interpretation based on the inferences drawn in different studies from the Northeastern Himalaya, India with CONISS Cluster analysis. The ascending order of the scale represents intensification of the monsoon

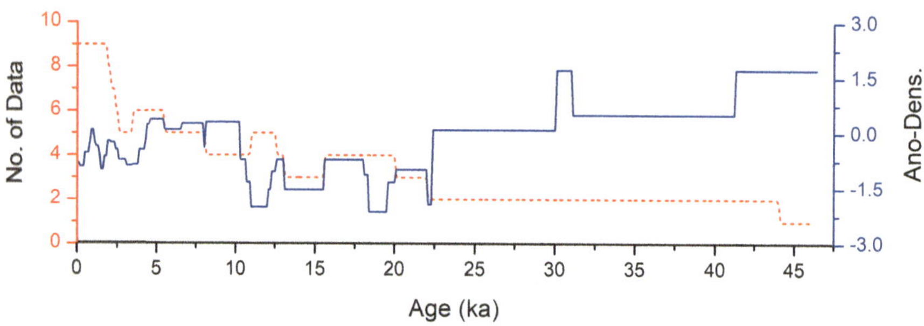

Fig. 3 Synthesis of the palaeoclimatic records represented by anomalous density function (blue line) and the temporal availability of the number of data points (climatic records; red line) from the northeastern Indian Himalaya

This allows us to have a broad understanding of the palaeoclimate of the area and will help in further correlations as a whole (Figs. 2, 3 and Tables 1 and 2).

4.2 Ordination Analysis

According to Gauch and Gauch (1982), ordination analysis primarily endeavours to represent sample and species relationships as faithfully as possible in a reduced-dimensional space. It can be used as a collective term for multivariate analysis used to find patterns and combinations in the data that are otherwise difficult to interpret. Ordination analysis can be classified into the constrained and unconstrained ones. To find which ordination method should be applied for a dataset, the rule of thumb, introduced by Lepš and Šmilauer (2003), is used. The present study involved, calculating the detrended correspondence analysis (DCA) followed by measuring the length of the first DCA axis. The length of the first axis decides whether a linear or unimodal based method is to be applied where a gradient length > 3 SD (standard deviation) suggests that a unimodal based method of direct gradient analysis (canonical correspondence analysis) should be selected to understand the relationship between response and explanatory variable. However, <3 SD gradient length indicates that a linear-based method of direct gradient analysis (redundancy analysis) to be selected. Further, principal component analysis (PCA) has been used to identify the patterns between samples (response variables) and determine minimum variables responsible for those patterns (explanatory variables). Here, redundancy analysis (RDA) is used to examine the direct relationship between the response variables and explanatory variables. Based on p-values, obtained from the RDA, we have tried to quantitatively assess correlation of the changes in the driving factors with the variability in meteorological parameters over the NEIH (Fig. 4 and Table 3). To understand which of these driving factor(s) are responsible for causing the change in temperature and precipitation in a predictable manner, we have used the Granger-causality test.

4.3 Granger-Causality Test

The Granger causality test is a widely used statistical analysis to determine whether one time series can cause the other time series, or not (Granger 1969). Although it is called causality test, it is in fact calculating the temporal correlation between two time series and hence it contain the directional information. Because of this directional information, it has some advantages compared to mere correlations between two time series. In a sense this Granger causality test is an attempt, in the original sense, to forecast given some available data at hand. According to Granger causality, if a signal $x(t)$ "Granger-causes" (or "G-causes") a signal $y(t)$, then past values of $x(t)$ should contain information that helps to predict y merely from the past values of x alone (Seth 2007). Actually this is not causality in a strict sense, and hence tones down to Granger causality that has implicit limitations in applying this method.

Table 3 Table showing the CRU grid information and results of the ordination analysis (RDA)

Identifier	Grid	Proportion explained		Constrained proportion		SOI		AO		PDO		IOD		AMO		TSI	
		PC1	PC2			r^2	p-value	r^2	p-value	r^2	p-value	r^2	p-value	r^2	p-value	r^2	p-value
27.25 N, 88.25 E	**G1**	0.3577	0.2677	0.1541	***											0.1271	0.001
					**	0.1288	0.002			0.0955	0.007			0.086	0.009		
					*												
27.25 N, 88.75 E	**G2**	0.3442	0.2682	0.1526	***	0.1409	0.001										
					**											0.1078	0.008
					*					0.0917	0.013			0.0937	0.012		
27.75 N, 88.25 E	**G3**	0.35	0.2655	0.1381	***	0.1442	0.001										
					**					0.1189	0.003					0.1009	0.007
					*												
					.									0.0559	0.074		
27.75 N, 88.75 E	**G4**	0.3486	0.2785	0.1493	***	0.1638	0.001										
					**					0.117	0.003						
					*									0.074	0.029	0.0901	0.011
28.25 N, 88.25 E	**G5**	0.349	0.2702	0.1331	***												
					**	0.131	0.003										
					*					0.0792	0.026					0.0782	0.019
					.									0.0537	0.077		
28.25 N, 88.75 E	**G6**	0.3522	0.2788	0.1378	***	0.1531	0.001										
					**					0.0952	0.005						
					*												
					.									0.0488	0.092	0.0521	0.077

28.75 N, 88.25 E	**G7**	0.3422	0.2742	0.1316	***	0.1315	0.001										
					**												
					*					0.0719	0.02					0.0638	0.035
28.75 N, 88.75 E	**G8**	0.3409	0.2964	0.1331	***	0.1591	0.001										
					**					0.0966	0.005						
					*												
					.									0.054	0.075		

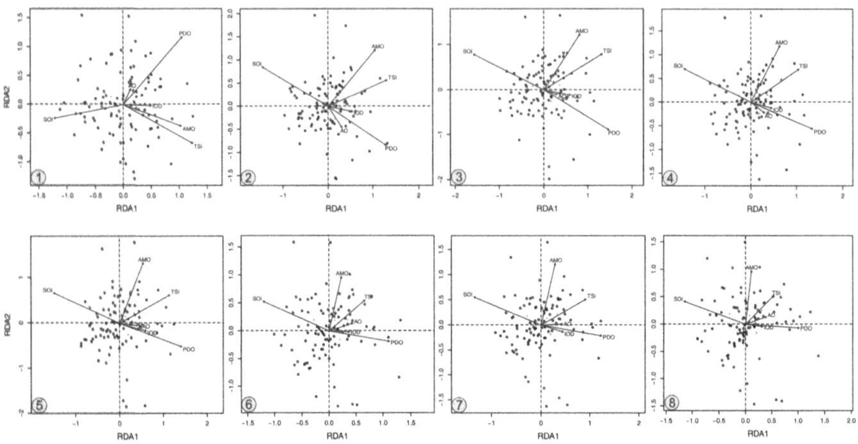

Fig. 4 Redundancy Analysis (RDA) showing the relationship between the response variables and explanatory variables for the CRU grids (please see Table 3 for grid information)

5 Results and Discussion

Understanding the natural systems and their sensitivities to different forcing factors on decadal to centennial time scales, require a perspective that can be achieved only from a better knowledge of past variability. Thus synthesizing observational datasets collected over decades enables us to identify broad trends in climatic variability over a region and correlations between the observational dataset and past records lays the foundation for reliable future climate model simulations. Keeping this in view, the present study attempts to provide a comprehensive analysis of Late Quaternary climate of monsoon dominated North Eastern Himalaya based on a compilation of 11 available (Figs. 2 and 3) palaeoclimatic records as follows.

- *Pre-LGM*: The available proxy climatic records from this region are scanty and more specifically for pre-LGM time period. Only two studies have identified and described the palaeoclimatic variability dating beyond 22 ka, in this region. Our synthesis show that more than 60% of these records inferred enhanced ISM during ~46–22 ka. The evidences of an overall humid climate (enhanced ISM) during this phase are reported from Darjeeling (Ghosh et al. 2015) and Ziro valley (Bhattacharyya et al. 2014). Even though the inferences are drawn only from two studies, they hold well with the $\delta^{18}O$ values from Bay of Bengal and also with the speleothem records from eastern China (Wang et al. 2001; Liu et al. 2020). This phase broadly corresponds to the relatively higher temperature and precipitation conditions recorded in the $\delta^{18}O$ values of the Guliya ice-core (Thompson et al. 1990, 1997; Shi et al. 2001). Although this time frame is characterized with an overall enhanced ISM, however significant variability in terms of ISM strength has been reported. For example, Ghosh et al. (2015) inferred a warm climate but weaker monsoon during ~41–31 ka while Bhattacharyya et al. (2014) reported an

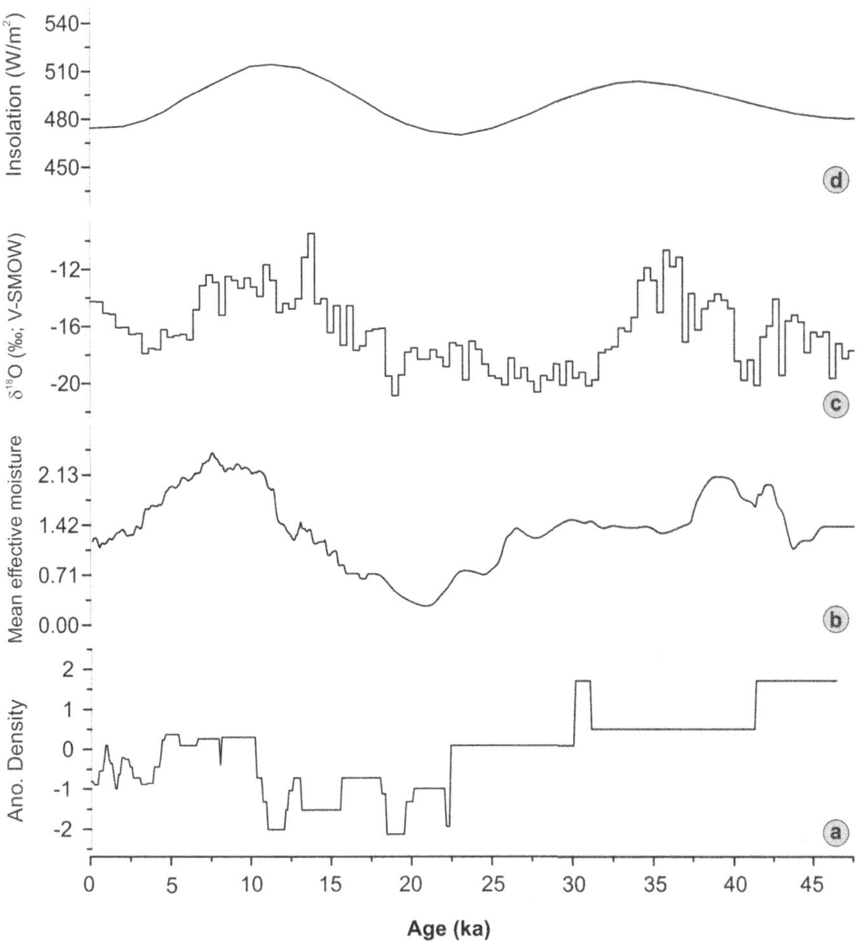

Fig. 5 (**a**) Comparison of the Indian summer monsoon anomalous density (ISM moisture) from the northeastern Indian Himalaya with other palaeoclimate proxy records viz., (**b**) Oxygen isotopic record of Guliya ice core (Thompson et al. 1997; enriched isotopic values indicate higher strengthen monsoon), (**c**) Mean effective moisture of the monsoon influenced central Asia (Herzschuh 2006) and, (**d**) Changes in insolation (June insolation 30°N) (Berger and Loutre 1991)

increased ISM during ~44–34 ka. However, the early MIS-2 showed a cooler climate with monsoon intensification in both Darjeeling as well as Ziro valley. Although the studies are in broad agreement locally as well as regionally, (Fig. 5) however the inferences maybe treated as tentative.

- *LGM:* The period between 22 and 18 ka witnessed reduced moisture conditions with ~80% of the records showing dry climate (weak monsoon) in the NEIH. The inference drawn from the palaeoclimatic records in Darjeeling (Ghosh et al. 2015) and Ziro valley further suggest low temperature conditions. However, another study from the Ziro Lake (Bhattacharyya et al. 2014) has suggested humid climatic conditions at ~21 ka with a progressive decline in ISM till 18 ka. The data is in conformity with the Bittoo Cave speoleothem $\delta^{18}O$ values suggesting reduced ISM precipitation during the LGM (Kathayat et al. 2017). These inferences are corroborated by the LGM glacier advance reported from all of the Himalayan region (Ali et al. 2013, 2019; Rana et al. 2019; Shukla et al. 2018; Murari et al. 2014 and references therein). Considering the sensitivity of Himalayan glaciers to precipitation and temperature changes, the LGM glacier advance has been attributed to the enhanced mid-latitude westerlies and associated temperature decline. Keeping the above facts in view, it has been suggested that the weakening of ISM during the LGM was influenced by decrease in solar insolation and increased snow cover on the Himalayan Tibetan orogen (Fig. 5; Herzschuh 2006; Owen et al. 2002; Thompson et al. 1997).

- *Last deglacial:* The last deglacial period has witnessed a gradual rise in temperature and revival of the ISM. Based on the limited data from the NEIH, it is observed that 50% of the available data (Chauhan and Sharma 1996; Ghosh et al. 2015) advocate for a moderate-wet climate while the remaining 50% (Ghosh et al. 2014; Bhattacharyya et al. 2014) propose dry climatic conditions during 18–13 ka. Similar inferences have been proposed by Rashid et al. (2011), suggesting a weak monsoon between 18.2 and 14.8 ka which further corroborated by lowest precipitation record inferred from the Hulu, Dongee and Sanbao speleothem records (Wang et al. 2001, 2006; Cheng et al. 2012 and references therein). Although a sharp increase in the monsoon strength is reported between 15 and 13 ka (Fig. 5; Herzschuh 2006), the present synthesis suggests drier climatic conditions (Ghosh et al. 2014; Bhattacharyya et al. 2014). Therefore the available data does not provide a clear last deglacial monsoon variability record may be due to poor age resolution and hence should be used with utmost care.
- *Early Holocene*: The beginning of the Holocene corresponds with the Younger Dryas (YD) cooling event. Our synthesis shows that about 80% of the climatic studies from the NEIH (during 13–11 ka) recorded dry to weak monsoon conditions (lower weightage values). There are multiple archives available from the region which suggests a cool and dry climate at the beginning of Holocene (Chauhan and Sharma 1996; Ghosh et al. 2014; Bera and Basumatary 2013; Bhattacharyya et al. 2014; Ali et al. 2018). This temperature decline along with low rainfall may be attributed to the weakening of the Atlantic Meridionial Overturning circulation during the YD (13–11 ka; Wang et al. 2001; Dansgaard et al. 1989; Alley 2000); the manifestations of which can be seen in the short term glacier advances throughout the Himalaya (Murari et al. 2014; Sati et al. 2014; Bisht et al. 2017; Shukla et al. 2018; Rana et al. 2019). The YD cooling in the region is also inferred in SST and $\delta^{18}O_{sw}$ values of Bay of Bengal (Liu et al. 2020; Kudrass et al. 2001), enriched $\delta^{18}O$ values from Hulu cave stalagmites (Wang et al. 2001) and the Dali Lake (Fan et al. 2018). Following this, the early Holocene period witnessed climatic amelioration with ~65% of the records (NEIH) showing moderate-wet climate conditions (11–8 ka). These inferences are in broad agreement with the synthesis of climate data from the ISM influenced

Asia, suggesting optimal moisture conditions during the early Holocene (11–7 ka; Herzschuh 2006). Majority of the records report intensified ISM in the region during this period (Chauhan and Sharma 1996; Ghosh et al. 2014; Ali et al. 2018). The inferences of gradual strengthening of the ISM can be validated from the detrital proxy record in the Central Himalaya (Bhushan et al. 2018). Other regional scale evidences that validate the post YD climate warming have been documented from both marine and terrestrial proxy records (Li et al. 2011; Hudson and Quade 2013). On the basis of depleted $\delta^{13}C$ values of proglacial lake sediments from the higher Sikkim Himalaya, Ali et al. (2018) have suggested enhanced ISM condition post YD. The above inferences corroborate well with the oxygen isotopic data from the Bay of Bengal and Andaman Sea (Rashid et al. 2011; Liu et al. 2020). Furthermore, the depleted $\delta^{18}O$ in speleothem records from northeast India (Berkelhammer et al. 2012; Dutt et al. 2015) and the Tianmen cave (Cai et al. 2012) also indicate insolation driven early Holocene strengthened ISM.

- *Mid Holocene*: The mid Holocene period has seen further wetness with more than 50% of the records suggesting moderate-wet to wet climatic conditions till ~4 ka BP in the Asian region (Herzschuh 2006). Our synthesis also demonstrate further wetness as 41% of the records show intensified ISM, while 26% suggest a moderate climate. These inferences are further supported by the terrestrial archives from the Central Himalayan region (Pant et al. 2005; Chakraborty et al. 2006; Ranhotra and Bhattacharyya 2010; Bhushan et al. 2018; Srivastava et al. 2017), suggesting increased ISM conditions. Regionally/ globally this time frame represents the Holocene climate optimum and is recorded in various climate reconstructions (Herzschuh 2006; Misra et al. 2019). Similarly, during this phase, Sharma and Chauhan (1994) observed an expansion of oak, appearance of broad-leaved temperate elements, such as *Juglans*, *Rhododendron* and *Ulmus* and inferred a period of optimal climate.
- *Late Holocene*: A sharp decline in the precipitation is observed at ~4 ka in the synthesis (Bera and Basumatary 2013; Ghosh et al. 2014; Mehrotra et al. 2019) and may broadly correspond to the 4.2 cooling event. This (4.2) marks the beginning of the recently proposed *Meghalayan* stage. The beginning of this stage is marked by a shift to heavier isotopic values at ~4.2 ka, reflecting an abrupt reduction in precipitation due to a weakening of the ISM across monsoon influenced Asia (Wang et al. 2005; Yang et al. 2011; Berkelhammer et al. 2012; Kathayat et al. 2017) as well as other continents (Mayewski et al. 2004; Staubwasser and Weiss 2006; Schimpf et al. 2011; Peck et al. 2015; Chase et al. 2017; Weiss 2017; Bailey et al. 2018; Walker et al. 2018). Ghosh et al. (2015) have suggested rainfall minima (weak monsoon), from the Darjeeling foot hills during 4.3–3.5 ka. However, Bera and Basumatary (2013) have reported a warm but dry climate from the foot Hills of Arunanchal Pradesh. Ali et al. (2018) have also reported a below average, but gradually increasing ISM since 4 ka in the higher Sikkim Himalaya. Further improvement in the moisture conditions is observed in the synthesis with ~40% records showing moderate to wet ISM. The increasing trend in the ISM precipitation during the last two millennia is

coeval with other records from the Himalayan region as well as the adjacent marine archives. Although, this time frame encompasses the little ice age (LIA), however it is interesting to note that, majority of the studies have not reported any climate change/variability at around LIA. Only Sharma and Chauhan (2001) have suggested a colder a drier climate in the Kupup lake during ~400–200 years BP. However, Ali et al. (2018) have suggested that no drop in ISM precipitation during the LIA in the region may be attributed to the dual interactions of ITCZ and strong mid latitude westerlies (Sinha et al. 2011; Dixit et al. 2014).

5.1 Investigating the Forcing Factors

Efficacy of climate drivers: To understand the nature and extent of the above mentioned forcing factors and their correlation with the climate variables, ordination analysis has been used. The climate (precipitation and temperature) data for past 118 years has been extracted from Climatic Research Unit (CRU TS.4.04; 0.5 latitude x 0.5 longitude; 1901–2018 CE; Harris et al. 2020) and used as response variables. The average annual temperature and precipitation of both summer and winter months from 8 grids that cover the entire NEIH were used. Furthermore, data of the seven driving factors under the same time frame was extracted and used as explanatory variable. Since ENSO-Modoki is a recent phenomenon compared to our timescales (been active only for a few decades), it has not been taken into consideration for ordination analysis. Thus for ordination only used six climatic drivers i.e. TSI, SOI, AMO, PDO, IOD and AO were used. The results obtained from the DCA, showed homogeneity in the data, as the length of the maximum gradient for all these grids was less than 3 SD. In order to find some relationship within the response variables we used PCA. More than 70% of the variance is explained by the first two components of the PCA, for all the grids. Besides this, since we assumed the forcing factors to be the explanatory variables, we tried find their relationship with the response (climatic) variables using RDA. For the present study, the amount of constrained proportion explained by the explanatory variables ranged between 13–15%. The maximum proportion (15.41%) was explained from the first grid, while the minimum proportion explained from the seventh grid (13.16%). Based on the r^2 and p-values from the RDA (Table 3) the two forcing factors are found to be causing maximum variability in the climatic variables, are Southern Oscillation Index (SOI) and the Pacific Decadal Oscillation (PDO). Other factors like TSI and AMO also show significant correlation but are not consistent for all the grids.

Causing factors: Annual rainfall (cumulative) and temperature (average) during 1901 to 2000 ($n = 100$) of 8 mentioned grids were used as time series representing effects. Annual mean TSI, AMO, IOD, PDO, AO, El-Nino, ENSO-Modoki and SOI from the same time period are considered as potential causal time series. Differenced time series ($x_{i+1}-x_i$, $i = 1,2, \ldots n$) data for both effects (response variables) and causes (explanatory variables) were used to make them stationary (detrended and deseasonalised) using *diff()* method implemented in *forecast* package (Hyndman et al. 2018). Another important parameter for the Granger causality test is to find appropriate lag order and this parameter is used to do regression analysis on itself (causal time series) with different lags so that the effect time series can be forecasted to a certain degree. It is called vector autoregressive model, VAR(q) with q lags. This lag parameter was optimized using two independent methods within R packages in R platform (R Core Team 2020). One was custom function and another one made use of *VARselect ()* method from *vars* package and both optimized the lag order i.e., q based on the Akaike Information Criteria, AIC (Pfaff 2008a, b). Lag order obtained from each procedure was used in doing Granger causality test. The procedure in testing whether y cause x, is to have a NULL hypothesis that state, "x do not G-causes y" and calculate F-ratio and p-value. If the p-value is less than 5% i.e. p-value < 0.05, the NULL hypothesis should be rejected. Rejecting NULL hypothesis means, "x do G-causes y". To make that statement, the other way around (another NULL hypothesis of "y do not G-causes x") should be accepted as well at the same time which will yield a higher p-value. With x and y are Drivers and Responses, respectively the following generalized NULL hypotheses are given below,

H_{0_1}: Drivers do not cause Responses.
H_{0_2}: Responses do not cause Drivers.

Rejections (p-value < 0.05) of the first NULL hypothesis (H_{0_1}) imply that drivers cause responses (precipitation and temperature). But it doesn't stop here and demand the other way round shouldn't be inferred. So the second NULL hypothesis (H_{0_2}) should be accepted as well (p-value > 0.05). To ascertain a driver cause a response, it is required to reject the H_{0_1} (p-value < 0.05) and accept H_{0_2} (p-value > 0.05). The same exercise was done with the lag order parameter that was optimized by the second procedure as well. There are two G-causal test namely GCT_1 and GCT_2 where the lag order was optimized based on two procedures, respectively. The results are given in Fig. 6. The top of Fig. 6 is showing the G-causal relationship with precipitation and the bottom figure is showing with temperature. The right arrow indicates that either GCT_1 or GCT_2 yielded G-cause and the left arrow indicates that both GCT_1 and GCT_2 yielded reverse G-cause i.e. accepting the H_{0_1} (p-value > 0.05) and rejecting H_{0_2} (p-value < 0.05). So in a sense the left arrow is stronger than the right arrow. The arrowless squares indicate unresolved where both the NULL hypotheses were either accepted or rejected. Granger causality test was implemented using *grangertest()* method in *lmtest* package (Zeileis and Hothorn 2002).

G-causality test suggests that IOD G-causes the precipitation change in Grids 1, 2, 4, 5 and 7, and AMO G-causes the precipitation change in Grid 2 since 1901 (Fig. 6). AMO G-causes temperature change in all the Grids, IOD in Grids 5–7, PDO in Grids 1–3 and Grid 5, and El Nino events G-causes temperature change in Grid 7. ENSO-Modoki is found to be G-caused by the change in temperature in Grids 2, 4, 5 and 8 (Fig. 6). Values of F-ratio and p-values for those significant G-causal links are

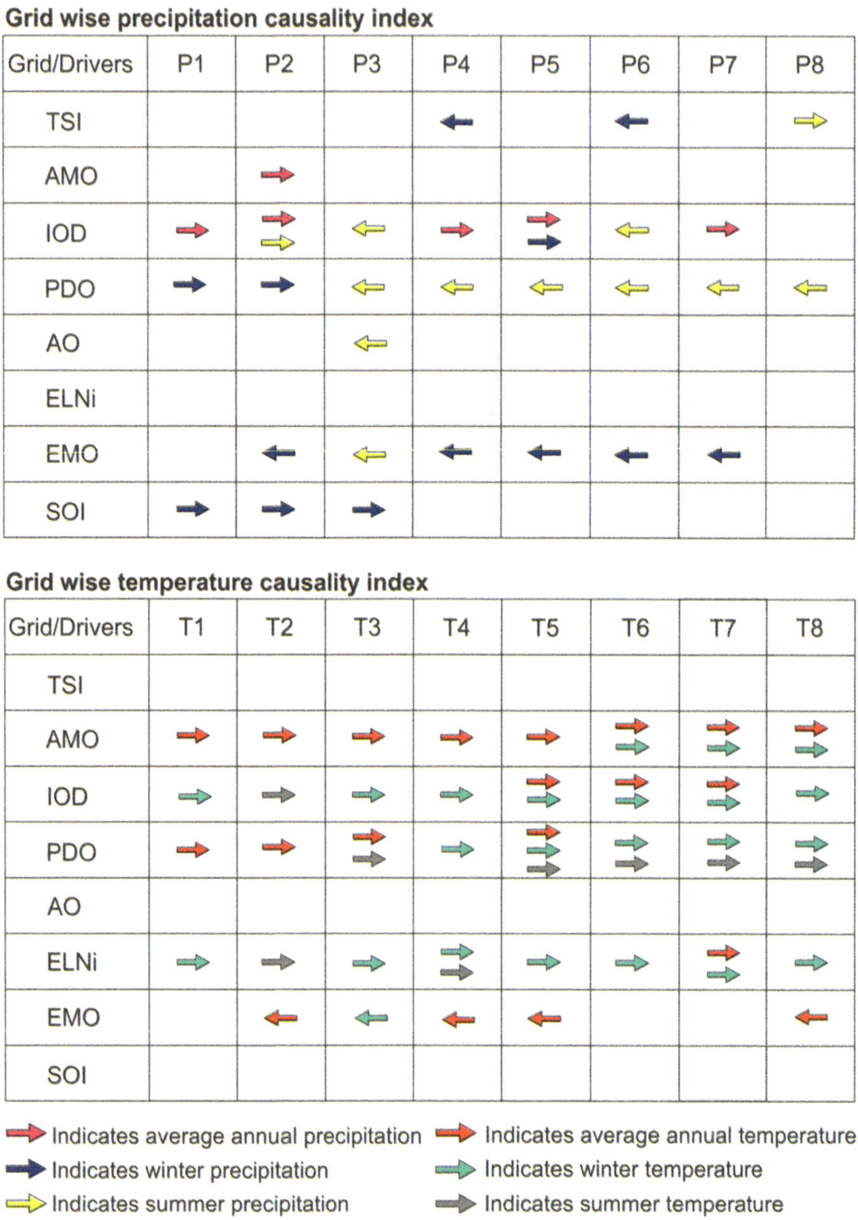

Fig. 6 Granger-test for understanding the causality factor with respect to precipitation (top) and temperature (bottom). TSI (Total Solar Irradiance), AMO (Atlantic Multidecadal Oscillation), IOD (Indian Ocean Dipole), PDO (Pacific Decadal Oscillation), AO (Arctic Oscillation), ELNi (El Nino Southern Oscillation), EMO (ENSO-Modoki), SOI (Southern Oscillation Index)

Table 4 Granger-causality; values of F-ratio and p-values for significant G-causal links (H_{0_1}: Drivers do not cause Responses)

Grid	Driver	F-ratio	P-value
WM_1	PDO	4.947	0.02847637
	SOI	7.538989994	0.007208995
ST_1	IOD	7.730317098	0.000788438
	ELNINO	7.305347492	0.001130263
SM_2	IOD	2.083762363	0.025282423
WM_2	PDO	3.981443192	0.048838398
	SOI	5.921628541	0.016806452
ST_2	IOD	7.140488	0.00131
	ELNINO	3.174475	0.077959
WM_3	SOI	4.908913742	0.029081862
ST_3	IOD	7.602335342	0.000879852
	ELNINO	3.080639482	0.08241948
WT_3	PDO	2.762017202	0.032620699
ST_4	IOD	8.144292958	0.000553907
	PDO	4.035735557	0.047354549
	ELNINO	5.545129739	0.005307474
WT_4	ENSO MODIKI	2.35368143	0.047541849
WM_5	IOD	2.157178251	0.029136702
ST_5	IOD	6.977694186	0.00150862
	PDO	4.814043856	0.030641627
	ELNINO	4.124646288	0.019218931
WT_5	PDO	2.595243817	0.041920421
ST_6	AMO	2.800876253	0.030766133
	IOD	2.628067665	0.039903424
	PDO	5.249963101	0.024133619
	ELNINO	3.923790345	0.023121292
WT_6	PDO	2.120969383	0.044152132
ST_7	AMO	3.04726834	0.021216196
	IOD	3.151918052	0.018115174
	PDO	4.365119333	0.039326794
	ELNINO	4.197794746	0.017970932
WT_7	PDO	2.514133146	0.017832055
SM_8	TSI	2.088237561	0.040087253
ST_8	AMO	2.719400399	0.034782177
	IOD	3.220825679	0.016324753
	AO	4.748050907	0.031778794
	ELNINO	3.642718534	0.029984449
WT_8	PDO	2.652435248	0.012911531

resolved, are given in Table 4. As TSI is apparently not causing, the same analysis was done using TSI as the driver and all the other drivers as responses. For this, we assumed TSI doesn't have a direct causal effect on the precipitation and temperature but via secondary causes such as AMO, IOD, PDO, AO, ENSO, ENSO-Modoki and

SOI. However the G-causal relationships of TSI remain unresolved as it has not shown any G-causal relationship with the climatic driving indices too. It is intriguing that neither the climate responses nor the climate driving indices have a significant temporal correlation with TSI. It should be noted that solar irradiance has an indirect causal connection with the climate via its interaction with the middle atmosphere, then with lower atmosphere and then with oceans, and it takes centenary to millennial timescales, but might not have manifested in decadal timescales as of our data (Haigh 2001). Further, the lag parameter also related to the timescales of the processes that we study and our maximum lag parameter is 25 years. So to understand the real causal relationship of TSI with other secondary causes (intermediaries) and climate parameters, a long term records are to be used. Using seasonal climate parameters and performing the G-causality test against respective climate driver indices that were chosen based on the correlation coefficients might be shedding more light into the better causal connections, and that is the assignment for future.

As a remark, it is not the true causality as mentioned above but it helps to know the directional effect on one time series and hence it helps from where mere correlations stop helping. Comparing the results that are obtained from multivariate statistics and G-causality test, TSI has shown to be causing the variability in the climatic parameters but no significant temporal correlation is observed and the usage of our short term data might be one among the possible reasons.

6 Conclusion

- Our reconstructed synthesis suggests that the pre-LGM (21–18 ka; Last glacial maximum) ISM was moderately higher; however, the conditions became drier during the LGM (weak ISM).
- A moderate to low ISM can be suggested during the last deglacial (~18–12 ka) period. Subsequently, an intensification of the ISM is observed that lasted till ~4 ka.
- Majority of the studies and our synthesis suggest an abrupt weakening of the ISM during ~4–3 ka and is followed by climatic amelioration.
- It is observed that even though the resolution of the data used for the synthesis is coarse it still correlates well with the local and region high resolution climate records.
- Based on the r^2 and p values from the RDA the two forcing factors are found to be causing maximum variability in the climatic variables, are Southern Oscillation Index (SOI) and the Pacific Decadal Oscillation (PDO).
- Other factors like TSI and AMO also show significant correlation but are not consistent for all the grids.
- G-causality test suggests that IOD G-causes play a dominant role in precipitation change in the area.
- While, AMO G-causes temperature change in the area, with significant influence of IOD and El Nino events in some parts.

Acknowledgments The authors are thankful to the Director, Birbal Sahni Institute of Palaeosciences, Lucknow, India for providing infrastructural facilities.

References

Ali SN, Biswas RH, Shukla AD, Juyal N (2013) Chronology and climatic implications of late quaternary glaciations in the Goriganga valley, central Himalaya, India. Quat Sci Rev 73:59–76

Ali SN, Dubey J, Ghosh R, Quamar MF, Sharma A, Morthekai P, Dimri AP, Shekhar M, Arif M, Agrawal S (2018) High frequency abrupt shifts in the Indian summer monsoon since younger Dryas in the Himalaya. Sci Rep 8(1):1–8

Ali SN, Dubey J, Morthekai P, Sharma A, Singh R, Prizomwala S (2019) Climate forcing and the initiation of glacier advance during MIS-2 in the North Sikkim Himalaya, India. J Asian Earth Sci 174:381–388

Alley RB (2000) The younger Dryas cold interval as viewed from Central Greenland. Quat Sci Rev 19(1–5):213–226

Arrawatia ML, Tambe S (2011) Biodiversity of Sikkim: exploring and conserving a global hotspot. Information and Public Relations Department, Gangtok, Sikkim

Ashok K, Feba F, Tejavath CT (2019) The Indian summer monsoon rainfall and ENSO. Mausam 70(3):443–452

Bailey HL, Kaufman DS, Sloane HJ, Hubbard AL, Henderson ACG, Leng M, Meyer H, Welker JM (2018) Holocene atmospheric circulation in the central North Pacific: a new terrestrial diatom and δ18O dataset from the Aleutian Islands. Quat Sci Rev 194:27–38

Benn DI, Owen LA (1998) The role of the Indian summer monsoon and the mid-latitude westerlies in Himalayan glaciation: review and speculative discussion. J Geol Soc 155(2):353–363

Bera SK, Basumatary SK (2013) Vegetation history and monsoonal fluctuations during the last 12,500 years BP inferred from pollen record at Lower Subansiri Basin, Assam, Northeast India. Palaeobotanist 62:1–10

Berger A, Loutre MF (1991) Insolation values for the climate of the last 10 million years. Quat Sci Rev 10(4):297–317

Berkelhammer MB, Sinha A, Stott L, Cheng H, Pausata FSR, Yoshimura K (2012) An abrupt shift in the Indian monsoon 4000 years ago. Geophys Monogr Ser 198:75–87

Bhattacharyya A, Mehrotra N, Shah SK, Basavaiah N, Chaudhary V, Singh IB (2014) Analysis of vegetation and climate change during late Pleistocene from Ziro valley, Arunachal Pradesh, eastern Himalaya region. Quat Sci Rev 101:111–123

Bhattacharyya A, Sharma J, Shah SK, Chaudhary V (2007) Climatic changes during the last 1800 yrs BP from paradise lake, Sela Pass, Arunachal Pradesh, Northeast Himalaya. Curr Sci 93(7): 983–987

Bhushan R, Sati SP, Rana N, Shukla AD, Mazumdar AS, Juyal N (2018) High-resolution millennial and centennial scale Holocene monsoon variability in the higher Central Himalayas. Palaeogeogr Palaeoclimatol Palaeoecol 489:95–104

Bhutiyani MR, Kale VS, Pawar NJ (2010) Climate change and the precipitation variations in the northwestern Himalaya: 1866–2006. Int J Climatol 30(4):535–548

Bisht P, Ali SN, Rana N, Singh S, Sundriyal YP, Bagri DS, Juyal N (2017) Pattern of Holocene glaciation in the monsoon–dominated Kosa Valley, central Himalaya, Uttarakhand, India. Geomorphology 284:130–141

Bookhagen B, Burbank DW (2006) Topography, relief, and TRMM-derived rainfall variations along the Himalaya. Geophys Res Lett 33(8)

Bookhagen B, Burbank DW (2010) Toward a complete Himalayan hydrological budget: spatio-temporal distribution of snowmelt and rainfall and their impact on river discharge. J Geophys Res Earth 115(F3)

Cai Y, Zhang H, Cheng H, An Z, Edwards RL, Wang X, Tan L, Liang F, Wang J, Kelly M (2012) The Holocene Indian monsoon variability over the southern Tibetan Plateau and its teleconnections. Earth Planet Sci Lett 335:135–144

Chakraborty S, Bhattacharya SK, Ranhotra PS, Bhattacharyya A, Bhushan R (2006) Palaeoclimatic scenario during Holocene around Sangla valley, Kinnaur northwest Himalaya based on multi proxy records. Curr Sci 00113891:91

Chakravarty S, Ghosh SK, Suresh CP, Dey AN, Shukla G (2012) Deforestation: causes, effects and control strategies. Glob Perspect Sustain For Manag 1:1–26

Chase BM, Chevalier M, Boom A, Carr AS (2017) The dynamic relationship between temperate and tropical circulation systems across South Africa since the last glacial maximum. Quat Sci Rev 174:54–62

Chauhan MS, Sharma C (1996) Late Holocene vegetation of Darjeeling (Jore-Pokhari) eastern Himalaya

Chen F, Xu Q, Chen J, Birks HJB, Liu J, Zhang S, Jin L, An C, Telford RJ, Cao X, Wang Z (2015) East Asian summer monsoon precipitation variability since the last deglaciation. Sci Rep 5(1): 1–11

Cheng H, Sinha A, Wang X, Cruz FW, Edwards RL (2012) The global paleomonsoon as seen through speleothem records from Asia and the Americas. Clim Dyn 39(5):1045–1062

Dansgaard W, White JWC, Johnsen SJ (1989) The abrupt termination of the younger Dryas climate event. Nature 339(6225):532–534

Dixit Y (2020) Regional character of the "global monsoon". Oceanography 33(2):56–64

Dixit Y, Hodell DA, Petrie CA (2014) Abrupt weakening of the summer monsoon in Northwest India~ 4100 yr ago. Geology 42(4):339–342

Dutt S, Gupta AK, Clemens SC, Cheng H, Singh RK, Kathaya G, Edwards RL (2015) Abrupt changes in Indian summer monsoon strength during 33,800 to 5500 years BP. Geophys Res Lett 42(13):5526–5532

Fan J, Xiao J, Wen R, Zhang S, Wang X, Cui L, Liu Y, Li H, Yue J (2018) The manifestation of the younger Dryas event in the East Asian summer monsoon margin: new evidence from carbonate geochemistry of the Dali Lake sediments in northern China. The Holocene 28(7):1082–1092

Finkel RC, Owen LA, Barnard PL, Caffee MW (2003) Beryllium-10 dating of Mount Everest moraines indicates a strong monsoon influence and glacial synchroneity throughout the Himalaya. Geology 31(6):561–564

Gadgil S, Kumar KR (2006) The Asian monsoon—agriculture and economy. In: The Asian monsoon. Springer, Berlin, pp 651–683

Gauch HG, Gauch HG Jr (1982) Multivariate analysis in community ecology, vol 1. Cambridge University Press, Cambridge

Ghosh R, Paruya DK, Khan MA, Chakraborty S, Sarkar A, Bera S (2014) Late quaternary climate variability and vegetation response in Ziro Lake Basin, eastern Himalaya: a multiproxy approach. Quat Int 325:13–29

Ghosh R, Bera S, Sarkar A, Paruya DK, Yao YF, Li CS (2015) A ~50 ka record of monsoonal variability in the Darjeeling foothill region, eastern Himalayas. Quat Sci Rev 114:100–115

Ghosh R, Biswas O, Paruya DK, Agrawal S, Sharma A, Nautiyal CM, Bera M, Bera S (2018) Hydroclimatic variability and corresponding vegetation response in the Darjeeling Himalaya, India over the past~ 2400 years. Catena 170:84–99

Govil P, Naidu PD (2011) Variations of Indian monsoon precipitation during the last 32 kyr reflected in the surface hydrography of the Western Bay of Bengal. Quat Sci Rev 30(27–28): 3871–3879

Granger CW (1969) Investigating causal relations by econometric models and cross-spectral methods. Econometrica 37:424–438

Grujic D, Coutand I, Bookhagen B, Bonnet S, Blythe A, Duncan C (2006) Climatic forcing of erosion, landscape, and tectonics in the Bhutan Himalayas. Geology 34(10):801–804

Gupta AK, Anderson DM, Overpeck JT (2003) Abrupt changes in the Asian southwest monsoon during the Holocene and their links to the North Atlantic Ocean. Nature 421(6921):354–357

Gupta AK, Das M, Anderson DM (2005) Solar influence on the Indian summer monsoon during the Holocene. Geophys Res Lett 32:L17703

Gupta AK, Mohan K, Das M, Singh RK (2013) Solar forcing of the Indian summer monsoon variability during the Ållerød period. Sci Rep 3(1):1–5

Haigh JD (2001) Climate variability and the influence of the sun. Science 294:2109–2111

Harris I, Osborn TJ, Jones P, Lister D (2020) Version 4 of the CRU TS monthly high-resolution gridded multivariate climate dataset. Sci Data 7(1):1–18

Hasson S, Pascale S, Lucarini V, Böhner J (2016) Seasonal cycle of precipitation over major river basins in south and Southeast Asia: a review of the CMIP5 climate models data for present climate and future climate projections. Atmos Res 180:42–63

Herzschuh U (2006) Palaeo-moisture evolution in monsoonal Central Asia during the last 50, 000 years. Quat Sci Rev 25(1–2):163–178

Hudson AM, Quade J (2013) Long-term east-west asymmetry in monsoon rainfall on the Tibetan plateau. Geology 41(3):351–354

Hyndman RJ, Athanasopoulos G, Bergmeir C, Caceres G, Chhay L, O'Hara-Wild M, Petropoulos F, Razbash S, Wang E, Yasmeen F (2018) Forecast: forecasting functions for time series and linear models

Kathayat G, Cheng H, Sinha A, Yi L, Li X, Zhang H, Li H, Ning Y, Edwards RL (2017) The Indian monsoon variability and civilization changes in the Indian subcontinent. Sci Adv 3(12): e1701296

Kaushal N, Breitenbach SF, Lechleitner FA, Sinha A, Tewari VC, Ahmad SM, Berkelhammer M, Band S, Yadava M, Ramesh R, Henderson GM (2018) The Indian summer monsoon from a speleothem δ18O perspective—a review. Quaternary 1(3):29

Kessarkar PM, Purnachadra Rao V, Naqvi SWA, Karapurkar SG (2013) Variation in the Indian summer monsoon intensity during the Bølling-Ållerød and Holocene. Paleoceanography 28(3): 413–425

Kudrass HR, Hofmann A, Doose H, Emeis K, Erlenkeuser H (2001) Modulation and amplification of climatic changes in the northern hemisphere by the Indian summer monsoon during the past 80 ky. Geology 29(1):63–66

Koul V, Parekh A, Srinivas G, Kakatkar R, Chowdary JS, Gnanaseelan C (2018) Role of ocean initial conditions to diminish dry bias in the seasonal prediction of Indian summer monsoon rainfall: a case study using climate forecast system. J Adv Model Earth Syst 10(3):603–616

Kumar KK, Soman MK, Kumar KR (1995) Seasonal forecasting of Indian summer monsoon rainfall: a review. Weather 50(12):449–467

Lepš J, Šmilauer P (2003) Multivariate analysis of ecological data using CANOCO. Cambridge University Press, Cambridge

Li Q, Lu H, Zhu L, Wu N, Wang J, Lu X (2011) Pollen-inferred climate changes and vertical shifts of alpine vegetation belts on the northern slope of the Nyainqentanglha Mountains (central Tibetan plateau) since 8.4 kyr BP. The Holocene 21(6):939–950

Liu S, Ye W, Chen MT, Pan HJ, Cao P, Zhang H, Khokiattiwong S, Kornkanitnan N, Shi X (2020) Millennial-scale variability of Indian summer monsoon during the last 42 kyr: evidence based on foraminiferal Mg/Ca and oxygen isotope records from the central Bay of Bengal. Palaeogeogr Palaeoclimatol Palaeoecol 562:110112

Mayewski PA, Rohling EE, Stager JC, Karlén W, Maasch KA, Meeker LD, Meyerson EA, Gasse F, van Kreveld S, Holmgren K, Lee-Thorp J, Rosqvist G, Rack F, Staubwasser M, Schneider RR, Steig E (2004) Holocene climate variability. Quat Res 62:243–255

Mehrotra N, Shah SK, Basavaiah N, Laskar AH, Yadava MG (2019) Resonance of the '4.2 ka event' and terminations of global civilizations during the Holocene, in the palaeoclimate records around PT Tso Lake, Eastern Himalaya. Quat Int 507:206–216

Ménégoz M, Gallée H, Jacobi HW (2013) Precipitation and snow cover in the Himalaya: from reanalysis to regional climate simulations. Hydrol Earth Syst Sci 17(10):3921–3936

Meyer MC, Hofmann CC, Gemmell AM, Haslinger E, Häusler H, Wangda D (2009) Holocene glacier fluctuations and migration of Neolithic yak pastoralists into the high valleys of Northwest Bhutan. Quat Sci Rev 28(13–14):1217–1237

Meyer MC, Aldenderfer MS, Wang Z, Hoffmann DL, Dahl JA, Degering D, Haas WR, Schlütz F (2017) Permanent human occupation of the central Tibetan plateau in the early Holocene. Science 355(6320):64–67

Mishra PK, Anoop A, Schettler G, Prasad S, Jehangir A, Menzel P, Naumann R, Yousuf AR, Basavaiah N, Deenadayalan K, Wiesner MG (2015) Reconstructed late quaternary hydrological changes from Lake Tso Moriri, NW Himalaya. Quat Int 371:76–86

Misra P, Tandon SK, Sinha R (2019) Holocene climate records from lake sediments in India: assessment of coherence across climate zones. Earth Sci Rev 190:370–397

Murari MK, Owen LA, Dortch JM, Caffee MW, Dietsch C, Fuchs M, Haneberg WC, Sharma MC, Townsend-Small A (2014) Timing and climatic drivers for glacier advance across monsoon-influenced regions of the Himalayan-Tibetan orogen. Quat Sci Rev 88:159–182

O'Neill AR, Badola HK, Dhyani PP, Rana SK (2017) Integrating ethnobiological knowledge into biodiversity conservation in the eastern Himalayas. J Ethnobiol Ethnomed 13(1):1–14

Overpeck J, Anderson D, Trumbore S, Prell W (1996) The southwest Indian monsoon over the last 18000 years. Clim Dyn 12(3):213–225

Owen LA, Finkel RC, Caffee MW (2002) A note on the extent of glaciation throughout the Himalaya during the global last glacial maximum. Quat Sci Rev 21(1–3):147–157

Pant GB, Parthasarathy SB (1981) Some aspects of an association between the southern oscillation and Indian summer monsoon. Arch Meteorol Geophys Bioclimatol Ser B 29(3):245–252

Pant RK, Basavaiah N, Juyal N, Saini NK, Yadava MG, Appel E, Singhvi AK (2005) A 20-ka climate record from central Himalayan loess deposits. J Quat Sci 20(5):485–492

Parthasarathy B, Yang S (1995) Relationships between regional Indian summer monsoon rainfall and Eurasian snow cover. Adv Atmos Sci 12(2):143–150

Peck VL, Allen CS, Kender S, McClymont EL, Hodgson DA (2015) Oceanographic variability on the West Antarctic peninsula during the Holocene and the influence of upper circumpolar deep water. Quat Sci Rev 119:54–65

Pfaff B (2008a) VAR, SVAR and SVEC models: implementation within R package vars. J Stat Softw 27(4):1–32

Pfaff B (2008b) Analysis of integrated and cointegrated time series with R. Springer, New York

Prasad S, Anoop A, Riedel N, Sarkar S, Menzel P, Basavaiah N, Krishnan R, Fuller D, Plessen B, Gaye B, Röhl U (2014) Prolonged monsoon droughts and links to Indo-Pacific warm pool: a Holocene record from Lonar Lake, Central India. Earth Planet Sci Lett 391:171–182

R Core Team (2020) R: a language and environment for statistical computing. R Foundation for Statistical Computing, Vienna, Austria. https://www.R-project.org/

Rana N, Sharma S, Ali SN, Singh S, Shukla AD (2019) Investigating the sensitivity of glaciers to climate variability since the MIS-2 in the upper Ganga catchment (Saraswati valley), Central Himalaya. Geomorphology 346:106854

Ranhotra PS, Bhattacharyya A (2010) Holocene palaeoclimate and glacier fluctuations within Baspa valley, Kinnaur, Himachal Pradesh. J Geol Soc India 75(3):527–532

Rashid H, England E, Thompson L, Polyak L (2011) Late glacial to Holocene Indian summer monsoon variability based upon sediment records taken from the Bay of Bengal. Terr Atmos Ocean Sci 22(2):2

Rasmusson EM, Carpenter TH (1983) The relationship between eastern equatorial Pacific Sea surface temperatures and rainfall over India and Sri Lanka. Mon Weather Rev 111(3):517–528

Ruddiman WF (1997) Tropical Atlantic terrigenous fluxes since 25,000 yrs BP. Mar Geol 136(3–4):189–207

Sarkar A, Ramesh R, Somayajulu BLK, Agnihotri R, Jull AJT, Burr GS (2000) High resolution Holocene monsoon record from the eastern Arabian Sea. Earth Planet Sci Lett 177(3–4): 209–218

Sati SP, Ali SN, Rana N, Bhattacharya F, Bhushan R, Shukla AD, Sundriyal Y, Juyal N (2014) Timing and extent of Holocene glaciations in the monsoon dominated Dunagiri valley (Bangni glacier), Central Himalaya, India. J Asian Earth Sci 91:125–136

Schimpf D, Kilian R, Kronz A, Klaus S, Spötl C, Wörner G, Deininger M, Mangini A (2011) The significance of chemical, isotopic and detrital components in three coeval stalagmites from the super humid southern most Andes (53°S) as high-resolution paleo-climate proxies. Quat Sci Rev 30:443–459

Schulz H, von Rad U, Erlenkeuser H (1998) Correlation between Arabian Sea and Greenland climate oscillations of the past 110,000 years. Nature 393(6680):54–57

Seth A (2007) Granger causality. Scholarpedia 2(7):1667

Sharma C, Chauhan MS (1994) Vegetation and climate since last glacial maximum in Darjeeling (Mirik Lake), Eastern Himalaya

Sharma C, Chauhan MS (2001) Late Holocene vegetation and climate of Kupup (Sikkim), Eastern Himalaya, India. J Palaeontol Soc India 46:51–58

Shi Y, Yu G, Liu X, Li B, Yao T (2001) Reconstruction of the 30–40 ka BP enhanced Indian monsoon climate based on geological records from the Tibetan Plateau. Palaeogeogr Palaeoclimatol Palaeoecol 169(1–2):69–83

Shrestha UB, Gautam S, Bawa KS (2012a) Widespread climate change in the Himalayas and associated changes in local ecosystems. PLoS One 7(5):e36741

Shrestha D, Singh P, Nakamura K (2012b) Spatiotemporal variation of rainfall over the central Himalayan region revealed by TRMM Precipitation Radar. J Geophys Res Atmos 117(D22)

Shrestha AB, Agrawal NK, Alfthan B, Bajracharya SR, Maréchal J, Oort BV (2015) The Himalayan climate and water atlas: impact of climate change on water resources in five of Asia's major river basins. The Himalayan Climate and Water Atlas: impact of climate change on water resources in five of Asia's major river basins

Shukla T, Mehta M, Jaiswal MK, Srivastava P, Dobhal DP, Nainwal HC, Singh AK (2018) Late quaternary glacier advance history of monsoon-dominated Dingad basin, central Himalaya, India. Quat Sci Rev 181:43–64

Sikka DR (1980) Some aspects of the large scale fluctuations of summer monsoon rainfall over India in relation to fluctuations in the planetary and regional scale circulation parameters. Proc Indian Acad Sci Earth Planet Sci 89(2):179–195

Sinha A, Cannariato KG, Stott LD, Li HC, You CF, Cheng H, Edwards RL, Singh IB (2005) Variability of southwest Indian summer monsoon precipitation during the Bølling-Allerød. Geology 33(10):813–816

Sinha A, Berkelhammer M, Stott L, Mudelsee M, Cheng H, Biswas J (2011) The leading mode of Indian Summer Monsoon precipitation variability during the last millennium. Geophys Res Lett 38(15)

Sirocko F, Sarnthein M, Erlenkeuser H, Lange H, Arnold M, Duplessy JC (1993) Century-scale events in monsoonal climate over the past 24,000 years. Nature 364(6435):322–324

Srivastava P, Agnihotri R, Sharma D, Meena N, Sundriyal YP, Saxena A, Bhushan R, Sawlani R, Banerji US, Sharma C, Bisht P (2017) 8000-year monsoonal record from Himalaya revealing reinforcement of tropical and global climate systems since mid-Holocene. Sci Rep 7(1):1–10

Staubwasser M, Weiss H (2006) Holocene climate and cultural evolution in late prehistoric–early historic West Asia. Quat Res 66:372–387

Tao SY (1987) A review of recent research on the east Asian summer monsoon in China. Monsoon Meteorol 1987:60–92

Thompson LG, Mosley-Thompson E, Davis ME, Bolzan JF, Dai J, Klein L, Gundestrup N, Yao T, Wu X, Xie Z (1990) Glacial stage ice-core records from the subtropical Dunde ice cap, China. Ann Glaciol 14:288–297

Thompson LO, Yao T, Davis ME, Henderson KA, Mosley-Thompson E, Lin PN, Beer J, Synal HA, Cole-Dai J, Bolzan JF (1997) Tropical climate instability: the last glacial cycle from a Qinghai-Tibetan ice core. Science 276(5320):1821–1825

Venkataraman K, Sivaperuman C (2018) Biodiversity hotspots in India. In: Indian hotspots. Springer, Singapore, pp 1–27

Walker M, Head MJ, Berkelhammer M, Björck S, Cheng H, Cwynar L, Fisher D, Gkinis V, Long A, Lowe J, Newnham R (2018) Formal ratification of the subdivision of the Holocene series/epoch (quaternary system/period): two new global boundary Stratotype sections and points (GSSPs) and three new stages/subseries. Episodes 41(4):213–223

Wang YJ, Cheng H, Edwards RL, An ZS, Wu JY, Shen CC, Dorale JA (2001) A high-resolution absolute-dated late Pleistocene monsoon record from Hulu Cave, China. Science 294(5550): 2345–2348

Wang Y, Cheng H, Edwards RL, He Y, Kong X, An Z, Wu J, Kelly MJ, Dykoski CA, Li X (2005) The Holocene Asian monsoon: links to solar changes and North Atlantic climate. Science 308: 854–857

Wang X, Auler AS, Edwards RL, Cheng H, Ito E, Solheid M (2006) Interhemispheric anti-phasing of rainfall during the last glacial period. Quat Sci Rev 25(23–24):3391–3403

Webster PJ (1987) The variable and interactive monsoon. In: Fein JS, Stephens PL (eds) Monsoons. Wiley-Interscience, New York, pp 269–330

Webster PJ, Yang S (1992) Monsoon and ENSO: selectively interactive systems. Q J R Meteorol Soc 118(507):877–926

Webster PJ, Magana VO, Palmer TN, Shukla J, Tomas RA, Yanai M, Yasunari T (1998) Monsoons: Processes, predictability, and the prospects for prediction. J Geophys Res 103:14451–14510

Weiss H (2017) 4.2 ka BP mega drought and the Akkadian collapse. In: Weiss H (ed) Megadrought and collapse. From early agriculture to Angkor. Oxford University Press, Oxford, pp 93–160

Wu R (2017) Relationship between Indian and east Asian summer rainfall variations. Adv Atmos Sci 34(1):4–15

Xavier PPK, Marzin C, Goswami B (2007) An objective definition of the Indian summer monsoon season and a new perspective on the ENSO—monsoon relationship. QJR Meteorol Soc 764: 749–764

Yang B, Bräuning A, Dong Z, Zhang Z, Keqing J (2008) Late Holocene monsoonal temperate glacier fluctuations on the Tibetan Plateau. Glob Planet Chang 60(1–2):126–140

Yang TN, Lee TQ, Meyers PA, Song SR, Kao SJ, Löwemark L, Chen RF, Chen HF, Wei KY, Fan CW, Shiau LJ (2011) Variations in monsoonal rainfall over the last 21 kyr inferred from sedimentary organic matter in Tung-Yuan Pond, southern Taiwan. Quat Sci Rev 30(23–24): 3413–3422

Zeileis A, Hothorn T (2002) Diagnostic checking in regression relationships. R News 2(3):7–10. https://CRAN.R-project.org/doc/Rnews/

Quaternary Climate of Narmada Valley: A Case Study on Understanding Provenance, Weathering and Depositional Environment Using Alluvium Geochemistry from Tawa River Basin, Hoshangabad District, Madhya Pradesh

Shradha Shukla, Jayshree Meshram, Chhaya Minz, Hemraj Suryavanshi, and Shubrasuchi Sarkar

Abstract The geochemical characterization of fine clastic stream sediment samples from both fluvial (Older Floodplain and Younger Floodplain) and erosional (Piedmont Slope, Pediplain) morphological units, of Tawa river basin, Hoshangabad District, Madhya Pradesh was attempted for understanding the Quaternary climate of Narmada Valley. The statistical evaluation and analysis of the chemical data suggests that Narmada alluvium is chemically immature derived from relatively un-weathered source, with average CIA value of 72.83 comparable with global shale standards.

The major and trace elemental ratios further indicate that these sediments are least affected by post depositional chemical weathering and the elemental characteristics are defined by the detrital mineralogy. The possible detrital phases in Narmada valley alluvium includes clay minerals (illites, smectite), phyllosilicates (muscovite) and the quartz, opaque's (Fe-Ti Oxides), phosphate minerals and heavy minerals including Zircon, Sphene (?), Monazite (?), Allnite (?) etc. The attenuation and fractionation of Heavy metals and other mineral phases elucidated through elemental

---

S. Shukla (✉)
Geological Survey of India, Gandhinagar, Gujarat, India
e-mail: shradha.sukla@gsi.gov.in

J. Meshram
Geological Survey of India, Hyderabad, Telangana, India

C. Minz
Geological Survey of India, Ranchi, Jharkhand, India

H. Suryavanshi
Geological Survey of India, NMH-II, Nagpur, Maharashtra, India

S. Sarkar
Geological Survey of India, Bhopal, Madhya Pradesh, India

B. Phartiyal et al. (eds.), *Climate Change and Environmental Impacts: Past, Present and Future Perspective*, Society of Earth Scientists Series,
https://doi.org/10.1007/978-3-031-13119-6_7

signatures of stream sediment samples (alluvium) from the Older, Younger Floodplain and Pedimont slope and Pediplain horizons of Narmada valley is attributed sedimentary fractionation, flow hydraulics in present day depositional environment.

Keywords Narmada alluvium · Climate · Alluvial geochemistry

1 Introduction

Earth had history of cyclic climatic variability with ubiquitous temporal and spatial implications across the geological time span. Considering the development and evolution of earth and its environment as a continuous and ongoing process, several researchers are involved in paleo climatic reconstructions using different proxies (Franckea et al. 2020). The terrestrial and aquatic ecosystems, belonging to the past and present day environment, act as repositories to preserve environmental imprints and the source characteristics. The quaternary fluvial sediments i.e. alluvium is a robust tool for studing of gradual environmental variability due to supergene processes. The elemental signatures of fluvial sediments acts as an indicator for provenance, tectonic setting and the degree of weathering (Martin et al. 1978; Potter 1978; Stallard and Edmond 1983; Wu et al. 2011).

The geochemical characterization of stream sediment samples from the major river basins for example Ganga (Singh 2010), Bramhputra (Rahman et al. 2020) Yangtze, Tongtian and Jinsha Rivers (Wu et al. 2011) and Amazon River (Vital and Stattegger 2000) and their detailed interpretation for provenance, weathering pattern and the depositional environment are no exception.

The Narmada basin of peninsular India possesses thick quaternary sediments and has witnessed the Pleistocene and Holocene climatic variation (De Terra and Paterson 1939b; Khartri 1961; Biswas and Dassarma 1984; Tiwari and Bhai 1997). It is also significant due to the discovery of fossils of pre historic man (Sonakia 1984; Khan and Sonakia 1992), vertebrates (Dassarma and Biswas 1978; Biswas and Dassarma 1982; Salahuddin 1987) and the remnants of lithic industry (Khartri 1961, 1962; Bain 2015).It is worth mentioning here that the central part of Narmada Valley was extensively studied in several perspectives including archeology, vertebrate paleontology, geomorphology and quaternary sedimentology. But for the first time the baseline geochemical data generated through stream sediment geochemistry of alluvium from Tawa basin, in parts of Hoshangabad district, Madhya Pradesh is utilized for understanding the provenance, weathering pattern, effect of sedimentary recycling and fluvial fractionation in peninsular India, hitherto the impact of variable present day depositional environment of Narmada Valley.

2 Location

The present study area is bounded by latitude 22°30′ to 22°45′ and longitude 77°45′ to 78°00′ with elevation ranging from 314 to 228 m having the north westerly slope (Fig. 1). This area is lying in the central part of Narmada river valley near the mouth of Tawa river basin (one of the major tributary of Narmada river) at the confluence of humid subtropical to savannah wet and dry climatic regime of present day climate (Fig. 1).

3 Geology

The pioneer work on identification and characterization of Narmada Quaternary sequence was done by Theobald 1860; Medlicott 1873; De Terra and Paterson 1939a, b; Khartri 1961. The De Terra and Paterson (1939b) had classified Narmada Alluvium in the three Groups viz. Lower, Upper and the Cotton Soil or Regur Group. According to him the base for both the lower and upper groups is defined

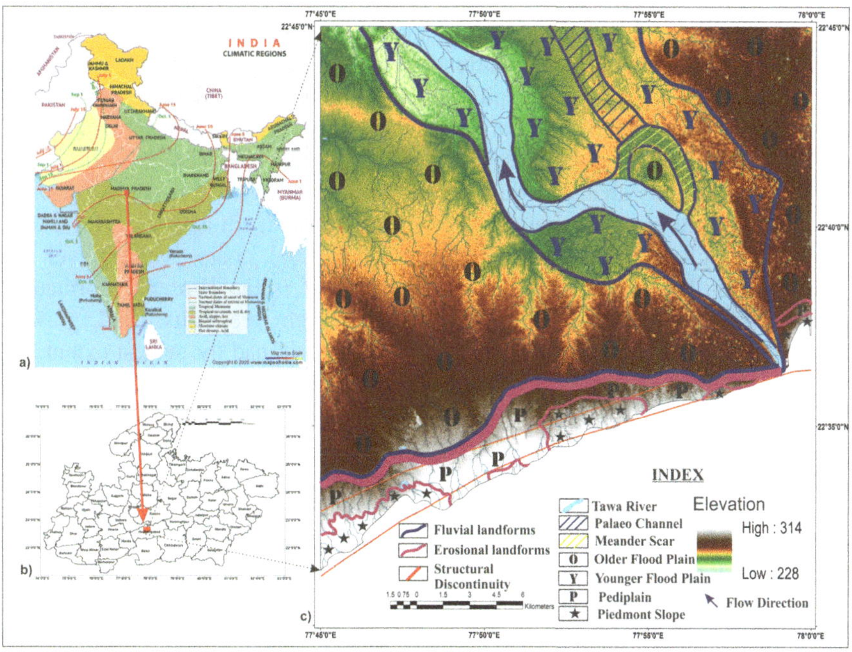

Fig. 1 The Drainage Map of the study area overlaid on the digital elevation model generated using CARTOSAT data showing location of geomorphic units with reference to the present climate of India. (Inset: Climate Map of India {source: www.mapofindia.com} & District map of Madhya Pradesh showing the location of present study area)

by a basal gravel and is overlain by pinkish or orange colored concretionary clays and silts. The first documentation of cyclic fining upward sequence and alternating fluvial and inter-fluvial climatic stages in Narmada Quaternary is given by De Terra and Paterson (1939b). He had also compared the fluvial and inter-fluvial cycles recorded in central India with the glacial and interglacial stages of Himalayas, and correlated with Pleistocene climate transition.

Khartri (1961) had described the stratigraphic sequence of Narmada Valley in a different version and had classified these formations with the archaeological connotations (Table 1). The detailed characterization of fluvial and inter-fluvial climate of Narmada valley was also attempted by Khartri (1961). Subsequently, another worker, Biswas and Dassarma (1984), had further reclassified the Narmada quaternary sequence and identified three phases of alluvial fills, each separated by disconformities having different mammal assemblages (Table 1). Biswas and Dassarma (1984) had also identified Pleistocene and Holocene environmental transitions within Narmada Valley and had recorded two cycles of humid to arid climate variations, based on the sedimentary structures and the type of litho-section observed. The comparative stratigraphy of Narmada valley with interpretations on prevailing climatic conditions given by different workers is given in Table 1.

The latest litho-stratigraphy classification of Narmada alluvium in Central part of the valley was given after Tiwari and Bhai (1997). They had classified the alluvium exposed within the present study area in lithologically five distinct formations (Table 1 and Fig. 2a). As per their classification scheme, the laterite gravels and coarse sand define the oldest Formation and are named Pilikankar Formation. The overlying formation is the Surajkund Formation and is characterized by yellow silty clay, red sandy silts, yellow, brown to steel grey silty clays, gravel and conglomerate. This is occupying the higher elevation and occurring at western and northeastern parts within the study area. The Surajkund Formation is overlain by Baneta Formation characterized by moderately calcareous brown sand, silt and silty clays, dark carbonaceous clays, brown and red sand with charcoal, coarse sand, gravel and conglomerate. The subsequently overlying lithounit is the Hirdepur Formation, indentified by the presence of highly calcareous light grey silts, sand, gravels, calcareous sandstone and conglomerate. This formation had the widest extent and is occupying both the erosional and the depositional terraces at varying locales within the study area. The Ramnagar Formation is the most recent alluvium having the restricted extent along the present river channels. It is non-oxidized and devoid of calcification and is characterized by pebbly sand, medium to fine sand and silt (Fig. 2a). Tiwari and Bhai (1997) has conotated the Middle Pleistocene age for the Pilikankar and Surajkund Formations whereas the Upper Pleistocene age for the Hirdepur and Baneta Formations (Table 1). The Holocene age is given for the Ramnagar Formation after Tiwari and Bhai (1997). Shukla et al. (2021) has reported REE's and heavy metals enrichment in Ramnagar Formation of Narmada basin through geochemical mapping.

Table 1 The comparative stratigraphy of the Narmada valley (Central India) after Khartri (1961), Biswas and Dassarma (1984), and Tiwari and Bhai (1997)

After Khartri (1961)				Biswas and Dassarma (1984)			Tiwari and Bhai (1997)				
Period		Geological formations	Climate		Litho unit	Age	Chrono stratigraphy	Morpho stratigraphy	Tephra stratigraphy	Formation	Paleo climate
Holocene		Black cotton soil Yellowish Brown silt with concretions	Inter-fluvial present day conditions	Narmada upper deposit (5 m)	Black cotton soil	Lower Holocene	Holocene	Depositional	–	Ramnagar Fm	Present climate
					Gray silt						
					Pedocal layer						
					Grey sand with lime concretions						
				≈ ≈ ≈ ≈ ≈ ≈ ≈≈	Disconformity	≈ ≈ ≈ ≈ ≈ ≈ ≈≈					
Pleistocene	Upper	Deposition of cross-bedded sand (fossils) Cemented sandy conglomerate (fossils)	Current sluggish but water still in plenty Fluvial Water level more than 30′ higher than the present	Narmada middle deposit (25 m)	Pinkinsh clay and silt with lime concretions Grit and sand (upper part yellow Lower part Grey) Boulder bed	Upper Pleistocene	Upper Pleistocene	Depositional	Volcanic ash (transported)	Hirdepur Fm	Warm and semiarid
	≈ ≈ ≈ ≈ ≈	Unconformity	≈ ≈ ≈ ≈ ≈ ≈ ≈	≈ ≈ ≈ ≈ ≈ ≈ ≈≈	Disconformity	≈ ≈ ≈ ≈ ≈ ≈ ≈≈		–	Volcanic ash	Baneta Fm	Warm and semiarid
	Middle	Boulder conglomerate (fossils) Red greasy clay	Inter-fluvial Present day conditions. Humidity continued	Narmada lower deposit (15 m)	Deep red clay and silt often mottled deep red gritty layer Deep red Boulder bed		Middle Pleistocene		–	Surajkund Fm	
				≈ ≈ ≈ ≈ ≈ ≈ ≈≈	Disconformity	≈ ≈ ≈ ≈ ≈ ≈ ≈≈			–	Pilikankar Fm	Warm and humid
		Laterite (?)	Fluvial		Laterite						
					Basement rock						

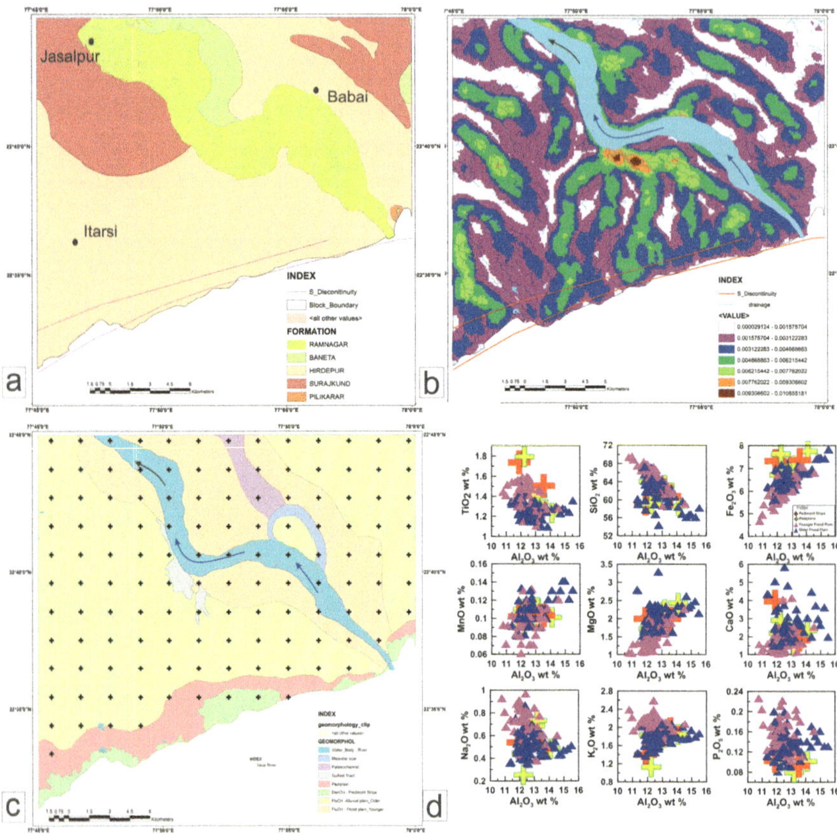

Fig. 2 (**a**) The Geological Map on 1:50,000 scale of part of Topographic sheet number 55F/14 (source: Map and Cartography Division, GSI Bhopal). (**b**) Drainage Density Map of the study area generated using stream network of part of Topographic sheet number 55F/14. (**c**) The Geomorphological Map on 1:50,000 scale of part of Topographic sheet number 55F/14 (source: BHUKOSH). (**d**) Bivariate Hraker plots of Major Oxides (wt %) vs Al_2O_3 (wt %) for the stream sediment samples of Narmada valley alluvium

4 Geomorphology

Synoptically the present study area is occupied by morphological units of fluvial and erosional geomorphic origin. Tawa River is flowing across the study area from southeast to northwest and the overall drainage is sub-dendritic to sub-parallel with drainage density upto 0.01 (Fig. 2b). The majority landscape is covered by the fluvial geomorphic units including the Older Floodplains, Younger Floodplains, gullied tracks, meandering scars, paleo-channel and the present river bed of Tawa river whereas the Piedmont slopes, peniplains originated by the denudational and erosional process occupy the higher reaches and define the southern boundary of Narmada quaternary's (Tiwari and Bhai 1997; Khan and Aziz 2016). The geomorphological map compiled after GSI-BHUKOSH data set on 1:50,000 scale is given in Fig. 2c. These geomorphic units are represented by different litho- stratigraphic horizons of Narmada quaternary succession exposed in the present study area. The Older Floodplains are composed of the Hirdepur and Surajkund Formations whereas the Younger Floodplains comprises Baneta and Ramnagar Formations. The Pilikankar Formation and the part of Hirdepur Formation exposed in the upper reaches along southern part of the study area comprises the Piedmont slopes and the peniplains. The most dominant units of the study area are Older Floodplains, Younger Floodplains exposed in the form of erosional and depositional terraces.

5 Methodology

Under the realm of National Geochemical Mapping Programme (NGCM) of Geological Survey of India, the systematic, grid wise (1 km × 1 km) sampling on 1: 50,000 scale was carried out in the Survey of India, Topographic sheet No. 55F/14, Hoshangabad District, Madhya Pradesh. The sampling media is the stream sediments of the 1st or 2nd order streams, or the other smaller channel(s), here referred as ‘0’ order or slope wash representing the maximum catchment area within the particular grid (1 km × 1 km) with the grain size ranging from silt to clay.

The collected samples were processed as per the Standard Operating Procedure (SOP) NGCM of Geological Survey of India. The processed 120 mesh size samples of four adjoining grids was thoroughly homogenized by coning and quartering and then composited for representing 2 km × 2 km grid. The details of sampling and processing are given in Shukla et al. (2014). All the adequate measures were adapted to avoid anthropogenic contamination and creation of a geochemical baseline data with the actual representation of the underling geology.

In the present study the geochemical data of 128 numbers of composite stream sediment samples, from 512 sq. km area, collected over the quaternary succession (alluvium), developed at different geomorphic horizons is utilized for geochemical characterization of Narmada Valley and holistic interpretation for climate variability. It is to be emphasized that theses surface samples were collected at different elevations. Thus out of these 128 samples, 74 samples represented the Older Floodplain, 46 samples represented the Younger Floodplain, 06 samples belong to the Pediplain and 02 samples to the pedimont slopes (Fig. 2c).

These samples were analyzed at Chemical Lab, Geological Survey of India, Central Region, Nagpur for the Major Oxide, trace and REE’s. The major oxides and trace elements were analyzed using the M/S Panalytical, X-Ray Fluorescence Spectrometer, using the standards GSD-1 to GSD-9 for calibration. The REE’s and RM were analyzed through ICP-MS, Perkin Elmer Sciex, model no. ELAN-6100 and the instrument was calibrated using the standards GSD-2, GSD-9, GSD-10, GSD-12. The precision of both the instrument (XRF & ICPMS) was ascertained by the repeat analysis of 5% samples. The precision is better than ±5% for major oxides and ± % for trace elements including REEs.

6 Results

6.1 Major Oxides

The elemental signatures are the best keys for understanding any sediment's evolutionary history, if devoid of alteration, weathering and digenesis. In the present study the major and trace element composition of surface samples (stream sediments), deposited at different geomorphic units of Tawa basin, Narmada valley, were statistically evaluated and interpreted.

The Silica (SiO_2) abundance within the analyzed samples ranges between 54.35% and 69.59% and that of Aluminium (Al_2O_3) varies from 10.76% to 15.52%. The abundance ranges of other major oxides varies from 4.67% to 7.84% (Fe_2O_3), 1.10% to 1.80% (TiO_2), 0.06% to 0.14% (MnO), 1.04% to 3.30% (MgO), 1.07% to 5.86% (CaO), 0.26% to 0.97% (Na_2O), 1.13% to 2.59% (K_2O) and 0.08% to 0.23% (P_2O_5). The complete analysis of all the 128 samples is given in online supplementary material-I and the summary statistics of the major and trace elemental abundances is presented in Table 2.

The bivariate statistics of Major Oxides in samples of Narmada alluvium shows that Al_2O_3 has a strong negative correlation with SiO_2, TiO_2, and scattered negative correlation with Na_2O and P_2O_5 hence the inverse relation between clay minerals and the quartz, opaque's (Fe-Ti Oxides), phosphate minerals (monazite) is observed. Simlarly strong positive correlation of Al_2O_3 with MgO, MnO, Fe_2O_3 and scattered slight positive correlation with CaO is also recorded, further validating presence for clay minerals and phyllosilicates within the Narmada Alluvium (Fig. 2d).

The K_2O/Al_2O_3 ratio varies from 0.00 to 0.40, for clay minerals and 0.50 to 0.90, for feldspars (Cox et al. 1995; Dar et al. 2020), the value of K_2O/Al_2O_3 ratio in our samples ranges from 0.09 to 0.22 with an average of 0.15 which also elucidates the dominance of clay minerals within the sediments deposited in the Tawa basin of Narmada valley.

Further to ascertain the compositional maturity of the stream sediment samples of Narmada valley the Log SiO_2/Al_2O_3 versus Log Fe_2O_3/K_2O plot after (Herron 1988) and Barium versus Strontium plot after Floyd et al. (1989) were used. The samples fall within the field of wacke, shales and Fe rich shales (Fig. 3a, b), indicating chemical immaturity of these sediments, and their derivation from relatively un-weathered source (Taylor and McLennan 1985; McLennan et al. 1993).

6.2 Trace Elements

The High Field Strength Elements (HFSE) such as Y, Sc, Th, U, Zr, Nb, Hf and Rare Earth Elements (REE's) have a wide range of abundances in the stream sediment samples of Narmada basin. The concentration ranges (in ppm) are 25 and 41 for Y, 241 to 868 for Zr, 11 to 24 for Nb, 10 to 21 for Th, 7.57 to 23.68 for Hf and 2.31 to

Table 2 The abundance range of major, trace and REEs within the surface samples of quaternary sediments (Alluvium) from Tawa basin, Hoshangabad district Madhya Pradesh

	Older flood plain (number of samples 74)			Younger flood plain (number of samples 46)			Pediplain (number of samples 06)			Pedimont slope (number of samples 02)		
Elements	Maximum	Minimum	Average	Maximum	Minimum	Average	Maximum	Minimum	Average	Maximum	Minimum	Average
SiO_2	67.32	54.35	61.83	69.59	58.71	65.24	64.11	58.68	61.55	60.82	59.08	59.95
Al_2O_3	15.52	11.34	12.78	13.76	10.76	12.31	14.10	12.11	12.97	13.53	11.87	12.70
Fe_2O_3	7.84	5.56	6.55	7.15	4.67	6.00	7.75	6.74	7.21	7.38	7.35	7.37
MnO	0.14	0.08	0.11	0.13	0.06	0.10	0.11	0.09	0.10	0.10	0.10	0.10
MgO	3.30	1.50	2.05	2.09	1.04	1.50	2.43	1.73	2.03	2.00	2.00	2.00
CaO	5.86	1.45	2.54	4.16	1.07	1.77	2.91	1.49	2.14	3.96	1.82	2.89
Na_2O	0.87	0.35	0.53	0.97	0.38	0.71	0.73	0.26	0.53	0.54	0.48	0.51
K_2O	2.20	1.38	1.78	2.59	1.68	2.08	1.93	1.13	1.74	1.97	1.40	1.69
TiO_2	1.55	1.10	1.27	1.59	1.10	1.40	1.80	1.24	1.41	1.74	1.51	1.63
P_2O_5	0.20	0.08	0.13	0.23	0.10	0.15	0.15	0.08	0.11	0.10	0.09	0.10
Ba	557	239	377	530	361	434	349	245	311	313.00	293.00	303
Co	29	13	20	24	10	18	30	18	24	26.00	25.00	26
Cr	138	45	79	87	21	52	102	56	80	72.00	63.00	68
Cu	64	33	46	49	28	37	60	42	51	61.00	51.00	56
Ga	19	14	16	18	15	16	18	16	17	17.00	14.00	16
Nb	21	11	18	24	18	21	21	15	18	19.00	19.00	19
Ni	63	34	50	55	25	39	61	41	53	48.00	42.00	45
Pb	51	7	18	25	15	19	19	15	17	20.00	16.00	18
Rb	102	63	86	102	77	91	83	54	76	85.00	59.00	72
Sc	19	9	12	14	7	10	14	10	13	13.00	12.00	13
Sr	151	74	98	136	79	105	112	60	87	108.00	83.00	96
Th	21	10	14	21	13	16	13	11	12	13.00	11.00	12
V	177	130	147	161	121	144	202	148	166	187.00	161.00	174

(continued)

Table 2 (continued)

	Older flood plain (number of samples 74)			Younger flood plain (number of samples 46)			Pediplain (number of samples 06)			Pedimont slope (number of samples 02)		
Elements	Maximum	Minimum	Average	Maximum	Minimum	Average	Maximum	Minimum	Average	Maximum	Minimum	Average
Y	37	25	31	41	29	35	33	25	29	29.00	29.00	29
Zn	112	52	65	83	51	62	76	60	68	67.00	66.00	67
Zr	640	241	396	868	327	547	484	301	379	356.00	345.00	351
Be	2.46	1.06	1.76	2.48	1.20	1.68	2.33	1.00	1.51	1.09	1.07	1
Ge	2.01	1.20	1.56	1.90	1.35	1.55	1.61	1.46	1.57	1.51	1.46	1
Hf	19.69	7.57	10.82	23.68	10.44	15.86	14.11	8.22	10.72	10.83	10.60	11
Ta	2.80	1.12	1.34	1.94	1.23	1.54	1.90	1.15	1.55	1.45	1.43	1
U	4.30	2.31	2.92	5.00	2.77	3.67	3.39	2.33	2.90	2.89	2.89	3
La	84.84	32.23	42.27	82.59	40.71	54.91	50.64	32.35	38.62	40.23	34.99	37.61
Ce	169.14	67.33	87.26	163.09	84.11	110.90	102.34	64.54	77.09	80.45	70.15	75.30
Pr	19.52	7.85	10.10	18.63	9.92	12.98	12.14	7.63	9.19	9.80	8.27	9.04
Nd	70.30	28.97	37.57	66.83	36.30	47.62	43.74	28.62	34.06	36.46	31.05	33.76
Sm	12.08	5.66	7.20	12.09	6.92	8.93	8.31	5.68	6.66	7.12	6.14	6.63
Eu	1.68	1.12	1.43	1.79	1.34	1.57	1.48	1.26	1.41	1.49	1.40	1.45
Gd	9.83	5.21	6.49	10.22	6.06	7.72	7.55	5.28	6.10	6.33	5.75	6.04
Tb	1.43	0.85	1.04	1.61	1.01	1.24	1.25	0.83	1.02	1.04	0.97	1.01
Dy	7.71	5.11	6.19	9.43	6.08	7.29	7.24	4.66	5.95	6.12	5.95	6.04
Ho	1.48	0.97	1.17	1.84	1.14	1.40	1.42	0.88	1.13	1.14	1.10	1.12
Er	4.36	2.79	3.41	5.53	3.19	4.12	4.16	2.49	3.26	3.29	3.28	3.29
Tm	0.69	0.42	0.54	0.83	0.51	0.65	0.61	0.40	0.50	0.52	0.49	0.51
Yb	4.40	2.81	3.44	5.55	3.31	4.23	3.84	2.60	3.21	3.31	3.21	3.26
Lu	0.71	0.43	0.54	0.89	0.53	0.67	0.62	0.41	0.52	0.54	0.53	0.54

Σ REE	387.54	162.96	208.64	378.24	203.24	264.23	245.34	157.80	188.72	197.83	173.29	185.56
$(La/Lu)_N$	13.80	6.48	8.37	11.95	7.29	8.79	9.62	6.23	8.10	7.94	7.13	7.54
$(La/Sm)_N$	4.54	3.53	3.78	4.41	3.62	3.95	4.00	3.54	3.74	3.68	3.65	3.66
$(Gd/Lu)_N$	1.84	1.24	1.48	1.71	1.24	1.43	1.62	1.26	1.47	1.35	1.44	1.40

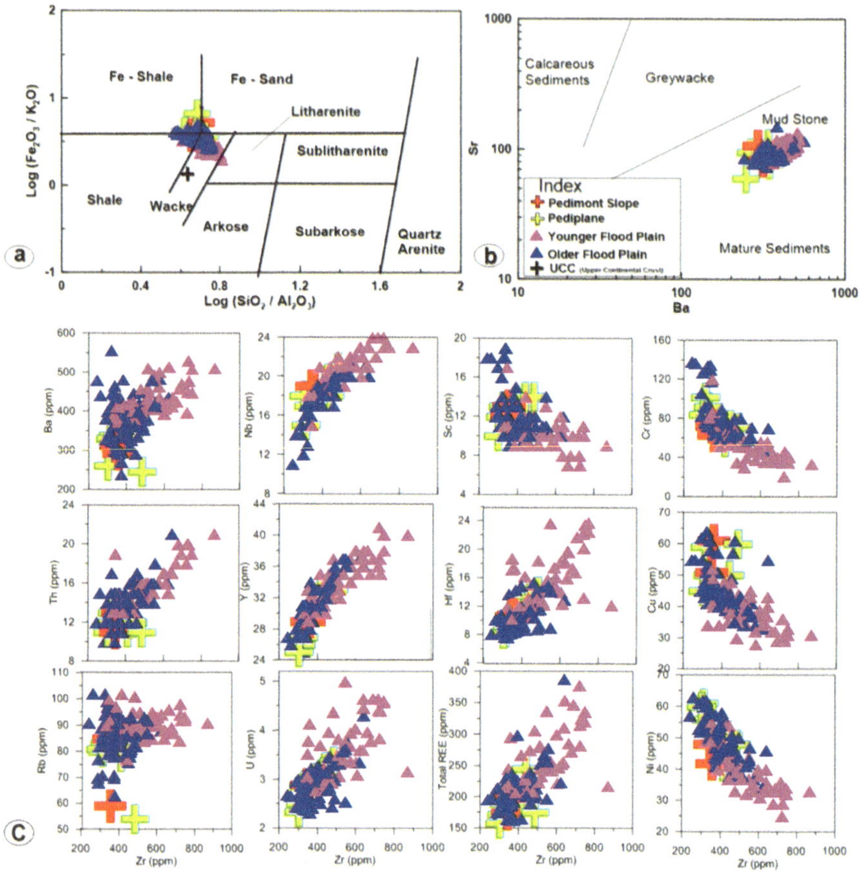

Fig. 3 Compositional maturity plots for the stream sediment samples of Narmada valley alluvium showing the immature nature of the sediments. (**a**) Log SiO_2/Al_2O_3 versus Log Fe_2O_3/K_2O after (Herron 1988). (**b**) Barium versus Strontium plot after Floyd et al. (1989). The values of UCC are after (Taylor and McLennan 1985). (**c**) Bivariate plots of Trace Elements versus Zircon in ppm after Hraker for the stream sediment samples of Narmada valley alluvium

5 for U. Owing to the high solubility the Large Ion Lithophile Elements (LILE's) i.e. Rb, Sr, Ba, and Pb also have a wide range of abundance varying from 54 to 102 for Rb, 60 to 151 for Sr, 239 to 557 for Ba and 7 to 51 for Pb. The concentration ranges for the compatible trace elements are 21 to 138 for Cr, 25 to 63 for Ni,10 to 30 for Co, 25 to 41 for V, 07 to 19 for Sc (values in ppm). The complete analysis of all the 128 samples is given in online supplementary material-I and the summary statistics of the major and trace elemental abundances within different geomorphic units of Narmada valley is presented in Table 2.

The Bivariate XY plots of Zircon (Zr) vs. trace elements (Ba, Th, Rb, Nb, Y, U, Sc, Hf, ∑REE, Cr, Cu and Ni were plotted to visualize the effect of hydraulic fractionation, and have a synoptic understanding on the source composition of these clastic sediments. These plots indicates that Zircon (Zr) had shown strong positive correlation with Ba, Nb, Th, Y, Hf, U and Σ REE except scattered positive correlation with Rb (Fig. 3c). Whereas Sc, Cr, Cu and Ni had shown strong negative correlation with Zr (Fig. 3c).The REE, Zr, Ba, Rb, Nb, Th and U include LILE (Large ion Lithophile Elements) and HFSE (High Field Strength Elements) associated with felsic minerals, and their significant correlation with each other and enrichment with alumina actually suggests the predominance of felsic minerals and clay minerals in the detrital mineralogy of Narmada Alluvium. The clustered and collinear variation with overlapping abundance ranges of the HFSE and LILE in stream sediments from Older Floodplain, Younger Floodplain, Pediplain and Pedimont slopes of Narmada Valley suggests the effect of sedimentary fractionation and hydraulic sorting.

Another diagram based on the ratio of Ba and Sr within the clastic sediments given after Floyd et al. (1989) had suggested that the Narmada Alluvium belongs to the immature category of sediments having the characters similar to that of mudstones (Fig. 3b).

6.3 Rare Earth Elements

Synoptically in the Chondrite normalised REE plots (Fig. 4a) samples from Narmada valley alluvium displays LREE enrichment with nearly flat HREE pattern and strong negative Eu anomaly, whereas in the North American Shale Composite (NASC) normalized plots these samples displays a slight (1.2–1.8 times) enriched flat pattern with negative Ce, Nb, Er and Eu anomalies (Fig. 4b). The NASC values are after Gromet et al. (1984).

Rare Earth Element composition of the samples collected from different geomorphic units have unique ratios. The sediment samples collected from the Pedimont slope shows the LREE enrichment that varies from 147 to 169 times of Chondrite with the average of 158 times and the HREE enrichment ranging from 19 to 21 times of Chondrite with the average of 19 (Fig. 4a and Table 2), indicating highly varied protoliths. For the same group of samples, their $(La/Lu)_N$ varies from ~7.13 to ~7.94 with an average of 7.53, $(La/Sm)_N$ varies from 3.65 to 3.68 with an average of 3.66 and $(Gd/Lu)_N$ varies from 1.35 to 1.44 with an average of 1.40.

Similar pattern with slightly more LREE enrichment ranging from 136 to 213 times of Chondrite with the average of 163 times and that of the HREE enrichment ranging from 15 to 25 times of Chondrite with the average of 19 is displayed by the samples collected from the Pediplains (Fig. 4a and Table 2). For the same group of samples, their $(La/Lu)_N$ varies from ~6.23 to ~9.62 with an average of 8.10, $(La/Sm)_N$ varies from 3.54 to 4.00 with an average of 3.74 and $(Gd/Lu)_N$ varies from 1.26 to 1.62 with an average of 1.47.

The other set of samples collected from the Older and Younger Flood plain also had identical Chondrite normalized REE patterns but with slightly enriched and varying abundance ranges. The LREE enrichment with in samples of Older

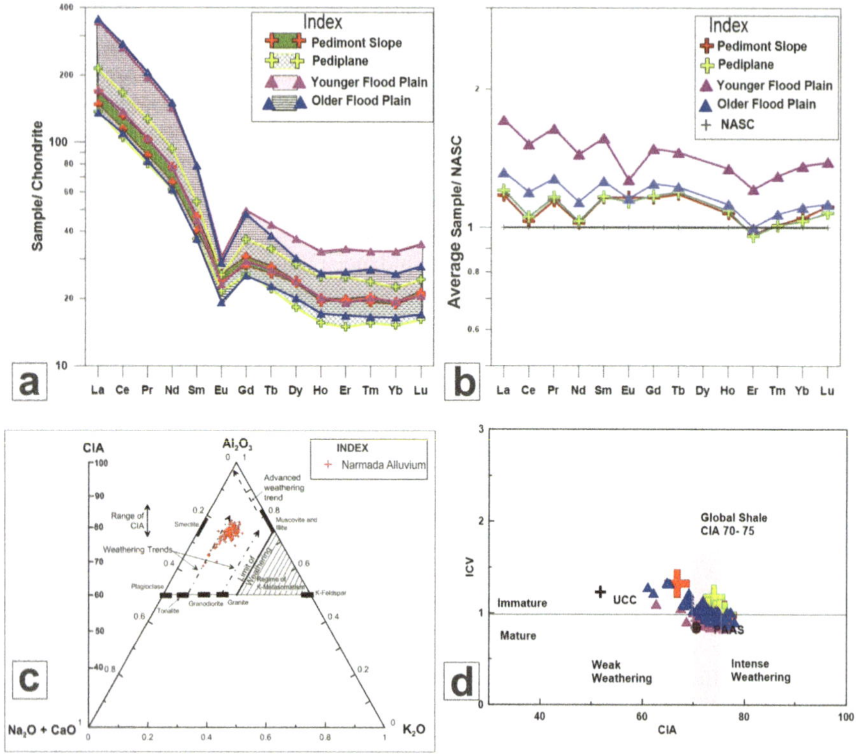

Fig. 4 (**a**) The Chondrite Normalized REE plots for the stream sediment samples of Narmada valley alluvium. The chondrite values are after Taylor and McLennan (1985). (**b**) The NASC Normalized REE plots for the stream sediment samples of Narmada valley alluvium. The NASC values are after Gromet et al. (1984). (**c**) A-CN-K ternary diagram (after Nesbitt and Young 1984) showing weathering trends for the stream sediment samples of Narmada valley alluvium. The red plus is the symbol for Narmada Alluvium. (**d**) The ICV (Index of Compositional Variability) versus CIA (Chemical Index of Alteration) plot for the stream sediment samples of Narmada valley alluvium (after Long et al. 2012). UCC (Upper Continental Crust) and PAAS (Post-Archean Australian Shale) values after Taylor and McLennan (1985) are included for comparison

Floodplain ranges from 136 to 357 times of Chondrite with the average of 179 times and that of the HREE enrichment ranges from 16 to 28 times of Chondrite with the average of 20 whereas LREE enrichment with in samples of Younger Floodplain ranges from 171 to 348 times of Chondrite with the average of 231 times and that of the HREE enrichment ranges from 19 to 35 times of Chondrite with the average of 24 (Fig. 4a). Within these samples of Older Floodplain the $(La/Lu)_N$ varies from ~6.48 to ~13.80 with an average of 8.37, $(La/Sm)_N$ varies from 3.53 to 4.54 with an average of 3.78 and $(Gd/Lu)_N$ varies from 1.24 to 1.84 with an average of 1.48 whereas in Younger Floodplain the $(La/Lu)_N$ varies from ~7.29 to ~11.95 with an average of 8.79, $(La/Sm)_N$ varies from 3.62 to 4.41 with an average of 3.95 and $(Gd/Lu)_N$ varies from 1.24 to 1.71 with an average of 1.43 (Table 2).

6.4 Calulations for Weathering Indices and Index of Compositional Variability

The Chemical weathering indices i.e. Ruxton ratio (R), Weathering Index of parker (WIP), Vogt's Residual Index (V), Chemical index of Alteration (CIA), Chemical index of Weathering (CIW), Plagioclase Index of Alteration (PIA) and Silica-Titania Index (STI) (Table 3) were calculated for the stream sediment samples of quaternary alluvium collected from the different geomorphic units of tawa basin, Narmada valley using the major oxides data. The mathematical equations used for the computation of all the enumerated indices along with the results are given in Table 3.

The Ruxton ratio (R) is a simple weathering ratio that considers alumina as immobile and relates the silica loss to the total elemental loss during weathering (Ruxton 1968; Chittleborough 1991). Weathering Index of parker (WIP) is another robust indicator based on the mobility and proportions of alkali and alkaline earth metals given after Parker (1970), Eswaran et al. (1973), Hamdam and Bumham (1996). The WIP value for the quaternary sediments of central part of Narmada Valley in the present study area varies between 1666.61 and 3217.01 (Table 3).

Vogt (1927) and Roaldset (1972) had given another well known geochemical ratio for accessing the maturity of residual sediments and the extent of weathering in quaternary sediments called as Vogt's Residual Index (V). The quaternary sediments from this part of Narmada Valley has the V value ranging from 1.67 to 4.96 (Table 3).

The most extensively used alteration ratio for plaeo-climatic reconstruction and provenance determination of clastic sediments is the Chemical index of Alteration (CIA) (Nesbitt and Young 1982, 1984; Fedo et al. 1995; Maynard et al. 1995). The equation used for calculation is given in Table 3 here all the oxides are in molar proportion and CaO represents Ca in silicate fraction. The CIA value for the quaternary clastic sediments of the present study area ranges from 61.05 to 77.98 (Table 3) with an average of 72.83 which is comparable with the average CIA values for global shale (~70 to ~75) (Nesbitt and Young 1982). This reflects the presence of muscovite, illites and smectite within the Narmada alluvium. The samples of the present study area when plotted on $Al_2O_3 - CaO + Na_2O - K_2O$ (A-CN-K) (Nesbitt and Young 1982, 1984; Fedo et al. 1996) follow a trend sub parallel to A-CN axis, within the limits of weathering (Fig. 4c).

Harnois (1988) had given another ratio i.e. Chemical index of Weathering (CIW) for determination of degree of weathering using the alumina, sodium and calcium. This ratio is independent of the plagioclase alteration, therefore Fedo et al. (1995) had given the Plagoiclase Index of Alteration (PIA) for understanding the impact of weathering of plagioclase on the elemental composition of quaternary clastics. The CIW value for the quaternary clastic sediments of the present study area ranges from 66.42 to 88.27 and the PIA varies between 63.18 and 86.35 (Table 3). For assessment of degree of chemical weathering on the weathered profile of metamorphic rock the Silica-Titania Index (STI) was given by de Jayawardena and Izawa (1994). The STI value for the Narmada alluvium ranges from 73.84 to 78.17.

Table 3 Summary of weathering indices for the samples of quaternary sediments (Alluvium) from Tawa basin, Hoshangabad disctict Madhya Pradesh

Index	Formula	Ideal trend of index up-profile (increase in weathering)	Optimum fresh value	Optimum weathered value	Allows Al mobility	Present study (Alluvium from Tawa basin, Hoshangabaddisctict Madhya Pradesh) Older flood plain	Younger flood plain	Pediplain	Pedimont slope	
R	SiO_2/Al_2O_3	Negative	>10	0	No	5.74	6.45	5.29	4.98	Maximum
						3.68	4.48	4.16	4.50	Minimum
						4.87	**5.32**	**4.76**	**4.74**	Average
WIP	(100) [(2 Na_2O/0.35) + (MgO/0.9) + (2K_2O/0.25) + (Cao/0.7)]	Negative	>100	0	Yes	2678.71	3217.01	2451.64	2335.04	Maximum
						1857.80	1983.70	1666.61	2219.43	Minimum
						2313.42	**2488.25**	**2222.22**	**2277.23**	Average
V	(Al_2O_3 + K_2O)/(MgO + CaO + Na_2O)	Positive	<1	Infinite	No	4.09	4.96	3.45	3.60	Maximum
						1.67	2.24	2.70	2.04	Minimum
						2.94	**3.74**	**3.15**	**2.82**	Average
CIA	(100) [Al_2O_3/(Al_2O_3 + CaO + Na_2O + K_2O)]	Positive	≤50	100	No	77.98	77.4	76.22	75.97	Maximum
						61.05	62.63	72.91	66.78	Minimum
						72.59	**73.05**	**74.65**	**71.38**	Average
CIW	(100) [Al_2O_3/(Al_2O_3 + CaO + Na_2O)]	Positive	≤50	100	No	86.10	88.27	85.60	85.42	Maximum
						66.42	71.21	79.47	72.50	Minimum
						80.81	**83.38**	**82.95**	**78.96**	Average
PIA	(100) [(Al_2O_3 − K_2O)/(Al_2O_3 + CaO + Na_2O − K_2O)]	Positive	≤50	100	No	84.45	86.35	83.58	83.35	Maximum
						63.18	66.64	77.84	69.92	Minimum
						78.41	**80.66**	**80.84**	**76.64**	Average
STI	(100) [(SiO_2/TiO_2)/((SiO_2/TiO_2) + (SiO_2/Al_2O_3) + (Al_2O_3/TiO_2))]	Negative	>90	0	No	78.17	77.77	76.65	74.98	Maximum
						73.84	75.55	73.85	74.21	Minimum
						76.50	**76.73**	**75.75**	**74.60**	Average

Modified after Prince and Velbel (2003)

The indexes were R, after Ruxton (1968); WIP after Parker (1970) and Harnois (1988); V after Vogt (1927) and Roaldset (1972); CIA after Nesbitt and Young (1982), CIW after Harnois (1988); PIA after Fedo et al. (1995); and STI after de Jayawardena and Izawa (1994)

The impact of Hydraulic sorting on the geochemical variability within the fine clastics was also quantified using the index of compositional variability (ICV) defined by the equation ICV = (Fe_2O_3 + K_2O + Na_2O + CaO + MgO + MnO + TiO_2)/Al_2O_3 (Cox et al. 1995). The ICV values for samples from Older Floodplain ranges from 0.87 to 1.34 with an average of 1.02 and that for samples from Younger Floodplain ranges from 0.83 to 1.10 with an average of 0.93.The samples from Pediplain had the ICV values ranging from 0.97 to 1.18 with an average of 1.04 and those from Pedimont Slope had ICV values ranging from 0.98 to 1.32 with an average of 1.15 (Fig. 4d).

7 Discussion

The elemental characters of fine clastics (alluvium) realistically reflects provenance, the degree of hydraulic sorting and intensity of weathering, witnessed during their transportation and deposition in present day climate. The geochemical data generated from stream sediments of Tawa basin, Narmada Valley is evaluated and interpreated for provenance delineation and understanding effect of quaternary climatic variability in syn-sedimentary deposition.

7.1 Influence of Sedimentary Sorting

The sedimentary fractionation and sorting are the dominant factors defining the mineral assemblage and hence the geochemical characters of fluvial sediments. The sedimentary fractionation resulted in the concentration of clay minerals (Al_2O_3), some titanium oxides (TiO_2), REE's and Zircon in the silt sized where as Quartz (SiO_2) along with other heavy minerals (Sphene, Illimanite, Zircon, Apatite etc.) in sand sized clastic sediments (McLennan et al. 1993; Cullers 1994a, b). The Major element ratios of Al_2O_3/SiO_2 vs. Fe_2O_3/SiO_2 in the stream sediments from Narmada Valley are plotted in the binary plot given after Wu et al. (2011), to visualize the effects of sorting. In order to have synoptic comparison the values of UCC, PAAS after Taylor and McLennan (1985) are also plotted (Fig. 5a). The stream sediment samples of the present study displays a collinear trend, indicating that the sedimentary sorting of Narmada valley sediments ranges from quartz to clay minerals. These samples have an intermediate composition in comparison to the Post Archean Australian Shale and the Upper continental crust. The sample from Pedimont Slope, Pediplain and Older Floodplain shows the obvious enrichment of Fe and Al in comparison those from the Younger Floodplain. This also indicates the predominance of clay minerals and phyllo-silicates in the stream sediments collected from Pedimont Slope, Pediplain and Older Floodplain and that of quartz in the Younger Floodplain.

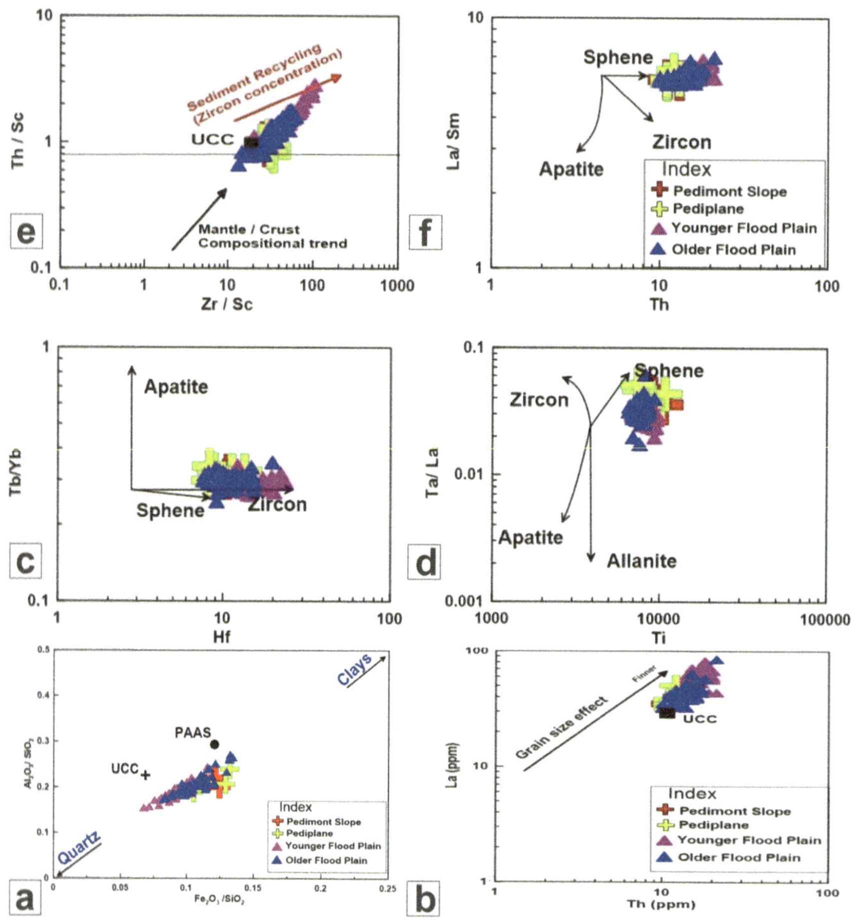

Fig. 5 (**a**) Al_2O_3/SiO_2 vs. Fe_2O_3/SiO_2 plot (after Wu et al. 2011) for the stream sediment samples of Narmada valley alluvium showing the effect of sedimentary sorting. The data of UCC and PAAS are from Taylor and McLennan (1985). (**b**) Bivariate plots of Th against La (ppm) of the samples from Narmada alluvium after Vital and Stattegger (2000). The values of UCC are after Taylor and McLennan (1985). Plots showing the effects of heavy mineral accumulation on selected elements and element ratios for the stream sediment samples of Narmada valley alluvium. (**c**) Tb/Yb vs. Hf; (**d**) Ta/La vs. Ti diagrams (after La Flèche and Camiré 1996). (**e**) Zr/Sc vs. Th/Sc, showing enrichment of zircon as opposed to compositional variations of the provenance; trends of compositional variations and sediment recycling from McLennan et al. (1993); (**f**) La/Sm vs. Th

The index of compositional variability (ICV) (Cox et al. 1995) is another robust indicator for quantifying impact of Hydraulic sorting's on the geochemical variability within the fine clastics. The average value of ICV for the quaternary sediments of Narmada alluvium is equivalent to 1, and it implies that this alluvium is geochemically immature and is weakly affected by weathering. The plot of ICV versus CIA after Long et al. (2012). Figure 4d shows the smooth positive relation between ICV and CIA. The decreasing values of average ICV form Pedimont Slope to Pediplain and subsequently to Older and Younger Floodplain indicate the progressive increase in the degree of hydraulic sorting.

McLennan et al. (1993) stated that the concentration of heavy mineral within clastic sediments is seldom uninfluenced by sedimentary sorting. Hence, the ratio of Th/ Sc had been proved as an index for provenance delineation and Zircon enrichment can be appreciated using the ratio of Zr/Sc (McLennan et al. 1993). Therefore, hydraulic sorting in Narmada valley stream sediments was ascertained using relation and ratios of the trace elements (Zr/Sc vs. Th/Sc, La/Sm vs. Th, Tb/Yb vs. Hf and Ta/La vs. Ti). The Zr/Sc vs. Th/Sc and Tb/Yb vs. Hf (Fig. 5c, e) shows that all the samples follow the trend of Zircon enrichment, hitherto the hydraulic sorting whereas the ratio plot of La/Sm vs. Th and Ta/La vs. Ti (Fig. 5d, f) shows clustering, suggestive that the concentration of Th and Ti are not influenced by the sedimentary fractionation. It is interesting to mention that maximum zircon enrichment (Fig. 5c, e) is observed in the samples from the Younger Floodplain. It is also recorded in Fig. 5e that Narmada alluvium is comparatively more enriched in Zircon than the Upper Continental Crust.

The similar evidence for sedimentary fractionation within the stream sediment samples from the Younger Floodplain, of Narmada Valley is recorded in the chondrite and NASC normalized REE plots (Fig. 4a, b). The samples collected over the pediplains have the lowest concentration for the REE's as displayed in the chondrite normalized plots (Fig. 4a). The REE's abundance in the stream sediment samples from Older and Younger Floodplain is further attenuated by the sedimentary fractionation and hydraulic sorting.

Likewise the effect of hydraulic sorting and the grain size variability was further confirmed by the bivariate plots of Th against La (ppm) (Fig. 5b); Th against Th/U (Fig. 6a) and Th/U against ΣLREE/ ΣHREE (Fig. 6b) after Vital and Stattegger (2000). McLennan et al. (1980) had utilized the coherent behavior of Lanthanum and Thorium for interpretation of provenance and grain size hydraulic sorting. The positive correlation between Thorium and Lanthanum is observed within the samples from Narmada Alluvium and the ratio plots of Th/U versus Th and ∑LREE/ ∑HREE versus Th/U hence further ascertains the predominance of sedimentary fractionation in defining the elemental characters of Narmada Valley alluvium (Fig. 6a, b).

7.2 *Degree of Weathering*

The computation of Chemical weathering indices including Ruxton ratio (R), Weathering Index of parker (WIP), Vogt's Residual Index (V), Chemical index of Alteration (CIA), Chemical index of Weathering (CIW), Plagioclase Index of Alteration (PIA) and Silica- Titania Index (STI) after Prince and Velbel (2003), validated that the quaternary sediments of Narmada Valley had witnessed the least amount of post depositional chemical weathering and belongs to immature category.

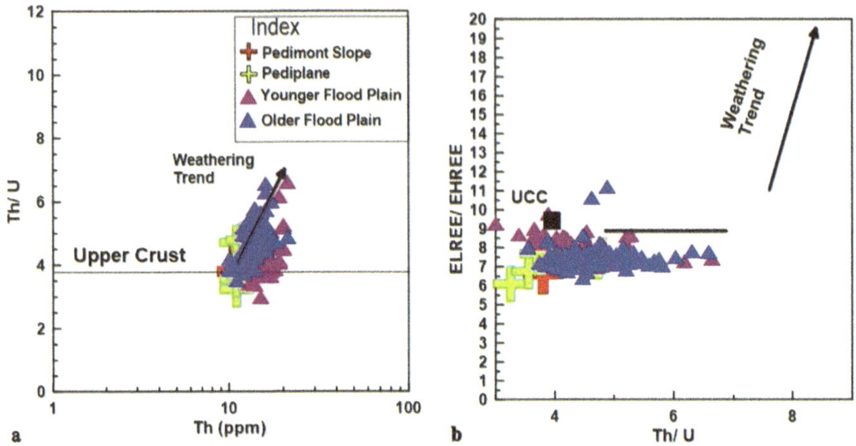

Fig. 6 Bivariate plots of (**a**) Th against Th/U and (**b**) Th/U against ΣLREE/ΣHREE of the samples from Narmada alluvium (after Vital and Stattegger 2000). The values of UCC are after Taylor and McLennan (1985)

The minimum value of STI, V, CIA, CIW and PIA is from the samples collected over the Pedimont Slope where as the maximum value for CIA, PIA is from the samples from Pediplains and that of CIW, V, WIP and STI are form Younger Floodplain (Table 3). Subsequently, from the ratio ranges of CIA and PIA it can be inferred that the sediments from piedmont slope's are least altered where as from the Pediplains are most altered. Besides average abundance of CIW also shows that the sediments from the piedmont slope are least altered where as those from the Younger Floodplain are most altered. The ranges of WIP in all the stream sediments falls in the range of unaltered samples, thus indicates that all theses samples preserves the source characters.

The SiO_2/Al_2O_3 and Fe_2O_3/K_2O ratio (Fig. 3a) for these stream sediment samples also provide similar inference, Hence It can further be interpreted that the geochemical signatures of these fine clastics are attributed by the syn-depositional process, thus preserves the provenance characters.

7.3 *Provenance*

The smooth and correlated trends of TiO_2, SiO_2, Fe_2O_3, MnO, MgO, Na_2O and P_2O_5 on Harker variation plots with Al_2O_3 as differentiating index reflects that all the samples of quaternary alluvium were genetically related and may be derived from the same/similar precursor. The positive correlation of iron, manganese, magnesium with aluminum (Fig. 3c) indicates predominance of clay minerals where as the negative correlation of silica, titanium, sodium and phosphorus with aluminum

(Fig. 3c) indicates the presence of phosphate (Monazite?), opaque's (ilmenite, sphene) or some amphiboles and quartz.

The multi-elemental correlation of the major and trace elements depicted existence of inverse relation between clay minerals, phyllosilicates and the quartz, opaque's (Fe-Ti Oxides), phosphate minerals (monazite). A significant correlation among elements of felsic linage i.e. Silica, Sodium, Potasium, Niobium, Yittrium, Zircon, Hafnium and that off iron, magnesium, manganese, cobalt, copper, chromium, nickel, scandium, vanadium (Fig. 3c) suggests about detrital origin for these elemental signatures. The presence of clay minerals within the sediments deposited in the Tawa sub-basin of Narmada valley is also ascertaind by K_2O/Al_2O_3 ratio and moderately significant correlation of Alumina with Iron, Manganese, Chromium, Copper, Gallium, Nickel, Scandium (Fig. 3c).

The REE, Zr, Ba, Rb, Nb, Th and U also shows a significant correlation among each other and positive enrichment with alumina, which eventually indicates the predominance of felsic minerals {Zircon (?), Sphene(?), Monazite (?)} and clay minerals in the detrital mineralogy of Narmada Alluvium. The major and trace element correlations therefore confirms that the elemental characteristics of the Narmada Alluvium are controlled by the dertital mineralogy and the visible compositional variation could be attributed to mineralogical and textural maturity induced by the flow hydraulics influenced by the depositional environment.

The Chondrite normalized enriched and fractionated LREE with flat HREE pattern and strong negative Eu anomaly (Fig. 4a) for the stream sediments of Narmada valley further attenuates the inference of sedimentary fractionation of highly varied protoliths. Cox et al. (1995) and Cullers (2000) had defined the ranges of index of compositional variability for provenence determination. The value of ICV > 0.84, denotes the dominance of rock forming minerals i.e. K-feldspars, plagioclase, pyroxenes and amphiboles whereas the ICV values <0.84 denotes the dominance of alteration products such as kaolinite, illite, and muscovite. Since the average ICV Value for the stream sediments collected over the Narmada Alluvium from Tawa Basin is 0.99, hence the dominance of rock forming minerals within the overall composition due to sedimentary fractionation in indicated.

To conclusively constrain the provenance using the major element characteristics the Discriminant function diagram (Roser and Korsch 1988) was utilized and all the quaternary sediments of Narmada Alluvium were plotted in the field of quartzose sedimentary provenance (Fig. 7a). The dominant source rock for Narmada Alluvium is Gondwana Supergroup of rocks, with minor amount of Deccan trap, and Betul Group of Rocks (Shukla et al. 2021). Hence the inference about quartzose sedimentary provenance fully substantiated with the major lithounits exposed in the upper reaches of the Tawa sub basin, Narmada valley.

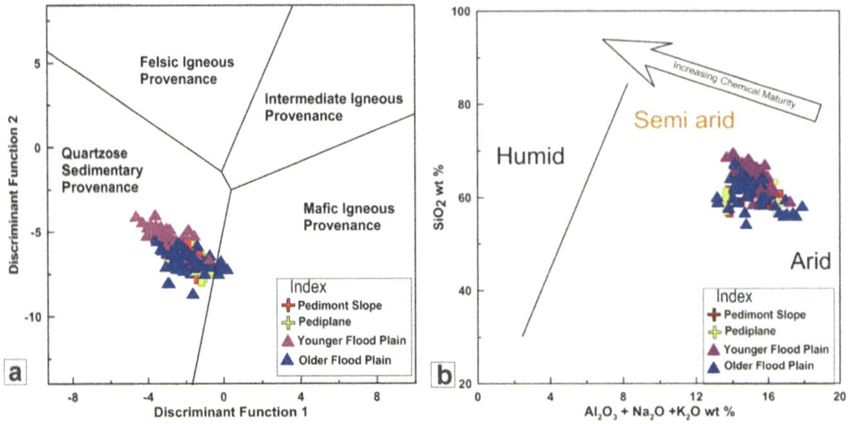

Fig. 7 (**a**) Discriminant function diagram of the Narmada alluvium (after Roser and Korsch 1988). Discriminant function 1 = (−1.773 × TiO_2%) + (0.607 × Al_2O_3%) + (0.76 × Fe_2O_3 T %) + (−1.5 × MgO%) + (0.616 × CaO%) + (0.509 × Na_2O%) + (−1.22 × K_2O%) + (−9.09). Discriminant function 2 = (0.445 × TiO_2%) + (0.07 × Al_2O_3%) + (−0.25 × Fe_2O_3 T %) + (−1.142 × MgO%) + (0.432 × Na_2O%) + (1.426 × K_2O%) + (−6.861). (**b**) Bivariate SiO_2 (wt. %) versus $Al_2O_3 + K_2O + Na_2O$ (wt. %) paleoclimate discrimination diagram. Fields after Suttner and Dutta (1986)

7.4 *Effect of Climate on Geochemistry of Geomorphic Units*

The major, trace and REE characteristics of stream sediments from Tawa basin, Narmada valley indicate that the Piedmont slope and Pediplain have similar geochemical characters with that of the Older Floodplain. A progressive chemical fractionation of Major, trace and REE's is recorded within the samples from Old and Younger Floodplain in Harkar Bivariate (Fig. 3c), Chondrite and NASC normalized REE patterns (Fig. 4a, b).

The ubiquitous presence of black cotton (regur) top soil in the present study area confirms the present day humid subtropical climatic conditions in the central part of the Narmada basin. Since the whole rock chemistry of any sedimentary succession either of present or past is influenced by the key environmental factors i.e. oxygen fugacity, humidity, CO_2 concentration, biological productivity etc., therefore Suttner and Dutta (1986) has given ar robust bivariate plot of SiO_2 wt. % versus ($Al_2O_3 + K_2O + Na_2O$) wt. % based on the weathering index of ancient sediments. This plot can provide useful inferences on present and past climate.

On this diagram our samples plot in the field of semiarid climate with sequential variation (Fig. 7b) from Older to Younger Floodplain. This geochemical variation can be inferred as the progressive climatic variability within Quaternary environment of Central India. Biswas and Dassarma (1984) and De Terra and Paterson (1939b) had already reported the cyclic climatic variations from Humid to semi-arid to arid, within the quaternary succession of Narmada Valley using the sedimentary structures. The present alluvium geochemistry of surface samples also corroborates the presence of climatic transitions from arid to semi arid within the depositional span of sediments of Older and Younger Floodplain from Tawa basin, Narmada valley.

Hence, the quaternary sediments of Narmada valley derived from the homogeneous precursor but deposited under different geomorphic horizons of fluvial system within intercontinental humid to semiarid climate of quaternary era, have uniqe elemental signatures with a limited overlap dependent on mineralogical and textural maturity viz. sedimentary fractionation and hydraulic sorting controlled by differential weathering and climatic variability in respective depositional environment.

8 Conclusions

The Alluvium Geochemistry of the stream sediment samples collected from distinct geomorphic horizons of Tawa river basin, Narmada valley suggests that these are chemically immature having unique elemental signatures with a limited overlap and derived from relatively un-weathered source. The degree of chemical weathering is different at each geomorphic horizons, but the pristine source and syn depositional characters are well preserved.

Narmada alluvium is comparatively more enriched in Zircon than Upper Continental Crust and has more enriched and fractionated LREE with flat HREE pattern when compared with Chondrite and North America Shale Composite (NASC). The major, trace and REE data along with ratio plots, confirms that geochemical signatures of all the fine clastics collected from Pediplain, Piedmont slope, Old and Younger Floodplain are controlled by detrital mineralogy with least to moderate amount of post-depositional chemical weathering.

The average CIA value for the clastic sediments from the present study area is 72.83 which is comparable with the average CIA values for global shale hence, confirms the presence of muscovite, illites, and smectite (Clay minerals) within the Narmada alluvium. Significant correlation among the elements associated with the felsic linage including Rare Earth elements and Hafnium, Uranium, Tantalum, Thorium, Zircon, Niobium, Silica in Narmada Alluvium indicates the felsic precursor and presence of heavy minerals including Zircon, Sphene (?), Monazite (?), Allnite (?) etc. The inverse relation between elements associated with clay minerals, phyllosilicates and the quartz, opaque's (Fe-Ti Oxides), phosphate minerals (monazite) indicates the effect of mineralogical and textural maturity induced by the flow hydraulics. The attenuation of heavy minerals and metals including Zircon, Th, La and REE's in quaternary alluvium is attributed to sedimentary fractionation and hydraulic sorting during deposition.

Interestingly it is also concluded that although the sedimentation of Narmada quaternary valley occurred from the homogeneous precursor in humid to semiarid climate between Pleistocene to Holocene time span but the fine clastic deposited under different geomorphic horizons of fluvial system have unique elemental signatures with a limited overlap dependent on mineralogical and textural maturity viz. sedimentary fractionation and hydraulic sorting controlled by differential weathering and climatic variability in respective depositional environment. Hence enrichment of heavy metals including REE's in Younger Floodplain in comparison to Older Floodplain of Narmada valley is attributed to changing present day climate.

Acknowledgments The assignment is executed as a part of NGCM work carried out under item No. 017/GCM/CR/MP/2013/014, Field Season 2013-14 of Geological Survey of India, Ministry of Mines. The authors gratefully acknowledge Director General, Geological Survey of India; Additional Director General and HOD, Geological Survey of India, Central Region, for providing necessary facilities technical, administrative and financial for successful completion of the research. The authors also express their heart full thanks to Late Shri A. K. Mathur, Project Director, for his keen supervision and motivation during execution of field work and data collection. The authors are also thankful to Chemical Lab, Nagpur for in time sample analysis and all the administrative and support system staff for their direct & indirect efforts in bringing up of this manuscript. Last but not the least authors also express kind acknowledgement towards the expert reviewer for his keen scrutiny and value comments for betterment of the manuscript. The competent authorities of GSI are also dully acknowledged for providing necessary approval for publication of this manuscript.

References

Bain WK (2015) A preliminary study on the Stone age artifacts of NetanKhedi, Sehore district of Central Narmada valley in Madhya Pradesh. Human Kind 11:31–49

Biswas S, Dassarma DC (1982) A new species of Stegodon from the alluvial deposits of the Narmada Valley in peninsular India. In: Proceedings of Neogene/Quaternary Boundary. Field Conference, India. Geological Survey of India, Kolkata, pp 15–20

Biswas S, Dassarma DC (1984) Quaternary alluvial deposits and associated fossil vertebrates of Narmada valley in Madhya Pradesh. Geological Survey of India. Unpublished Report. Accession No. GSI-CHQ-24454

Chittleborough DJ (1991) Indices of weathering for soil and paleosol formed on silicate rocks. Aust J Earth Sci 38:115–120

Cox R, Low DR, Cullers RL (1995) The influence of sediment recycling and basement composition on evolution of mud rock chemistry in the South-Western United States. Geochim Cosmochim Acta 59:2919–2940

Cullers RL (1994a) The controls on the major and trace element variations of shales, siltstones, and sandstones of Pennsylvanian–Permian age from uplifted continental blocks in Colorado to platform sediment in Kansas, USA. Geochim Cosmochim Acta 58:4955–4972

Cullers RL (1994b) The chemical signature of source rocks in size fractions of Holocene stream sediment derived from metamorphic rocks in the wet mountains region, USA. Chem Geol 113: 327–343

Cullers RL (2000) The geochemistry of shales, siltstones, and sandstones of Pennsylvanian-Permian age Colorado, U.S.A.: implications for provenance and metamorphic studies. Lithos 51:181–203

Dar SA, Khan KF, Mir AR (2020) Provenance and paleo-weathering of Paleoproterozoic siliciclastic sedimentary rocks of Bijawar Group, Sonrai Basin, Uttar Pradesh, India: using a geochemical approach. J Sediment Environ 5:399–413

Dassarma DC, Biswas S (1978) Fossil hippopotamus from Madhya Pradesh Geological Survey of India. News 9(6):7

De Jayawardena US, Izawa E (1994) A new chemical index of weathering for metamorphic silicate rocks in tropical regions: a study from Sri Lanka. Eng Geol 36:303–310

De Terra H, Paterson TT (1939a) Studies on the ice age in India and associated human cultures, vol 499. Caxmegie Insituation Publication, Washington

De Terra H, Paterson TT (1939b) Pleistocene chronology in the Narmada Valley of Central India. In: op. cit, pp 313–326
Eswaran H, Stoops G, Paepe D, R. (1973) A contribution to the study of soil formation of Isla Santa Cruz Galapegose. Pédologie 23:100–122
Fedo CM, Nesbitt HW, Young GM (1995) Unrevalling the effects of potassium metasomatism in the sedimentary rocks anspaleosols with implications for paleoweathering conditions and provenance. Geology 23:921–924
Fedo CM, Eriksson KA, Krogstad EJ (1996) Geochemistry of shales from the Archean (~3.0 ga) Buhwa Greenstone Belt, Zimbabwe: implications for provenance and source-area weathering. Geochim Cosmochim Acta 60:1751–1763
Floyd PA, Winchester JA, Park RG (1989) Geochemistry and tectonic setting of Lewisian Clastic Metasediments from the early proterozoic loch Maree Group of Gairloch, NW Scotland. Precambrian Res 45:203–214
Franckea A, Holtvoeth J, Codileana AT, Lacey JH, Bayon G, Dosseto A (2020) Geochemical methods to infer landscape response to quaternary climate change and land use in depositional archives: a review. Earth Sci Rev 207(103218):1–23
Gromet LP, Dymek RF, Haskin LA, Korotev RL (1984) The north American shale composite: its compilation, major and trace element characteristics. Geochim Cosmochim Acta 48:2469–3482
Hamdam J, Bumham CP (1996) The contribution of nutrients from the parent material in three deeply weathered soils of peninsular Malaysia. Geoderma 74:219–233
Harnois L (1988) The CIW index: a new chemical index of weathering. Sediment Geol 55:319–322
Herron MM (1988) Geochemical classification of terrigenous sands and shales from core or log data. J Sediment Petrol 58:820–829
Khan AA, Aziz M (2016) Geomorphology and sedimentation of the area around hominid locality Hathnora Narmada Rift Valley Central India. Int J Adv Res 4(12):2778–2804
Khan A, Sonakia A (1992) Quaternary deposits of Narmada with special reference to the hominid fossil. J Geol Soc India 39:147–154
Khartri AP (1961) Stone age and pleistocene chronology of the Narmada Valley (Central India). Anthropos 56(3/4):519–530
Khartri AP (1962) Mahadevian': an Oldowan pebble culture of India. Asian Perspect 6:186–197
La Flèche MR, Camiré G (1996) Geochemistry and provenance of metasedimentary rocks from the Archean Golden pond sequence (casa Berardi mining district, Abitibi sub province). Can J Earth Sci 33:676–690
Long X, Yuan C, Sun M, Xiao W, Wang Y, Cai K, Jiang Y (2012) Geochemistry and Nd isotopic composition of the early Paleozoic flysch sequence in the Chinese Altai, Central Asia: evidence for a northward-derived mafic source and insight into model ages in an accretionaryorogen. Gondwana Res 22:554–566
Martin JM, Thomas AJ, Van Gricken R (1978) Trace element composition of Zaire suspended sediments. Neth J Sea Res 12:414–420
Maynard JB, Sutton SJ, Robb LJ, Ferraz MF, Meyer FM (1995) Apaleosol developed on hydrothermally altered granite from the hinterland of the Witwaterrand basin: characteristics of a source of basn fill. J Geol 103:357–377
McLennan SM, Nancy WB, Taylor R (1980) Rare earth element–thorium correlations in sedimentary rocks, and the composition of the continental crust. Geochim Cosmochim Acta 44:1833–1839
McLennan SM, Hemming S, McDaniel DK, Hanson GN (1993) Geochemical approaches to sedimentation, provenance and tectonics. In: Johnsson MJ, Basu A (eds) Processes controlling the composition of clastic sediments. Geological Society of America, Boulder, CO, pp 21–40
Medlicott HB (1873) Note on celt found by Mr. Hacket in the fossiferous deposits of the Narmada Valley. Records GSI 3:49–57
Nesbitt HW, Young GM (1982) Early Proterozoic climates and plate motions inferred from major element chemistry of lutites. Nature 299:715–717

Nesbitt HW, Young GM (1984) Prediction of some weathering trends of plutonic and volcanic rocks based on thermodynamic and kinetic considerations. Geochim Cosmochim Acta 48: 1523–1534

Parker A (1970) An index of weathering for silicate rocks. Geol Mag 107:501–504

Potter PE (1978) Petrology and chemistry of modern big river sands. J Geol 86:423–449

Prince JR, Velbel MA (2003) Chemical weathering indices applied to weathering profiles developed on heterogeneous felsic metamorphic parent rocks. Chem Geol 202:397–416

Rahman MA, Chandra Das S, Pownceby MI, Tardio J, Alam MS, Zaman MN (2020) Geochemistry of recent Brahmaputra River sediments: provenance, tectonics, source area weathering and depositional environment. Fortschr Miner 10:813

Roaldset E (1972) Mineralogy and geochemistry of quaternary clays in Numedalarea southern Norway. Norsk Geolisk Tidsskrift 52:335–369

Roser BP, Korsch RJ (1988) Provenance signatures of sandstone-mudstone suites determined using discriminant function analysis of major-element data. Chem Geol 67:119–139

Ruxton BP (1968) Measures of the degree of the chemical weathering of rocks. J Geol 76:518–527

Salahuddin (1987) Late quaternary ecology, fauna and human culture of Central Narmada valley Madhya Pradesh. PhD. Thesis, Department of Archeology Deccan College, P.G.R.I. University of Poona, Pune

Shukla S, Meshram J, Minz C (2014) Final report on geochemical mapping in toposheet no. 55F/14 and parts of 55F/9, Hoshangabad District, Madhya Pradesh. Geological Survey of India. Unpublished Report. Accession no. GSI-22977

Shukla S, Meshram J, Minz C, Suryavanshi H, Sarkar S (2021) Reporting of REE's and heavy metals through geochemical mapping in Ramnagar formation of Narmada quaternary basin, Hoshangabad district. Madhya Pradesh. Indian J Geosci 75(1):50–70

Singh P (2010) Geochemistry and provenance of stream sediments of the Ganga River and its major tributaries in the Himalayan region, India. Chem Geol 269:220–236

Sonakia A (1984) The skull cap of early man and associated mammalian fauna from Narmada Valley alluvium Hoshangabad area, Madhya Pradesh, India. Records Geol Surv India 113(6): 159–172

Stallard RF, Edmond JM (1983) Geochemistry of the Amazon: 2. The influence of geology and weathering environment on the dissolved load. J Geophys Res 88:9671–9688

Suttner LJ, Dutta PK (1986) Alluvial sandstone composition and palaeoclimate. 1. Framework mineralogy. J Sediment Petrol v. 56 (3):329–345

Taylor SR, McLennan SM (1985) The continental crust: its composition and evolution. Blackwell, Oxford

Theobald W (1860) On the tertiary and alluvial deposits of the central portion of the Narbudda Valley. Mem Geol Surv India II:279–291

Tiwari MP, Bhai HY (1997) Quaternary stratigraphy of the Narmada valley. In: Quaternary geology of narmada valley: a multidisciplinary approach, vol 46. Geological Survey of India, Kolkata, pp 33–63

Vital H, Stattegger K (2000) Major and trace elements of stream sediments from the lowermost Amazon River. Chem Geol 168:151–168

Vogt T (1927) Sulitjelmafeltets geologi og petrografi. Norges Geologiske Undersokelse 121:1–560. (in Norwegian with English Abstract)

Wu W, Xu S, Lu H, Yang J, Yin H, Liu W (2011) Mineralogy, major and trace element geochemistry of riverbed sediments in the headwaters of the Yangtze, Tongtian River and Jinsha River. J Asian Earth Sci 40:611–621

Heterogeneity in Glacier Area Loss in Response to Climate Change in Selected Basins of Western Himalaya

Riyaz Ahmad Mir, Zahid Majeed, Rayees Ahmed, Sanjay K. Jain, Syed Towseef Ahmed, Muneer Ahmad Mukhtar, and Gowhar Farooq Wani

Abstract Himalayan glaciers are characterized to exhibit heterogeneous and variable changes in areal loss. To understand this erratic behavior of Himalayan glaciers six major sub-basins (upper Jhelum, Suru and Zanskar, Chandra, Spiti, Satluj and upper Indus basin) covering parts of western Himalayan region were selected for analysis. For this purpose, Landsat images (MSS, TM, ETM+, OLI/TIRS), Survey of India toposheets (SoI), high resolution Google Earth and Corona KH-4B images and ASTER DEM with a limited field check were used. The results indicated that the western Himalayan glaciers during last 5–6 decades exhibit a heterogeneous pattern of area loss that is falling within a range of 5–20% with a mean loss of 15%. The Satluj basin indicated a higher average area loss of 22.1% wherein the Tirungkhad and Baspa sub-basins indicated a loss of 26.1% from 1966 to 2011 and 18.1% from 1976 to 2011. The upper Indus basin also indicated a higher loss of 21.2% from 1962 to 2017 followed by the Jhelum basin indicating a higher area loss of 19.41% from 1990 to 2018. The Chandra basin indicated an area loss of 10.7% from 1962/1971 to 2013/2014/2015 whereas the Suru and Zanskar basins indicated a lower area loss of 5.7% from 1962/1971 to 2013/2015 followed by the Spiti basin indicating a lower average area loss of 4.3% from 1962/1965 to 2013/2014. During the same time period, the study observed statistically significant rising trends in T_{min}, T_{max}, and T_{avg} at a rate of 0.071 °C/year, 0.076 °C/year, 0.07 °C/year and 0.013 °C/year, 0.002 °C/year, 0.004 °C/year at Rakcham (Satluj basin) and Pahalgam (Jhelum basin) stations. The Manali station (near Chandra basin) indicated

R. A. Mir (✉)
Department of Earth Sciences, Indian Institute of Technology Roorkee, Roorkee, India

Geological Survey of India, Srinagar, Jammu and Kashmir, India

Z. Majeed · M. A. Mukhtar
Geological Survey of India, Srinagar, Jammu and Kashmir, India

R. Ahmed · S. T. Ahmed · G. F. Wani
Department of Geography and Disaster Management, University of Kashmir, Srinagar, India

S. K. Jain
Water Resource System Division, National Institute of Hydrology, Roorkee, India

B. Phartiyal et al. (eds.), *Climate Change and Environmental Impacts: Past, Present and Future Perspective*, Society of Earth Scientists Series,
https://doi.org/10.1007/978-3-031-13119-6_8

a significant rising trend for T_{min} (0.056 °C/year) and significantly declining trends for T_{max} and T_{avg}. Contrarily, the precipitation at the Rakcham and Manali station indicated a significant declining nature (−6.28 and −52.5 mm/year). The rising temperature is therefore considered as a major factor controlling the glacier area loss in the region. Thus, suitable strategies are required to be evolved timely to manage the Himalayan glacier resources that are under the threat of loss due to the recent climate warming and change. The higher areal loss in glacier resources may have concerning implications on future water availability in the region.

Keywords Glacier wastage · Climate change · Jhelum · Suru and Zanskar · Chandra · Spiti · Satluj · Indus · Water resources

1 Introduction

In the Himalayan mountain system, about 54,000 individual glaciers covering an area of 60,000 km^2 and volume of 6000 km^3 are reported to exist (Bajracharya and Shrestha 2011). In the Indian Himalayan region, about 9675 glaciers with an area of 37,000 km^2 and volume of 2000 km^3 are recorded (Raina and Srivastava 2008). These huge glacier resources affect significantly the river system surface runoff, recharge aquifers, sustain hydropower generation, agriculture, and ecosystems and finally contribute to sea-level fluctuations (Dyurgerov and Meier 2005; Mir et al. 2017; Majeed et al. 2020).

Generally, the glaciers and ice reserves are considered very sensitive to any climatic variations as the glaciers respond directly to the climate and therefore, serve as the best and early indicators of climatic variations. Likely in the Himalayan region, any increase in air temperature and changes in precipitation patterns influence the overall glacier dynamics and behavior (Benn et al. 2012; Mir 2021; Sood et al. 2021; Singh et al. 2021). All over in the world, the glaciers are reported to be in a state of shrinkage particularly in response to the rising temperature vis a vis climate change (Sorg et al. 2012; Mir et al. 2015a, b, 2017). As per the reports, the world's average surface temperature has increased between 0.3 °C and 0.6 °C over the past hundred years (Bajracharya et al. 2008). Similarly, an analysis of the temperature data over the Himalayan region has immensely revealed a warming pattern, albeit at diverse rates during different periods depending on the seasons and regions (Bhutiyani et al. 2007; Dash et al. 2007; Singh et al. 2008; Koul and Ganjoo 2010; Dimri and Dash 2012; Mir et al. 2015a).

In the western Indian Himalayan region, previous reports have suggested that the temperature is rising significantly and the precipitation in general is decreasing with intra-regional differences (Sontakke et al. 2009; Bhutiyani et al. 2010; Dimri and Dash 2012; Mir et al. 2015a; b; c). In response to increasing temperature and decreasing precipitation, the most obvious and clear impact is certainly the widespread melting and retreat of Himalayan glaciers (Kääb et al. 2007; Scherler et al. 2011; Swain et al. 2018; Majeed et al. 2020; Krishna et al. 2020). Majority of the earlier studies showed that the Himalayan glaciers have been receding and losing huge area and mass in response to climate change (Mayewski and Jeschke 1979; Mehta et al. 2011; Mir et al. 2013; Bahuguna et al. 2014; Mir et al. 2017, 2018; Mir 2018, 2021; Jain and Mir 2019) with a noticeable exception of the Karakorum region where some glaciers are advancing which is known as Karakorum anomaly (Hewitt 2005; Mir and Majeed 2018; Mir 2021).

Notably, the Himalayan glacier recession and change has been found to be heterogeneous and vary profusely from one region to another, from one glacier to another (Scherler et al. 2011; Kääb et al. 2015) probably due to the heterogeneous and varying climate across the region, variable morphological, topographical and geomorphic factors as well as the presence of variable supraglacial debris (Shukla et al. 2018; Mir et al. 2018). For instance, the Nathawat et al. (2008), reported that glaciers have lost 18% of area and retreated by 6–33 m/year from 1962 to 2001 in parts of the Zanskar valley. A recent study on five large glaciers of the Zanskar basin reported an area loss of 13–15% at a retreat rate of 8–19 mm/year of clean glaciers and 14–21% of area loss at a rate of 2–3 m/year of debris covered glaciers (Shukla and Qadir 2016). In contrast, Kulkarni et al. (2011) have reported an area loss of 9.2% (2.41 km^2/year) from 1962 to 2001 in the Zanskar basin. Similarly, Pandey and Venkataraman (2013) has carried out a detailed study of 15 large glaciers in the Chandra basin and have reported an average glacier retreat rate of 15.5 m/year from 1980 to 2010. Similarly, higher changes in glacier terminus and area have been reported for several glaciers in the Himalayan region e.g., Drung-Drung (9 m/year) in the Jammu and Kashmir region; Rekha-Samb (12 m/year) in Nepal; and Zemu (14 m/year), South Lhonak (42 m/year), and Rathong (18 m/year) in Sikkim (Kulkarni and Karyakarte 2014). Similarly, in other sub-basins of Himalayan region a heterogeneous and variable retreat and area loss has been recorded. But, in general, the Himalayan glaciers are retreating faster than the other glacier on the Earth, wherein the average rate of the loss is ~0.4% area per year (Bolch et al. 2010, 2012; Bhambri et al. 2011, 2012; Kamp et al. 2011; Kulkarni et al. 2011; Chand and Sharma 2015; Mehta et al. 2011; Nainwal et al. 2008; Govindha Raj 2010; Mir and Majeed 2018).

Thus, keeping the above facts in view, this study attempts to understand, decode and bracket the variable and heterogeneous glacier area loss in few selected catchments of the Indus River System, covering parts of western Himalaya. The glacier area loss has also been compared with the observed temperature and precipitation changes analyzed for 3 metrological stations located in Satluj basin, near Chandra basin and in Jhelum basin of the study area. For understanding the dynamics and behavior of the selected glaciers, the high resolution (available partly), medium and coarse resolution Landsat satellite data and other ancillary data sources with a limited field check have been used.

2 Study Area

The Indus River system, located in western Himalayan region, is one of the greatest river systems with its origin from the Mount Kailash in Tibet on the northern side of the Himalaya at an altitude of 5486 m amsl (Fig. 1). This transboundary Indus River system has a total area of 1.12 million km^2 distributed between Pakistan (47%), India (39%), China (8%) and Afghanistan (6%) (Mir 2016; Singh 2017). In India, the drainage area of the Indus River system lying in the states of Ladakh, Jammu and Kashmir, Himachal Pradesh, Punjab, Rajasthan, Haryana and Chandigarh is approximately 4,40,000 km^2, which is nearly 14% of the total area of the country (Mir 2016; Singh 2017). Besides it, at least 300 million people are estimated to be dependent on the Indus River system for their livelihood (Mir et al. 2017; Shukla et al. 2018; Singh 2017). The Indus River is fed by about 24 tributaries out of which 8 tributaries are defined as the major tributaries. The major tributaries such as the Jhelum, Chenab, Ravi, Satluj and Beas rivers are east flowing while as the Shyok, Kabul, Gomal and Gilgit rivers flow towards west and north, respectively (Mir 2016; Singh 2017).

In this study, the selected basins include the Jhelum basin, Suru and Zanskar basins, Chandra basin, Spiti basin and Satluj basin covering the parts of Greater

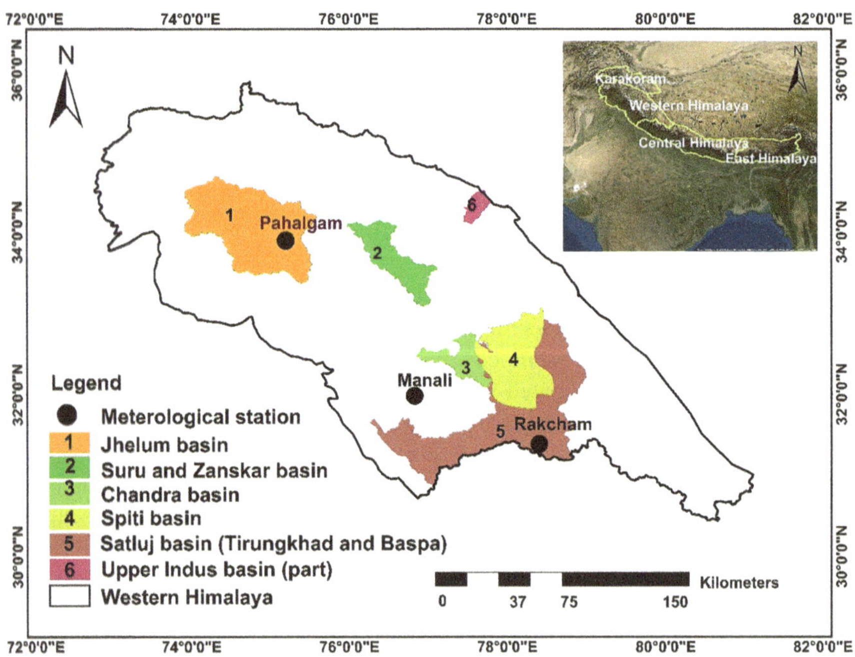

Fig. 1 Location map of the selected basins of Indus River system for the study located in western Himalaya

Himalayan Range (GHR). The study area also includes the Phyang, Ganglas, Rong, Khalsar and No. 5 catchments covering the parts of Ladakh Mountain Range (LMR) of upper Indus basin. The glacial systems of LMR are located at the interface between rapidly receding glaciers of the GHR and advancing glaciers of Karakoram Mountain Range (KMR). For the Jhelum basin, Suru and Zanskar basins, Chandra basin and Spiti basin only few major glaciers have been analyzed whereas, for the Satluj basin represented here by its highly glaciated south-western 2 sub-catchments i.e., Tirungkhad and Baspa basins, an inventory of 32 and 107 glaciers have been analyzed. For the upper part of Indus basin, an inventory of 90 glaciers of sub-catchments of Phyang, Ganglas, Rong, Khalsar and No. 5, etc. have been analyzed.

The climate of the Western Himalayan (WH) region is dependent on precipitation occuring during the winter and early spring months (December to April). The precipitation is produced by the Western Disturbances (WD) originating from the Mediterranean region. In this region, the snowfall starts in early November and stops by the end of April, depending on the altitude of the area. The Jhelum basin witnesses temperate climate. It receives the annual precipitation of 1100 mm (Mir et al. 2016). The temperature varies between −6 °C and 30 °C (Mir and Jeelani 2015). The Suru-Zanskar valley has a distinct climatic characteristic due to its location in the shadow zone of Great Himalaya (Mamgain and Sastry 1975; Klimes 2003). The study region has cold sub arid type of climate. The temperatures varies between mean minimum temperatures of −15 °C to −35 °C. In this area, the precipitation of about 250 mm is received. The Chandra basin is in monsoon-arid transition zone and receives precipitation of about 514 mm. The Spiti and Satluj basins receive the mean annual precipitation of 456 mm. The basins experience large variations in the climatic conditions ranging from the sub-tropical climate at the bottom of the Satluj valley to the alpine in the upper reaches/parts of which are perpetually under snow. The LMR lies in the semi-arid climatic regime produced as a result of rain shadow effect of the HMR and KMR that are affected by two distinct climatic regimes. For example, in the HMR about 80% of precipitation is produced by the summer monsoon whereas the westerly dominated KMR receives two-thirds of its precipitation through winter westerlies (Bookhagen and Burbank 2010).

3 Materials and Methods

In this study, more than 60 numbers of images have been used for the glacier mapping available during different time periods as well as for different glaciers and selected basins. Depending upon the data availability and suitability, only a limited number of satellite data images were used for the glacier mapping and delineation. The details of the data sources used are given in (Table 1). The ancillary and satellite data images used include the SoI toposheets, declassified high-resolution Corona KH-4A images, Landsat satellite images such as Multispectral Sensor (MSS), Thematic Mapper (TM), Landsat 7 Enhanced Thematic Mapper

Table 1 Multi-temporal ancillary and satellite imagery data sources used for the study

Selected basin	Benchmark Glaciers	Satellite/ sensor	Date of acquisition	Scale (m) and spatial resolution
Jhelum basin	Kolahoi, Bodhpathri, Shishram, Harmukh, Nehnar	SoI Toposheet	1962	1:50000
		Landsat MSS	24-10-1980	60
		Landsat 5 (TM)	13-09-1990	30
		Landsat 5 (TM)	16-10-1992	28.5
		Landsat 5 (TM)	09-10-2000	15, 30
		Landsat 5 (TM)	18-06-2000	28.5
		Landsat 7 (ETM+)	30-09-2001	28.5
		Landsat 5 (TM)	17-10-2010	30
		Landsat 8 (OLI)	17-10-2014	30
		Sentinel-2A	15-09-2018	10, 20
Suru and Zanskar basins	Parkhachik, Pensilungpa, Shafat, Dalung, Durung Drung, Hagshu, Padam	SoI Toposheet	1962	1:50000
		MSS	11-10-1975	90
		Corona KH-4B	28-09-1971	8
		Corona KH-4B	25-08-1972	8
		Landsat 5 TM	04-07-1992	30
		Landsat 7 ETM+	16-07-1999	15, 30
		Landsat 7 ETM+	16-12-1999	15, 30
		Landsat 7 ETM+	13-10-1999	15, 30
		Landsat 7 ETM+	01-10-2003	15, 30
		Landsat 7	10-05-2003	15, 30
		Landsat 5 TM	29-09-2011	15, 30
		Landsat 8 OLI	14-07-2013	15, 30

(continued)

Table 1 (continued)

Selected basin	Benchmark Glaciers	Satellite/ sensor	Date of acquisition	Scale (m) and spatial resolution
		Landsat 8 OLI	28-11-2013	15, 30
		Google Earth	2012 & 2015	15, 2.5
Chandra basin	Samundar Tapu, Batal, Sonapani, Bara Shigri	MSS	15-11-1972	80
		Corona KH-4B	28-09-1971	8
		Landsat 5 TM	09-10-1989	30
		Landsat 7 ETM+	16-11-1999	15, 30
		Landsat 7 ETM+	13-10-1999	15, 30
		Landsat 8 OLI	28-11-2013	15, 30
		Google Earth	2015	15, 2.5
Spiti basin	Parang, Nisti, Padma, Pin	SoI Toposheet	1962, 1965	1:50000
		Corona KH-4B	24-09-1965	8
		Landsat 7 ETM+	01-10-2003	15, 30
		Landsat 5 TM	19-10-2001	15, 30
		Landsat 8 OLI	28-11-2013	15, 30
		Google Earth	2013	15, 2.5
Satluj basin	Glaciers of Baspa basin	MSS	02-11-1976	60
		Landsat (TM)	11-11-1992	30
		Landsat (ETM+)	08-10-2000	30
		LISS III	16-09-2007	27
		Landsat (TM)	13-09-2011	30
	Glaciers of Tirungkhad basin	SoI Toposheet	1966	1:50000
		Mss	01-10-1979	79

(continued)

Table 1 (continued)

Selected basin	Benchmark Glaciers	Satellite/ sensor	Date of acquisition	Scale (m) and spatial resolution
		Landsat (TM)	21-10-1990	30
		Landsat (TM)	13-09-2011	30
Upper Indus basin	Glaciers of Ladakh mountain range (part)	SoI Toposheet	1962	1:50000
		MSS	21-11-1976	60
		MSS	18-06-1977	60
		MSS	18-07-1978	60
		Landsat (TM)	16-09-1998	30
		Landsat (TM)	31-10-2000	15
		Landsat (TM)	03-08-2011	30
		Landsat (OLI/TIRS)	06-10-2017	30

(ETM+), Landsat 8 Operational Land Imager (OLI) (OLI) etc. In addition, high-resolution images provided through virtual Google Earth have also been used to extend the time period to recent years and to use as a recent source of information. The ASTER DEM has also been used for the glacier mapping and other attribute estimation. The Corona KH4B and TM images were available from United States Geological Survey website (USGS; http://edcsns17.cr.usgs.gov/NewEarthExplorer/) whereas, a subsets of Google Earth™ (GE) images were selected and saved from virtual Google Earth. ASTER DEM data was available freely from the website as http://www.jspacesystems.or.jp/ersdac/GDEM/E/index.html) for download and use. All the data images were processed in the GIS platform.

For the estimation of glacier area, the co-registration and ortho-rectification of the data images was carried out separately for the each selected basin wherever required. It is because, the Landsat data series mostly used in this analysis is provided orthorectified (Mir and Majeed 2018). Therefore, as per the availability either the TM/ETM image were used as reference image for the coregistration and rectification of other images using the UTM projection system of WGS-84 datum. The coregistration was done based on ground control/tie points in GIS platform. The available topographical maps were also scanned and georeferenced in the GIS platform. The high resolution, Corona images with minimal snow cover were used generally to extract the historic extent of the glaciers wherever, this data was available. The available datasets with minimum snow cover/or cloud cover and easily visible features were used for glacier mapping. For the present study area, such type of data is generally available during September to November months of the year. Therefore, the choice of data was also limited to these 3 months. The glacier extents were manually delineated through visual inspections from all the satellite datasets based on the false color composite (RGB-532). The manual digitization was adopted because, many glaciers are debris covered and were carefully delineated to avoid wrong inclusion of debris covered parts. Based on the glacier outlines of the different years, the area vacated by the glaciers were calculated in GIS platform. For certain glaciers, the length was also estimated to determine the glacier retreat (Mir and Majeed 2018). The details of the methodology for glacier mapping are discussed in detailed in Mir et al. (2017, 2018), Mir and Majeed (2018), Mir (2021). In addition, a limited field check was also carried out at few glaciers during 2013, 2014, 2016 and 2018 using the Garmin GPS map 76CSX for the accuracy assessment of the glacier mapping based on satellite data sources (Majeed et al. 2016; Krishna et al. 2017; Mir and Majeed 2018; Mukhtar and Majeed 2020).

Furthermore, for the climate change analysis, the metrological data of Rakcham station located in Satluj basin (extreme western part of western Himalaya) and Manali station located in Beas basin near to Chandra basin (central part of the western Himalaya) and Pahalgam station located in Lidder catchment of Jhelum basin (extreme eastern part of western Himalaya) have been used. For Rakcham and Manali station, the data was procured from Bhakra Beas Management Board (BBMB) whereas, for the Pahalgam station, the data was procured from Indian Meteorological Department (IMD), Srinagar. The temperature data was available from 1985 to 2008 for Rakcham station, 1986 to 2011 for Manali station and 1990 to 2018 for Pahalgam station. The precipitation for Rakcham and Pahalgam was also available for the same time period as temperature respectively, but, for the Manali station, the precipitation data was available from 1989 to 2010. The climatic data almost coincides with the same time period for which the glacier area loss has been assessed. The trend analysis of the climatic data was carried out for annual periods using the statistical tests i.e., Mann Kendall test and Sen's Slope estimator. The climate data was also analyzed using simple linear regression technique. The details of the methods are discussed in detail in Mir et al. (2015b, 2017) and Jain and Mir (2019).

3.1 Uncertainty in the Study

Using the buffer method (Granshaw and Fountain 2006; Mir et al. 2017, 2018), the uncertainty in glacier area estimated varied between the 0.47 km^2 (4.7%) for high resolution Corona images and 0.75 km^2 (7.5%) for Landsat images. For Google Earth (GE) images, an error of 0.47 km^2 (4.7%) was determined. For this analysis, a buffer size of 7.5 m for Corona, 8.3 m for GE images (registration error) and 7.5 m (digitizing error) for all other Landsat images was selected. Furthermore, the accuracy assessment based on the comparison of few bench mark points and certain glacial features at the selected glaciers observed in the field revealed an error of

2–5%. Additionally, the high-resolution Google Earth images were also used to check the accuracy by repeating the digitization of a few selected glaciers. This method resulted in an accuracy of approximately $\leq$2%. In nutshell, an average mapping uncertainty of (5%) was considered for this study (Mir 2021). Furthermore, the quality of the climatic data was assessed and checked statistically as well as visually.

4 Results

In this study, six major sub-basins (i.e., Jhelum, Suru and Zanskar, Chandra, Spiti, Satluj and parts of Upper Indus basin) witnessing variable and different microclimatic regimes were selected to understand the glacier area loss, its heterogeneity and variability during the last 3–6 decades. In case of Jhelum basin, 5 glaciers were selected as benchmark and representative glaciers for the study. Similarly, in case of Suru and Zanskar basins 7 glaciers, Chandra basin 5 glaciers and Spiti 5 glaciers were selected for the study. However, in case of Baspa and Tirungkhad sub-basins of Satluj basin, an updated glacier inventory of 32 and 107 glaciers were used for the study. Similarly, for the upper Indus basin, 5 small sub catchments of Ganglas, Phyang, Khalsar, Rong and No. 5 draining parts of LMR were selected. The detailed results are given in Table 2. The details of the analysis are discussed below.

4.1 Glacier Area Loss in Jhelum Basin

For the representation of the upper Jhelum basin, 5 major glaciers were selected as benchmark. The glaciers include the Kolahoi, Nehnar, Shishram, Bodpathri and Harmukh as shown in (Fig. 2). The Kolahoi glacier drains Lidder and Sind catchments. From 1979 to 2016, the whole Kolahoi glacier has lost an area of 1.73 km^2 (13.63%). However, it is also divided into the K1 and K2 parts for change analysis as per drainage pattern. From 1979 to 2016, the K1 unit revealed an area loss of 1.07 km^2 (11.66%) at a rate of 0.02 km^2/year, whereas, the K2 unit showed a higher area loss of 0.67 km^2 (18.7%) at a rate of 0.018 m^2/year. From 1979 to 2016, the length of K1 unit has retreated by 0.78 km (13.6%) at a rate of 22 m/year whereas; the K2 unit's length has exhibited a lower retreat of 0.39 km (9.8%) at a rate of 11 m/year. The Nehnar glacier is located in Sind catchment. It has lost an area 0.40 km^2 (20.9%) at the rate of 0.01 km^2/year from 1.91 km^2 (1990) to 1.51 km^2 (2018). A total retreat of 246 m at a rate of 8.7 m/year from 1990 to 2018 is recorded for this glacier.

The Shishram glacier is located in Lidder catchment. It is a simple basin mountain glacier with clean ice surface. This glacier has lost an area of 0.91 km^2 (16.33%) at a rate of 0.03 km^2/year from 1990 to 2018. The glacier length has retreated from 2240 m (1990) to 2060 m (2018) at the rate of 6.42 m/year. The Bodpathri glacier is one of the important glaciers of Panjtarni Group of glaciers located in Sind catchment. The total area and length of the glacier in the year 2018 was estimated to be

Table 2 Glacier area loss in the selected basins of western Himalaya

S. no.	Selected basins and benchmark glaciers		Glacier area (km^2) (year)		Area change (km^2)	Area change (%)	Rate of change (km^2 a^{-1})
1	Jhelum basin	Kolahoi	12.74 (1979)	11.0 (2016)	1.74	13.63	0.047
		Nehnar	1.91 (1990)	1.51 (2018)	0.40	26.4	0.014
		Shishram	5.57 (1990)	4.66 (2018)	0.91	16.3	0.03
		Bodpathri	2.5 (1990)	1.91 (2018)	0.59	23.6	0.021
		Harmukh	3.5 (1990)	2.90 (2018)	0.6	17.14	0.02
2	Suru and Zanskar basins	Parkachik glacier	49.5 (1971)	48.8 (2015)	0.7	1.5	0.015
		Pensilungpa glacier	10.90 (1962)	10.78 (2015)	0.12	1.1	0.017
		Shafat glacier	17.38 (1962)	16.69 (2015)	0.69	3.97	0.063
		Dalung glacier	13.7 (1962)	11.09 (2015)	2.64	19.24	0.05
		Durung Drung	54.15 (1971)	53.43(2013)	0.72	1.3	0.031
		Hagshu glacier	52.12 (1962)	50.94(2015)	1.18	0.023	0.0003
		Padam glacier	26.7 (1962)	23.36 (2015)	3.4	12.7	0.06
3	Chandra basin	Samundar Tapu glacier	70.84 (1971)	63.03 (2013)	7.81	11.1	0.26
		Batal glacier	4.78 (1962)	4.65 (2014)	0.13	2.71	0.043
		Bara Shigri glacier	112.52 (1962)	111.12 (2013)	1.4	1.2	0.020
		Sonapani glacier	23.62 (1971)	17.09 (2016)	6.53	27.6	0.613
4	Spiti basin	Parang glacier	4.96 (1962)	4.48 (2013)	0.48	9.6	0.15
		Padma glacier	5.78 (1962)	5.67(2013)	0.11	1.9	0.002
		Nisti glacier	4.38 (1965)	4.20 (2014)	0.18	4.10	0.0038
		Pin glacier	11.17 (1965)	11.02 (2014)	0.15	1.34	0.003
5	Satluj basin	Tirungkhad Basin (32 glaciers)	112 (1966)	82 (2011)	30.0	26.7	0.66
		Baspa Basin (107 glaciers)	227.4 (1976)	186.2 (2011)	41.2	18.1	1.18
6	Upper Indus basin	Ladakh Mt. range (90 glaciers)	24.1 (1962)	19.0 (2017)	5.1	21.2	0.092

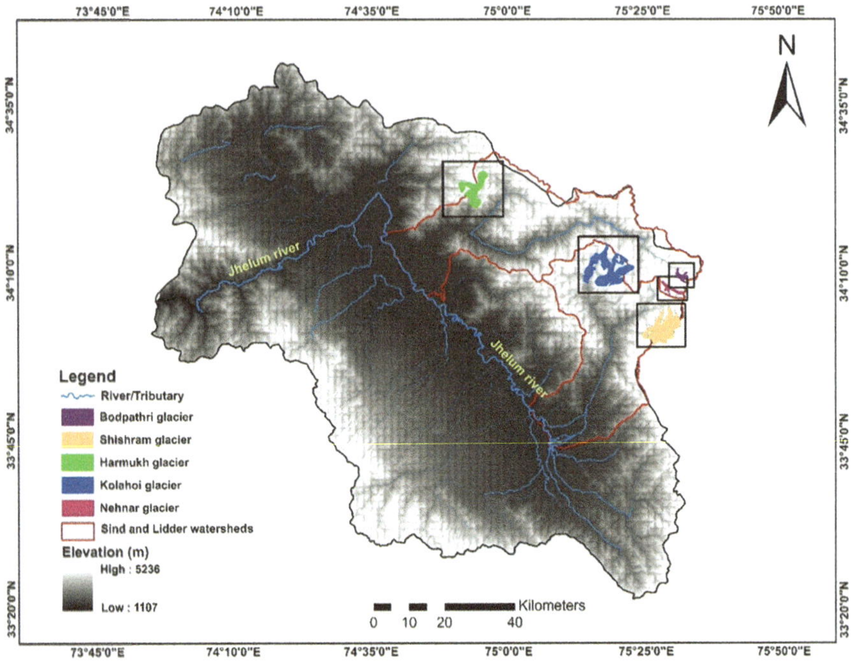

Fig. 2 Location map of the benchmark glaciers of Jhelum basin

1.91 km^2 and 2560 m respectively. The glacier has lost an area of 0.59 km^2 (23.6%) from 1990 to 2018 whereas, the length has retreated by 205 m at a rate of 7.30 m/year. The Harmukh glacier is also located in Sind catchment. It is the simple basin clean valley glacier with the total area and length of 2.90 km^2 and 3025 m respectively.

The glacier has lost an area of around 0.6 km^2 (17%) from 1990 to 2018 at a rate of 0.02 km^2/year, whereas the length has decreased from 3245 to 3025 m with a total loss of 220 m during this period. Overall, based on these representative glaciers, an average area loss of 0.848 km^2 (19.41%) at a rate of 0.03 km^2/year during last 3 decades from 1979/1990 to 2016/2018 is observed. The loss in area varies from 13.63% (Kolahoi Glacier) to 26.4% (Nehnar Glacier) in the Jhelum basin. The retreat rate is also variable and heterogeneous in this basin.

4.2 Glacier Area Loss in Suru and Zanskar Basin

In the Suru and Zanskar basins, 7 major glaciers have been selected as representative and benchmark. The selected glaciers include the Parkhachik, Pensilungpa, Shafat, and Dalung located in Suru basin whereas, the Durung Drung, Hagshu and Padam glaciers located in Zanskar basin (Fig. 3). The Parkachik glacier (Kangriz) is a large compound basin valley type glacier. The glacier is 15 km in length and covers an

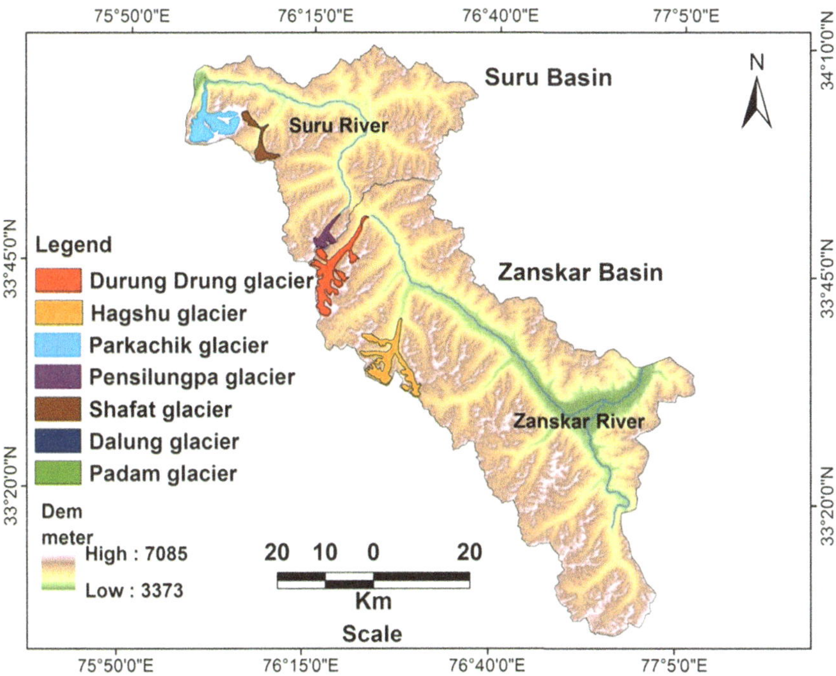

Fig. 3 Location map of benchmark glaciers of Suru and Zanskar basins

area of 49.65 km^2. The glacier from 1971 to 2015, has retreated by 127 m i.e. (0.75%) at a rate of 2.9 m/year with a simultaneous decrease in area from 49.5 to 48.8 km^2 i.e. 740 m^2 (1.5%) at a rate of 74 m^2/year. The Pensilungpa glacier is a compound basin valley glacier type in the upper catchment of Suru valley. It is about 8 km in length and covers an area of 18.56 km^2. From 1962 to 2015, the glacier has shown a retreat of 394.35 m at an annual retreat of 7.44 m/year. The glacier has lost an area of 0.12 km^2 (1.1%) at a rate of 0.002 km^2/year. The Shafat glacier is 8 km long, compound basin valley type glacier. During the time from 1962 to 2015, it has retreated by 1398.44 m, with an average annual retreat of 26.39 m/year. Simultaneously, it has lost an area of 0.69 km^2 (3.97%) at a rate of 0.013 km^2/year. The glacier has receded by 519 m, 658 m, 221 m from 1962 to 1975, 1975 to 1999, and 1999 to 2015 respectively at a rate of 40, 47, and 8 m/year respectively.

The Dalung glacier covers an area of 13.7 km^2 during 1962 which has decreased to 11.0 km^2 (2015). Overall, the glacier has shrunk by 2.6 km^2 (19.2%) at a rate of 0.05 km^2/year during the last 53 years period (5 decades). The length of glacier was 7.93 km during 1962 and has decreased gradually to 6.77 km during 2015. Overall, the total retreat of 1240 m (15.7%) from 1962 to 2015 at a rate of 23.3 m/year was found for Dalung glacier.

The Durung Drung glacier is the second longest glacier in Himalaya after Siachen glacier. This glacier from1971 to 2013 has lost an area of 0.72 km^2 (1.3%) at a rate of

0.017 km^2/year. During the same time period, the glacier has retreated by 550 m at a rate of 13.09 m/year. The glacier has retreated by 321 m/year, 228 m/year from 1971–2004 to 2004–2013 respectively at a rate of 10 m/year and 25 m/year respectively. The Hagshu glacier is 17 km long, covering an area of 58 km^2, compound valley glacier located in Zanskar basin. The glacier from 1962 to 2015 has lost an area of 1.18 km^2 (0.023%). It has retreated by about 1583 m at a rate of 29.87 m/year. The Padam glacier is 20 km long and the area of glacier has decreased considerably from 26.7 km^2 during 1962, to 23.3 km^2 during 2015. It has decreased by 3.4 km^2 (21.7%) from 1962 to 2015 at a rate of 0.06 km^2/year. The length of glacier was observed to be 12.6 km during 1962, which has gradually decreased to 9.91 km (2015). For this glacier, the total retreat observed was 2690 m (21.3%) from 1962 to 2015 at a rate of 50.7 m/year. Overall, based on these representative glaciers, an average area loss of 1.35 km^2 (5.7%) at a rate of 0.03 km^2/year during the last 5–6 decades from 1962/1971 to 2013/2015 is observed. The loss in area varies from 0.03% (Hagshu Glacier) to 19.2% (Dalung Glacier) in the Suru and Zanskar basin. The retreat rate is also variable for selected glaciers.

4.3 Glacier Area Loss in Chandra Basin

For the representation of Chandra basin, 4 major glaciers were selected as benchmark. The glaciers include the Samundar Tapu, Batal, Bara-Shigri and Sonapani as shown in (Fig. 4). The Samundar Tapu glacier is a large compound basin valley type glacier covering an area of 95 km^2. From 1971 to 2013, the glacier has lost an area of 7.8 km^2 (11.1%) at a rate of 0.26 km^2/year. From 1971 to 2013, it has retreated by 1266 m at a rate of 24.82 m/year. The Batal glacier is a simple basin type glacier located on the right bank of Chandra River. From 1962 to 2013, the glacier has lost an area of 0.13 km^2 (2.71%) at a rate of 0.04 km^2/year. Batal glacier has retreated by 375 m from 1962 to 2014 at a rate of 7.21 m/year. The Bara Shigri glacier is a large compound basin valley type glacier and the largest glacier of Himachal Pradesh, covering an area of about 136 km^2. Bara Shigri showed a retreat of about 1717 m at a rate of recession of 32.4 m/year. The Bara Shigri glacier has lost an area of 1.4 km^2 (1.2%) from 1962 to 2013. The Sonapani glacier has lost an area of 6.53 km^2 (27.6%) from 1971 to 2016. The glacier has retreated by 202 m at a rate of 11.2 m/year from 1971 to 1989, 31 m at a rate 2.8 m/year during 1989–2000 and 596 m at rate of 37.3 m/year during 2000–2016. During the recent time period from 2000 to 2016, the rate of recession has increased rapidly. Due to this rapid recession of the glacier a major tributary towards its lower left side has also detached from it. Overall, based on these representative glaciers an average area loss of 3.96 km^2 (10.7%) at a rate of 0.23 km^2/year during last 5–6 decades from 1962/1971 to 2013/2014/2015 is observed. The loss in area varies from 1.2% (Bara Shigri Glacier) to 27.6% (Sonapani Glacier) in the Chandra basin. The rate of recession is also variable during different time periods for the selected glaciers.

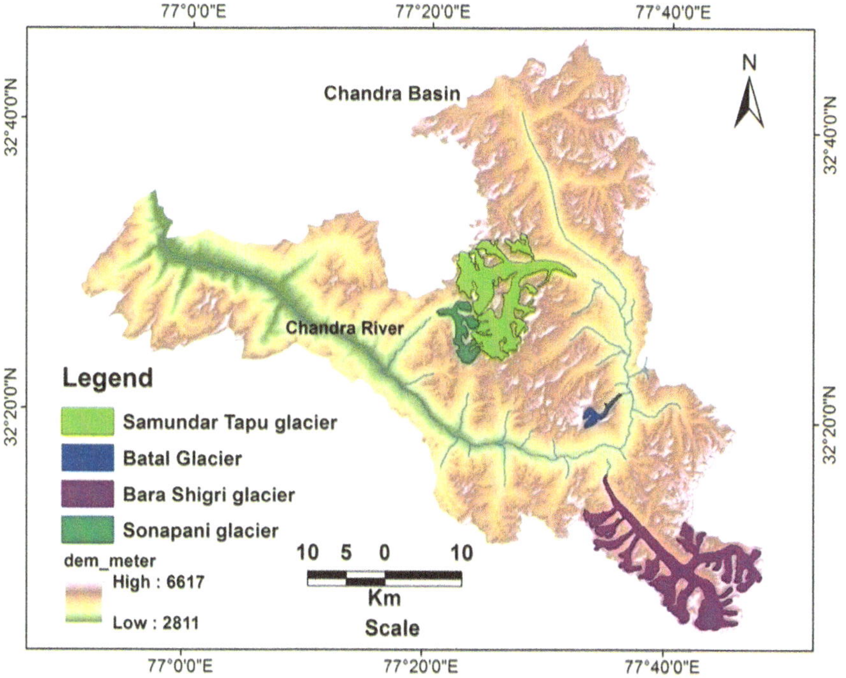

Fig. 4 Location map of benchmark glaciers of Chandra basin

4.4 Glacier Area Loss in Spiti Basin

The Spiti basin forms the upper part of the Satluj basin. However, the data of only 4 glaciers was available for this sub-basin and hence, it has been separately discussed. The studied glaciers include the Parang, Padma, Nisti and Pin glaciers as shown in (Fig. 5). The Parang glacier is 3.85 km long, covering an area of 4.48 km^2. From 1962 to 2013, the glacier has lost an area of 0.48 km^2 (9.6%) at a rate of 0.15 km^2/year. Similarly, from 1962 to 2013, it has shown a retreat of 563 m at a rate of 11.04 m/year. The Padma glacier is small, compound basin valley type glacier about 5.2 km long, covering an area of 7 km^2. From 1962 to 2013, the glacier has lost an area of 0.11 km^2 (1.9%) at a rate of 0.002 km^2/year. From 1962 to 2013, it has shown a total retreat of 355 m at a rate of 6.95 m/year. The Nisti glacier is simple basin type valley glacier about 4.4 km long, covering an area of 5.92 km^2. From 1965 to 2013, the glacier has lost an area of 0.18 km^2 (4.1%) at a rate of 0.003 km^2/year. Between 1965 and 2014, Nisti glacier has retreated by 712 m with an average annual rate of 14.53 m/year. The Pin glacier is simple basin valley type glacier about 7.15 km long and it covers an area of 10.25 km^2. From 1965 to 2013, the glacier has lost an area of 0.15 km^2 (1.34%) at a rate of 0.003 km^2/year. During 1965 to 2014, Pin glacier has shown a total retreat of 621 m at a rate of 12.67 m/year. Overall, based on these representative glaciers an average area loss of 0.23 km^2 (4.3%) at a

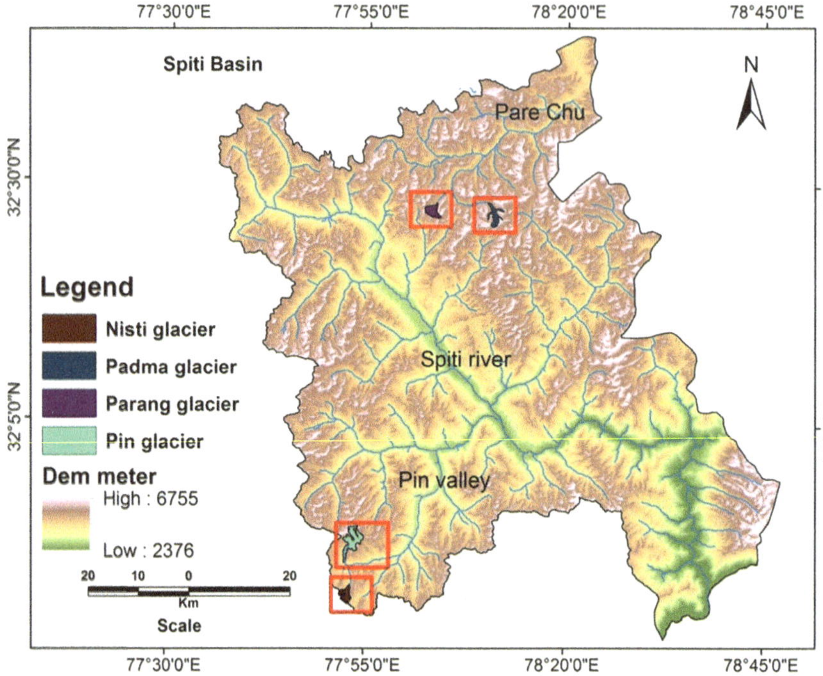

Fig. 5 Location map of benchmark glaciers in Spiti basin

rate of 0.04 km^2/year during last 6 decades from 1962/1965 to 2013/2014 is observed. In the Spiti basin, the loss in area varies from 1.3% (Pin Glacier) to 9.6% (Parang Glacier). The glacier retreat was also found variable and heterogeneous.

4.5 Glacier Area Loss in Satluj Basin

For the representation of the Satluj basin, 2 highly glaciated sub-basins (i.e., Baspa and Tirungkhad basins) located towards its south-eastern corner were selected as benchmark for the study. For the Baspa and Tirungkhad basins, an inventory of 109 and 32 glaciers was created to study and represent the Satluj basin. The location map of the sub-basins and its studied glaciers is shown in Fig. 6.

(a) **Tirungkhad basin glaciers:** The Tirungkhad basin lies to the south-east of the Satluj basin. Lambar glacier is one of the main glaciers of the Tirungkhad basin with an area of 15.3 km^2 and length of 6.6 km. An inventory of 32 glaciers was generated for this basin (Fig. 7). The glaciers covered an area varying from 0.4 to 15.3 km^2 (Lambar glacier). In this basin also, the glaciers have lost the area significantly from 1966 to 2011. The total glacial area calculated for 1966 was

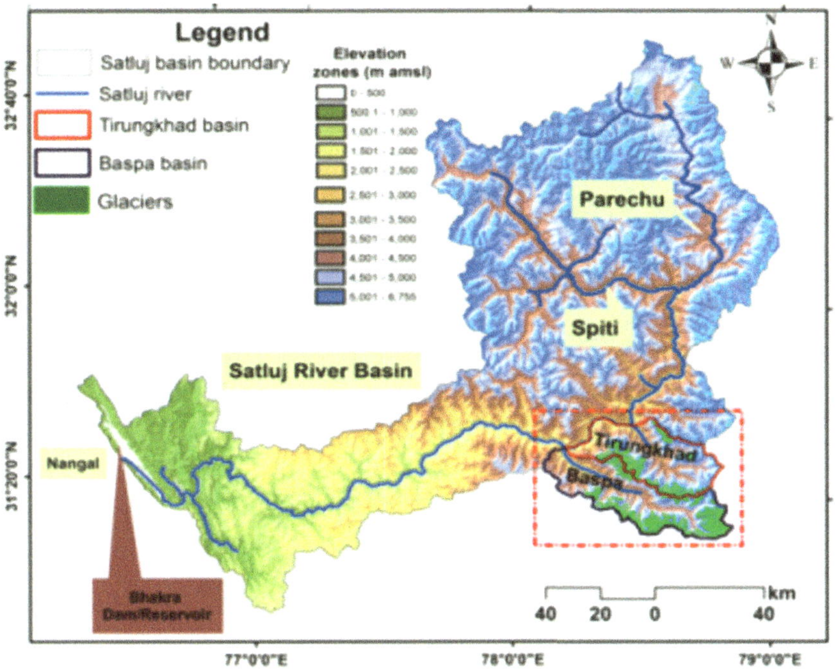

Fig. 6 Location map of Tirungkhad and Baspa basins with glaciers as benchmark for the Satluj basin

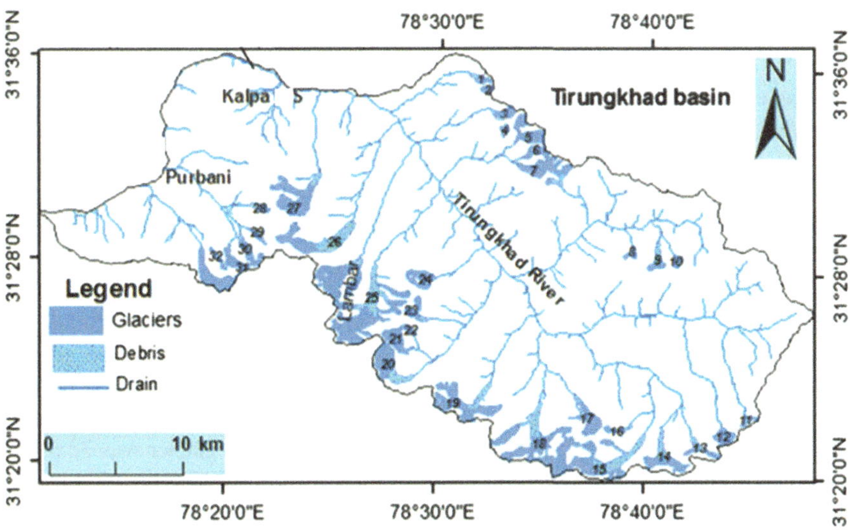

Fig. 7 Location map of glaciers inventoried (2011) in Tirungkhad basin as benchmark for Satluj basin

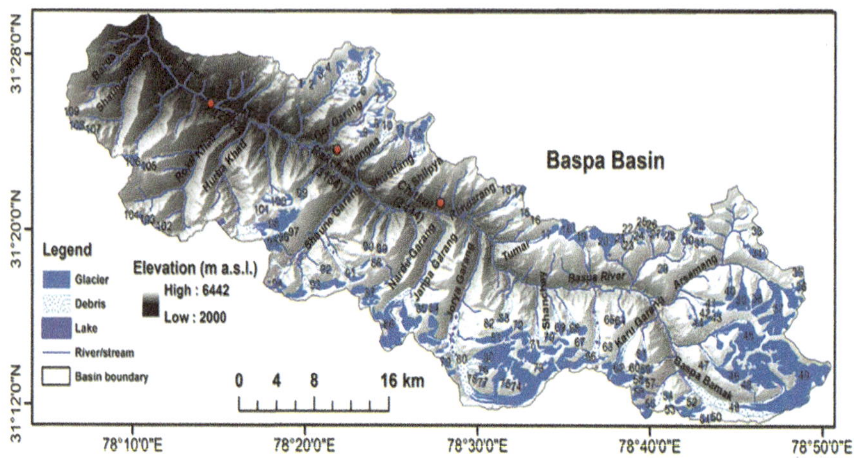

Fig. 8 Location map of glaciers inventoried (2011) in the Baspa basin as benchmark for Satluj basin. The elevation in the background is represented by the ASTER DEM

112 km^2, 91 km^2 in 2000, 87 km^2 in 2006 and 82 km^2 in 2011. The glaciers have lost an area from 112 km^2 in 1966 to 82 km^2 in 2011, i.e., a reduction of 29.1 km^2 or 26.1%. An analysis of length revealed that the length of glaciers retreated about 31.9 km i.e., 26% from 1966 to 2011. The glacier length changed by 17.6 km from 1966 to 2000, 6.6 km from 2000 to 2006 and 7.6 km from 2006 to 2011. Overall, the rate of retreat of glacier length was observed to be 0.71 km (0.70% a^{-1}) from 1966 to 2011. Among the glaciers from 1966 to 2011, the changes in the length varied from 0.10 to 1.46 km.

(b) **Baspa basin glaciers:** In the Baspa basin, the main glaciers present are Baspa Bamak, Shaune Garang, Jorya Garang and Karu glaciers. An inventory of 109 glaciers covering an area of 187.0 km^2 was produced for this basin (Fig. 8). The total area of these glaciers cover about 17% of entire basin area. The Baspa Bamak, Gour Joriya are large glaciers in this basin having an area of 32.3 0.6 and 26 km^2 each. The glacier length ranged from 0.30 0.01 km (G-1) to 17.8 0.8 km (G-49). Glaciers having a length of 2 km are prevalent in the basin. Analysis of glaciers total area during different time periods indicated a continuous and significant loss in glacier area as well as in length. During 1976 the glaciers (no. 109) covered an area of 227.4 km^2 with the largest glacier (Baspa Bamak, G-49) covering an area of 35 km^2.During 2011 the glaciers (no. 97) covered an area of 186.2 km^2 in which the maximum size of largest glacier was 32.3 km^2 (G-49). The glacier area has shrunken from 227.4 km^2 (1976) to 207.8 km^2 in 1992, 198.1 in 2000, 192.1 km^2 in 2006 and to 186.2 km^2 by 2011. Overall, an aerial loss of 41.2 km^2 (18.1%) at a rate of decrease of 1.18 km^2/year (0.52%) was observed during last 3 decades (36 years) from 1976 to 2011. The retreat analysis of 33 large compound valley type glaciers with well defined long and clearly observable tongues indicated that the length change varied from 0.87 km at the rate of 17.2 m/year to 0.60 km at the rate of 24.8 m/year. Overall, from 1976 to 2011, the mean length/snout retreat was found to be 615 m at a rate of 19 m/year in this basin.

4.6 *Glacier Area Loss in Upper Indus Basin (LMR)*

To represent the upper Indus basin, 5 small sub-catchments covering parts of the LMR and draining into the Indus River as well as its major tributary Shyok River were selected as benchmark. In this area, an inventory of 90 glaciers having an area larger than 0.008 km^2 were mapped (Fig. 9). The glaciers covered a total area of 21.1 km^2 which is about 2.6% of the total area of study area. The size of the glaciers varies from a maximum area of 0.91 km^2 (G. No. 41) and a minimum area of 0.008 km^2 (G. No. 9). In this study only 47 glaciers having an area greater than

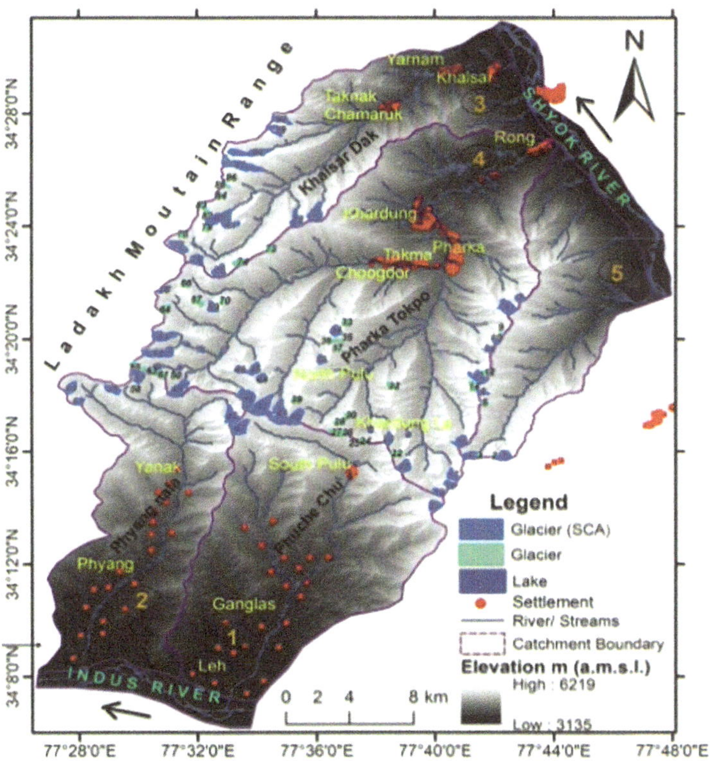

Fig. 9 Location map of glaciers inventoried (2017) in sub-basins 1 and 2 draining the southwestern slopes of the Ladakh Mountain range and the sub-basins 3, 4 and 5 draining the northeastern slope of the LMR covering the parts of Upper Indus Basin

0.12 km^2 have been analyzed to understand the glacier changes. From the change detection studies, it has been observed that during 1962, the total area of glaciers was 24.1 km^2. However, during a recent year of 2000, the glacier area has decreased to 21.7 km^2. Similarly, the glacier area has further reduced to 19.9 and 19 km^2 during 2011 and 2017 respectively. A reduction in glacier area of about 21.2% (5.1 km^2) at a rate of 93 m^2/year is observed from 1962 to 2017.

5 Discussion

5.1 Heterogeneity in Glacier Area Loss in the Selected Basins

During recent years, a number of studies have been carried out in western Himalaya suggesting a very heterogeneous and variable glacier area changes (Berthier et al. 2007; Wagnon et al. 2007; Brahmbhatt et al. 2012; Nathawat et al. 2008; Dobhal et al. 2008; Kulkarni et al. 2010; Pithan 2011; Bolch et al. 2012; Mir et al. 2014a, b, 2017, 2018; Murtaza and Romshoo 2017; Jain and Mir 2019; Mir 2021). Thus, in this study, an attempt to decode, bracket and understand this heterogeneous nature and behavior of glacier area loss in the selected basins of western Himalaya has been carried out based on the major representative and benchmark glaciers as discussed above. For this purpose, ancillary data (SoI maps), Landsat data series (MSS, TM, ETM, ETM+, OLI), aerial photo imageries (Corona), Google Earth images, ASTER DEM in conjunction with a limited field check at certain selected glaciers have been used (Fig. 10). The satellite data sources used have variable resolutions and have been used as per the availability and suitability. The glacier area loss and recession of few glaciers of the selected basins is shown in Fig. 11. The glacier area loss for each selected glacier representing its corresponding basin has been averaged to understand average glacier area loss in the region. After averaging the average glacier loss of selected basins, it is observed that the western Himalayan region in general has lost an average area of 15% during the last 3/5–6 decades. However, the glacier area loss has been very heterogeneous and variable in nature in its different sub-basins and for different glaciers (Fig. 12). For instance, observations here revealed that the average glacier area loss varies from 4.24% (Spiti basin) to 26.7% (Tirungkhad basin). The Spiti basin is the upper part of the Satluj basin whereas; the Tirungkhad basin is the south-eastern sub-basin of the Satluj basin located in the western Himalayan region. Similarly, the Suru and Zanskar basins revealed lower percentage of area change (5.69%) followed by the Chandra basin that revealed a loss of 10.65%. For the Satluj basin an average area loss of 22.1% is observed based on its average glacier area loss in Tirungkhad and Baspa sub-basins respectively. The lower percentage of area loss in case of Spiti, Suru and Zanskar basins and Chandra basin may be attributed to the selection of only few major glaciers wherein the chances of snow contribution through snow avalanches and high-altitude range of accumulation zone are more. Thus, the higher contribution of snow overall leads to a reduced reflection and response of its dynamic changes to the climatic variations in the region.

Fig. 10 Field photograph showing (**a, b**) snout and terminus of Bodpathri glacier observed in field during 2018, (**c**) lateral moraine of Bodpathri glacier during 2018, (**d**) Nehnar glacier snout and terminus in Jhelum basin during 2019, (**e**), lateral moraine and lake near Samundar Tapu glacier in Chandra basin during 2013, (**f**) snout and terminus of Padma glacier in Spiti basin during 2013, (**g**) snout/terminus and proglacial lake of Darung Drung glacier in Zanskar basin during 2014 and (**h**) snout of Shafat glacier Suru basin during 2014, of western Himalaya

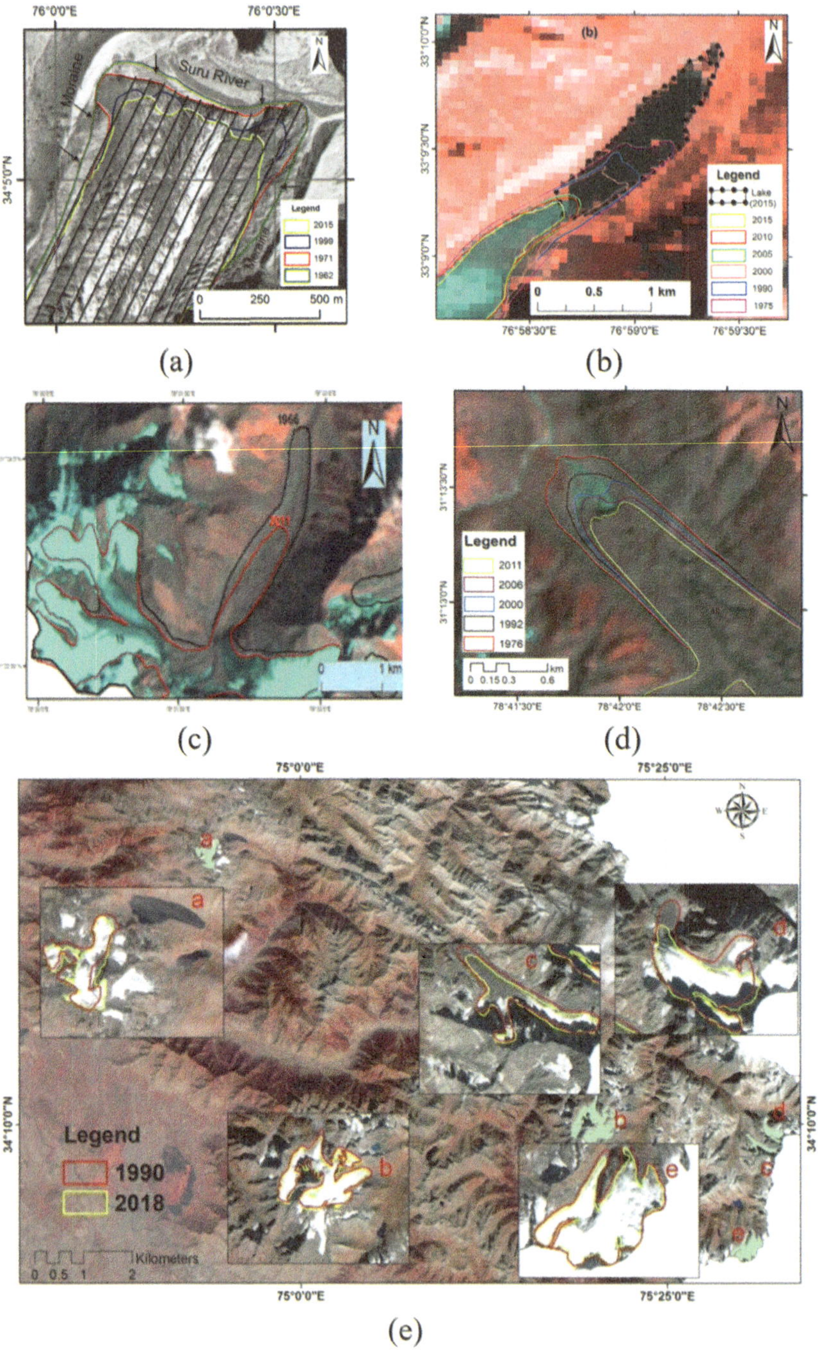

Fig. 11 Glacier recession and area loss of (**a**) Parkachik glacier in Suru basin (back ground corona image, 1971) (**b**) Padam glacier in Zanskar basin (background image TM 2015), (**c**) G-19 in Tirungkhad basin (background TM 2011), (**d**) Baspa Bamak glacier in Baspa basin (background

Contrary to this, the Jhelum basin revealed a higher loss of 19.41% based on the major selected benchmark glaciers. However, in case of Baspa, Tirungkhad and upper Indus basin (parts), the higher glacier area loss of 18.10%, 26.7% and 21.2% is based on glacier inventories, wherein the glaciers of all sizes, above a thresh hold of 0.008 km^2 in area have been considered for change detection analysis. It is a general reporting in the Himalayan region that the small glaciers lose more area than the large glaciers (Mir et al. 2017; Mir 2021). Therefore, the higher area loss in these basins is attributed particularly to the higher area loss of small glaciers in the region. In addition to it, the presence of supraglacial and proglacial lakes is also highly affecting the glacier changes and its dynamics (Jain and Mir 2019; Ahmed et al. 2021a, b; Ahmed et al. 2022). For examples, the Dalung glacier in Suru and Padam glacier in Zanskar basin revealed high area loss of 19.2% and 12.7% in comparison to other glaciers located within the same basin which on the other hand indicated a very lower area changes (Table 2). Moreover, the resolution of data images and maps, local climatic conditions, related orographic, topographic and geomorphic factors, nature and thickness of debris cover, contribution through snow avalanches and tributary glaciers and other related factors etc. may also be controlling the heterogeneous nature of glacier area loss and its dynamics in this region. The detailed results are given in (Table 3).

5.2 *Comparison of Glacier Area Loss with Previous Studies*

A variable and heterogeneous pattern of glacier area loss has been observed in the selected basins of the western Himalayan region in this study. In order to further understand the dynamics and behavior of glacier area loss in these selected basins, the results have been compared for each catchment/or study basin with few recent studies (published 2012 onward). The analysis revealed that the observed heterogeneous area loss is almost in line with the previous/and recent studies carried out in this area. For instance, in the Jhelum basin, the glacier area loss of 19.4% is relatively less than the glacier area loss reported by recent studies (e.g., Romshoo et al. 2015, 2020; Murtaza and Romshoo 2017; Shafiq et al. 2020). But the results are in line and similar with two recent studies wherein a glacier area loss of 17.92% and 22.9% has been reported. Recently, another study revealed that the total glacier area in the Jhelum basin has reduced by 20% from 85.25 km^2 in 1990 to 68.17 km^2 in 2020 at the rate of 0.56 km^2/year which is similar to present study results (Ahmad et al. 2022). Similarly, the Suru and Zanskar basin glacier area loss of 5.69% is relatively less than a glacier area loss of 13.85%, 12.91% and 19.85% as reported by previous studies (Rashid and Majeed 2018; Taloor et al. 2019; Rai et al. 2013). But

Fig. 11 (continued) image TM 2011), (**e**) Harmukh, Kolahoi, Nehnar, Bodpathri and Shishram glaciers in Jhelum basin, of western Himalaya

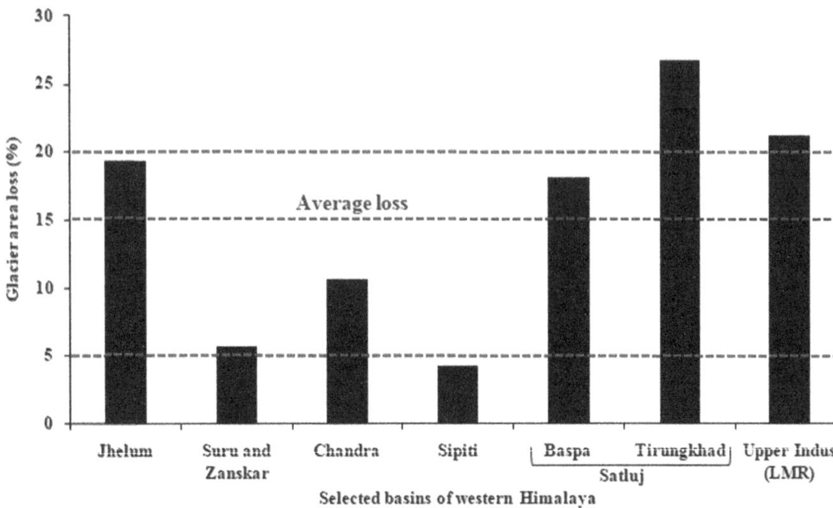

Fig. 12 Average and heterogeneous glacier area loss in the selected basins of western Himalaya

Table 3 Average glacier area loss in the selected basins of western Himalaya during last 3/5–6 decades

S. no	Selected basins	Area change (km^2)	Average area change (%)	Rate of change (km^2/year)
1	Jhelum	0.85	19.41	0.03
2	Suru and Zanskar	1.35	5.69	0.03
3	Chandra	3.97	10.65	0.23
4	Spiti	0.23	4.24	0.04
5	Baspa	41.20	18.10	1.18
	Tirungkhad	30.00	26.70	0.66
6	Upper Indus (LMR)	5.10	21.20	0.09

the area loss is in line with the area loss of 6% reported by Shukla et al. (2020)—a recent study. However, in this basin, Ghosh et al. (2014), have reported only 2% of glacier area loss which is relatively low than the present study also.

For Chandra basin, a glacier area loss of 10.65% as reported in this study is almost similar to the area loss of 12.24% and 8.27% reported for the Sutridhaka and Chotta Shigri glaciers (Sharma et al. 2016) which may be attributed to the fact that only few major glaciers have been studied in this study also. However, Pandey and Venkataraman (2013) have reported a little glacier area loss of 2.5% only in the Chandra basin. But, Garg et al. (2017) have also reported a variable glacier area loss for different glaciers such as 20.3% for Sakchum glacier and 0.72% for Bara Shigri glacier in this area. Sahu and Gupta (2020) have also reported a loss of 4.89% in the Chandra basin which is relatively lower than the present reported area loss. For the Spiti basin, a glacier area loss of 4.24% as found in this study is almost similar to the glacier area loss of 4% and 5.94% reported for the nearby Miyar basin and Parvati glacier in this area (Patel et al. 2018; Chand et al. 2020). For the Satluj basin, an average loss of 22.4% is almost similar to the glacier area loss of 23.8% and 18.09% reported for Baspa basin—major and highly glaciated sub-basin of Satluj basin (Singh et al. 2021; Gaddam et al. 2016). However, the area loss as reported in the present study is higher than a loss of 5.6% reported for Tirungkhad basin by a recent study mainly based on aerial photographs as base maps (Mandal and Sharma 2020). For the parts of Upper Indus basin, a glacier area loss of 21.2% is almost similar to a loss of 16.7% to the glacier area loss reported for Drass valley falling in the same region. But, comparatively the Machoi and Gaya glaciers have indicated a higher glacier area loss than the present results (Majeed et al. 2021; Rashid et al. 2021). However, Chudley et al. (2017) and Ali et al. (2017) have reported relatively little loss of 12.5% in this area as compared to the present study.

Overall, from this analysis, it is observed that the glacier area loss of the present study falls almost within the range of variable and heterogeneous glacier area loss reported for different basins/or sub-basins and different glaciers studied either individually or in groups or inventoried before change detection studies by different workers. But, interestingly, it is important to note that the glacier number, glacier size, type and resolution of data sources etc. affects considerably the output results vis a vis percentage of changes in glaciers and its number, size in addition to morphological, topographic and other microclimatic factors. For example, based on high resolution data sources Mandal and Sharma (2020) have reported considerably low area loss in Tirungkhad basin glaciers. Similar results of lower percentage of area loss have been obtained in case of Parkachik glacier in this study also based on the high resolution data sources such as Corona data sources. Additionally, higher percentage of area loss in case of few glaciers in this study may also be attributed to the use of SoI maps for which a higher inaccuracy has been reported by previous studies (Bhambri et al. 2011; Mir et al. 2017). Furthermore, it is notable that for Gaya glacier and Nehnar glacier an extremely higher area loss of around 45% and 50% during last 5–6 decades has been reported. This higher area loss is about three times higher than the average glacier loss of 15% (range: 5–20%) observed for the western Himalayan glaciers in this study. These observations reflect that certain glaciers in the region may be losing area out of range as found in this study. Thus, to further decode this out of range behavior of certain glaciers, further detailed studies utilizing high resolution data sources and a dense network of datasets of glaciers covering all the basins may be taken up in future in the region. The detailed results are given in Table 4.

5.3 *Observed Climate Change and Its Comparison with Previous Studies in the Area*

Previous studies have revealed that the changes in glacier resources in general is caused due to the warming pattern of climate and changing nature of precipitation (Shekhar et al. 2010; Mir et al. 2017; Jain and Mir 2019). Therefore, to understand

Table 4 Comparison of average glacier area loss in the selected basins of western Himalaya during last 3/5–6 decades with the previous studies in the area

Basin	Study area	Study period	Area change (km^2)	Change (%)	Rate of change (km^2/year)	Source
Jhelum basin	**6 glaciers**	**1990–2020**	**0.85**	**19.41**	**0.03**	**Present study**
	Kashmir basin	1980–2018	29.3	28.82	0.77	Romshoo et al. (2020)
	9 glaciers	1980–2013	5.20	17.92	0.15	Murtaza and Romshoo (2017)
	Nehnar glacier	1962–2017	1.4	50.35	0.02	Shafiq et al. (2020)
	Lidder valley	1962–2013	11.49	28.89	0.22	Marazi and Romshoo (2018)
	Kolahoi glacier	1962–2014	3.18	22.99	0.06	Shukla et al. (2017)
	Lidder valley	1962–2013	12.66	27.46	0.24	Romshoo et al. (2015)
Suru and Zanskar basins	**7 glaciers**	**1962/ 1971–2015**	**1.35**	**5.69**	**0.03**	**Present study**
	Zanskar valley	1962–2001	15	2	0.38	Ghosh et al. (2014)
	Suru-sub basin	1971–2017	32	6	0.69	Shukla et al. (2020)
	Drang Drung glacier	1971–2017	11.03	13.85	0.23	Rashid and Majeed (2018)
	Drang Drung glacier	1976–2017	9.09	12.91	0.22	Taloor et al. (2019)
	Doda valley	1962–2001	72.13	19.85	1.84	Rai et al. (2013)
Chandra basin	**4 glaciers**	**1962/ 1971–2016**	**3.97**	**10.65**	**0.23**	**Present study**
	Chandra and Bhaga basin	1980–2010	9.4	2.5	0.31	Pandey and Venkataraman (2013)
	Chandra basin	1971–2016	31.3	4.89	0.69	Sahu and Gupta (2020)
	Sutridhaka glacier	1962–2013	3.09	12.24	0.06	Sharma et al. (2016)
	Batal glacier	1962–2013	0.29	5.89	0.005	Sharma et al. (2016)
	Sakchum glacier	1993–2014	3.17	20.30	0.15	Garg et al. (2017)
	Chotta Shigri glacier	1993–2014	1.26	8.27	0.06	Garg et al. (2017)

(continued)

Table 4 (continued)

Basin	Study area	Study period	Area change (km^2)	Change (%)	Rate of change (km^2/year)	Source
	Bara Shigri glacier	1993–2014	0.92	0.72	0.04	Garg et al. (2017)
Sipiti basin	**4 glaciers**	**1962–2013/2014**	**0.23**	**4.24**	**0.04**	**Present study**
	Miyar basin	1989–2014	9	4	0.36	Patel et al. (2018)
	Paravati glacier	1965–2018	2.2	5.94	0.04	Chand et al. (2020)
Satluj basin	**Tirungkhad and Baspa basin glaciers**	**1966/1976–2011**	**36.6**	**22.4**	**0.92**	**Present study**
	Baspa basin	2000–2018	38.52	18.09	2.14	Singh et al. (2021)
	Tirungkhad Basin	1965–2018	1.95	5.58	0.03	Mandal and Sharma (2020)
	Baspa basin	1962–2014	41.3	23.87	0.79	Gaddam et al. (2016)
Upper Indus basin (Ladakh mountain range)	**90 glaciers**	**1962–2017**	**5.10**	**21.20**	**0.09**	**Present study**
	Ladakh mountain range	1991–2014	45.3	12.8	1.81	Chudley et al. (2017)
	Upper Indus basin	1996–2014	11.9	12	0.66	Ali et al. (2017)
	Machoi glacier	1972–2019	1.88	29	0.04	Rashid et al. (2021)
	Drass valley	1962–2013	31.25	16.63	0.65	Koul et al. (2016)
	Gaya glacier	1969–2019	0.42	45.6	0.008	Majeed et al. (2021)

the status of temperature and precipitation in this region, the available climate data of 3 stations was analyzed. The meteorological data of Pahalgam station from 1990 to 2018 revealed that T_{min}, T_{max} and T_{avg} are increasing in the area. More interestingly, the T_{min} and T_{avg} are increasing very significantly. The T_{min} is increasing at a rate of 0.013 °C/year whereas; T_{avg} is increasing at a rate of 0.004 °C/year. The T_{max} is also increasing but, the trend is insignificant statistically. The precipitation has shown a very insignificant increasing trend in this basin. The results are given in the Table 5. The linear trends of temperature and precipitation are shown in Fig. 13a–d. The meteorological data of Manali station revealed that T_{min} is increasing very

Table 5 Statistical parameters of temperature and precipitation of Pahalgam (Jhelum basin), Manali (Near Chandra basin) and Rakcham (Satluj basin) stations located in western Himalaya. (Notes: Significance of trends at 95% confidence level is represented by Mann-Kendall (Zs) statistics, the slope of trend is given by Sens slope (Qi) and probability of the trend is represented by the p-value. The significant trends are given by bold font values)

Meteorological station	Climate variable		Zs	Qi	P
Pahalgam	Temperature	Tmin	**3.882**	**0.013**	**0.001**
		Tmax	0.937	0.002	0.348
		Tavg	**2.044**	**0.004**	**0.40**
	Precipitation		0.131	0.001	0.895
Manali	Temperature	Tmin	**2.207**	**0.056**	**0.027**
		Tmax	**−2.692**	**−0.227**	**0.007**
		Tavg	**−3.046**	−0.083	0.002
	Precipitation		**−2.48**	**−52.50**	**0.01**
Rakcham	Temperature	Tmin	**2.086**	**0.071**	**0.036**
		Tmax	**2.825**	**0.076**	**0.004**
		Tavg	**2.729**	**0.07**	**0.026**
	Precipitation		**−2.88**	**−6.28**	**0.004**

significantly a rate of 0.056 °C/year from 1986 to 2011. However, T_{max} and T_{avg} revealed a declining trend during the same time period. The declining trends in T_{max} are also significant statistically. The T_{max} has also a dominant control on the declining trend of T_{avg} in the area. Interestingly, the precipitation has shown a very significant declining trend in this area. The results are given in the Table 4. The linear trends of temperature and precipitation are shown in Fig. 13e–h.

Similarly, the estimation of trends indicated that the temperature for Rakcham station has increased from 1985 to 2008 significantly. The trend analysis indicated an increasing pattern in temperature especially in T_{min}. A remarkable phase of cooling persisting from 1985 to 1997 was observed in T_{avg}, T_{max}, and T_{min}. After the 1997s, the temperature has increased, and a more fluctuated pattern of warming was observed. The T_{avg}, T_{max} and T_{min}, have increased almost at a similar rate of 0.070 °C, 0.071 °C and –0.076 °C/year thereby indicating an insignificant variation in magnitude of changes. Contrary to temperature, the annual precipitation indicated a decreasing pattern for Rakcham station with relatively higher interannual fluctuations. The linear trends of temperature and precipitation are shown in Fig. 13i–l. Overall, the interannual variability in trends of temperature and precipitation may be attributed to regional differences in orography, topography and other related factors in the region. Furthermore, as per the previous studies, the conversion of solid to liquid precipitation has been reported in the region and that may be the probable reason that the Rackham station is showing decreasing trends in precipitation whereas, increasing trends are observed at Pahalgam station. The Manali station, however, indicated a significant declining trend in precipitation. Thus, it is inferred that local orographic and other topographic factors may also be controlling this heterogeneous behaviour of precipitation pattern in the region. Therefore, a detailed study in this direction is recommended by utilizing large and dense data sets.

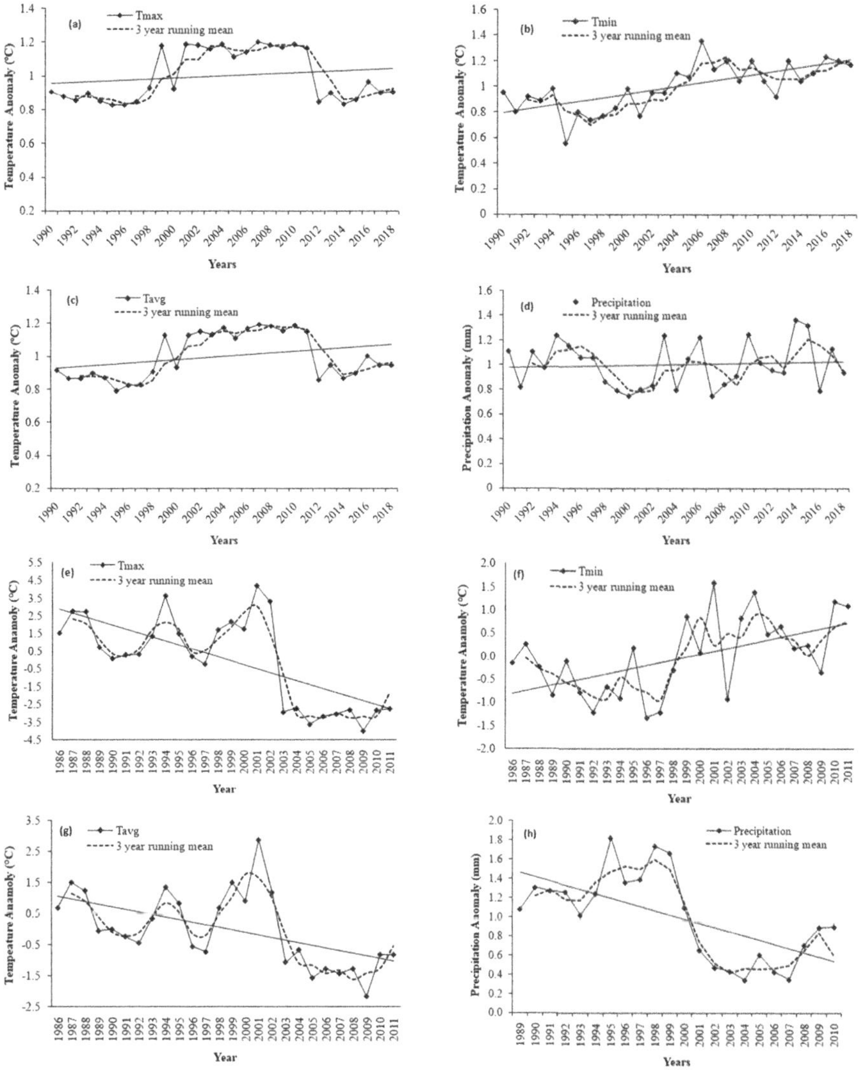

Fig. 13 Trends and three-year moving average of annual temperature (**a**) T_{max}, (**b**) T_{min}, (**c**) T_{avg} and (**d**) annual precipitation anomalies of Pahalgam station located in Jhelum basin, (**e**) T_{max}, (**f**) T_{min}, (**g**) T_{avg} and (**h**) annual precipitation anomalies of Manali station located near Chandra basin and (**i**) T_{max}, (**j**) T_{min}, (**k**) T_{avg}, and (**l**) annual precipitation anomalies of Rakcham station located in Satluj basin, of western Himalaya

Overall, the present results are in line with a number of previous studies carried out in the region (e.g., Bhutiyani et al. 2010; Dar et al. 2014; Dad et al. 2021; Mir et al. 2015a, b, c; Shafiq et al. 2018, 2019; Romshoo et al. 2020). For instance, in the upper Indus river basins around Ladakh region, a recent study has demonstrate an increase in annual temperatures of about 0.162 °C (0.0013 °C/year) from 1901 to

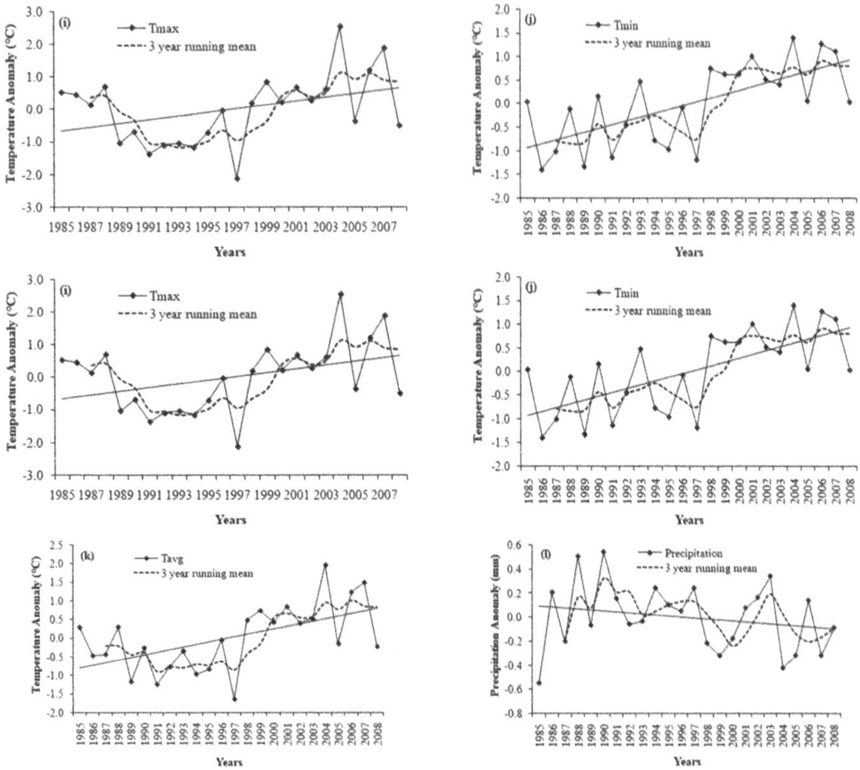

Fig. 13 (continued)

2017 along with a decline in annual precipitation slightly by 0.032 mm/year (Lone et al. 2019). Similarly, an analysis of meteorological data from Leh station clearly indicates a rise of 1 °C for winter and 0.5 °C for summer months during last 35 years with a decline in precipitation from November to March (Angmo and Mishra 2009). Chevuturi et al. (2018) reported a warming trend with reduced precipitation in/ over Leh (Ladakh) areas during last decades. Over last 28 years, a small increase in annual mean temperature of 0.426 °C/decade prior to 1995 with a significant increase at a rate of 0.375 °C decade at Drass station has been reported (Koul et al. 2016). Similarly, in the Zanskar and Suru basins, an increase in the mean annual temperature (T_{max} and T_{min}) of 0.77 °C (0.25 and 1.3 °C) has been reported from 1901 to 2017 (Shukla et al. 2020). Reports have suggested that the minimum as well as maximum temperatures are increasing significantly while the precipitation is also increasing insignificantly from 1961 to 2017 (Rashid and Majeed 2018). Another recent study suggested that the temperature as well as the liquid precipitation have increased from 1900 to 2020 in the Zanskar basin (Taloor et al. 2021). In the Jhelum basin, a significant increasing trend in annual mean maximum temperature at a rate of 0.03 °C/year with a consistent decrease in the annual precipitation at an alarming rate of about 7.9 mm/year from 1980 to 2014 has been recorded (Shafiq et al. 2019).

The historical average annual, minimum, and maximum temperatures at Pahalgam station have increased by 1.61 °C and 0.95 °C with an insignificant decrease in the total annual precipitation from 1980 to 2018 (Murtaza and Romshoo 2017; Romshoo et al. 2020).

In the Chandra basin, a study has reported that the summer mean temperature has increased significantly at a rate of 0.02 °C/year whileas, the winter precipitation has showed decreasing trend at a rate of 4.4 mm/year from 1961/1951 to 2015 (Sahu and Gupta 2020). Kaushik et al. (2020) also reported an increase of 0.021 °C/year in annual temperature and reduction of 2.74 mm/year in annual precipitation from 1979 to 2017. In the Satluj basin, Mir et al. (2017) have reported a rising trend at a rate of 0.076 and 0.071 °C/year of the mean annual T_{min} and T_{max} in Baspa basin western Himalaya from 1976/1985 to 2008. Mir et al. (2015a, b, c) have also reported significant positive trends in temperature particularly T_{min} from 1985 to 2008 in the Satluj basin. The decreasing trends in snowfall were explained as a result of increasing trends in temperature particularly T_{min} in this basin. In another study, from 1984 to 2008, the mean annual temperature (T_{max} and T_{min}) has been reported to have increased significantly, accompanied with a fall in snow water equivalent (SWE) and rainfall in this area (Mir et al. 2014). Similarly, Mandal et al. (2016) have reported a significant increase in average temperature for the entire country and huge variability in precipitation controlling the glacier health in the Spiti basin, western Himalaya. Another study revealed that the mean annual temperature from 1995 to 2015 combined with no significant trend of precipitation from 1948 to 2015 has probably resulted in accelerated mass loss of the glaciers in Spiti region (Tawde et al. 2017).

Thus, from these results, it can be concluded that the increase in temperature is the main fundamental factor controlling the glacier health in the region. The precipitation variability also pays a significant role in the maintaining the glacier health, but, the nature of precipitation is highly erratic and variable as found in this study.

5.4 *Limitations of the Study*

In this study, an attempt to bracket and decode the heterogeneous and variable glacier area loss in the western Himalayan region has been focused using mainly the satellite data sources with a limited field check. But, there are certain limitations associated with the study and the approach adopted. For instances,

1. The remotely sensed images were used successfully to observe the changes in area for monitoring purposes. But, relying on coarse (80 m) and medium (30 m) resolution images of many Landsat datasets (such as MSS, TM, and ETM) may be inadequate. Although, this limitation has been tried to overcome using the high resolution Corona images, but, this dataset was also not available for all the glaciers studied in this area. Similarly, the area changes estimated using the old SoI maps (1960s) may also be providing a major limitation as previous studies

have reported a higher inaccuracy associated with this only old available dataset (Bhambri et al. 2011; Mir and Majeed 2018). Therefore, as per availability, the MSS or Corona images has been used as a base map rather than SoI map to make sure higher accuracy of glacier area estimation for the considered base years.
2. In this study, only a few glaciers (4 to or 7 glaciers) as per the availability of the suitable dataset have been selected as benchmark and representative to understand glacier area changes for large basins such as Zanskar, Spiti etc. This approach may be giving unfiltered/or incomplete information. Therefore, all the glaciers (glacier inventory) present in any basin shall be taken up to overcome this limitation to attain a filtered signal and a complete understanding of glacier area loss and changes.
3. The availability of climate data at a variable temporal scale from only three stations located at lower elevation than the glaciers in the study area may also be providing inadequacy to infer about the climate change in the area. So, to overcome limitation, a large and dense climate dataset shall be analyzed to gain a concrete evidence of the climate change in the area.

Nevertheless, the results obtained in this study have been compared with a number of previous studies that were found to be in tune with them and hence, the results provided in this study can be considered very significant and useful for carrying out any future studies and developing any monitoring and mitigation strategies in the region.

6 Conclusion

This study based on few benchmark and representative glaciers located in few major selected basins of western Himalayan region found heterogeneity in glacier area loss in response to recent climate change during last few decades in the area. The study suggested that the heterogeneous and variable glacier area loss in the western Himalayan region is most dominantly controlled by significant temperature rising and warming. The T_{min} is considered the main, active and predominant factor controlling the glacier area loss and recession. It is because the T_{min} is showing a highly significant increasing trend at all the stations under study. However, the precipitation is showing an erratic changing pattern. During the last 3/5–6 decades, the selected basins of western Himalaya indicated a variable and heterogeneous glacier area loss and pattern. For example, the Jhelum basin indicated a higher average glacier area loss of 19.41%, followed by the Satluj basin that indicated a loss of 22.1% and upper Indus basin indicating a loss of 21.2% respectively. The Chandra basin also indicated a relatively higher area loss of 10.7%. However, Suru and Zanskar basins and Spiti basin indicated a lower area loss of 5.7% and 4.3% respectively.

Overall, it is observed that during last 5–6 decades, the western Himalayan glaciers are losing area heterogeneously within a range of 5–20% with an average area loss of 15%. During the same time period, climate data analysis indicated significant rising trends in T_{min}, T_{max}, and T_{avg} at Rakcham at a rate of (0.071 °C/year, 0.076 °C/year and 0.07 °C/year) and Pahalgam (0.013 °C/year, 0.002 °C/year, 0.004 °C/year) stations. However, the temperature at Manali station indicated a significant rising trend for T_{min} (0.056) and significant declining trends for T_{max} and T_{avg}. The precipitation at the Rakcham and Manali stations indicated significant declining nature at a rate of (−6.28 mm/year and −52.5 mm/year). Overall, the declining nature of the glacier resources in response to climate change and warming is expected to affect the water resources and other related sectors in future in this region. The glaciers of the selected basins are vital source to the headwaters of Indus River basin and also support the hydro-power generation, irrigation, domestic uses and tourism water supplies in the downstream areas. Thus, a continuous monitoring, assessment and management of these water resources is very imperative for which timely strategies are suggested to develop for the holistic development and management of these glacier resources.

Acknowledgments First, thanks are due to NASA/USGD for making the Landsat datasets freely available under the umbrella USGS web server, So I, BBMB and IMD for providing hydrometeorological data. The authors are grateful to ESS for giving this opportunity to publish this work in their special volume.

References

Ali I, Shukla A, Romshoo SA (2017) Assessing linkages between spatial facies changes and dimensional variations of glaciers in the upper Indus Basin, western Himalaya. Geomorphology 284:115–129

Angmo T, Mishra S (2009) Impacts of climate change in Ladakh and Lahaul & Spiti of the western Himalayan region. Energy and climate change in cold regions of Asia

Ahmed R, Wani GF, Ahmad ST, Sahana M, Singh H, Ahmed P (2021a) A review of glacial lake expansion and associated glacial lake outburst floods in the Himalayan region. Earth Syst Environ 5(3):695–708

Ahmed R, Wani GF, Ahmad ST, Mir RA, Almazroui M, Jain SK, Ahmed P (2021b) Spatiotemporal dynamics of glacial lakes (1990–2018) in the Kashmir Himalayas, India using RemoteSensing and GIS. Discov Water 1(1):1–17

Ahmad ST, Ahmed R, Wani GF, Sharma P, Ahmed P, Mir RA, Alam JB (2022) Assessing the status of glaciers in Upper Jhelum Basin of Kashmir Himalayas using multi-temporal satellite data. Earth Syst Environ 6(2):375–389

Ahmed R, Ahmad ST, Wani GF, Mir RA, Ahmed P, Jain SK (2022) High resolution inventory and hazard assessment of potentially dangerous glacial lakes in upper Jhelum Basin, Kashmir Himalaya, India. Geocarto Int:1–32

Bajracharya SR, Shrestha BR (2011) The status of glaciers in the Hindu Kush-Himalayan region. International Centre for Integrated Mountain Development (ICIMOD)

Bahuguna IM, Rathore BP, Brahmbhatt R, Sharma M, Dhar S, Randhawa SS, Ajai (2014) Are the Himalayan glaciers retreating? Curr Sci 106:1008–1013

Bajracharya SR, Mool PK, Shrestha BR (2008) Global climate change and melting of Himalayan glaciers. Melting glaciers and rising sea levels: impacts and implications, pp 28–46

Benn DI, Bolch T, Hands K, Gulley J, Luckman A, Nicholson LI et al (2012) Response of debris-covered glaciers in the Mount Everest region to recent warming, and implications for outburst flood hazards. Earth Sci Rev 114(1–2):156–174

Berthier E, Arnaud Y, Kumar R, Ahmad S, Wagnon P, Chevallier P (2007) Remote sensing estimates of glacier mass balances in the Himachal Pradesh (Western Himalaya, India). Remote Sens Environ 108(3):327–338

Bhambri R, Bolch T, Chaujar RK, Kulshreshtha SC (2011) Glacier changes in the Garhwal Himalayas, India 1968–2006 based on remote sensing. J Glaciol 57(203):543–556

Bhambri R, Bolch T, Chaujar RK (2012) Frontal recession of Gangotri glacier, Garhwal Himalayas, from 1965 to 2006, measured through high resolution remote sensing data. Curr Sci (00113891) 102(3):489–494

Bhutiyani MR, Kale VS, Pawar NJ (2007) Long-term trends in maximum, minimum and mean annual air temperatures across the northwestern Himalaya during the twentieth century. Clim Chang 85(1):159–177

Bhutiyani MR, Kale VS, Pawar NJ (2010) Climate change and the precipitation variations in the northwestern Himalaya: 1866–2006. Int J Climatol 30(4):535–548

Bolch T, Menounos B, Wheate R (2010) Landsat-based inventory of glaciers in western Canada, 1985–2005. Remote Sens Environ 114(1):127–137

Bolch T, Peters J, Yegorov A, Pradhan B, Buchroithner M, Blagoveshchensky V (2012) Identification of potentially dangerous glacial lakes in the northern Tian Shan. In: Terrigenous mass movements. Springer, Berlin, pp 369–398

Bookhagen B, Burbank DW (2010) Toward a complete Himalayan hydrological budget: spatiotemporal distribution of snowmelt and rainfall and their impact on river discharge. J Geophys Res Earth 115(F3)

Brahmbhatt RM, Bahuguna IM, Rathore BP, Kulkarni AV, Nainwal HC, Shah RD, Ajai (2012) A comparative study of deglaciation in two neighbouring basins (Warwan and Bhut) of Western Himalaya. Curr Sci 103:298–304

Chand P, Sharma MC (2015) Glacier changes in the Ravi basin, North-Western Himalaya (India) during the lastfour decades (1971–2010/13). Glob Planet Chang 135:133–147

Chand P, Jain SK, Thakur HP, Kumar S, Sharma MC (2020) Recessional pattern and surface elevation change of the Parvati Glacier, North-Western Himalaya (1965-2018) using remote sensing. Int J Remote Sens 41(24):9360–9392

Chevuturi A, Dimri AP, Thayyen RJ (2018) Climate change over Leh (Ladakh), India. Theor Appl Climatol 131(1):531–545

Chudley TR, Miles ES, Willis IC (2017) Glacier characteristics and retreat between 1991 and 2014 in the Ladakh range, Jammu and Kashmir. Remote Sens Lett 8(6):518–527

Dad JM, Muslim M, Rashid I, Reshi ZA (2021) Time series analysis of climate variability and trends in Kashmir Himalaya. Ecol Indic 126:107690

Dar RA, Rashid I, Romshoo SA, Marazi A (2014) Sustainability of winter tourism in a changing climate over Kashmir Himalaya. Environ Monit Assess 186(4):2549–2562

Dash SK, Jenamani RK, Kalsi SR, Panda SK (2007) Some evidence of climate change in twentieth-century India. Clim Chang 85(3):299–321

Dimri AP, Dash SK (2012) Wintertime climatic trends in the western Himalayas. Clim Chang 111(3):775–800

Dobhal DP, Gergan JT, Thayyen RJ (2008) Mass balance studies of the Dokriani glacier from to, Garhwal Himalaya, India. Bull Glaciol Res 25:9–17

Dyurgerov MB, Meier MF (2005) Glaciers and the changing earth system: a 2004 snapshot, vol 58. Institute of Arctic and Alpine Research, University of Colorado, Boulder

Gaddam VK, Kulkarni AV, Gupta AK (2016) Estimation of glacial retreat and mass loss in Baspa basin, Western Himalaya. Spatial Inf Res 24(3):257–266

Garg PK, Shukla A, Tiwari RK, Jasrotia AS (2017) Assessing the status of glaciers in part of the Chandra basin, Himachal Himalaya: a multiparametric approach. Geomorphology 284:99–114

Ghosh S, Pandey AC, Nathawat MS, Bahuguna IM (2014) Contrasting signals of glacier changes in Zanskar valley, Jammu & Kashmir, India using remote sensing and GIS. J Indian Soc Remote Sens 42(4):817–827
Govindha Raj KB (2010) Remote sensing-based hazard assessment of glacial lakes: a case study in Zanskar basin, Jammu and Kashmir, India. Geomat Nat Haz Risk 1(4):339–347
Granshaw FD, Fountain AG (2006) Glacier change (1958–1998) in the north cascades national park complex, Washington, USA. J Glaciol 52(177):251–256
Hewitt K (2005) The Karakoram anomaly? Glacier expansion and the 'elevation effect, Karakoram Himalaya. Mount Res Dev 25(4):332–340
Jain SK, Mir RA (2019) Glacier and glacial lake classification for change detection studies using satellite data: a case study from Baspa basin, western Himalaya. Geocarto Int 34(4):391–414
Kääb A, Frauenfelder R, Roer I (2007) On the response of rock glacier creep to surface temperature increase. Glob Planet Chang 56(1–2):172–187
Kääb A, Treichler D, Nuth C, Berthier E (2015) Brief communication: contending estimates of 2003–2008 glacier mass balance over the Pamir–Karakoram–Himalaya. Cryosphere 9(2): 557–564
Kamp U, Byrne M, Bolch T (2011) Glacier fluctuations between 1975 and 2008 in the greater Himalaya range of Zanskar, southern Ladakh. J Mt Sci 8(3):374–389
Kaushik S, Rafiq M, Joshi PK, Singh T (2020) Examining the glacial lake dynamics in a warming climate and GLOF modelling in parts of Chandra basin, Himachal Pradesh, India. Sci Total Environ 714:136455
Klimes L (2003) Life-forms and clonality of vascular plants along an altitudinal gradient in E Ladakh (NW Himalayas). Basic Appl Ecol 4(4):317–328
Koul MN, Ganjoo RK (2010) Impact of inter- and intra-annual variation in weather parameters on mass balance and equilibrium line altitude of Naradu Glacier (Himachal Pradesh), NW Himalaya, India. Climatic Change 99(1):119–139
Koul MN, Bahuguna IM, Rajawat AS, Ali S, Koul S (2016) Glacier area change over past 50 years to stable phase in Drass valley, Ladakh Himalaya (India). Am J Clim Chang 5(01):88
Krishna K, Majeed Z, Pramanik S (2017) Long-term monitoring of mass balance of Hamtah glacier, Lahaul and Spiti district, Himachal Pradesh – on expedition basis. (GL/ NR/HQ /2016/060 GSI, Un-Published Reports)
Krishna K, Kumar P, Mishra R, Majeed Z (2020) Glacier fluctuation in Suru and Zanskar basins using ground-based observations and remote sensing technique. India J Geosci 74
Kulkarni AV, Karyakarte Y (2014) Observed changes in Himalayan glaciers. Curr Sci 106:237–244
Kulkarni AV, Rathore BP, Singh SK (2010) Distribution of seasonal snow cover in central and western Himalaya. Ann Glaciol 51(54):123–128
Kulkarni AV, Rathore BP, Singh SK, Bahuguna IM (2011) Understanding changes in the Himalayan cryosphere using remote sensing techniques. Int J Remote Sens 32(3):601–615
Lone SA, Jeelani G, Deshpande RD, Mukherjee A (2019) Stable isotope (δ18O and δD) dynamics of precipitation in a high altitude Himalayan cold desert and its surroundings in Indus river basin, Ladakh. Atmos Res 221:46–57
Majeed Z, Krishna K, Mukhtar MA, Kumar P (2016) Generation of primary baseline field-data andalso some repeat monitoring data from glaciers in the gap areas of remote tribal districts of Kargil and Leh in Jammu & Kashmir and Lahaul & Spiti district in Himachal Pradesh. Accession No. NRO-22281-GSI, Un-published Reports
Majeed Z, Mukhtar MA, Mir RA, Kumar P, Krishna K (2020) Sonapani glacier recession over a century from 1906–2016, Chandra basin, Himachal Himalaya. J Geol Soc India 95(1):36–44
Majeed U, Rashid I, Sattar A, Allen S, Stoffel M, Nüsser M, Schmidt S (2021) Recession of Gya glacier and the 2014 glacial lake outburst flood in the trans-Himalayan region of Ladakh, India. Sci Total Environ 756:144008
Mamgain VD, Sastry MVA (1975) Palaeontologica Indica. Geol Surv India XLIII 1–10:1–33
Mandal ST, Sharma MC (2020) Spatial changes in glaciers between 1965 and 2018 in Tirungkhad watershed, Upper Sutlej Basin, Himachal Pradesh. Earth Syst Environ 4(2):427–438

Mandal A, Ramanathan AL, Angchuk T, Soheb M, Singh VB (2016) Unsteady state of glaciers (Chhota Shigri and Hamtah) and climate in Lahaul and Spiti region, western Himalayas: a review of recent mass loss. Environ Earth Sci 75(17):1–12
Marazi A, Romshoo SA (2018) Streamflow response to shrinking glaciers under changing climate in the Lidder Valley, Kashmir Himalayas. J Mt Sci 15(6):1241–1253
Mayewski PA, Jeschke PA (1979) Himalayan and trans-Himalayan glacier fluctuations since AD 1812. Arct Alp Res 11(3):267–287
Mehta M, Dobhal DP, Bisht MPS (2011) Change of Tipra glacier in the Garhwal Himalaya, India, between 1962 and 2008. Prog Phys Geogr 35(6):721–738
Mir RA (2016) Geoinformatics for cryospheric studies under changing climate in western Himalaya, Ph. D Thesis, Indian Institute of Technology Roorkee, India
Mir RA (2018) Recent changes of two parts of Kolahoi glacier and its controlling factors in Kashmir basin, western Himalaya. Remote Sens Appl Soc Environ 11:265–281
Mir RA (2021) Remote sensing-based assessment of glacier resources in parts of Ladakh Mountain range, a trans-Himalayan region, book title, water, cryosphere and climate change in the Himalayas; a geospatial approach. Springer, New York
Mir RA, Majeed Z (2018) Frontal recession of Parkachik glacier between 1971-2015, Zanskar Himalaya using remote sensing and field data. Geocarto Int 33(2):163–177
Mir RA, Jain SK, Saraf AK, Goswami A (2013) Glacier volume changes and their climatic causes in Tirungkhad Basin located in Western Himalaya. J Remote Sens GIS 4(2):35–46
Mir RA, Jain SK, Saraf AK, Goswami A (2014a) Glacier changes using satellite data and effect of climate in Tirungkhad basin located in western Himalaya. Geocarto Int 29(3):293–313
Mir RA, Jain SK, Saraf AK, Goswami A (2014b) Detection of changes in glacier mass balance using satellite and meteorological data in Tirungkhad basin located in Western Himalaya. J Indian Soc Remote Sens 42(1):91–105
Mir RA, Jain SK, Saraf AK, Goswami A (2015a) Accuracy assessment and trend analysis of MODIS-derived data on snow-covered areas in the Sutlej basin, Western Himalayas. Int J Remote Sens 36(15):3837–3858
Mir RA, Jain SK, Saraf AK, Goswami A (2015b) Decline in snowfall in response to temperature in Satluj basin, western Himalaya. J Earth Syst Sci 124(2):365–382
Mir RA, Jain SK, Saraf AK (2015c) Analysis of current trends in climatic parameters and its effect on discharge of Satluj River basin, western Himalaya. Nat Hazards 79(1):587–619
Mir RA, Jain SK, Jain SK, Thayyen RJ, Saraf AK (2017) Assessment of recent glacier changes and its controlling factors from 1976 to 2011 in Baspa basin, western Himalaya. Arct Antarct Alp Res 49(4):621–647
Mir RA, Jain SK, Lohani AK, Saraf AK (2018) Glacier recession and glacial lake outburst flood studies in Zanskar basin, western Himalaya. J Hydrol 564:376–396
Mir, R. A., Jeelani, G. H. (2015) Textural characteristics of sediments and weathering in the Jhelum River basin located in Kashmir Valley, western Himalaya. Journal of the Geological Society of India, 86(4), 445–458.
Mukhtar MA, Majeed Z (2020) Glaciological studies carried out by GSI in the UT J and K and UT Ladakh. GSI, Part X Misc. Pub. No. 30
Murtaza KO, Romshoo SA (2017) Recent glacier changes in the Kashmir alpine Himalayas, India. Geocarto Int 32(2):188–205
Nainwal HC, Negi BDS, Chaudhary M, Sajwan KS, Gaurav A (2008) Temporal changes in rate of recession: evidences from Satopanth and Bhagirath Kharak glaciers, Uttarakhand, using total station survey. Curr Sci 86:653–660
Nathawat MS, Pandey AC, Rai PK, Bahuguna IM (2008) Spatio-temporal dynamics of glaciers in Doda valley, Zanskar range, Jammu and Kashmir, India. In: Proceedings of the International Workshop on Snow, Ice, Glacier and Avalanches, vol 256264. IIT Bombay, Mumbai
Pandey P, Venkataraman G (2013) Changes in the glaciers of Chandra–Bhaga basin, Himachal Himalaya, India, between 1980 and 2010 measured using remote sensing. Int J Remote Sens 34(15):5584–5597

Patel LK, Sharma P, Fathima TN, Thamban M (2018) Geospatial observations of topographical control over the glacier retreat, Miyar basin, Western Himalaya, India. Environ Earth Sci 77(5): 1–12
Pithan F (2011) A model study of the energy and mass balance of Chhota Shigri glacier in the Western Himalaya, India. Cryosphere Discuss 5(1):95–129
Rai PK, Nathawat MS, Mohan K (2013) Glacier retreat in Doda Valley, Zanskar Basin, Jammu & Kashmir, India. Univ J Geosci 1:139–149
Raina VK, Srivastava D (2008) Glacier atlas of India. Geological Society of India, Bangalore
Rashid I, Majeed U (2018) Recent recession and potential future lake formation on Drang Drung glacier, Zanskar Himalaya, as assessed with earth observation data and glacier modelling. Environ Earth Sci 77(12):1–13
Rashid I, Majeed U, Najar NA, Bhat IA (2021) Retreat of Machoi Glacier, Kashmir Himalaya between 1972 and 2019 using remote sensing methods and field observations. Sci Total Environ 147376
Romshoo SA, Dar RA, Rashid I, Marazi A, Ali N, Zaz SN (2015) Implications of shrinking cryosphere under changing climate on the streamflows in the Lidder catchment in the Upper Indus Basin, India. Arct Antarct Alp Res 47(4):627–644
Romshoo SA, Fayaz M, Meraj G, Bahuguna IM (2020) Satellite-observed glacier recession in the Kashmir Himalaya, India, from 1980 to 2018. Environ Monit Assess 192(9):1–17
Sahu R, Gupta RD (2020) Glacier mapping and change analysis in Chandra basin, Western Himalaya, India during 1971–2016. Int J Remote Sens 41(18):6914–6945
Scherler D, Bookhagen B, Strecker MR (2011) Spatially variable response of Himalayan glaciers to climate change affected by debris cover. Nat Geosci 4(3):156–159
Shafiq M, Ahmed P, Mir AA, Hassan H (2018) Trend analysis of winter precipitation over Kashmir valley from 1980–2016
Shafiq MU, Rasool R, Ahmed P, Dimri AP (2019) Temperature and precipitation trends in Kashmir Valley, north western Himalayas. Theor Appl Climatol 135(1):293–304
Shafiq MU, Islam ZU, Bhat IA, Ahmed P (2020) Spatio-temporal behaviour of Nehnar Glacier from 1962 to 2017, Jhelum basin, Kashmir Himalayas, India. Phys Geogr 41(6):517–536
Sharma P, Patel LK, Ravindra R, Singh A, Mahalinganathan K, Thamban M (2016) Role of debris cover to control specific ablation of adjoining Batal and Sutri Dhaka glaciers in Chandra Basin (Himachal Pradesh) during peak ablation season. J Earth Syst Sci 125(3):459–473
Shekhar MS, Chand H, Kumar S, Srinivasan K, Ganju A (2010) Climate-change studies in the western Himalaya. Ann Glaciol 51(54):105–112
Shukla A, Qadir J (2016) Differential response of glaciers with varying debris cover extent: evidence from changing glacier parameters. Int J Remote Sens 37(11):2453–2479
Shukla A, Ali I, Hasan N, Romshoo SA (2017) Dimensional changes in the Kolahoi glacier from 1857 to 2014. Environ Monit Assess 189(1):1–18
Shukla A, Garg PK, Srivastava S (2018) Evolution of glacial and high-altitude lakes in the Sikkim, eastern Himalaya over the past four decades (1975–2017). Front Environ Sci 6:81
Shukla A, Garg S, Mehta M, Kumar V, Shukla UK (2020) Temporal inventory of glaciers in the Suru sub-basin, western Himalaya: impacts of regional climate variability. Earth Syst Sci Data 12(2):1245–1265
Singh DS (ed) (2017) The Indian Rivers: scientific and socio-economic aspects. Springer, New York
Singh P, Umesh KH, Kumar N (2008) Modelling and estimation of different components of streamflow for Gangotri Glacier basin, Himalayas/Modélisation et estimation des différentescomposantes de l'écoulement fluviatile du bassin du Glacier Gangotri, Himalaya. Hydrol Sci J 53:309–322
Singh DK, Gusain HS, Dewali SK, Tiwari RK, Taloor AK (2021) Analysis of snow dynamics in Beas river basin, western Himalaya using combined terra–aqua MODIS improved snow product and in situ data during twenty-first century. Water, cryosphere, and climate change in the Himalayas: a geospatial approach, 115

Sontakke NA, Singh HN, Singh N (2009) Monitoring physiographic rainfall variation for sustainable management of water bodies in India. In: Springer MKJ (ed) Natural and anthropogenic disasters: vulnerability, preparedness and mitigation. Springer, Dordrecht, The Netherlands, pp 293–331

Sood V, Gupta S, Gusain HS, Singh S, Taloor AK (2021) Topographic controls on subpixel change detection in western Himalayas. Remote Sens Appl Soc Environ 21:100465

Sorg A, Bolch T, Stoffel M, Solomina O, Beniston M (2012) Climate change impacts on glaciers and runoff in Tien Shan (Central Asia). Nat Clim Chang 2(10):725–731

Swain AK, Mukhtar MA, Majeed Z, Shukla SP (2018) Depth profiling and recessional history of the Hamtah and Parang glaciers in Lahaul and Spiti, Himachal Pradesh, Indian Himalaya. Geol Soc Lond Spec Publ 462:35–49

Taloor AK, Kotlia BS, Jasrotia AS, Kumar A, Alam A, Ali S, Kouser B, Garg PK, Kumar R, Singh AK, Singh B (2019) Tectonoclimatic influence on landscape changes in the glaciated Durung Drung basin, Zanskar Himalaya, India: a geospatial approach. Quat Int 507:262–273

Taloor AK, Kothyari GC, Manhas DS, Bisht H, Mehta P, Sharma M et al (2021) Spatio-temporal changes in the Machoi glacier Zanskar Himalaya India using geospatial technology. Quat Sci Adv 100031

Tawde SA, Kulkarni AV, Bala G (2017) An estimate of glacier mass balance for the Chandra basin, western Himalaya, for the period 1984–2012. Ann Glaciol 58(75pt2):99–109

Wagnon P, Linda A, Arnaud Y, Kumar R, Sharma PV (2007) Four years of mass balance on Chhota Shigri Glacier, Himachal Pradesh, India, a new benchmark glacier in the western Himalaya. J Glaciol 53:603–611

Proglacial Landscape Transformations in Arctic, Ny-Alesund Area, Svalbard: Paraglacial Processes and Climate Warming During Late Quaternary

Sharat Dutta, Mohd Sadiq, and Amit Dharwadkar

Abstract The Arctic region is very sensitive to climate change and the transformations from largely ice covered to relatively ice free has been rapid during the Holocene. This rapid warming in the Arctic has a significant impact on the development and evolution of landforms. In glaciated Arctic region (glacial valleys), abrupt climatic shifts are captured in the landforms developed and their modification from glacial to proglacial (through glacier retreat). These transformations are evident in the form of systematic progression from glacier snout, morainic ridges (evidence of past glacial advance/retreat), proglacial lakes and drumlins (recent glacial melt) as well as permafrost and outwash plains. The paraglacial processes and related geomorphic and sedimentary archives in the vicinity of present glaciers margin are evidence of enhanced climatic amelioration and associated sediment transfer. The evolution of the landforms, in response to varying climate, provide the long-term proxy records which are invaluable for assessment of climate change trends. Proper characterization of these landforms in field and understanding of various sedimentary processes are important to realize the spatio-temporal implications of such geomorphic transformations. The distinctive landforms distribution in Ny-Alesund area is studied, documented and interpreted to deduce the drastic changes in palaeoclimatic regimes. In the proglacial areas, evidences of enhanced solifluction processes within the diamictite deposits are clear indication of enhanced proglacial sediment readjustment through permafrost degradation. Extensive solifluction lobes mark transformation from proglacial to paraglacial system modification that controls the meltwater sediment transportation and sediment aggradation in outwash plain. Characterization of the geomorphic attributes in vicinity of present glacial margin as evidences of such transformations that clearly indicate significant warming during the Late Quaternary. This study provides a fresh outlook to understand the

S. Dutta (✉)
Polar Studies Division, Geological Survey of India, Faridabad, India
e-mail: sharat.dutta@gsi.gov.in

M. Sadiq · A. Dharwadkar
Polar Studies Division, National Center of Excellence in Geoscience Research, Geological Survey of India, Faridabad, Haryana, India

B. Phartiyal et al. (eds.), *Climate Change and Environmental Impacts: Past, Present and Future Perspective*, Society of Earth Scientists Series,
https://doi.org/10.1007/978-3-031-13119-6_9

proglacial-paraglacial transition in this part of Arctic region associated with concomitant sediment readjustment processes in time and space.

Keywords Paraglacial process · Proglacial · Late Quaternary · Glacial foreland · Geomorphology

1 Introduction

After withdrawal of glacier ice, the glacially conditioned sediments are available of reworking by non-glacial processes like transport through meltwater channels and slope readjustment. The processes are frequently clubbed together as "Paraglacial" processes. Sensu stricto the paraglacial concept is defined as "nonglacial processes that are directly conditioned by glaciation" (Church and Ryder 1972). This concept of sediment readjustment was originally put forward to explain rapid redistribution of metastable glacial sediment by fluvial and debris flows after onset of deglaciation (Ryder 1971a, b). The concept has been applied extensively to define sediment readjustments in variable sedimentary regimes like hillslope sediment adjustment (Owen 1991; Ballantyne and Benn 1994; Ballantyne 1995; Watanabe et al. 1998; Curry 1999), glacier marginal areas (Matthews et al. 1998; Etzelmüller 2000), fluvial/lacustrine environment (Ryder 1971a; Church and Ryder 1972; Jackson et al. 1982; Church and Slaymaker 1989), marine coastal environment (Forbes and Syvitski 1994; Forbes et al. 1995). Nonetheless, studies indicate climatic transition and enhanced period of sediment reworking. The studies were further supported by sedimentary transformations through fluvial activity and slope processes for shorter time scale (Jackson et al. 1982; Eyles et al. 1988; Harrison 1991; Wright 1991). The relevance of such paraglacial processes in deciphering the post-glacial climatic changes is obviously the objective of geoscientists to quantify the post glacial rapid palaeoclimatic changes. The decadal studies to deduce glacial fluctuation and palaeoclimatic history suggests appreciable depression in Equilibrium Line of Altitude (ELA) during Last Glacial maxima and advocate climatic amelioration during early Holocene (Inge Svendsen and Mangerud 1992 and reference therein). The transformation from largely ice-covered area to ice free has been relatively rapid since Last Glacial Maxima (LGM). This rapid warming in the Arctic has a significant impact on the development and evolution of landforms in glacial foreland. The evidences of such climatic amelioration are available in glacier forelands (proglacial areas) of present glacial environments in the form of sedimentary and geomorphic archives. The climatic shifts are captured in evolution of the landforms from glacial to proglacial conversion and subsequent modification through paraglacial processes. The present study is focussed on identification and documentation of various landforms in Ny-Alesund, Svalbard. The present study focussed to study archives of paraglacial processes and related geomorphic changes in the vicinity of present glaciers margin of selected glaciers in Ny-Alesund, Svalbard, Arctic to decipher palaeoclimatic variation. The absolute chronology of landforms through Optically Stimulated Luminescence (OSL) dating.

2 Study Area Geomorphology

The study area lies in western part of Svalbard archipelago, situated within the Arctic circle (Fig. 1a) and constituted of four major Islands, viz. Spitsbergen, Nortdaustlandet, Edigeoya and Barentsoya surrounded by Arctic Ocean, Barents Sea, Norwegian Sea and Greenland Sea (Fig. 1b). The Svalbard area mostly constituted of either permanently ice covered (glaciers and ice sheet) with marginal seaward and hilltop land exposures seasonally. The Ny-Alesund area exposes tongue shaped Peninsula with Kongsfjorden in the north (Fig. 1c). Several glaciated valleys occupied by both surging as well as non-surging glaciers (Fig. 1c), proglacial

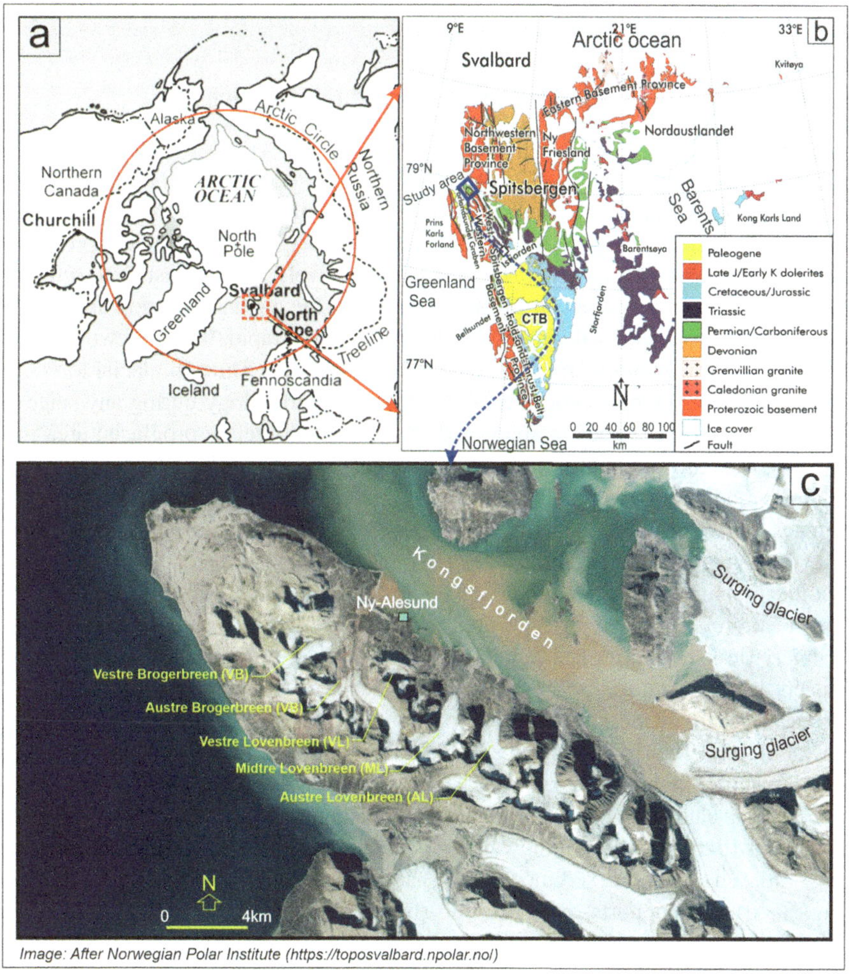

Fig. 1 (**a**) Arctic circle. (**b**) Geology and location of Svalbard (modified after Gee and Tebenkov 2004). (**c**) Location of Ny-Alesund (Image source in **b**, **c** Norwegian Polar Institute; web-https://toposvalbard.npolar.no/)

foreland (Fig. 2a, b), morainic ridges (recessional, longitudinal and end moraines), outwash plain (Fig. 2b) within small region forms terrestrial and maritime geomorphology. The bay facing glacier Valleys and the extent of arcuate end moraine ridges demarcate the proglacial areas from present glacier snout and outwash plain (Figs. 1c and 2b). The morainic ridges are dissected by meltwater stream (glacier streams and permafrost degradation melt or seasonal snow melt) (Fig. 2c). Characteristically, the present proglacial region exhibits variable degree of gully erosion and channel density, whereas, the outwash plain with gentle slope (~2°) towards Kongsfjorden bay shows prolific distributary channel network and braided channel pattern (Fig. 2d). Dissection and erosion of end moraine ridges forms Low angle alluvial fans (around 2°–3° slope) fed by proglacial sediments merges the proglacial area with outwash plain (Fig. 2b, d, e, and f).

3 Proglacial Morphology and Paraglacial Process Dynamics

In the Glacial foreland and other related environments, the concept of paraglacial processes is being introduced with new and distinct geomorphological processes (Church and Ryder 1972) which encompasses relatively rapid adjustment of deglaciated landscapes to nonglacial conditions through the enhanced action and amplitude of subaerial surficial processes. These include rapid fluvial reworking of glacigenic sediment and permafrost changes on exposure. Thus, in glacial foreland, the degree of operational processes and landscape morphology during any stage of "paraglacial period" varies in space and can be used to define proglacial area into zones. In the study area, the exposure of glacial foreland to subaerial conditions led to rapid modification of landforms through action of snowmelt channels, slope changes due to permafrost degradation and wind action. The proglacial sedimentary zone is divided into two on the basis of morphology of landscape, paraglacial modification and phases of degradation under 'paraglacial period'. **Zone-I:** marginal proglacial area consists of deglaciated valley floor typically shows less erosional channels (less reworking). The recent glacial retreat is evidenced near snout of glaciers leaving almost flat ground moraine surface with headward gully erosion (towards present glacier) through seasonal snowmelt, permafrost degradation, and/or glacial melt (Fig. 3a). The enhanced glaciofluvial activity led to meltwater channel draining outwash plain (Fig. 3b). The paraglacial sediment readjustment is key process that catalysed through glacial/snow melt leaving behind remnant glacial pavements (flat or gentle sloping valley floor) with limited gully erosion in this zone (Fig. 3a). The surficial sediment debris and kettle depressions are also evident (Fig. 3a–d) that supports enhanced sub-surface melting of ice lenses (permafrost degradation). However, within Zone-II, relatively more pronounced meltwater reworking of glacigenic sediment is evident. The former valley floor which presently shows Kame field is extensively modified into solifluction lobes, collapse cone,

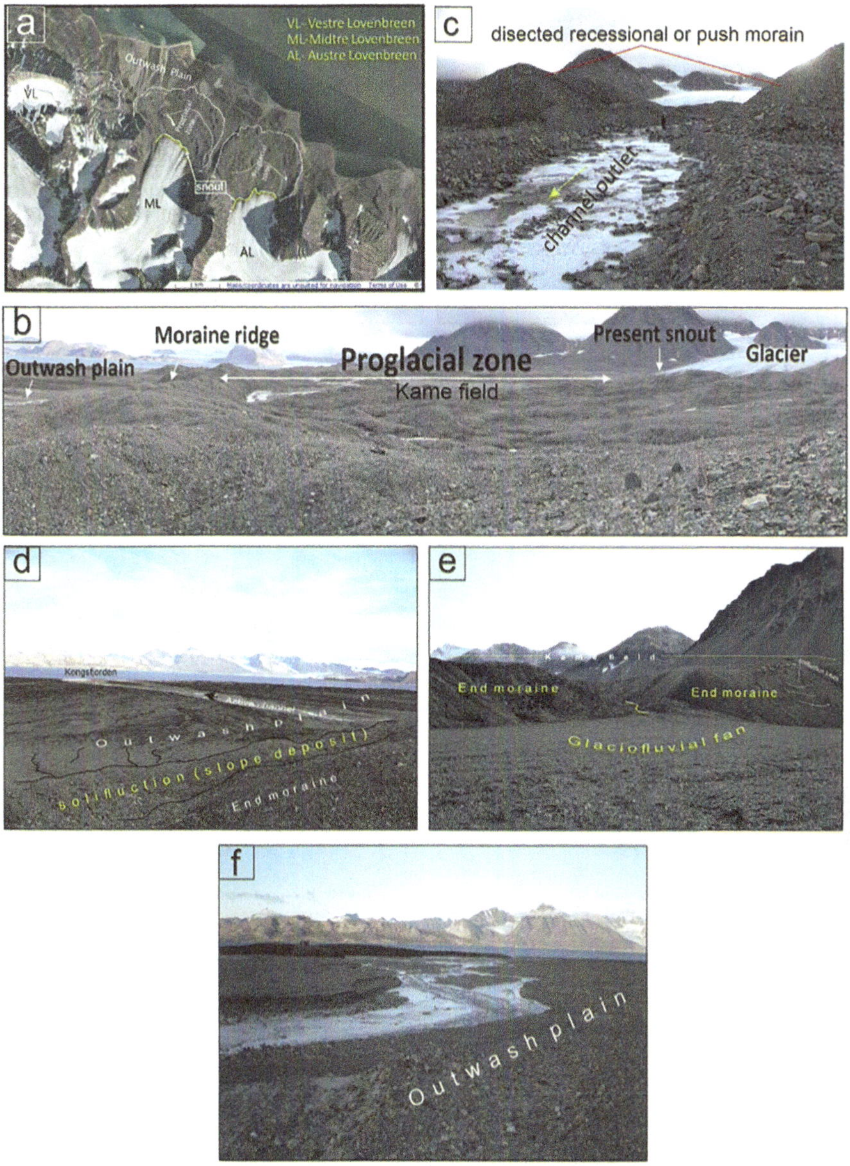

Fig. 2 (**a**) Present snout and proglacial foreland zones of Midtre Lovenbreen (ML) and Austre Lovenbreen glaciers, Ny-Alesund, Svalbard, Arctic (Image source: Norwegian Polar Institute; web-https://toposvalbard.npolar.no/). Note the arcuate end moraine demarcating proglacial zone from outwash plain. (**b**) Geomorphic features of ML proglacial area showing Kame field (Hummocky undulating surface), present ice front (snout), end moraine ridge and outwash plain. (**c**) Outlet channel morphology in present day condition. Note the dissected recessional moraine ridge. (**d**) Gentle sloping outwash plain with braided meltwater channels. (**e**) Dissected end moraine ridge and development of glacio-fluvial fan. (**f**) Outwash plain and braided channel debouching into Kongsfjorden

Fig. 3 (**a**) Satellite image showing proglacial zone-I and II of ML glacier Arctic (Image source: Norwegian Polar Institute; web-https://toposvalbard.npolar.no/). Note the flat glacial pavement with feeble headward erosion facilitated by glacial melting in Zone-I and extensive erosional channels in Zone-II with Hummocky surface. (**b**) Meltwater channel and kettle lakes in proglacial Zone-I. (**c**) Hummock (Kame field) of ML glacier in Zone-II number of solifluction lobes and kettle hole. (**d**) Kettle lake in proglacial Zone-I of ML glacier. (**e**) End Moraine modification through collapse cone. (**f**) Kettle hole in proglacial Zone-II. (**g**) Thermokarst process in Zone-II of ML glacier

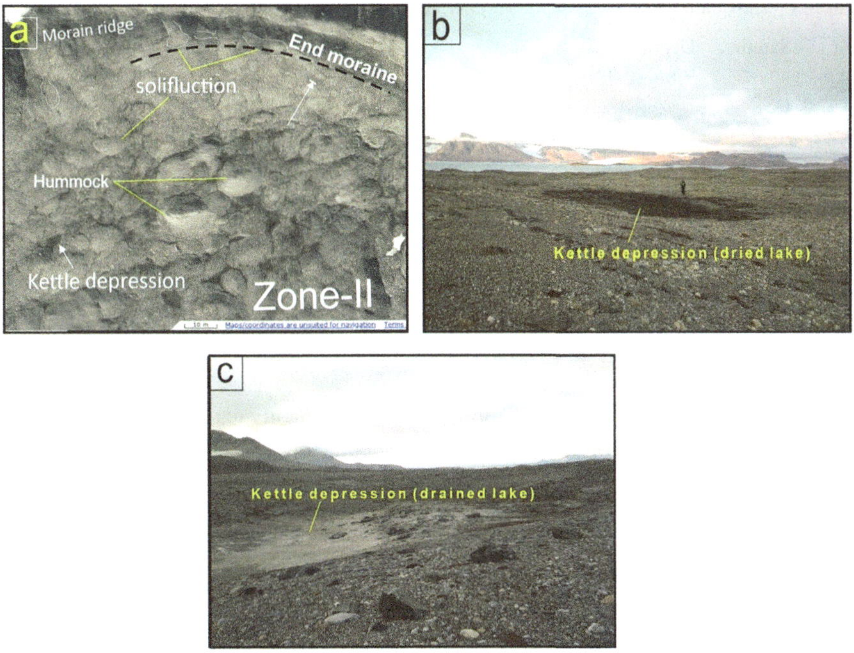

Fig. 4 (**a**) Satellite image showing extensive solifluction activity and developed Kame field in Zone-II of ML glacier Arctic (Image source: Norwegian Polar Institute; web-https://toposvalbard.npolar.no/). (**b**, **c**) Kettle depression in Zone-II of ML glacier

kettle holes (active layer deformation) and thermokarst activity (Figs. 2b, 3c, e, f, g, and 4a). Subglacial ice melting formed uneven bad land topography that resembles hummocks (Fig. 2b). Proglacial lakes and depressions (drained lake remnant) are also characteristic landforms (Fig. 4a–c). The end moraine ridges modified by collapse of debris (collapse cone). The unstable or metastable glacially conditioned sediment is being reworked to small streams feeding outwash plain.

4 Material and Methods

The research work was part of Indian Scientific Expedition to Arctic during 2018–2019 (summer). And taken up for geomorphological mapping of study area and understanding of paraglacial processes in glacial foreland of selected glaciers viz. *Vestre Lovenbreen (VL), Midtre Lovenbreen (ML), Austre Lovenbreen (AL) and Vestre Brogerbreen (VB)*. The spatial distribution of associated landforms was mapped initially through remote sensing and subsequently field traverses were taken for validation. With systematic field traverses, geomorphological attributes of landforms (location, elevation, slope etc.) were recorded. Optically Stimulated

Table 1 The quartz OSL age, equivalent dose (De) and annual dose rate

Sr. n o	Sample no	Depth (m)	U (ppm)	Th (ppm)	K %	Equivalent dose (De) in gray	Dose rate (Gray/ka)	Age (ka)
1	OSL-M-1	5	2.37	9.19	1.49	62 ± 2	2.4 ± 0.1	26 ± 2
2	OSL-M-2	1	2.30	9.19	1.90	55 ± 2	2.8 ± 0.1	19 ± 1
3	RM-5	1	2.77	9.34	2.43	139 ± 5	3.4 ± 0.2	41 ± 3
4	RM-6	7	1.99	8.17	1.68	59 ± 3	2.4 ± 0.1	24 ± 3
5	A7 (S6)	0.8	3.78	10.62	1.5	54 ± 5	2.89 ± 0.2	19 ± 2
6	A9 (S12)	3	3.54	3.56	0.92	162 ± 6	1.9 ± 0.1	87 ± 6

Luminescence (OSL) sampling of moraine ridge sediments was essential part of the study to establish chrono-stratigraphy of deglaciation period and associated paraglacial processes for palaeoclimatic correlation/inferences. Four OSL samples were collected which were processed in OSL laboratory, Nation Center for Excellence in Geoscience Research, (NCEGR), Geological Survey of India that equipped with automated Riso TL/DA 15 & 20 readers for quartz OSL dating. Samples collected during earlier studies were used for interpretation. Samples from the middle part of the pipe were processed in subdued red light for the removal of carbonate and organic matter using 1 N HCl and 30% H_2O_2, respectively. The 90–150 micron size grains were hand-sieved from which quartz grains were extracted by density separation using sodium polytungstate solution (Sp gr. 2.58). The grains were etched for 80 min in hydrofluoric acid (HF) to remove the outer 20 micron layer affected by alpha radiation. The HF treatment also removed any feldspar contaminations. The grains were treated with HCl (for about 30 min) and re-sieved. The purity of the etched quartz (i.e., feldspar contaminations) was tested using infrared stimulated luminescence (IRSL). The quartz grains were mounted as a mono layer (about 3 mm diameter) on stainless steel disc (10 mm diameter) using silicon oil for luminescence measurements. A Risoe TL/OSL Reader (model 15–20A) fitted with Sr 90 radiation source was used to determine the radiation energy received by the sample after its burial, that is, palaeodose or equivalent dose (ED). The single aliquot regeneration (SAR) protocol (Murray and Wintle 2000) was used for ED determination. The OSL measurements were carried out using 220 °C preheat for 10 s, 200 °C cut heat, laboratory irradiation from a calibrated beta source (Sr/Yr 90) and 40 s blue light stimulation. The ED values (Table 1) were calculated using the initial 0.8 s integral of the OSL by Duller's Analyst software.

5 Chronology of Deglaciation: Glacial Foreland Exposure Period

We determined absolute chronology of glacier advance and recession of selected glacier namely Midtre Lovenbreen (ML) and Vestre Brogerbreen (VB) on the basis of OSL dating (Table 1) using quartz as dosimeter to constrain 'paraglacial period' and associated processes post deglaciation. The OSL dating of glacigenic sediment section near present snout (OSL-RM-5) of Midtre Lovenbreen (ML) glacier gave an OSL age of 41 ± 3 ka. The sample was collected from sandy bed exposed along sub-glacial channel in Zone-I. This surface represents the older glacial valley floor deglaciated during end phase of Marine Isotopic stage-III (MIS-III interglacial) (Fig. 5a, b). Whereas, an OSL sample (OSL-RM-6) collected from medium sand horizon from recessional moraine ridge in Zone-II gave depositional age of 24 ± 2 ka (MIS-III and II transition) (Fig. 5a, b). On the other hand, end moraine ridge of same glacial foreland gave an OSL burial age of 19 ± 2 ka (Fig. 5a, b). Another sample from basal part of end moraine ridge gave burial age of 26± 2 ka (Fig. 5a, b). During earlier studies in the same area, OSL ages from moraines deposits of VB and AB glaciers is also constrained. OSL age of farthest moraine deposit ~4 km from present snout of VB glacier gave burial age of 19 ± 2 ka. Whereas, recessional moraine near present snout position gave 87 ± 6 ka. The older age in the bottom and younger age at the top in case of ML end moraine ridge deposit suggests atleast two distinct phases of glacier advance and overlapping glacial till deposition (Fig. 5a, b). The younger ages of recessional moraine as compared to end moraine further supports the glacial retreat in the study area.

6 Discussion and Conclusion

6.1 Glacial Foreland, Paraglacial Period and Landform Adjustment

The retreat of glacier exposes loose valley-floor glacigenic deposits that usually composed of stacked lateral, terminal and ground moraines (Mattson and Gardner 1991; Ballantyne and Benn 1994, 1996). The studied proglacial morphology and evidence of paraglacial activity in Vestre Brogerbreen (VL), Midtre Lovenbreen (ML) and Aautre Lovenbreen (AL) glaciers, Ny Alesund depicts Late Quaternary warmer phases of rapid paraglacial sediment readjustment. The morphological evidence constitutes glacially conditioned sedimentary deposits and associated valley floor morphology which is susceptible for reworking through nonglacial processes that formed a disequilibrium paraglacial land system in glacial foreland. Such disequilibrium conditions after glacial retreat are marked by changes in the rate and pattern of sediment transfer (Clark 1987; Church and Ryder 1972; Ballantyne 2002). In the study area where cold and dry environments prevail in present, the fluvial

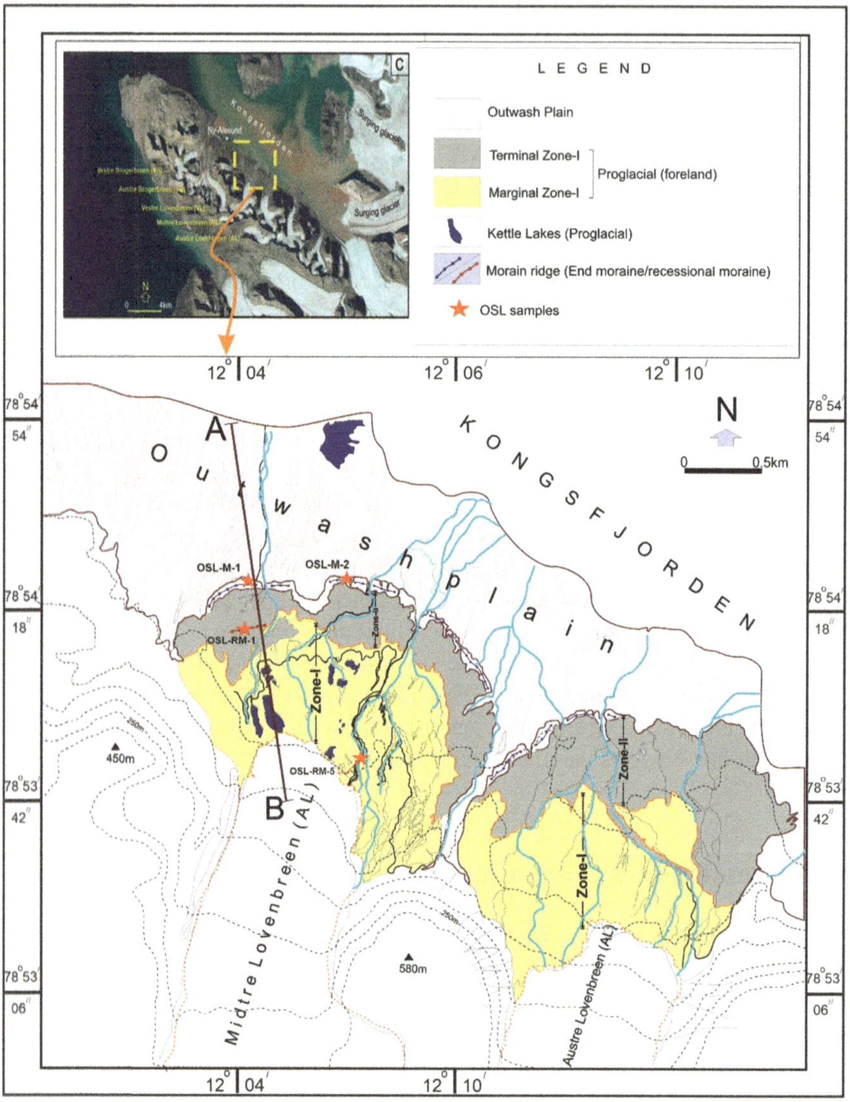

Fig. 5 (**a**) Map showing geomorphic features of proglacial foreland (Zone-I and II) of ML glacier and OSL sample locations. (**b**) Schematic model of ML glacial advance and retreat during past ~40 ka time period

processes are likely to be weak compared to temperate environments of dominated fluvial processes. Therefore, the paraglacial processes over exposure of glacial valley floor or glacier foreland are supposed to be delayed (Fitzsimons 1996) and limited gully erosion through meltwater stream (seasonal snowmelt and/or glacial ice melt) is likely dominant process. Under such conditions, evidences of enhanced

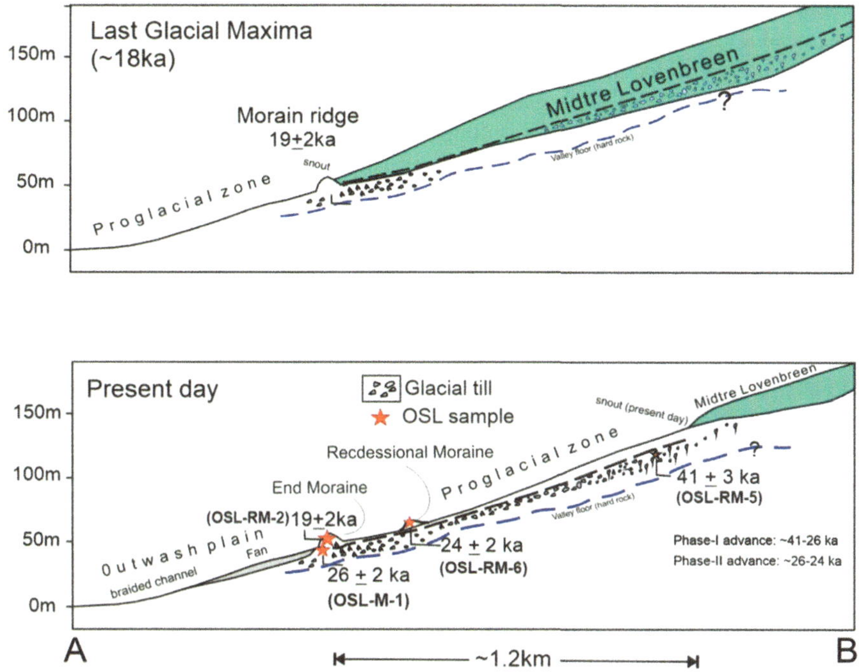

Fig. 5 (continued)

paraglacial processes after onset of glacier retreat indicative of climatic warming. The chronological data suggests, Vestre Brogerbreen (VL) glacier witnessed maximum advance during Last Glacial maxima (LGM; ~19 ka). Followed by fast recession exposing glacial foreland susceptible for paraglacial process operation. The older ages of glacial sediment near the snout of VB glacier represents older valley floor ground moraine surface. Pre-LGM advances are evidenced by recessional moraine during warmer warm phase around 26 ka (terminal MIS-III: Interglacial period). This suggests replenishment of paraglacial system through sediment rejuvenation on repeated glacial advances and retreat during Late Quaternary and near completion of paraglacial period and peneplanation of metastable landforms thorough reworking. In the ML and AL glacier foreland, though recessional moraine represents valley floor rejuvenation upto end moraine ridge indicating glacier advance during ~19 ka (OSL-M-1. from top of deposit) and ~26 ka (OSL M-2; from bottom of exposed end moraine), the paraglacial processes were operational post LGM with rapid modification of landforms. The proglacial domain of ML and AL glaciers shows enhanced sediment reworking owing to climatic warming. The onset of permafrost degradation and active layer seasonal freeze and thaw (periglacial changes: solifluction, gelifluction, thermokarst etc.) played catalysing role in enhancement of paraglacial processes. The melting of ice lenses within glacigenic sediments produces bad land surface morphology that resemble hummocky microform. Based on geomorphology and erosional landforms, the studied proglacial domains of selected glaciers are sub divided into two. Zone-I: marginal proglacial area showing characteristic headward erosion. Such headward erosion is result of subglacial ice melt or permafrost degradation owing to sub aerial exposure of glacigenic sediment after glacier retreat (Fig. 5b). The limited erosion with preserved trains of glacier drift sediments (flat gentle topography) accounts for initiation of paraglacial adjustment. Deep erosional gullies are dominant that indicates segregated ice melting within the sediment upto present permafrost table beneath. Proglacial lakes are evidence of glacial ice melt owing to climate warming. In zone-II, extensive reworking is evidenced. The original glacial debris modified in the form of small mounds appears as hummocky ground (Fig. 2b). Such hummocky appearance and bad land topographic features are well described in glacial forelands (Gravenor and Kupsch 1959; Clayton 1967; Boulton 1967, 1972; Clayton and Moran 1974; Eyles 1983; Paul 1983) and indicative of stage of paraglacial readjustment towards establishment of geomorphic landscape equilibrium. Although, various explanations are projected viz. chaotic erosion by subglacial meltwater (Shaw et al. 1996; Munro and Shaw 1997), subglacial upheaving by active ice, forming hummock like features along with drumlins (Aario 1977) and tectonics under subglacial environment (Aber et al. 1989). In the studied glaciers, the gravity driven collapse of glacigenic debris lying over stagnant ice on melting best explains the observed morphology. Although the outwash deposits are derived from glacial streams, the morainic sediment dissected through snowmelt or headward erosional streams indicative of enhanced period of melting and consequent sediment reworking and landform transformation during Late Quaternary. The moraine ridge slopes reworked through debris cones, slumps, solifluction under periglacial changes. Thermal-contraction cracks within till deposits indicative of active layer deformation and permafrost lowering on account of climate warming. All these processes have played significant role in total sediment yield through paraglacial processes and carving of glacial foreland as well as outwash geomorphology. The paraglacial concept apprehended by Church and Ryder (1972) that emphasises rapid rate of sediment adjustment in deglaciated landscapes in distinct warm climatic phases through the enhanced operation of a wide range of subaerial processes explains the proglacial landscape evolution in study area.

7 Conclusion

The geomorphological field observation and remote sensing image analysis supplemented with OSL based chronology depicts following inferences on paraglacial processes and landscape development on account of palaeoclimatic variations

1. The study area glaciers Vestre Brogerbreen (VB) and Midtre Lovenbreen (ML) depicts two phases of paraglacial transformation and sediment readjustment under warmer phase of MIS-III and post LGM (~18 ka) time period

2. The onset of deglaciation post LGM 'paraglacial period' responded in delayed paraglacial readjustment in glacial foreland of VB, ML and AL glacier owing to permafrost stability of metastable sedimentary archives.
3. Periglacial (permafrost as well as non-permafrost) sedimentary processes played decisive role in enhancement of paraglacial modification and indicative of fast climate warming since LGM

Acknowledgments The authors extend their thanks to the Director General, CHQ, GSI, ADG and Head MIV, ADG and HoD, NR Lucknow, Deputy Director General, GSI Faridabad for technical and administrative support and permission to publish this work. Sincere thanks to Director, NCPOR, Goa for providing permission to participate in Indian Arctic Expeditions. Thanks to Dr. Radhakrishna Chunchekar and Mohd. Atif Raza, TL-OSL laboratory, NCEGR, GSI, Faridabad for providing OSL dating data. Sincere thanks to the reviewers and editors for valuable comments and suggestions that have helped to improve the manuscript. Thanks are also due to the officers of the Polar Studies Division, GSI Faridabad for suggestions.

References

Aario R (1977) Classification and terminology of morainic landforms in Finland. Borras 6:87–100

Aber JS, Croot DG, Fenton MM (1989) Glaciotectonic landforms and structures. Kluwer Academic Publishers, Dordrecht, p 200

Ballantyne CK (1995) Paraglacial cone formation on recently deglaciated terrain. The Holocene 5: 25–33

Ballantyne CK (2002) Paraglacial geomorphology. Quat Sci Rev 21(18):1935–2017

Ballantyne CK, Benn DI (1994) Paraglacial slope adjustment and resedimentation following recent glacier retreat, Fabergstolsdalen, Norway. Arct Alp Res 26:255–269

Ballantyne CK, Benn DI (1996) Paraglacial slope adjustment during recent deglaciation and its implications for slope evolution in formerly glaciated environments. In: Anderson MG, Brooks S (eds) Advances in hillslope processes, vol 2. Wiley, Chichester, pp 1173–1195

Boulton GS (1967) The development of a complex supraglacial moraine at the margin of Sorbreen, Ny Friesland, Vestspitsbergen. J Glaciol 6:717–735

Boulton GS (1972) Modern arctic glaciers as depositional models for former ice sheets. J Geol Soc Lond 128:361–393

Church MA, Ryder JM (1972) Paraglacial sedimentation: a consideration of fluvial processes conditioned by glaciation. Geol Assoc Am Bull 83:3059–3071. https://doi.org/10.1130/0016-7606(1972)83[3059:PSACOF]2.0.CO;2

Church M, Slaymaker O (1989) Disequilibrium of Holocene sediment yield in glaciated British Columbia. Nature 337:452–454

Clark MJ (1987) The alpine sediment system: a context for glaciofluvial processes. In: Gurnell AM, Clark MJ (eds) Glaciofluvial sediment transfer—an alpine perspective. Wiley, Chichester, UK, pp 9–29

Clayton L (1967) Stagnant glacier features of the Missouri Coteau in North Dakota. In: Clayton L, Freers TF (eds) Glacial geology of the Missouri Coteau and adjacent areas. North Dakota Geological Survey Miscellaneous Series, vol 30. U.S. Geological Survey, Reston, pp 25–46

Clayton L, Moran SR (1974) A glacial process-form model. In: Coates DR (ed) Glacial geomorphology. Binghamton State University of New York Publications in Geomorphology, New York, pp 89–119

Curry AM (1999) Paraglacial modification of slope form. Earth Surf Process Landf 24:1213–1228

Etzelmüller B (2000) Quantification of thermo-erosion in pro-glacial areas – examples from Svalbard. Z Geomorphol 44:343–361

Eyles N (1983) Modern Icelandic glaciers as depositional models for 'hummocky moraine' in the Scottish highlands. In: Evenson EB (ed) Tills and related sediments. Balkema, Rotterdam, pp 47–59

Eyles N, Eyles CH, McCabe AM (1988) Late Pleistocene subaerial debris-flow facies of the Bow Valley, near Banff, Canadian Rocky Mountains. Sedimentology 35:465–480

Fitzsimons SJ (1996) Paraglacial redistribution of glacial sediments in the Vestfold Hills, East Antarctica. Geomorphology 15:93–108

Forbes DL, Syvitski JPM (1994) Paraglacial coasts. In: Carter B, Woodroffe CD, IGCP Project 274 (eds) Coastal evolution: late quaternary shoreline morphodynamics. Cambridge University Press, Cambridge

Forbes DL, Orford JD, Carter RWG, Shaw J, Jennings SC (1995) Morphodynamic evolution, self-organization, and instability of coarse-clastic barriers on paraglacial coasts. Mar Geol 126:63–85

Gravenor C, Kupsch WO (1959) Ice-disintegration features in western Canada. J Geol 67:48–64

Harrison S (1991) A possible paraglacial origin for the drift sheets in upland Britain. Quat Newsl 64:14–18

Inge Svendsen J, Mangerud J (1992) Paleoclimatic inferences from glacial fluctuations on Svalbard during the last 20 000 years. Clim Dyn 6:213–220. https://doi.org/10.1007/BF00193533

Jackson LE, MacDonald GM, Wilson MC (1982) Paraglacial origin for terraced river sediments in Bow Valley, Alberta. Can J Earth Sci 19:2219–2231

Matthews JA, Shakesby RA, Berrisford MS, McEwen LJ (1998) Periglacial patterned ground in the Styggedalsbreen glacier foreland, Jotunheimen, southern Norway: microtopographical, paraglacial and geochronological controls. Permafr Periglac Process 9:147–166

Mattson LE, Gardner JS (1991) Energy exchanges and ablation rates on the debris covered Rakhiot glacier. Pakistan Z f Gletscherk Glazialgeol 25(1):17–32

Munro M, Shaw J (1997) Erosional origin of hummocky terrain in south-central Alberta. Geology 25:1027–1030

Murray AS, Wintle AG (2000) Luminescence dating of quartz using an improved single-aliquot regenerative-dose protocol. Radiat Meas 32(1):57–73

Owen LA (1991) Mass movement deposits in the Karakoram Mountains: their sedimentary characteristics, recognition and role in Karakoram landform evolution. Z Geomorphol 35: 401–424

Paul MA (1983) The supraglacial land system. In: Eyles N (ed) Glacial geology: an introduction to engineers and earth scientists. Pergamon Press, Oxford, pp 71–90

Ryder JM (1971a) Some aspects of the morphometry of paraglacial alluvial fans in south-central British Columbia. Can J Earth Sci 8(10):1252–1264. https://doi.org/10.1139/e71-114

Ryder JM (1971b) The stratigraphy and morphology of paraglacial alluvial fans in south-central British Columbia. Can J Earth Sci 8:279–298

Shaw J, Rains B, Eyton R, Weissling L (1996) Laurentide subglacial outburst floods: landform evidence from digital elevation models. Can J Earth Sci 33:1154–1168

Watanabe T, Dali L, Shiraiwa T (1998) Slope denudation and supply of debris to cones in Landtang Himal, Central Nepal Himalaya. Geomorphology 26:185–197

Wright MD (1991) Pleistocene deposits of the South Wales coalfield and their engineering significance. In: Forster A, Culshaw MG, Cripps, Little A, Moon CF (eds) Quaternary engineering geology, vol 7. Geological Society, Engineering Geology Special Publications, London, pp 441–448

Impact of Changing Climate Over Polar Ice Sheet: A Case Study from Larsemann Hills, East Antarctica

Pradeep Kumar, Abhishek Verma, Deepak Gajbhiye, Vikash Chandra, Ajanta Goswami, and Sharat Dutta

Abstract Changing climate and its impact on the polar cryosphere is one of the prime concern of researchers and decision makers. Present study is focused on the on ground insitu observation on impact of changing climate over the Larsemann Hills area situated in Prydz Bay region, East Antarctica. The study area covering 216 km^2 is bounded by the Polararboken ice streaming the west and Dålk glacier in the east. Annual and monthly accumulation/ablation rates have been estimated with the help of 149 stakes. Movement of the ice sheet is monitored at regular interval with the help of stake displacement using Differential-GPS. Snow/ice thickness at the ice sheet margin has been estimated using Ground Penetrating Radar. Subsequently surface mass balance has been calculated for the region. All the stakes in the study area show accumulation except the SN1 stake farm which is located towards the margin of the bay. Snow accumulation increases away from the shore due to lesser melting at high altitude. Observations indicate more than 3.5 times average accumulation during 2019–2020 compared to previous year with a high degree of confidence. The average velocity of the ice sheet based on the SN1 stake farm was calculated to be 0.78 m/year. In general, higher velocities during the study period were recorded at the margins of Polararboken and Dålk Glaciers. The general direction of the ice sheet movement, in the study area, is towards NNW to North. The net Surface Mass Balance for 2019–2020 is estimated to be 80.304204×10^6 m^3 SWE and specific surface mass balance is estimated 0.37 m SWE. The study shows that there is an increase in snow thickness as a result of positive mass balance as compared to previous years. The positive mass balance trend in the area is confirmed using different methodologies and data.

Keywords Larsemann Hills · Climate change · Glaciology · Mass balance · Ice loss

---

P. Kumar (✉) · A. Verma · D. Gajbhiye · V. Chandra · S. Dutta
Polar Studies Division, Geological Survey of India, Faridabad, India
e-mail: sharat.dutta@gsi.gov.in

A. Goswami
Department of Earth Science, Indian Institute of Technology Roorkee, Roorkee, India

B. Phartiyal et al. (eds.), *Climate Change and Environmental Impacts: Past, Present and Future Perspective*, Society of Earth Scientists Series,
https://doi.org/10.1007/978-3-031-13119-6_10

1 Introduction

Global warming is determined by the increase in the combined surface air and sea surface temperature averaged for the global extension and for above 30 years' period of the observations. The average warming increased by the 0.87 °C during the decade of 2006–2015 period with the reference to the preindustrial level (likely between 0.75 °C and 0.99 °C) (Allen et al. 2018). The volume of the glaciers is decreasing for more than two decades. The Greenland and Antarctica is losing mass. Glaciers are also shrinks worldwide (Box et al. 2006; Monaghan et al. 2006). The Global mean sea level (GMSL) is rising and the accelerating and the major contributors are glaciers and ice sheet (Meredith et al. 2019). The rising temperature in the Antarctica is prime concerned for the sustainability of the Antarctic cryosphere. The assessment and monitoring of the Antarctic ice sheet with changing climatic parameters is important for the valuation of health of the ice sheet in terms of ice loss and sea level rise.

East Antarctic ice sheet is the largest fresh water reservoir with convergence area of 14 million sq. km. The thickness of the ice sheet varies from few meters at periphery to few (~4 km) km inward. Any variation in the ice thickness will provide inputs to understand the effect of regional and global climate change. In the past few decades substantial warming has resulted in melting of ice shelf as well as Antarctic ice sheet and glaciers. According to a study, Prior to 2012, Antarctica lost ice at a steady rate of 76 billion tons per year—a 0.2 mm per year contribution to sea level rise. However, since then there has been a sharp, threefold increase—between 2012 and 2017 Antarctica lost 219 billion tons of ice per year, a 0.6 mm per year sea level contribution (Shepherd et al. 2018).

Ground based studies were conducted in an area covering 220 sq. km in Larsemann Hills, East Antarctica during Indian Antarctic Expeditions. The Larsemann Hills (69°20′S to 69°30′S, 75°55′E to 76°30′E), with an area of approximately 40 sq. km, is located on the Ingrid Christensen Coast of Princess Elizabeth Land, East Antarctica. To the northeast of the Larsemann Hills are the Rauer Islands and the Vestfold Hills and to the west-southwest are the Bolingen Islands and the Amery ice shelf (Fig. 1). Five research stations namely the Progress I and II (Russia), Law-Racovita (Australia-Romania), Zhongshan (China) and Bharati Research Base (India) are located in this area. Dålk Glacier (69°26′S 76°27′E) is eight nautical miles (~15 km) long glacier, draining into the southeast part of Prydz Bay between the Larsemann Hills and Steinnes whereas Polarårboken Glacier (69°36′S76°0′E) is three nautical miles (~5.6 km) long glacier situated in northeast of Stein Islands, draining westward into the north part of Publications Ice Shelf. Broknes and Storness peninsulas are two largest peninsulas while Fisher, Grovness, Mcleode, Soloman and Manning are some of the larger islands of this area (Fig. 2).

The study area lies about 5 km south-east of Bharati Research Base, on Polar ice sheet and is located along the coast of South Grovness and Broknes (Fig. 2). The stakes have been installed between Dalk (Approx. 2.5 km south-east of Progress-II Station) Glacier and Polararboken (Approx. 12 km south-west of Bharati Station)

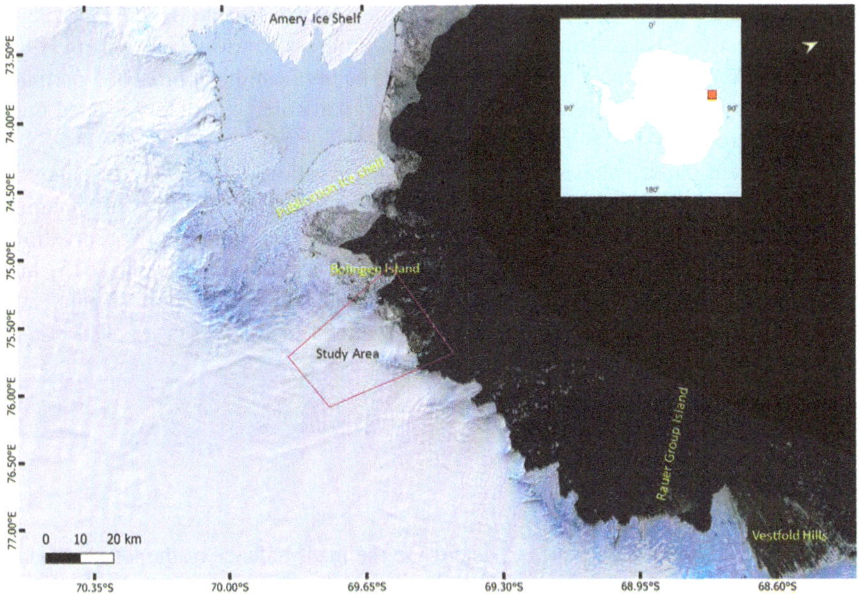

Fig. 1 Regional map of the study area. Overlaid by satellite image. (Source: NASA)

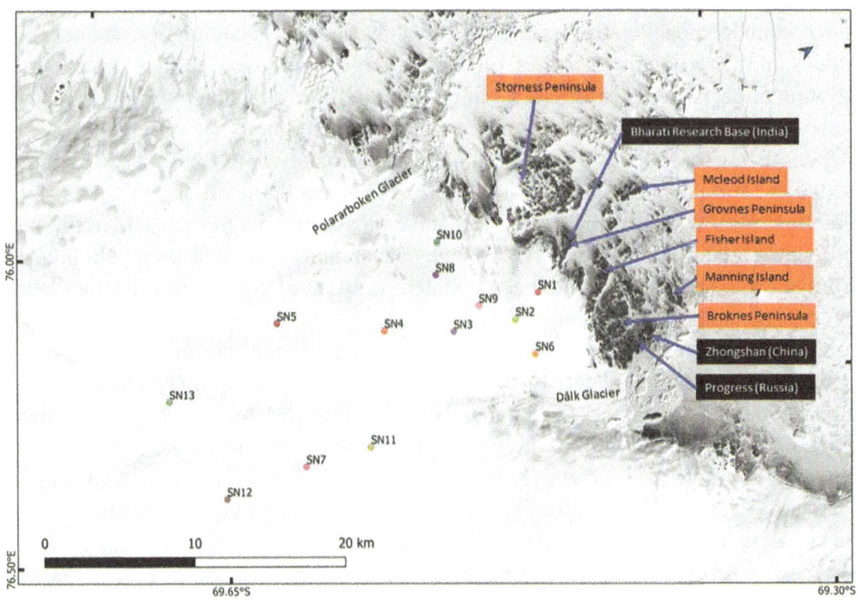

Fig. 2 Stake Position in study area Larsemann Hills, East Antarctica

Glacier. Bharti station can be reached from Cape Town to Bharati (Prydz Bay) either by ship directly or by air from Cape Town via Progress Runway (Larsemann Hills) and Novo Airbase (near Schirmacher Oasis). The area can be approached partially on foot or by Pisten Bully, whereas some parts, particularly remotely located ones can be covered only by helicopter.

The current paper is based on the ground observations of glacial parameters in combination with local meteorological parameters to assess the effect of changing climate over the ice sheet in the study area. The study incorporates the observations and analysis of the longest monitored stake farm (SN1) (Shah et al. 2015) and subsequent stake networks as reported in unpublished GSI Report (Kumar and Dutta 2020). The relationship between observations of the stakes are also briefly discussed.

2 Methodology

One of the important techniques to determine the mass balance of the ice sheet is to study the accumulation/ablation using stake networks. This traditional glaciological technique employs regular (annual) measurements of exposed lengths of fixed stakes or stake networks. The exposed heights of the individual stakes are compared with preceding years exposed-length data and accordingly determine the accumulation/ablation in the area. Positive mass balance values indicate accumulation and negative values indicate ablation. Stakes often submerge below ice surface when snow accumulation is high or they may fall after extensive ablation. In both cases old stakes are replaced with new stakes to maintain continuity of observations. The stakes are anchored in ice up to a depth of approximately 2 m so that stakes protrude above the surface and remain visible even after heavy snowfall or snow accumulation due to snow deposition. Individual stake measurements from a stake farm have been averaged for representative changes in annual ablation/accumulation. Data from several stake farms have been collated to arrive at cumulative changes in the entire area under observation.

Atmospheric parameters such as wind, temperature etc. during the observation period (annual) were collected from Automatic Weather Station (AWS) fixed near Bharti Station situated at Larsemann Hills by India Meteorological Department (IMD) and correlated with accumulation data to assess the response of ice sheet to changing weather parameters (Kumar and Dutta 2020). Measurement of surface velocity yields important information about the movement of ice sheet to evaluate glacial dynamics. For this purpose, precise positions of the stakes installed over ice sheet were recorded using Differential Global Positioning System (DGPS) (Leica make). These positions are compared with corresponding position during previous year to yield the magnitude and direction of annual displacement of the ice sheet surface (Swain and Chandra 2017; Kumar and Dutta 2020).

3 Observation and Results

3.1 Accumulation/Ablation

The observations for snow accumulation/ ablation patterns in the Larsemann Hills area were initiated in 2013–2014. The effect of changing climate on ice sheet at regional scale requires continuous and long term monitoring to decipher the trends and relative patterns of changes in the ice-sheet. Snow accumulation/ablation were measured from the stake farm (SN01) established during the year of 2013–2014 are given in Table 1 (Shah et al. 2015; Swain and Chandra 2017). The temperature data along with the monthly accumulation data have been analyzed using regression analysis by Ordinary Least Square (OLS) model (Kumar and Dutta 2020). The model is based on the regression equation:

$$y = b_0 + b_1 x_1 + \varepsilon \tag{1}$$

Where $x_1$is dependent variable (here, amount of accumulation/ablation)

b_0 is constant and b_1 is correlation coefficient of the independent variable x_1.

ε is the error $= \frac{\rho}{\sqrt{n}}$

The confidence level was also calculated for the data and establish the authenticity and accuracy of the measurements. The standard error of calculation was also computed for the data. Calculated margin of error is given by the following formula:

$$\overline{x} \pm t_{n-1,\alpha/2} \frac{s}{\sqrt{n}} \tag{2}$$

Where $\frac{s}{\sqrt{n}}$ is standard error, $t_{n\ -\ 1,\ \alpha/2}$ is the t-table value for margin of error.

Confidence level have been estimated using the

$$\overline{x} \pm x_{\alpha/2} \frac{\rho}{\sqrt{n}} \tag{3}$$

The high value of the R^2 (0.78) shows a strong correlation between the monthly temperature data of February (Table 2) and accumulation/ablation value measuring in March (Table 1). The regression model also shows the strong R^2 value with the temperature data for the month of February. The temperature data for the February 2019 is not available. Hence, the mean value of the Feb-2018 and Feb-2020 data have been utilized for the analysis. The data for March shows very poor correlation. Monthly surface temperature of February month shows negative correlation between temperature and accumulation/ablation (Fig. 3). It is usually observed that higher the temperature, lesser is the accumulation in the region. The OLS model which is a type of linear least square method for estimating the unknown parameters in a linear regression model, also shows significance to the data by higher F-Statistic (Table 3).

Table 1 Accumulation (+)/ablation (−) in cm for SN1 Stake Network, Larsemann Hills, East Antarctica as recorded in the month from 2014–2015 to 2019–2020

Accumulation/Ablation (cm)						
Stake no.	2014–2015	2015–2016	2016–2017	2017–2018	2018–2019	2019–2020
Period	March	March	March	March	March	March
1	−3.5	84.5	−28.0	19.0	−13.0	−4.0
2	−3.5	84.0	−32.5	4.0	3.0	8.0
3	−8.5	67.5	−34.0	21.0	−8.0	−2.0
4	−3.0	67.5	−20.5	3.0	−3.0	−9.0
5	−7.5	73.0	−33.5	8.0	6.0	−8.0
6	−28.5	68.5	−32.0	21.0	13.0	−5.0
7			−23.5	22.0	1.0	−6.0
8	−4.0	75.0	−21.0	15.0	−2.0	1.0
9	−18.5	86.5	−29.0	22.0	−4.0	−4.0
10	−2.5	89.5	−36.0	41.0	−25.0	−5.0
11	−5.0	87.5	−26.5	30.0	−43.0	24.0
12	3.0	78.0	−26.0	6.0	8.0	12.0
13	−14.0	84.0	−35.0	29.0	−5.0	−20.0
14	6.5	70.5	−21.0	20.0	−8.0	−9.0
15	−14.5	64.0	−20.5	16.0	−2.0	−7.0
16	−16.0	67.0	−33.0	21.0	−4.0	−6.0
17	−23.0	83.5	−23.5	23.0	−12.0	−3.0
18	2.5	74.0	−25.5	43.0	−21.0	−14.0
19	−1.5	80.5	−27.0	38.0	−23.0	−1.0
20	8.5	87.5	−19.0	18.0	−5.0	6.0
21	−10.0	84.5	−15.5	23.0	5.0	−1.0
22	−6.0	80.5	−24.5	18.0	−4.0	0.0
23	−1.0	75.5	−22.5	20.0	−8.0	8.0
24	−8.0	77.5	−27.5	9.0	5.0	0.0
25			−31.5	21.0	−5.0	5.0
Mean	−6.9	77.8	−26.7	20.4	−6.2	−1.6
Standard deviation	9.1	7.8	5.6	10.3	12.0	8.9
Std error	1.9	1.6	1.1	2.1	2.4	1.8

The constant value −125.9584 and coefficient −46.1099 was used from the model analysis to estimate the correlation in the Eq. (1). The red line in the Fig. 3 was derived from the equation. The ε has been estimated to be 0.29 °C. The annual accumulation/ablation range have also been estimated which is given in Table 4.

The data from the stake farms installed in the marginal area of ice sheet in 250 × 250 m grid pattern was considered as sample data for the whole region. The denser the stake network, more the accuracy of the model and extrapolation of the time series data. Using these data, the future accumulation-ablation pattern may be predicted. Different variables like temperature, precipitation, wind speed and

Table 2 Average monthly surface temperature data for Larsemann hills area

Year	Jan.	Feb.	Mar.	Apr.	May	Jun.	Jul.	Aug.	Sept.	Oct.	Nov.	Dec.
2014	1.3	−2.8	−10	−14.9	−10.7	−18.8	−18.4	−16.4	−12.2	−10.3	−2.7	0
2015	0.8	−2.1	−8.3	−15.1	−21.4	−14.9	−22	−17.2	−15.8	−12.4	−4.8	0.4
2016	0	−4	−10.4	−13.7	−15.2	−20.2	−18.6	−14.5	−20.6	−11.3	–	–
2017	0.8	−2.3	−8.6	−15.5	−12.6	−19.4	−16.7	−17.6	−18.2	−13.4	−4.3	−1
2018	−0.4	−3.4	−9.4	−9.8	−14.4	−13.5	−13.4	−12.9	−17.2	−12.1	−7.1	−0.3
2019	−0.2	−3.05[a]	−7.3	−13.6	−16.7	–	−17.6	−13.6	–	−11.1	−3.4	0.5
2020	1.8	−2.7	−11	−11.2	−12.5	−13.1	–	–	–	–	–	–
Mean		−2.9										
Std. dev		0.71										

[a]Statistically estimated for calculations. Source: https://ramadda.data.bas.ac.uk

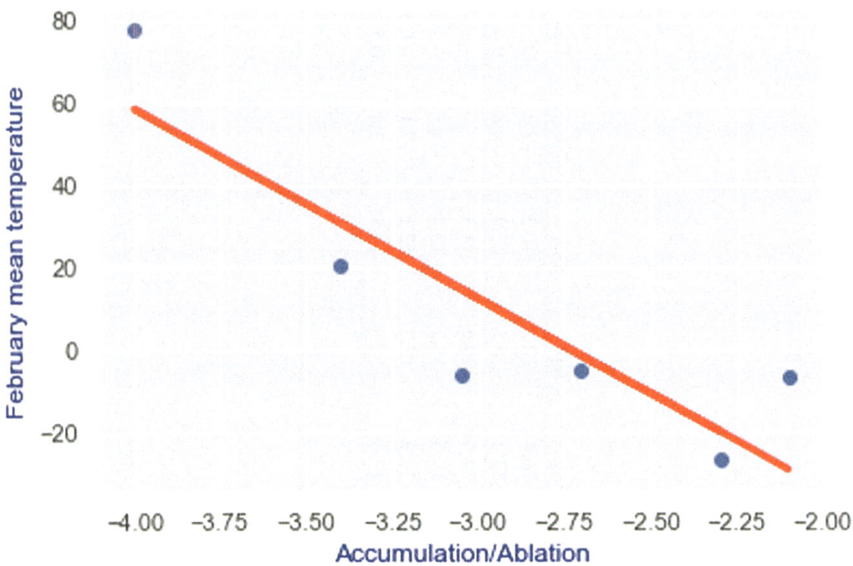

Fig. 3 Graph showing strong correlation between accumulation/ablation of March and monthly temperature data of February month

Table 3 Ordinary Least Square (OLS) regression result of accumulation/ablation and surface temperature data of February month since 2014

OLS Regression results	
Dep. variable: Accum Mar	R-squared: 0.784
Model: OLS	Adj. R-squared: 0.731
Method: Least Squares	F-statistic: 14.55
Date: Wed, 08 Jul 2020	Prob (F-statistic): 0.0189
Time: 16:39:48	Log-Likelihood: −25.021
No. observations: 6	AIC: 54.04
Df residuals: 4	BIC: 53.63
Df model: 1	
Covariance type: nonrobust	
coef std err t	P>|t| [0.025 0.975]
const −125.9584 36.210 −3.479 0.025 −226.494 −25.423	
Avg_feb_temp	−46.1099 12.087 −3.815 0.019 −79.668 −12.552
Omnibus: nan	Durbin-Watson: 0.757
Prob(Omnibus): nan	Jarque-Bera (JB): 0.583
Skew: 0.337	Prob(JB): 0.747
Kurtosis: 1.630	Cond. No. 15.3

wind direction can be used for fine-tuning of the model. During the current study, only the surface air temperature data has been analyzed. Though the duration of observational data is restricted in nature to assess and predict the changing climate,

Table 4 Annual accumulation/ablation range of the region is shown below with the 90% confidence

	Annual range of accumulation/ablation value in cm with the 90% confidence calculated using 'Student—t' value.					
Year	2014–2015	2015–2016	2016–2017	2017–2018	2018–2019	2019–2020
Lower range	−9.4	75.7	−28.2	17.7	−9.3	−3.9
Higher range	−4.4	79.9	−25.3	23.2	−2.9	0.8

an attempt has been made to understand the trends and patterns under current conditions.

It is observed that there is ablation during the month of December 2019 (Mean Value −6 cm) (Table 5). There is net accumulation between March and June 2020 (mean value 5.5 cm) during initial winter season. The huge accumulation may be attributed to drift snow in the region. The seasonal data during summer and winter are comparable and show consistent relationship during the corresponding months (Table 5). The regression analysis derived R^2 value shows strong correlation (0.818) between the accumulation/ablation value and corresponding monthly temperature data (Table 6). On other hand, it is evident that the surface air temperature of the region is having the great impact on the value of accumulation/ablation. In other words, it is also inferred that the lag in the cause and the effect is not much which suggests that the fall or rise in surface temperature affects the accumulation/ablation pattern during the next month. The F-Statistic shows that the model is less significant to predict the future trend using this OLS model. The number of observations in large time span would increase the significance of the model for predicting future trend. The analysis shows b_0 and b_1 is −6.05 and −1.68 respectively for the Eq. (1) (Fig. 4). The error factor in the independent variable is 3.22 °C. The data is little negatively skewed due to the temperature of the December month which is making outlier. If the outlier is removed, the R^2 value increases to 0.996 (Fig. 5 and Table 7). The high value may because of the only factor to affect during the Polar winter time is temperature it is also estimated that there should not be any precipitation value which can affect the correlation. It is also observed that after removing the outlier value (December 2019 data) the model becomes very significance with the F-Statistic value 242.6 (Table 7). Since only three observations were considered, the model is not much reliable for future predictions apart from providing a template for adding more data.

The stakes other than SN1 is installed in year 2018. Hence we have only 2 year of the annual accumulation-ablation pattern. The 2 year of the data cannot be analyzed for the regression analysis. Annual and monthly pattern and the value of accumulation-ablation is calculated.

The study starts since 2013 with installation of the stakes. In order to expand the study area more stake networks installed during 2017–2018 field season. During the

Table 5 Monthly data of exposed height and ablation (−ve value) /accumulation (+ve values) of SN1 network

Period	Nov–Dec. 19	Mar–Apr. 20	Apr–May. 20	May–Jun. 20
Temperature during observation period (°C)	0.5	−11.2	−12.5	−13.1
Stake 1	−4.8	25.7	0.9	1.6
2	−4.8	0.0	2.6	28.4
3	−7.2	1.1	21.4	17.4
4	−1.2	4.3	23.1	18.9
5	−1.2	3.2	18.9	25.3
6	−12	−2.1	29.1	22.1
7	−18	11.8	19.7	36.3
8	−4.8	4.3	15.4	4.7
9	−9.6	4.3	8.6	14.2
10	−12	0.0	29.1	−15.8
11	−2.4	9.6	14.6	44.2
12	3.6	9.6	11.1	15.8
13	−4.8	3.2	24.9	31.6
14	−10.8	13.9	4.3	−1.6
15	−3.6	7.5	15.4	17.4
16	−12	−1.1	25.7	25.3
17	−2.4	6.4	8.6	28.4
18	−1.2	1.1	23.1	23.7
19	−13.2	13.9	18.0	20.5
20	−7.2	9.6	10.3	26.8
21	−13.2	5.4	15.4	20.5
22	−1.2	3.2	18.9	36.3
23	−7.2	−1.1	18.0	12.6
24	−1.2	−2.1	24.9	15.8
25	3.6	5.4	14.6	42.6
Mean	**−6.0**	**5.5**	**16.7**	**20.5**
Standard deviation	5.5	6.4	7.8	13.7

period, 12 stake networks have been installed in the study area. For annual estimation of accumulation/ablation, January month data have been selected due to availability of the seamless data in that month every year (Kumar and Dutta 2020).

4 Discussion

Polar ice sheet of Antarctica, particularly that in East Antarctica is in focus today as there is a large uncertainty in estimation of the mass gain or loss in last 20 years (Hanna et al. 2013). Different techniques of estimation of mass balance of the Polar

Table 6 OLS regression result of monthly accumulation/ablation and corresponding monthly surface temperature data for 2019–20 observation period

OLS Regression Results	
Dep. variable: Accumulation/ablation	R-squared: 0.818
Model: OLS	Adj. R-squared: 0.727
Method: Least Squares	F-statistic: 9.002
Date: Sat, 11 Jul 2020	Prob (F-statistic): 0.0954
Time: 13:27:49	Log-Likelihood: −11.608
No. observations: 4	AIC: 27.22
Df residuals: 2	BIC: 25.99
Df model: 1	Covariance Type: nonrobust
coef std err t	P>|t| [0.025 0.975]
const	−6.0508 5.957 −1.016 0.417 −31.680 19.579
Temp	−1.6784 0.559 −3.000 0.095 −4.085 0.728
Omnibus: nan	Durbin-Watson: 2.012
Prob(Omnibus): nan	Jarque-Bera (JB): 0.569
Skew: −0.818	Prob(JB): 0.752
Kurtosis: 2.144	Cond. No. 20.5

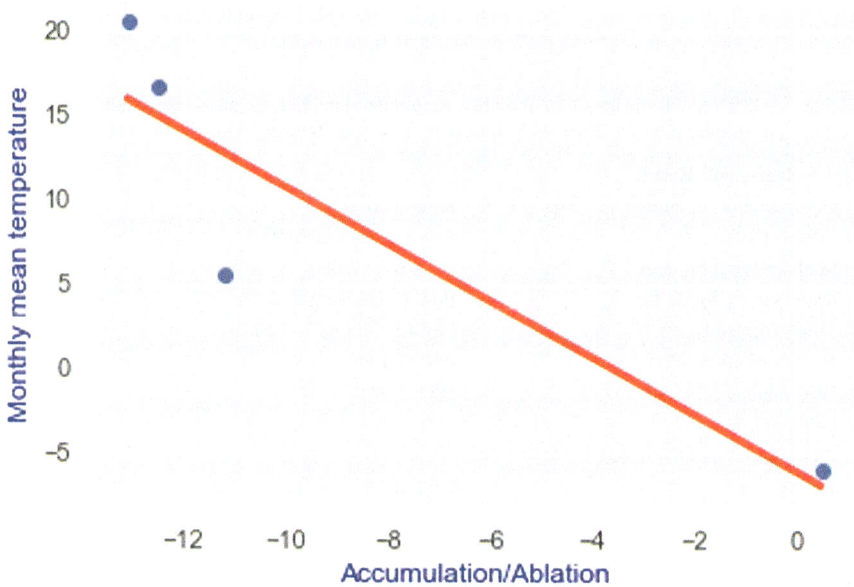

Fig. 4 Graph showing strong correlation between monthly accumulation/ablation value and monthly temperature

ice sheet viz. volumetric, space gravimetric or mass budget, particularly for East Antarctica, has provided different results and it is not certain that whether this sector of Antarctic ice sheet is actually gaining or losing mass. To employ these techniques,

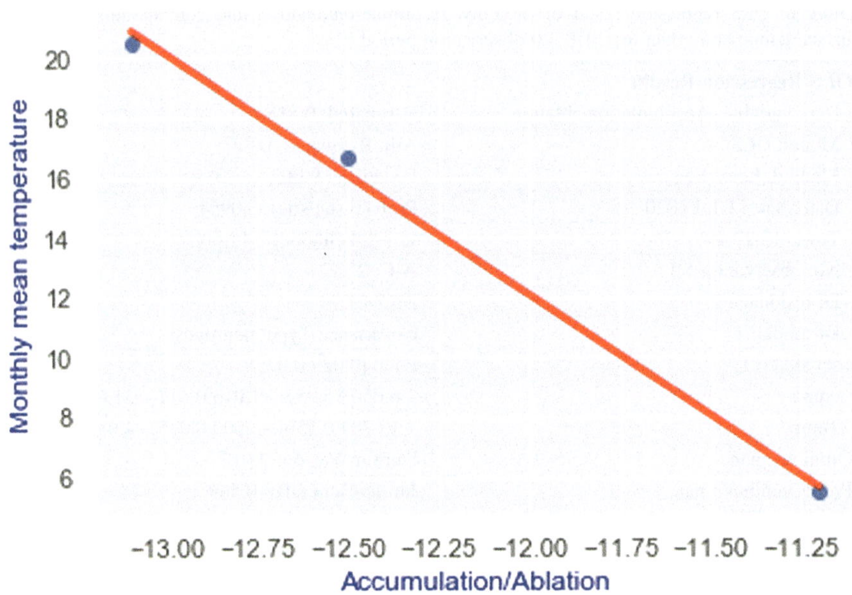

Fig. 5 Graph showing strong correlation between monthly accumulation/ablation value and monthly temperature after removing the outlier value from the dataset

Table 7 OLS regression result of monthly accumulation/ablation and corresponding monthly surface temperature data for 2019–2020 observation period after removing outlier value

OLS Regression Results	
Dep. variable: Accumulation/ablation	R-squared: 0.996
Model: OLS	Adj. R-squared: 0.992
Method: Least Squares	F-statistic: 242.6
Date: Sat, 11 Jul 2020	Prob (F-statistic): 0.0408
Time: 14:48:44	Log-Likelihood: −1.5719
No. observations: 3	AIC: 7.144
Df residuals: 1	BIC: 5.341
Df model: 1	
Covariance type: nonrobust	
coef std err t	P>|t| [0.025 0.975]
const	−84.2237 6.334 −13.298 0.048 −164.702 −3.745
Temp	−8.0257 0.515 −15.576 0.041 −14.573 −1.479
Omnibus: nan	Durbin-Watson: 2.913
Prob(Omnibus): nan	Jarque-Bera (JB): 0.445
Skew: 0.572	Prob(JB): 0.801
Kurtosis: 1.500	Cond. No. 192

radar altimetry (Davis et al. 2005; Zwally et al. 2005; Pritchard et al. 2009), laser altimetry (Jay Zwally et al. 2011), Gravity Recovery and Climate Experiment (GRACE) satellite data (Velicogna 2009; Pritchard et al. 2009) and ground data on the net ice accumulation on the ice sheets with estimates of discharge across the grounding line (Rignot and Kanagaratnam 2006) are required to be undertaken. The ice velocity and dynamic changes in the Dalk glacier front mainly derived from Gaofen-2 satellite and Polar Hawk-1 UAV suggest that, during 92 days of observation period, the front of Dalk Glacier advanced about 35–40 m, with an average ice velocity of ~0.5 m per day (Zhou et al. 2014). However, ground validation of the same has not been reported so far. The ice velocity and dynamic changes in the Dalk glacier front mainly derived from Gaofen-2 satellite and Polar Hawk-1 UAV suggest that, during 92 days of observation period, the front of Dalk Glacier advanced about 35–40 m, with an average ice velocity of ~0.5 m per day (Zhou et al. 2014). However, ground validation of the same has not been reported so far. Therefore, it is imperative to record and verify the ice dynamics of these fast flowing glaciers and the surrounding region, particularly to understand the ice flow velocity in different parts of the Polar ice sheet and the glaciers, measure the glacier and ice sheet thickness and thereby understand the stress pattern developed due to the ice movement leading to formation of crevasses (Swain 2020).

Geological Survey of India initiated glaciological studies in Antarctica since 2nd Indian Scientific Expedition to Antarctic (ISEA) (Kaul et al. 1985) in Queen Maud Land. Regular monitoring of snow accumulation/ablation studies was started from 4th ISEA (Singh et al. 1988) and continued till present. A summary of the accumulation/ablation of snow of two decades in the Schirmacher Oasis region was summarized by Chaturvedi et al. (2005). They highlighted the rising trend of deposition of snow on the ice shelf near India Bay region. Small-scale glaciological observations are being taken by different countries in the Ingrid Christensen coast. India initiated glaciological studies in this part of East Antarctica by establishing a stake network farm during the 33rd Indian Antarctic Expedition (2013–2014) The stakes fixed on the Polar ice sheet are being monitored annually for snow accumulation/ablation. GPR profiling was also carried out in Quilty Bay and on Polar ice cap to the south of Grovnes peninsula and bed rock interface was detected at around 60 m depth (Shah et al. 2017).

The present paper deals with the assessment of impact of climate change through integrated glaciological studies of Polar ice sheet between Polararboken and Dalk Glaciers, Larsemann Hills, East Antarctica, undertaken in 2019–2020. The study was conducted with the objective of recording ice sheet health since the establishment of stake networks in the study area. The assessment of the health of the ice sheet includes the monitoring of accumulation/ablation pattern for long time record. The monthly as well as annual observations were recorded during the study (Tables 1 and 5). The ice movement was also monitored using DGPS and an average annual velocity of 1.08 m/year was observed during 2017 to 2018 and 0.78 m/year during 2018 and 2019 (Kumar and Dutta 2020).

The accumulation/ablation pattern has been correlated with the temperature data and the correlation is quantified using different statistical parameters and equations.

The confidence level of the data has also been estimated. All the data range have been presented with the 90% of confidence. With increasing confidence, data range gets wider and has less accuracy in the prediction. Hence, to keep the data range significant, confidence level was kept at 90%. The accumulation/ablation data used average values of the stake networks distributed evenly in the study area and mean value was calculated along with the standard deviation. Some of the stake network show Gaussian distribution of the accumulation/ablation data with the short standard deviation. Both the variables, accumulation/ablation as well as temperature, were analyzed using regression analysis by Ordinary Least Square (OLS) model (Eq. 1). The longest available data duration, for both annual and monthly variation was recorded at the oldest stake network (SN1) installed in 2013–2014. Hence the observed data from this network shows good correlation with surface temperature (Kumar and Dutta 2020). Accumulation/ablation have been much affected by the temperature compared to other meteorological parameters (Kumar and Dutta 2020). Anomalous accumulation pattern was observed in the data of March–April 2020 period reflecting snow drift due to prevailing wind pattern. This observation warrants measurement of more parameters like precipitation and wind speed near the stake farm and opens new scope for study dimension.

The ice sheet movement was estimated using DGPS in static mode at some interval. The ice-streams bounding the study area in east and west show higher velocities as compared to the central part of the ice sheet. The velocity data shows that the SN10 and SN13 stake networks moved very fast due to their proximity to the ice stream flow line (Kumar and Dutta 2020). The SN1 which was northern most stake network shows very small displacement due to obstruction faced in the form of exposed rocky mass in the north (Kumar and Dutta 2020).

5 Conclusion

The stake network SN1 shows minor ablation during (2018–2019) though it followed the biannual repeat sine pattern with since the inception of the monitoring program in 2013–2014 (Table 1). The ablation pattern was much affected by the temperature in the region. The correlation between the annual temperature of February month showed good correlation with the accumulation/ablation value of the march month. The accumulation/ablation variation over the ice sheet indicate a lag of about 1 month as compared to the surface temperature. The accumulation/ablation value range could be calculated with 90% confidence level which could be used for the prediction of the future trends with reasonable accuracy. Stake networks SN1, SN2, SN4, SN7, SN9, SN11, SN12, SN13 shown ablation during the November–December month (Fig. 1). Some of the stake network like SN1, SN2, SN3, SN8, SN9 and SN10 have shown huge accumulation during the month of May 2020 (Kumar and Dutta 2020). The ice velocity responsible for the ice loss from the margin of the ice sheet is an important parameter reflecting the impact of changing climate (Kumar and Habib 2018).The average velocity of the ice sheet was observed to be 0.78 m/year. Higher velocities were observed at the margin of the ice streams (Polararboken Glacier and Dalk Glaciers). The general direction of the ice sheet movement in the area was towards NNW to North (Kumar and Dutta 2020). The direction of velocity followed the sub-ice topography in the region. The net surface mass balance for 2019–2020 is estimated to be 80,304,204 m^3 SWE and specific surface mass balance for study period 2019–2020 is 0.37 m SWE. These studies would help in estimating the long term mass balance in the area along with the ice dynamics. The studies would provide a model which would realistically reflect the ice sheet behavior trend in future.

Acknowledgments Authors extend their gratitude to Director General, Geological Survey of India for providing the opportunity to work in Antarctica. The authors are grateful to Additional Director General and National Mission Head – IV, GSI under whose aegis the project was approved and initiated. Authors want to record their gratitude to Dr. S. P. Shukla and Amit Dharwadkar, Director, Polar Studies Division, GSI, Faridabad for overall supervised guidance. The authors wish to thank one and all officers and staff in Polar Studies Division, GSI, Faridabad for providing necessary help and co-operation. We thank NCAOR, Goa for providing logistic support during the 39th ISEA. Thanks are also due to Shri. Debdip Chakravarti, Station Leader 39th ISEA, Bharati Station, East Antarctica. Also thanks are due to Shri. Pawan Sirodkar, Lab Assistant and his team, Communication officer Sh. Ankush and chef Sh. Jitendra for the kind generosity and help during the field work. We sincerely thank Shri. M. Javed Beg, Programme Director (Logistics) and Dr. Shailendra Saini, Scientist, E and Voyage Leader, 39th ISEA, NCAOR, Goa. Authors extend their thanks to Dr. Rahul Mohan, Group Director and Scientist F, NCPOR for his continuous support and push for this publication.

References

Allen MR, Dube OP, Solecki W, Aragón-Durand F, Cramer W, Humphreys S, Kainuma M, Kala J, Mahowald N, Mulugetta Y, Perez R, Wairiu M, Zickfeld K (2018) Global warming of 1.5°C. An IPCC special report on the impacts of global warming of 1.5°C above pre-industrial levels and related global greenhouse gas emission pathways, in the context of strengthening the global response to the threat of climate change. In: Framing and context. IPCC, Geneva

Box JE, Bromwich DH, Veenhuis BA, Bai LS, Stroeve JC, Rogers JC, Steffen K, Haran T, Wang SH (2006) Greenland ice sheet surface mass balance variability (1988-2004) from calibrated polar MM5 output. J Clim 19(12):2783–2800. https://doi.org/10.1175/JCLI3738.1

Chaturvedi A, Beg MJ, Singh A, Ravindra R (2005) Two decades of observations over polar ice cap margin in Schirmacher range of central Dronning Maud Land Antarctica. In: Symposium of Antarctic glaciology (ISAG-7) Milano, Italy, 7

Davis CH, Li Y, McConnell JR, Frey MM, Hanna E (2005) Climate change: snowfall-driven growth in East Antarctic ice sheet mitigates recent sea-level rise. Science 308(5730): 1898–1901. https://doi.org/10.1126/science.1110662

Hanna E, Navarro FJ, Pattyn F, Domingues CM, Fettweis X, Ivins ER, Nicholls RJ, Ritz C, Smith B, Tulaczyk S, Whitehouse PL, Jay Zwally H (2013) Ice-sheet mass balance and climate change. Nature 498(7452):51–59. https://doi.org/10.1038/nature12238

Jay Zwally H, Li J, Brenner AC, Beckley M, Cornejo HG, Di Marzio J, Giovinetto MB, Neumann TA, Robbins J, Saba JL, Yi D, Wang W (2011) Greenland ice sheet mass balance: distribution of increased mass loss with climate warming; 2003-07 versus 1992-2002. J Glaciol 57(201): 88–102. https://doi.org/10.3189/002214311795306682

Kaul MK, Chakraborty SK, Raina VK (1985) Experiment on artificial augmentation of ablation on the shelf ice, Antarctica, vol 2. Technical Publication, New Delhi

Kumar P, Dutta S (2020) Assessment of impact of climate change through integrated glaciological studies of polar ice sheet between Polararboken and Dalk glaciers, Larsemann Hills, East Antarctica. Unpublished Report, Geological Survey of India, July

Kumar P, Habib Z (2018, September) Ice sheet dynamics from Schirmacher oasis to Wohthat montains, CDML, East Antarctica and their stress pattern. Unpublished Report, Geological Survey of India

Meredith MM, Sommerkorn S, Cassotta C, Derksen A, Ekaykin A, Hollowed G, Kofinas A, Mackintosh J, Melbourne-Thomas MMC, Muelbert G, Ottersen H, Pritchard, Schuur EAG (2019) IPCC Special report on the ocean and cryosphere in a changing climate. In: Polar regions. IPCC, New Delhi

Monaghan AJ, Bromwich DH, Fogt RL, Wang SH, Mayewski PA, Dixon DA, Ekaykin A, Frezzotti M, Goodwin I, Isaksson E, Kaspari SD, Morgan VI, Oerter H, Van Ommen TD, Van Der Veen CJ, Wen J (2006) Insignificant change in Antarctic snowfall since the international geophysical year. Science 313(5788):827–831. https://doi.org/10.1126/science.1128243

Pritchard HD, Arthern RJ, Vaughan DG, Edwards LA (2009) Extensive dynamic thinning on the margins of the Greenland and Antarctic ice sheets. Nature 461(7266):971–975. https://doi.org/10.1038/nature08471

Rignot E, Kanagaratnam P (2006) Changes in the velocity structure of the Greenland ice sheet. Science 311(5763):986–990. https://doi.org/10.1126/science.1121381

Shah MY, Ayemi K, Kumar P, Mandal A, Mallik RK (2015) Glaciological studies in the Larsemann Hills, East Antarctica for assessing the impact of climate change. Unpublished Report, Geological Survey of India

Shah MY, Ayemi K, Shrivastava P (2017) GPR survey and physical measurements of sea ice in Quilty Bay, Larsemann hills, East Antarctica and its correlation with local atmospheric parameters. J Geol Soc India 90:371–377. https://doi.org/10.1007/s12594-017-0726-4

Shepherd A, Ivins E, Rignot E, Smith B, van den Broeke M, Velicogna I, Whitehouse P, Briggs K, Joughin I, Krinner G, Nowicki S (2018) Mass balance of the Antarctic ice sheet from 1992 to 2017. Nature 558(7709):219–222. https://doi.org/10.1038/s41586-018-0179-y

Singh RK, Mukerji S, Srivastava D, Kaul MK (1988) Snow accumulation and ablation pattern on ice shelf near Dakshin Gangotri, Antarctica, and Development of fast Ice off Dakshin Gangotri. Fifth Indian Expedition to Antarctica, Scientific Report 5:189–204

Swain AK (2020) Glacier stress pattern as an indicator for climate change. In: Climate change and the white world. Springer, New York, pp 119–138. https://doi.org/10.1007/978-3-030-21679-5_9

Swain AK, Chandra V (2017) Glaciological monitoring around Grovness, Larsemann Hills, East Antarctica. Unpublished Report, Geological Survey of India

Velicogna I (2009) Increasing rates of ice mass loss from the Greenland and Antarctic ice sheets revealed by GRACE. Geophys Res Lett 36(19):2–5. https://doi.org/10.1029/2009GL040222

Zhou C, Deng F, Ai S, Wang Z, E, D. (2014) Determination of ice-flow velocity at the polar record glacier and Dalk glacier using DInSAR. Wuhan Daxue Xuebao (Xinxi Kexue Ban)/Geomatics and Information Science of Wuhan University 39(8):940–944. https://doi.org/10.13203/j.whugis20130116

Zwally HJ, Giovinetto MB, Li J, Cornejo HG, Beckley MA, Brenner AC, Saba JL, Yi D (2005) Mass changes of the Greenland and Antarctic ice sheets and shelves and contributions to sea-level rise: 1992-2002. J Glaciol 51(175):509–527. https://doi.org/10.3189/172756505781829007

Prevalent Climate Variables During Ablation Season Around Gangotri Glacier

Manohar Arora and Jatin Malhotra

Abstract Hydrological investigations for the glacierized basins play important role in the development of water sector of the country. The glacier monitoring is very less and very few glaciers are being continuously monitored in Indian Himalayas. The Himalayan cryosphere system is sustained and regulated by the interaction of temperature and precipitation forced through the orographic processes. In this study, an extensive meteorological analysis has been carried out for the data collected at Gangotri Glacier base station at Bhojwasa for four consecutive ablation seasons (2014–2017) from May to October every year. The observed meteorological parameters included rainfall, temperature, evaporation, relative humidity, wind speed and direction and sunshine hours.

Average monthly rainfall recorded for the months June, July, August and September are 34.8, 87.6, 54.6 and 32.6, respectively. During the complete ablation period the total rainfall and its distribution varies from year to year. Based on the collected 4 years of data the average seasonal rainfall for the Gangotri Glacier was observed to be about 221.2 mm. The average daily maximum and minimum temperatures over the ablation season were determined to be 15.7° C and 4.7° C, respectively, whereas average mean temperature was 10.2° C. The wind data was also analysed and it was found that on an average the daytime wind speeds are nearly 4 times stronger in magnitude than during the night. On the seasonal scale daily mean sunshine hours were observed to be 4.7 hours. Monthly total pan evaporation was observed to be 118, 125, 107, and 121 mm for the month of June, July, August and September respectively. The mean daily evaporation for the ablation season as a whole is observed to be 3.9 mm, which is comparable to the pan evaporation data for the stations situated at lower altitudes of the Himalayas.

Keywords Gangotri Glacier · Rainfall · Temperature · Evaporation and climatology

---

M. Arora (✉) · J. Malhotra
National Institute of Hydrology, Roorkee, India

B. Phartiyal et al. (eds.), *Climate Change and Environmental Impacts: Past, Present and Future Perspective*, Society of Earth Scientists Series,
https://doi.org/10.1007/978-3-031-13119-6_11

1 Introduction

Climate is one of the most significant factor that influences glacier dynamics. Anthropogenic changes in the atmosphere affects the health of a glacier by changing the temperature regime and is imitated in the retreat or advance of glaciers. Hence meteorological studies of glacier basins are very much necessary to understand climate-glacier close interactions (Arora et al., 2008).

Approximately two third of all of the world's fresh water is protected in glacier ice. The melt produced from Glaciers provide potable water for people in various places around the world, including the Himalayas. Glaciers respond to variations in climate, primarily snowfall and temperature, by changing in length and thickness therefore understanding the regimes of the glaciers is important. In comparison to other mountains of the world the meteorological records of the Himalayas are poorly documented. Difficult approaches, rough topographies and severe weather conditions are considered to be the major problem in creating a good database. The installation, operation and maintenance of the meteorological instruments is of utmost important in the high-altitude Himalayan region. Strengthening of hydro meteorological network in high-mountain areas is very much required for improvement in the database. The data analysis helps us to understand weather conditions, its control on the water availability of the region, snow and glacier melting, sediment transport mechanism, tourism development and also to give an understanding to the complications related to natural hazards, viz. flash floods, debris flows, cloudbursts, landslides and avalanches. Singh et al. (2005) analysed the meteorological data for the study area from year 2000 to 2003. Further, Koul and Ganjoo (2010) studied the impact of inter and intra annual variation in meteorological parameters on the health status of Naradu glacier. They observed that the trends in seasonal and monthly mean temperatures depicted an interesting shift of peak summer and winter season. The data also explained night warming during summer and winter and low temperatures during peak summer season and very cold during winter.

The Himalayan glaciers have most rugged topography coupled with extreme climatic conditions making them one of the most difficult environments in the world. Special efforts are needed to carry out meteorological studies in the glaciated regions. In glacier monitoring though radiation and temperature are important parameters for melt production but wind speed, rainfall and air moisture also play crucial role in melt generation. A conventional observatory was installed near the snout of the glacier and data collected has been analysed and presented in this paper.

2 Study Area

The Gangotri Glacier (Lat. $30^0 43'$ N–$31^0 01'$ N and Long. $79^0 00'$ E–$79^0 17'$ E) is the second largest glacier of the Indian Himalayas. The meltwater stream originating from the snout of the Gangotri Glacier (4000 masl) is known as the Bhagirathi River.

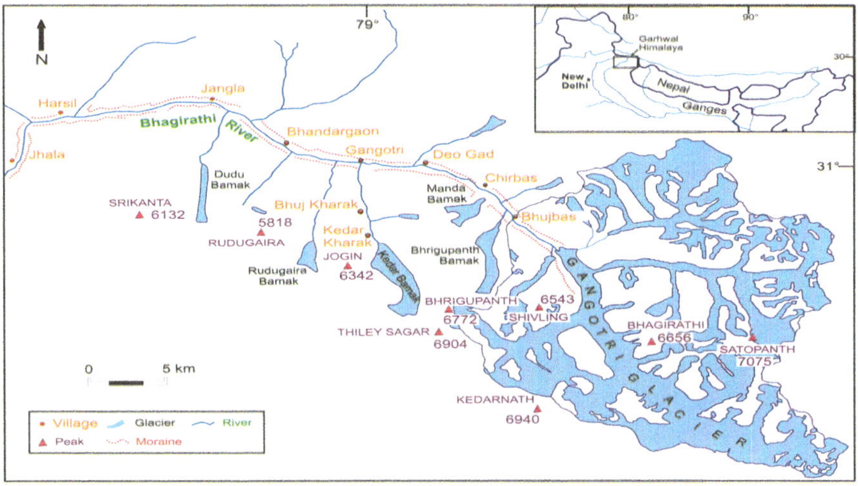

Fig. 1 Location of the Gangotri Glacier in the Garhwal Himalayas

The Gangotri Glacier system (most commonly known as Gangotri Glacier), is a group of many glaciers with the main Gangotri Glacier (length: 30.20 km; width: 0.20–2.35 km; area: 86.32 km^2) as its trunk. It is a temperate mountain valley glacier, which flows in the northwest direction. The major tributary glaciers of the Gangotri Glacier system are the Raktvarn, Swachand, Chaturangi, and Maiandi glaciers that merge with the stem glacier from the North-east, and the Meru, Ghanohim and Kirti that merge with the trunk glacier from the South-west. The altitudinal range of the glacier varies between 4000 to 7000 m. The total area of the Gangotri catchment is 556 sq. km out of which 228 sq. km is glaciated upto the established water discharge site (4 km downstream of glacier snout known as Gaumukh). The glacier snout is at about 18 km above from the Gangotri settlement. There are 238 glaciers of different sizes in the Bhagirathi basin which occupies about 12% of the basin area. Gangotri glacier is the largest glacier in the Bhagirathi basin. Figure 1 shows the location and area of the Gangotri Glacier with the gauging site.

3 Methodology

A conventional hydro-meteorological observatory was established at Bhojwasa site (~3800 ma.s.l.) about 4 km downstream to the snout of Gangotri Glacier for the collection of hydro-meteorological data during the ablation season (May–October) of the corresponding hydrological year. This observatory is well equipped with major meteorological instruments that includes ordinary rain gauge, self-recording rain gauge, maximum and minimum thermometers, dry and wet bulb thermometers, hygrograph, pan-evaporimeter, anemometer, wind vane and sunshine recorder.

These instruments are calibrated each year as per the standards and guidelines of Indian Meteorological Department (IMD) for the optimum results and to maintain the accuracy in measurements. Indian Standard timings (0830, 1130, 1430 and 1730) for hydro-meteorological data collection are used according to the practice followed by the IMD. The monitoring of hydro-meteorological parameters during the ablation season was carried out using conventional methods and Automatic Weather Station (AWS) at the Meteorological Observatory near the snout of Gangotri Glacier. The AWS continuously records the observations of various hydro-meteorological parameters throughout the winter season as well at a desired interval of time.

An AWS is also installed The hydro-meteorological records viz. air temperature, relative humidity, wind speed and direction, evaporation, rainfall and discharge for the ablation seasons (May–October) of 2014, 2015, 2016 and 2017 were collected and analyzed for the present study.

4 Result and Discussion

4.1 Air Temperature (T_a)

Air temperature tends to play a dominant role in snow/ice melting (Kesarwani et al., 2015; Kesarwani, 2015). In the Gangotri Glacier valley, Maximum and minimum temperature thermometers were placed in a Stevenson's screen at meteorological site. Daily maximum (T_x), minimum (T_n) and average (T_a) air temperatures observed near the snout of Gangotri Glacier for different years are shown in Fig. 2. The observed average air temperature records during the ablation seasons of 2014 (9.0 °C), 2015 (9.3 °C), 2016 (10.0 °C) and 2017 (9.4 °C) shows that the Gangotri Glacier valley experiences temperature ranging from 9.0 to 10 °C in a controlled manner (increasing trend from May to till July and then starts decreasing). The highest average temperature was observed 10 °C in 2016 whereas July 2015 (11.6 °C) was the warmest month during the observation period (Table 1). The frequency distribution analysis of diurnal air temperature specifies that maximum temperature occurred between 1200 and 1400 h (local time) while minimum during the 0000 to 0200 h (local time).

4.2 Relative Humidity (R_H)

The relative humidity (R_H) of the air is the ratio of the actual amount of moisture in the air to the saturated amount and its variability depends on the temperature and the pressure of the environment (Ahmad et al. 2017). In the Gangotri Glacier valley, dry bulb and wet bulb thermometers were placed in a Stevenson's screen. A psychrometric chart was used to calculate the relative humidity. The relative humidity tends to be lower during the day time and higher at night as a consequence of inverse

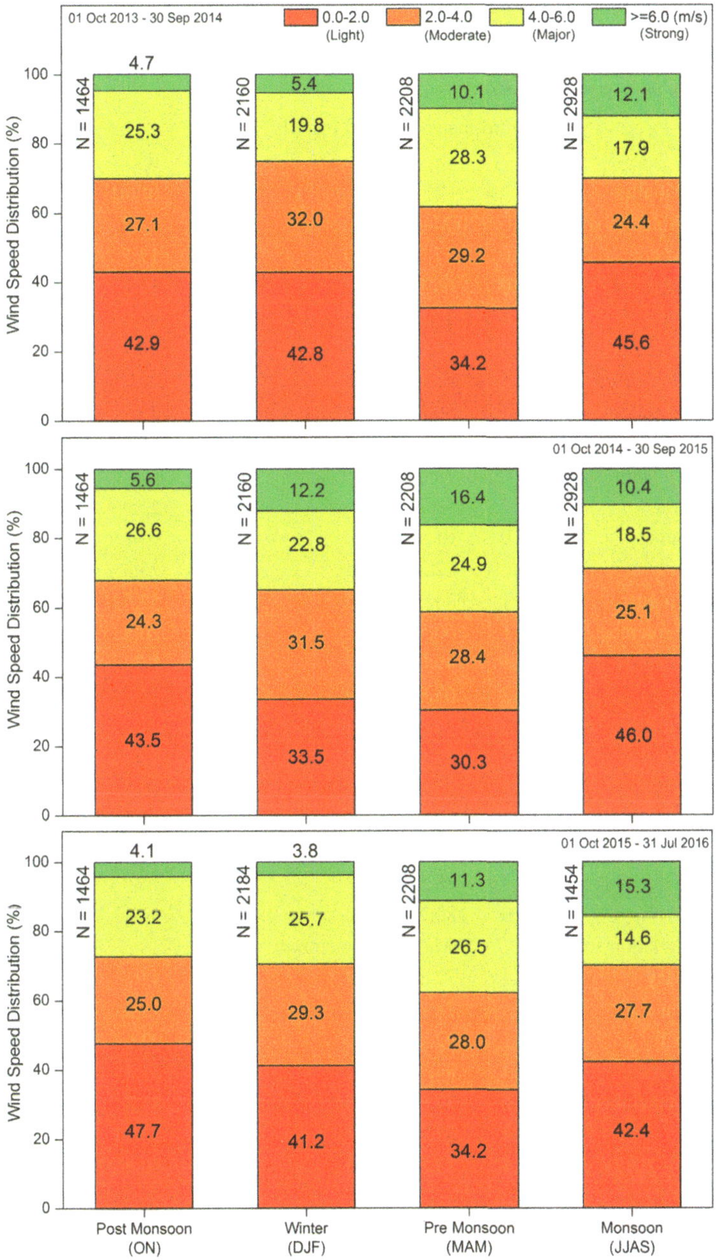

Fig. 2 Seasonal fluctuations in wind speed distribution (%) observed at meteorological site of the Gangotri Glacier valley. N = Total number of diurnal observations

Table 1 Calculated mean monthly air temperature (T_a), relative humidity (R_H), wind speed (W_S), evaporation (E), meltwater discharge (Q_{obs}) and total rainfall (R) from the observed records available for the ablation seasons (May-Oct) of 2014, 2015, 2016, and 2017. Corresponding unit for each variable is given in brackets

Year	Month	Hydro-meteorological records[¥]					
		T_a* (°C)	R_H* (%)	W_S* (kmh^{-1})	E* (mm)	R[†] (mm) / %	Q_{obs}* (mm d^{-1})
	Ablation season 2014 (20 May–14 October)						
2014	May	7.0	57.6	2.0	4.4	17.7	0.8
	June	11.0	66.5	3.3	4.4	12.3	5.5
	July	11.4	72.7	6.6	3.3	91.0	23.7
	August	10.4	66.0	6.5	5.1	49.2	15.5
	September	8.5	64.3	6.3	4.5	20.8	5.0
	October	5.2	59.4	6.4	3.4	2.6	2.0
		8.9	65.8	5.4	4.2	193.6	8.8
	Ablation season 2015 (13 May–09 October)						
2015	May	7.7	50.7	8.5	4.3	22.9	3.6
	June	9.0	61.9	7.0	4.1	90.8	6.6
	July	11.6	72.2	6.4	3.7	75.4	12.4
	August	11.1	73.0	6.5	3.4	52.5	13.6
	September	9.1	60.5	6.7	3.9	20.3	5.4
	October	7.1	49.0	6.7	3.4	1.3	3.6
		9.3	63.8	6.9	3.8	263.2	7.5
	Ablation season 2016 (17 May–30 September)						
2016	May	8.3	57.6	8.2	4.8	5.4	4.2
	June	11.2	75.6	8.4	4.6	10.2	9.5
	July	11.2	87.4	7.0	3.1	91.3	14.6
	August	10.1	91.0	6.6	2.5	85.1	12.4
	September	9.4	88.5	6.3	3.6	7.7	8.3
		10.0	82.7	7.2	3.7	199.7	9.8
	Ablation season 2017 (20 May–30 September)						
2017	May	7.4	77.8	6.9	2.7	4.9	5.2
	June	9.5	78.7	7.6	3.3	25.9	7.2
	July	11.1	92.4	6.9	2.4	92.6	16.6
	August	10.4	84.7	6.2	3.0	31.4	14.3
	September	8.6	77.6	6.0	2.7	81.5	6.0
		9.4	82.9	6.7	2.8	236.3	9.9
Mean* (2014–2017)		9.4	73.8	6.6	3.7	40.6/223.2	9.0

relationship with air temperature. Monthly and seasonal variation in relative humidity pattern showed that it ranged between 50.1 to 91.1% in the glacier valley due to the influence of Indian summer monsoon and local precipitation system (Table 1). Values of high humidity in the glacier valley indicated that high amount of precipitation and cloud cover whereas low was associated with clear sky. The average relative humidity for the ablation season of 2014 was calculated to be 65.8%, 63.8% for 2015, 82.7% for 2016 and 82.9% for 2017 (Table 1).

4.3 Wind Speed (W_S) and Direction (W_D)

In the high-altitude region, wind plays a fundamental role in local climatology and changing the heat budget of the system (Singh et al., 2005; Kesarwani, 2015). A cup-counter anemometer is used to measure the wind speed and direction. Diurnal analysis of wind speed (W_S) records suggest that its pattern is dissimilar for day and night and vary with change in season. Wind speed is observed to be higher during the day time due to stronger temperature deficit while it is minimum in the morning and night hours. Similarly, in each ablation season higher frequency of wind speed is observed during the pre-monsoon months except ablation season of 2014 (Table 1; Fig. 2). The highest average wind speed (8.7 km/h) was observed during 2016 ablation season followed by 8.0 km h^{-1} in 2017, 8.2 km h^{-1} in 2015 and 6.3 km h^{-1} in 2014. The average wind speed was calculated 7.8 km h^{-1} speed during the whole observation period (Table 1).

The direction of wind is primarily influenced by the topographical condition of valley which plays a significant role in the melting of snow/ice (Kesarwani et al., 2015). In the study area, wind flows from Chaukhamba group of peaks (7138 m a.s. l.) to the Gangotri Glacier valley and the maximum frequency of wind direction is observed North-West to South-East, and vice versa during the ablation season (Fig. 3). The frequency distribution analysis of wind direction records suggested that the rate of recurrence of anabatic wind (from valley to mountain) had higher frequency than the katabatic wind (from mountain to valley) during the monsoonal months (July, August and September), leading to increment of sensible heat causing more surface melting of snow/ice. However, as a consequence of residual winter snow cover during the months of May and June, the katabatic regime was higher than the anabatic in the region and as snow melted out the fraction increased towards the anabatic regime. On an average, the fraction of anabatic regime (65%) was higher than katabatic regime (35%) during all the ablation seasons, which showed that the glacierized area experienced much surface melting caused by turbulent heat fluxes.

4.4 Evaporation (E)

Evaporation play an important role in estimating the hydrological budget or streamflow modelling of a catchment. The primary forcing factors which are responsible for changing the evaporation dynamics are atmospheric temperature, wind speed and direction, net radiation and relative humidity (Singh et al., 2005). In Gangotri Glacier valley, analysis of the evaporation records suggested that it was

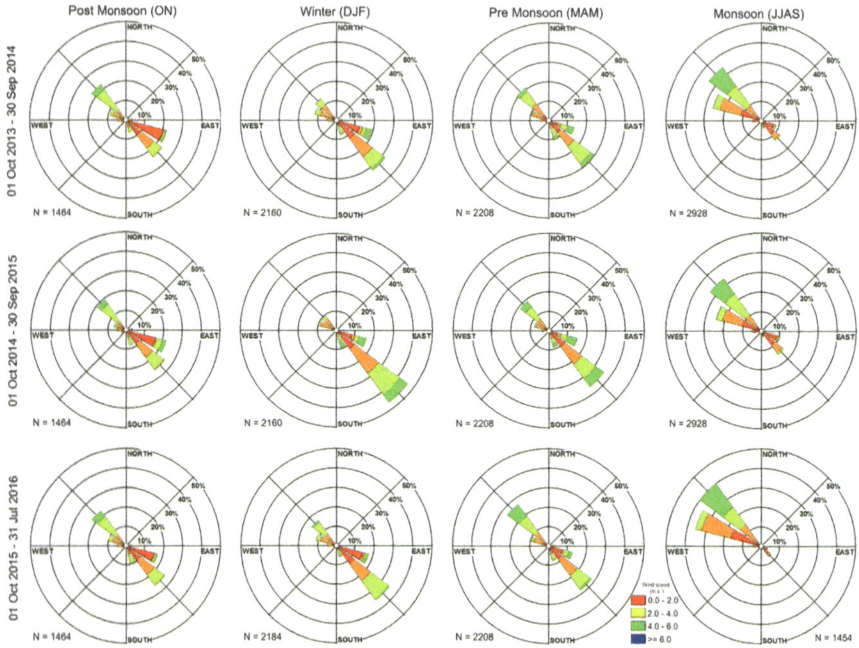

Fig. 3 Rose diagram showing seasonal fluctuations in wind direction and speed observed at meteorological site of the Gangotri Glacier valley. N = Total number of diurnal observations

generally high during the pre-monsoon months (May–June) due to favourable meteorological conditions such as clear sky, high sunshine hours, low relative humidity, high temperature and high wind speed (Table 1 and Fig. 2). During the monsoon months, it was observed to be low because of high moisture conditions, high cloud cover and associated less sunshine hours. The highest average evaporation was estimated 4.2 mm during the ablation season of 2014 followed by 3.8 mm in 2015, 3.7 mm in 2016 and 2.8 mm in 2017. However, average evaporation of the Gangotri Glacier valley was calculated to be 3.6 mm during the study period.

4.5 Rainfall (R)

The Indian summer monsoon produces heavy rainfall on the southern hill slopes of Himalaya (starting from East to West) during the summer months (June to October), whereas north-western Himalayan ranges receives less amount of precipitation (Singh et al., 2005; Loukas and Quick 1993). The Gangotri Glacier valley experiences orographic barrier and lies in the rain shadow region (leeward side) /northern part of the Central Himalaya, therefore, receives less amount of precipitation (< 300 mm) during the summer monsoon months (Singh et al., 2005; Srivastava, 2012; Singh and Kumar 1997). An ordinary rain gauge was used to collect the daily rainfall records. Analysis of observed rainfall records shows that its distribution pattern for monthly or seasonal scale is very dissimilar due to highly variable frequency of Indian summer monsoon in the glacier valley (Fig. 2 and Table 1). Generally, it was observed that the study area receives more than 75% annual rainfall between June and August (2014–79%; 2015–83%, 2016–93%) except in 2017 (63%). In 2017, September received second highest rainfall (34%) after July (39%) due to dominant pattern of Indian summer monsoon rainfall in this month (Table 1). A comparison of rainfall records indicated that highest total rainfall was received in 2015 (263.2 mm) followed by 236.3 mm in 2017, 199.7 mm in 2016 and 193.6 mm in 2014 (Table 1).

5 Conclusion

The climate science of the high-altitude, glacier regions is quite different from that in the plains. This paper highlights the results of detailed meteorological observations conducted for four-ablation seasons (May–September; 2014–2017) on the Gangotri Glacier. Results depicts fewer rainfall, lower air temperatures, relatively high wind speed, higher evaporation rates and moderately high humidity. Average seasonal (May–October) rainfall was about 221 mm. The daily rainfall observed scarcely exceeds 15 mm in the study area. About 80% rain events recorded daily rainfall of less than 5 mm. However, relatively good rainfall occurs in this area during July–September. The inter-seasonal variability of rainfall over the summer season is more. The average daily maximum and minimum temperatures over the summer season were computed to be 15.7°and 4.7 °C respectively, whereas average mean temperature was 10.2 °C. Diurnal variations in temperature indicates that generally maximum temperature is observed around 1400 h, while the minimum is observed at the early morning hours. The warmest month in all years was found out to be June. Diurnal temperature range was found to be more in May and October due to relatively cloud-free weather conditions, while it was lowest in the month of September generally due to presence of moist cloudy conditions. Results show that changes in minimum temperature are more substantial than the changes in maximum temperature. Strong winds observed during daytime would have contributed to high rate of evaporation even at high altitudes. Predominantly more sunshine hours in the region is also responsible for more evaporation. In general, high relative humidity was observed throughout the melt period.

References

Ahmad L, Kanth RH, Parvaze S, Mahdi SS (2017) Experimental agrometeorology: a practical manual. Springer International Publishing, p 159

Arora M, Singh P, Goel NK, Singh RD (2008) Climate variability influence on hydrological responses of a large Himalayan Basin. Water Resour Manag 22(10):1461–1475

Kesarwani K (2015) Energy balance and mass fluctuation of the Himalayan Glaciers: a case study of Chorabari Glacier, Garhwal Himalaya. Ph.D. thesis submitted to Kumaun University, Nainital, Uttarakhand, pp 236

Kesarwani K, Dobhal DP, Durgapal A, Mehta M (2015) High altitude meteorlogy and cloud cover conditions: a study from Chorabari Glacier, Central Himalaya, India. Himal Geol 36(2): 134–142

Koul MN, Ganjoo RK (2010) Impact of inter and intra annual variation in weather parameters on mass balance and equilibrium line altitude of Naradu Glacier, N W Himalaya, India. Clim Chang 99:119–139

Loukas A, Quick MC (1993) Rain distribution in a mountainous watershed. Nord Hydrol 24:225–242

Singh P, Kumar N (1997) Effect of orography on precipitation in the Western Himalayan region. J Hydrol 199:183–206

Singh P, Haritashya UK, Ramasastri KS, Kumar N (2005) Prevailing weather conditions during summer season around Gangotri Glacier. Curr Sci 88(5):753–760

Srivastava D (2012) Status report on Gangotri Glacier, Science and Engineering Research Board, Department of Science and Technology, New Delhi, Himalayan Glaciology Technical Report No. 3, 102p

Compacted Snow Dune Complexes in Antarctica and their Applicability as New Climate Change and Basement Tectonic Parameters

Anshuman Misra, K. S. Misra, and D. P. Dobhal

Abstract Compacted snow dunes were described for the first time as long stripes by pilots flying over eastern Antarctica. With the availability of satellite imagery, the interest has increased manifolds. Later Synthetic Aperture Radar (SAR) data provided nearly three-dimensional, enhanced relief pictures, of these dune complexes. Divergent views regarding their formation have been presented earlier. One group believed that they are erosional features imprinted on ice sheets due to strong katabatic winds, while the other one suggested that they have formed due to the accumulation of snow. Our interpretation of SAR and optical images over an area of 1,000,000 sq. km, suggests that they are a type of "fore-dunes", characterized by gentle wind-ward and steeper lee-ward sides. Deposited and compacted during intense climatic conditions. We believe that alternating wet and arid conditions prevailed during their formation and compaction. Wet periods were dominated by moisture-laden, low amplitude and long wavelength winds. This was accompanied by very heavy snowfall and accumulation as linear ridges perpendicular to the wind direction. Arid periods were marked by dry winds, compaction and development of wide-spread glazed surfaces on the lee-ward sides. Similarly, during the following period, with slightly less intense climatic conditions, snow sheets were deposited. Pile-ups along the high-rise regions later consolidated as ridges (sastrugi). These prominent climatic variations of continental proportions are mapped and seem to correspond with glacial and interglacial periods of the Quaternary Era. Prevailing as well as paleo-wind directions coincide with each other this indicates that there has been no marked change ever since they have formed. Furthermore, abrupt termination and shifting of compacted dunes, and formation of elongated valleys, filled by consolidated snow along parallel set of lineaments, are mapped in all the studied

A. Misra (✉)
Kumaun University, Nainital, India

K. S. Misra
University of Petroleum and Energy Studies, Dehradun, India

D. P. Dobhal
Wadia Institute of Himalayan Geology, Dehradun, India

B. Phartiyal et al. (eds.), *Climate Change and Environmental Impacts: Past, Present and Future Perspective*, Society of Earth Scientists Series,
https://doi.org/10.1007/978-3-031-13119-6_12

regions. This suggests that the lineaments are of basement origin and have propagated upward through the cover of ice and snow. We believe that the compacted snow dunes complexes and lineaments are unique markers and can serve as new parameters, to understand climatic changes and basement configuration of Antarctic continent.

Keywords Snow dunes · Snow sheets · Climate change · Lineaments · Basement tectonics · Radar images · Antarctica

1 Introduction

In the icy Antarctic continent, compacted snow dunes, as well as snow sheets, are perhaps the most important geomorphic features. Dune like pattern was identified, for the first time in eastern Antarctica (Swithinbank 1988). He conclusively differentiated them from clouds by exactly similar pattern, observed in multi-date Landsat imagery and described them as mega-dunes. Nothing much could be achieved from optical remote sensing because it was difficult to get images during dark six winter months, while during illuminated summer months the radiation from snow, ice and cloud reduces the perception and thus do not provide worthwhile images. Furthermore, due to the orbital parameter of Landsat, mostly images are of peripheral region, whereas, the snow dunes and snow sheets are dominantly located in the central part of Antarctica. Synthetic Aperture Radar (SAR) images have made an eloquent exposition of not only distribution of dune complexes but also features associated with the individual dunes. An excellent and detailed description of these mega-dunes is given by Fahnestock et al. (2000), They also estimated that more than 300,000 sq. km is occupied by dune fields in eastern Antarctica. However, as studies progressed many more areas with a prolific distribution of dunes are identified. Shuman et al. (2011) calculated that 900,000 sq. km area is occupied by dune complexes. During our studies, we interpreted mosaic of SAR data from Canadian Radars Atas well as European Space Agency's Sentinel 1 and 2 images and located three new areas. Two are located west of Amery bay and one very close to the south pole (Fig. 1). Furthermore, we have located a large peripheral area around three earlier reported regions. In this way, we added more than 100,000sq. km area and believe that more than 1,000,000sq. km area in eastern Antarctica is covered by the compacted snow dunes. Apart from this, we could locate extensive dune regions, either covered by snow sheets or obliterated later in glacier catchment. Misra and Dobhal (2015) based on their characteristic shape, size, distribution and wind direction, categorized them as 'foredunes'. As these foredunes are highly compacted, we prefer to describe them as compacted snow dunes.

It has been observed during the present study that the compacted snow dunes not only abruptly terminate, but also show shift along the set of lineaments. These set of lineaments are parallel to each other and along with the other set, form orthogonal pattern, as have been mapped in all the shield regions of the world. Furthermore, controlled elongated valleys portions are filled by consolidated snow sheets. They

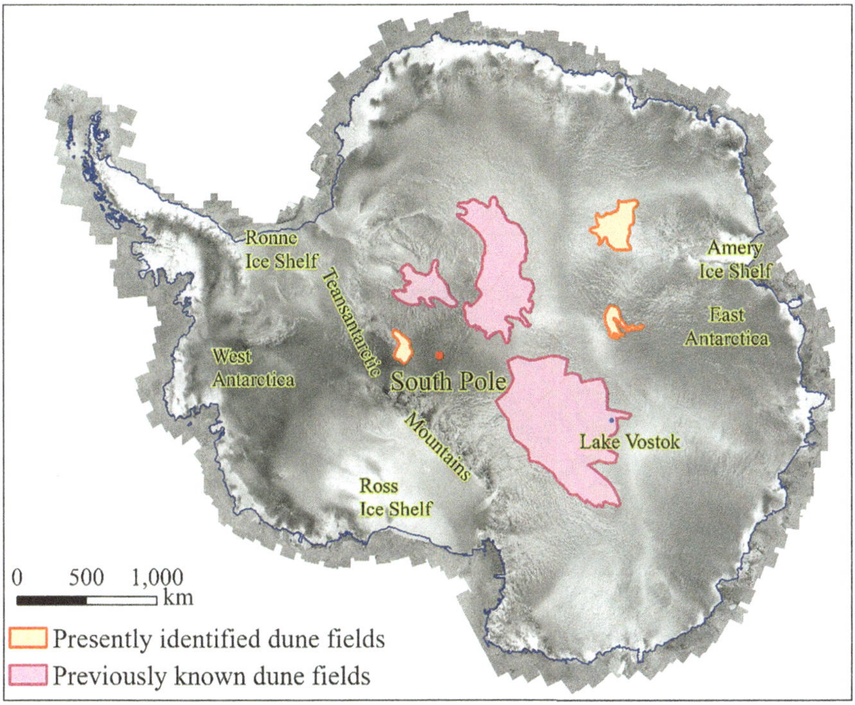

Fig. 1 Mosaic of Synthetic Aperture Radar data from Canadian Radarsat 1, under Antarctic Mapping Mission (Jezek 2008). It can be seen that three new areas are identified during the present study and extensions around known areas are also demarcated

have rough surfaces and therefore have lighter tones in radar images. The boundaries of snow sheets are also controlled by either of these set of lineaments.

2 Snow Dunes, Snow Sheets and their Formation

Geomorphologically compacted dune fields predominantly occupy plateau regions around highlands generally located between 2500 to 3500 m asl and have remarkable stability. Within the fields, the distribution of dunes is controlled by vertical tectonics. Uplifted linear blocks have well preserved dunes, while downward moved blocks for snow-filled valleys. These valleys are narrow in comparison to the uplifted blocks (Fig. 2).

Individual snow dunes are of low amplitude generally 2 to 4 m in height, very long, often extending for more than 100 km. They have a very regular shape and exactly the same pattern in all the regions. Although they are remarkably straight, a certain amount of sinuosity is invariably present. Apart from this swelling and pinching along their length as well as rounding of crests, is also commonly seen.

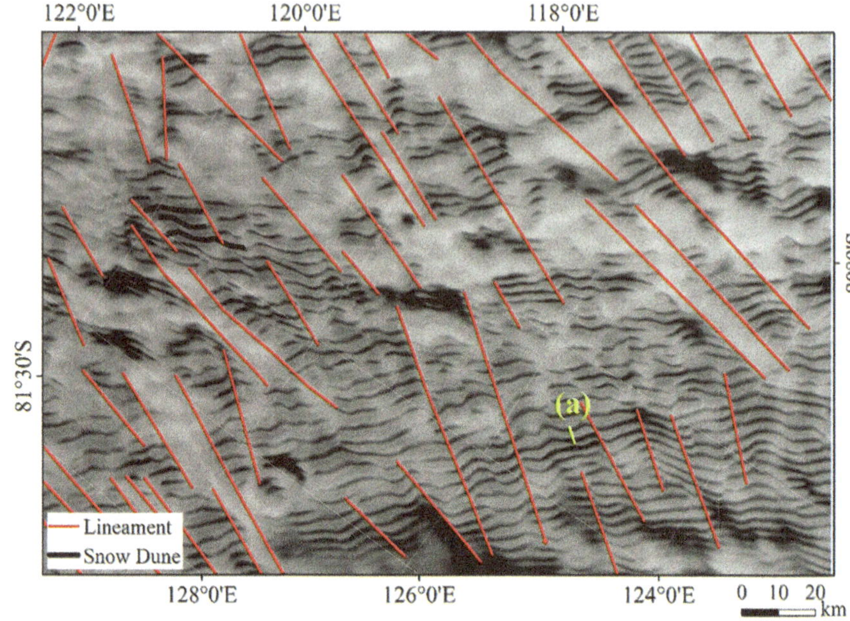

Fig. 2 Synthetic Aperture Radar image of part of eastern Antarctica, showing the compacted snow dunes. They are characterized by darker tones, linear shape and very regular pattern. Light tone areas are covered by consolidated as well as unconsolidated snow sheets. (**a**) Profile drawn by using REMA Virtual webpage, across compacted dunes, shows approximately 2 m height of individual dunes and almost same amount of depression between them. Interdune distance is about 1.5 km. in this case

Other unique features observed are that they have large lateral inter-dune separation, averaging from 2 to 3 km. These characteristics have led to several misconceptions. Their long and linear shape matches with the longitudinal (sief), dunes are seen in various desert regions as well as on planet Mars. These longitudinal dunes form along the wind direction and do not have gentle windward and steep leeward sides. Transvers dunes are mainly of three types namely fore dunes, parabolic dunes and barchans. Foredunes and parabolic dunes are formed by winds laden with moisture and thus they are mostly located close to the coastal regions, however, they can also form in interior regions. The primary requirement for their formation is, therefore, that the wind should be laden with moisture. A concerted effort is made during the present study, to describe these attributes and understand their formation during unique climatic conditions.

It has also been found that in many areas they have been obliterated by subsequent glaciation. In adjacent areas, however, they are not disturbed suggesting, their compaction before the onset of the next glaciation. They have formed perpendicular to the wind direction and correspond with coastal fore-dunes. This study has suggested that the accumulation of snow has taken place during interglacial periods and subsequently compacted during intense glacial cooling. Apart from dune complexes, there are vast regions covered by consolidated snow sheets, which seem to have formed during similar but less intense climatic conditions. Association with innumerable dunes forms vast complexes, while many others form smaller clusters. Interdune spaces within the clusters are much less (1 to 2 km) in comparison to dune complexes.

The first view regarding the formation of snow dunes was presented by Fahnestock et al. (2000). They suggested strong katabatic winds due to temperature gradient. Misra and Dobhal (2015) observed that snow ridges were deposited, later compacted during glacial period. However, our study of entire eastern Antarctica has brought out very regular dune patterns and does not support the idea of randomly distributed isolated hills as an obstruction to deposit these dunes. Prevailing katabatic winds are perpendicular to the longer axes of snow dunes (Dadic et al. 2013). Arcone and Jacobel (2009) recorded several-meter-thick co-sets with preserved bedding surfaces under the glazed lee-ward faces. Scarchilli et al. (2010) observed and measured that the maximum blowing snow is transported during March to August months in Antarctica. These extraordinary transport events alternated with compaction and formation of glazed surfaces. Profiles across pro-grading windward sides have shown more than 100 m thick layering continuing all along the length of the dunes. Shuman et al. (2011), derived elevation profiles from ICES at data and reported accretion on windward sides and an intricate relationship between accumulation rate and surface profiles. Meteorological data from the interior of eastern Antarctica indicates that the dune areas are characterized by katabatic wind flow having constant speed and uniform direction. Furthermore, trends of the dune crests are perpendicular to the directions of katabatic winds (Parish and Bromwich 1991). Fairly good evidence, collected from the field traverses, pit examination and other satellite measurements (Giovinetto 1963; Comiso 1994; Fahnestock et al. 2000) indicated that very coarse-grained firn exists in the dune fields. This model assumes that when the uppermost snow warms in the summer, tiny amounts of water vapor leave the snow grains and enter the air between them. During winter months the snowpack cools tremendously, to −62 °C or lower. Warmer air from summer is still present in the lower layers. Water vapor from the warm layer moves towards the colder snow at the surface and sticks to the snow grains.

During the present study most important support for the formation of compacted dune complexes has come from the geomorphic principles. It has been observed globally that depositional landforms have regular patterns while erosional ones are highly irregular and erratic. Geomorphic and glaciological evidence gathered during the present study, suggest that dune complexes have formed following two stages. In the first stage, the falling snow crystals are piled up by low amplitude, long wavelengths and moisture-laden winds, during wet interglacial periods. Later in the second stage consolidation and compaction has taken place due to intense cooling in intervening glacial periods. Metamorphism and growth of crystals continued during compaction. The accompanying winds were very dry, severe and have formed wind-polished glazed surfaces on the leeward sides. They also swept interdunal areas and caused progradation on the successive windward sides. Ekaykin et al. (2016) during field surveys found that the leeward sides of the dunes are characterized by reduced accumulation and increased concentration of $\delta^{18}O$ molecules likely due to post-deposition alteration of the snow isotopic content, while on the windward side the accumulation is more and their enrichment of $\delta^{16}O$ water isotopes. Radar and satellite images show features of compacted dunes that suggest alternating fine and coarse grains, forming rough and smooth surfaces. These surfaces are indiscriminately covered by an overburden of consolidated snow sheets. Wherever the thickness is less, clusters of snow dunes can be seen through them. Furthermore, these sheets cover well-formed dunes, it is visualized that they have accumulated under slightly different climatic conditions, separating them from dune fields due to less availability of moisture and short wavelength winds.

3 SAR Images, Mapping of Snow Dunes, Snow Sheets and Basement Tectonic Features

Synthetic Aperture Radar (SAR) can image through the cloud, mist cover and can be obtained even during the dark winter months. Furthermore, SAR measures surface roughness and presents nearly 3-dimensional view of the relief, along with dielectric constant useful to demarcate dry and wet condition of snow. Amount of reflected energy received by the sensors, is controlled largely by the look angle and surface roughness. Smooth surfaces reflect away most of the energy and thus have dark tones, while rough surfaces scatter energy, making them appear bright. It is therefore, the dune fields have alternating light and dark bands in SAR images.

Spectacular multi-date images of Antarctica were provided from C band of RADARSAT-1to estimate glacier surface velocities. The present study utilized modified high resolution radar mosaic data of 1997 and 2000 having 100 m resolution. Snow dune patterns in parts of the largest dune field have remained unchanged for 34 years (Fahnestock et al. 2000). Our study using Sentinel 2 data of 2021 has further indicated that there is no change thereafter, suggesting that the dune complexes are fairly stable for more than half a century. Geomorphological evidences such as the covering of dunes by stabilized snow sheets as well as later obliteration by moving ice in the glacial catchment, suggest their geological antiquity.

Earlier field traverses during the 1950s and 1960s across dune complexes in several sections could not recognize the overall dune pattern because of large lateral separation. They could only be recognized with the availability of satellite images providing a synoptic view. Shuman et al. 2011 conducted field studies during 2003–2009 and provided site-specific field data, utilized in the verification of our interpretation. The significant observations made by them include that the compacted snow dunes are generally 2 to 8 m high, 1 to 2 km wide ridges, up to 100 km long and laterally separated by 2 to 6 km. They also recognized progradation on windward sides and no accumulation on leeward sides. Interpretations from the Ground Penetrating Radar (GPR) data have also suggested that the glazed surfaces on compacted dunes have formed due to polishing done by the winds. All available multi-date satellite data sets both optical and SAR utilized during the present study are presented in Table 1.

Among the basement tectonic features the most important are the lineaments. They are easily recognized in remote sensing data sets because of linearity. Their mapping has provided sub-surface basement details in areas covered by ice sheets, intensely glaciarized regions with debris, moraine material, consolidated snow sheets and snow cover. They are conspicuous by abrupt termination of innumerable compacted snow dunes and dune complexes. Elongated depressions are also bounded by these lineaments and seem to have developed due to subsidence. These depressions are filled by consolidated snow sheets and are disposed in an orthogonal pattern.

4 Dune Complexes and Basement Tectonics in Antarctica

Lineaments are mapped from all the regions of the world including the most stable shield portions. They are quite obvious in satellite images by straight or curvilinear courses and association of several geomorphological and tonal characteristics. They have been correlated with basement structures. Slack (1981) discussed paleo-tectonic features in Power River Basin in Wyoming and demonstrated one-to-one correlation between vertical basement faults in exposed Archaean terrain to the areas covered by Cretaceous sedimentary sequence. He has also cited evidence for tectonic activity throughout Phanerozoic period. Shurr (1982) presented the upward propagation of basement faults to the surface through a column of sedimentary rocks in Montana and South Dakota provinces of USA. He also emphasized their bifurcation to form a zone of satellite lineaments. Upward propagation of basement faults with large vertical displacement and formation of Devonian reef complexes and continuous movement through the sedimentary sequences in Western Sedimentary Basin of Canada, Misra et al. (1991). They have identified basement structures in aeromagnetic data and displacements by changes in thickness and facies patterns. Misra and Misra (2013) observed in peninsular India, where basement faults can be seen extending from exposed Achaean terrain laterally and propagating through a thick column of Cretaceous Deccan basalt, in western India. From the above description it is clear the basement faults because of associated tectonic activity get propagated upward through the sedimentary and volcanic rocks. Basement tectonics was first recognized by Misra and Misra (2013) and conformed during the present study, suggest sudden truncation and vertical movement along the lineaments (Fig. 2). Furthermore, they concluded that glacial and aeolian processes have later accentuated the lineaments into transversely elongated zones. During detailed study, we have found that these zones are several kilometers wide and are filled by snow (Fig. 2a). Lighter tone zones suggest that the surfaces are fairly rough. Perhaps, the most interesting observation is that dunes show warping along transverse axes (Fig. 3). In certain regions, the compacted snow dunes are very well-formed and preserved; retain their original shape and size. In adjacent areas,

Table 1 List of satellites data sets

Sr.no	Data used
1.	Sentinel 1
	S1A_EW_GRDM_1SDH_20191226T105929_20191226T110021_030518_037E9F_16D5
	S1A_EW_GRDM_1SDH_20200502T163358_20200502T163502_032388_03BFFB_7065
	S1A_EW_GRDM_1SDH_20200505T165826_20200505T165930_032432_03C180_9E0D
	S1A_EW_GRDM_1SDH_20201114T110303_20201114T110349_035243_041D9E_5433
	S1A_EW_GRDM_1SDH_20201231T165832_20201231T165936_035932_04356A_32DD
	S1A_EW_GRDM_1SDH_20210322T163402_20210322T163506_037113_045E96_79E6
	S1A_EW_GRDM_1SDH_20210407T110300_20210407T110346_037343_04667B_3847
	S1A_EW_GRDM_1SSH_20170319T163337_20170319T163441_015763_019F3F_59A4
	S1A_EW_GRDM_1SSH_20190412T214527_20190412T214631_026762_030192_54E6
	S1A_EW_GRDM_1SSH_20210131T214539_20210131T214643_036387_044547_6EA8
	S1B_EW_GRDM_1SDH_20170925T105836_20170925T105948_007547_00D535_30C3
	S1B_EW_GRDM_1SDH_20180312T105833_20180312T105945_009997_012207_2BB9
	S1B_EW_GRDM_1SDH_20210329T104028_20210329T104132_026228_03215C_3545
	S1B_EW_GRDM_1SSH_20170326T104003_20170326T104107_004878_00885C_062F
2.	Sentinel 2
	L1C_T50CMT_A020719_20210222T232807
	L1C_T50CNS_A020547_20210210T224828
	L1C_T50CNT_A020733_20210223T225823
	L1C_T50CNU_A020547_20210210T224828
	L1C_T51CWP_A020719_20210222T232807
	L1C_T52CDT_A020719_20210222T232807
3.	Landsat 8
	LC08_L1GT_094114_20210115_20210308_01_T2
4.	Antarctic Mapping Mission, Mosaic
5.	Modified Antarctic Mapping Mission, Mosaic
6.	Antarctica DEM CryoSat
7.	Reference Elevation Model of Antarctica (REMA) DSM 100 m as well as 8 m virtual references

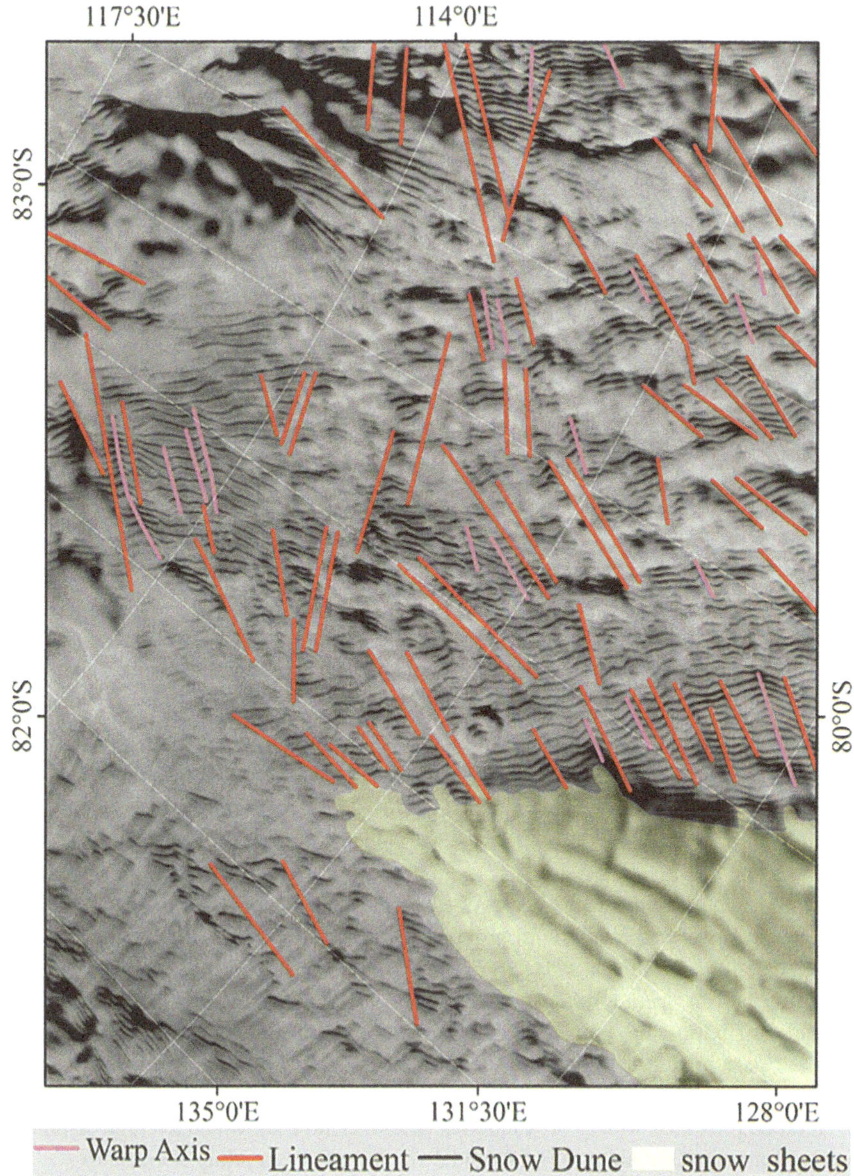

Fig. 3 Synthetic Aperture Radar image in part of eastern Antarctica. Compacted dune complexes can be seen as black stripes. Warping of dunes along several transverse axes can also be seen. Dune complexes are covered by snow sheets in SE portion, while snow dunes are obliterated by in SW portion

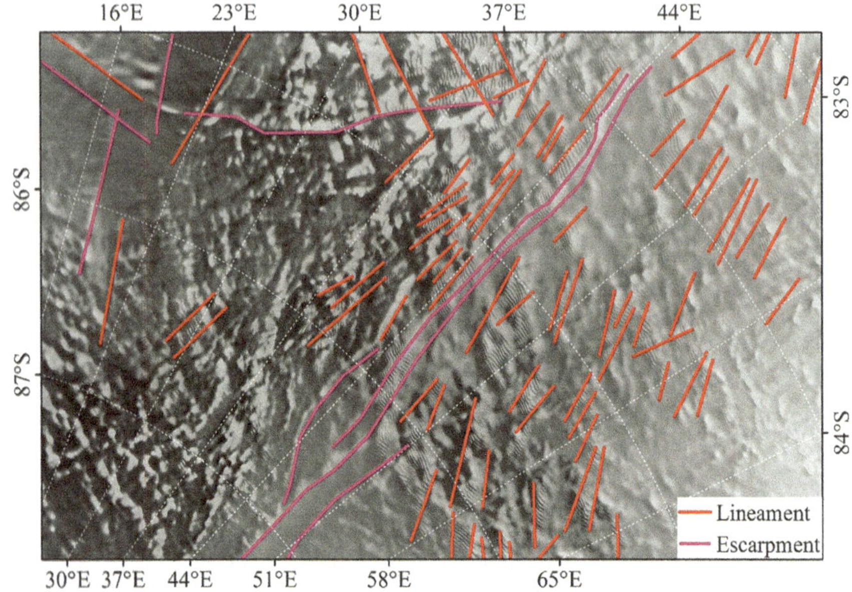

Fig. 4 Synthetic Aperture Radar image of eastern Antarctica, depicting several sets of lineaments. The important feature associated with the lineament is that there is vertical movement along them

however, the dune complexes are obliterated to different stages subsequently by erosional processes. Many a time these areas are also covered by later stabilized snow sheets as well as unconsolidated snow. These fields are presently coldest regions with temperatures dipping to −80 °C and dominated by strong winds with blowing snow.

Among several lineament sets, NW-SE and NE-SW trends are more prominent. In many cases, escarpments with pronounced down throw, are seen continuing for more than 400 km. Lineaments sets are present in almost entire Antarctic continent and affect all the geomorphic features develop during glacial history. Their widespread distribution and parallelism along with development of escarpments suggest that they are related to basement tectonics (Fig. 4). Measurements made by using Cryosat2 Digital Elevation Modal and profiles drawn across the escarpment by using Reference Elevation Model of Antarctica (REMA) have shown an average displacement of nearly 10 m along NE-SW and 5 m along E-W trending lineaments.

5 Dune Complexes and Climate Change Parameter

Misra and Dobhal (2015) emphasized that the compacted dunes represent the effects of alternating glacial and inter-glacial periods. These effects have been studied and documented in different parts particularly Europe and North America. Four major and more than a hundred little ice ages have been identified. The severity of unusual blizzards lasting for several months and dumping huge amount of snow are reported from North America in historical times. It can be visualized that these effects must have been more severe in the icy Antarctic continent. In large areas, the dune fields are covered by snow sheets. These sheets seem to have also followed two stages, corresponding with the dune formation however, certain variations such as moisture content, amplitude, and wavelength of corresponding winds. Glacial and interglacial periods, however, continued after the formation of dune complexes and snow sheets. This is evident by the development of enormous glacier catchments, where dune fields are later obliterated, suggesting that this event postdates dune formation. Wise et al. (2007) also reported older mega dune fields are covered by younger snow firn. They observed distinguishing features such as 2 to 4 km spacing, zebra-striped patterns different from modern dunes as relics of past climate.

6 Results

The exact nature of the compacted snow dunes in totality has emerged during the study. The dunes have gentle windward and steeper leeward sides and have formed perpendicular to the wind direction. Prevailing, as well as paleo-wind directions which have formed these dunes, match together. This suggests that the wind directions have not changed ever since their formation. They are the product of accumulation of snow and have not been carved on ice sheets as believed earlier by some scientists. Glazed surfaces, however, developed by wind polishing are present only on the leeward sides. During the present study, they have emerged as significant criteria for illustrating the paleo-climatic conditions. A certain unique set of events have played an important role in their accumulation and compaction. The study suggests that the dune complexes covered almost entire eastern part of the Antarctica, indicating that climatic conditions were of continental nature and were not confined locally. The dune complexes largely occupy elevated regions. The prevailing climatic conditions during the summer and winter months were similar to those, existing in the past, however, they were very severe and long, corresponding with glacial and interglacial periods during Quaternary era. The relentless winds were also severe when they formed the compacted snow dunes by moisture-laden blizzard winds. These intense conditions with enormous snowing and accumulation prevailed in almost entire eastern Antarctica. Stabilized snow sheets overlie the dune regions like an overburden. Many a time dune, as well as dune-like patterns, can be recognized where ever burden is thin. Weather conditions were similar to those prevailing presently; however, the magnitude was very high and also continued for a much longer time. Similarly, consolidated snow sheets have formed under slightly different weather conditions. Other intriguing features associated with the compacted snow dunes are their low height and fairly wide lateral separation. This could be because of amplitude and wavelength of winds prevailed during their formation.

Radar images have also been found to be extremely useful in mapping lineaments, representing upward continuations of nearly vertical basement faults. Two sets of parallel lineaments together form orthogonal pattern, similar to the other shield regions of the world. This study has elaborated disposition of lineaments in Antarctica for the first time.

7 Synthesis and Conclusions

The individual compacted snow dunes are remarkably similar to each other in height, width and linearity. Their disposition and geometric mutual relationship form a regular pattern. These patterns also have exactly the same symmetry throughout the vast stretch of eastern Antarctica. Any concept to explain their formation must therefore include all these aspects. Comparison between these unique snow dunes and desert dunes has brought out very many similarities and dissimilarities. Both of them are the product of aeolian (wind) action. The desert winds are dry and change direction with the season as well as between day and night. Polar winds prevailing over Antarctica are not affected by these factors and have uniform directional (katabatic) flow. However, during seasons and climatic conditions, their moisture content has varied. Severe snowing with moisture-laden winds (blizzards) has provided ideal conditions for their deposition and much cooler glacial (ice age) conditions favored compaction and metamorphism. Furthermore, the distribution of snow dune complexes suggests that they covered almost entire eastern Antarctica and their present-day separation is dependent on three factors, such as preferential preservation on highlands, later indiscriminate covering by snow sheets and selective destruction by glaciers. Extensive distribution and similarities have led us to believe that the event responsible for their formation was of continental extent. During the formation of desert dunes and sand sheets, particles of different sizes move by rolling and jumping (saltation) along the wind direction. Finer particles also move in suspended conditions depending on the wind velocity, extremely fine particles are lifted up to greater heights and transported hundreds of kilometers and settle down with the diminished wind velocity (loess deposits). In case of snow dunes, the falling crystals are being lighter, glided to long distances and it appears that this is mostly responsible for the large separation between the dunes. Furthermore, the amplitude and wavelength of flowing wind could have played vital role in geometric similarity and vast extent.

Mapping and monitoring glacio-morphic details in snow-clad mountainous regions are very difficult because of inclement weather and difficulties in getting proper logistic support. Satellite data obtained in visible and infrared regions also fails to provide sharp images due to high radiance values and persistent cloud cover. Furthermore, cover of fresh snow and dark winter months hamper collection of proper satellite imagery in Antarctica. SAR data highlights terrain roughness and also images subsurface features, such as facies characters of snow dunes and ice sheets to a certain extent.

Compacted snow dune complexes have formed mainly in two stages. During first stage enormous snowfall with high moisture content and severe winds accumulated snow in the form of fore-dunes and coincided with the wet inter-glacial period. This was followed by glacial cycle of metamorphism of accumulated snow to form compacted dunes. The disposition of snow dunes is having a very high degree of uniformity and regularity. A significant difference observed between the foredunes and the snow dunes are the spacing between them. The paleo-wind directions deduced from the study of compacted snow dunes and snowfields correspond with the prevailing wind direction.

The findings that the lineaments observed in satellite imagery are surface expressions of basement faults that have propagated upwards through the thick cover of ice sheets and snow, can significantly contribute to our understanding of the Geology, particularly to the tectonics of Antarctica. Occasionally only in peripheral areas, the rocks get exposed in summer months, when snow melts. Operating surface geophysical surveys are also very difficult in extreme climatic and logistic constraints. Even the optical remote sensing has not been successful in Antarctica because six dark winter months, persistent clouds cover other inclement weather conditions of haze and mist. Under these conditions, mapping of basement faults by SAR data has emerged as a powerful tool. Furthermore, the study has demonstrated that the lineaments exist in entire 1,000,000 sq. km. mapped areas in eastern Antarctica. Varying density and disposition of lineaments can also be utilized to interpret subsurface tectonic features (Misra et al. 1991).

Acknowledgements We wish to profusely thank Dr. Satish Tripathi, Formerly Deputy Director General, Geological Survey of India; Honorary Editor, Earth Science India and General Secretary, The Society of Earth Scientists; for inviting us to contribute this paper to forthcoming volume on Climate Change. Detailed discussions with Dr Rasik Ravindra, a veteran Antarctic Geologist; Panikkar Chair Professor, Ministry of Earth Sciences; Former Deputy Director General, Geological Survey of India and Director, National Centre for Polar and Ocean Research (NCPOR), were of great help in understanding the field characteristics of compacted dune complexes and associated climatic conditions. Canadian Government Laboratory Visiting Fellowship, provided to (KSM) to work in Radatsat project is thankfully acknowledged. We are also grateful to AMM-1 and MAMM for providing the SAR data for this study. Comments provided by the referee were helpful to improve the manuscript. Director, Wadia Institute Himalayan Geology, Dehradun and Head of the Geology Department, Kumaun University, Nainital, have very kindly provided all available facilities to carry out this work, is thankfully acknowledged.

References

Arcone SA, Jacobel RW (2009) Large scale metamorphism in East Antarctica: Radar evidence within firn and englacial profiles. American Geophysical Union, Fall Meeting 2009, abstract id. C52A-05

Comiso JC (1994) Surface temperatures in the polar regions from Nimbus 7 temperature humidity infrared radiometer. J Geophys Res 99(C3):5181–5200. https://doi.org/10.1029/93JC03450

Dadic R, Mott R, Horgan HJ, Lehning M (2013) Observations, theory, and modeling of the differential accumulation of Antarctic mega dunes. J Geophys Res Earth 118(4):2343–2353. https://doi.org/10.1002/2013JF002844

Ekaykin A, Eberlein L, Lipenkov V, Popov S, Scheinert M, Schroder L, Turkeev A (2016) Non-climatic signal in ice core records: lessons from Antarctic mega dunes. Cryosphere 10(3):1217–1227. https://doi.org/10.5194/tc-10-1217-2016

Fahnestock MA, Scambos TA, Shuman CA, Arthern RJ, Winebrenner DP, Kwok R (2000) Snow mega dune fields on the East Antarctic Plateau: extreme atmosphere-ice interaction. Geophys Res Lett 27(22):3719–3722. https://doi.org/10.1029/1999GL011248

Giovinetto M (1963) Glaciological studies on the McMurdo-South Pole traverse, 1960–1961. Ohio State Univ Inst Polar Stud Rep 7(7):1960–1961. Available at http://kb.osu.edu/dspace/handle/1 811/38737

Jezek KC (2008). The RADARSAT-1 Antarctic mapping project

Misra A, Dobhal DP (2015) Synthetic Aperture Radar (SAR) data in mapping and study of compacted snow dune complexes and active lineament tectonics in Antarctica. In XII International Symposium on Antarctic Earth Sciences (ISAES), NCAOR, Goa 504, S22–411, Abstract

Misra KS, Misra A (2013) Hydrocarbon exploration in sub-basalt basins around peninsular India. Search Discovery 50804:16–19

Misra KS, Slaney VR, Graham D, Harris J (1991) Mapping of basement and other tectonic features using SEASAT and thematic mapper in hydrocarbon-producing areas of the Western Sedimentary Basin of Canada. Can J Remote Sens 17(2):137–151

Parish TR, Bromwich DH (1991) Continental-scale simulation of the Antarctic katabatic wind regime. J Clim 4(2):135–146. https://doi.org/10.1175/1520-0442(1991)004<0135:cssota>2.0.co;2

Scarchilli C, Frezzotti M, Grigioni P, De Silvestri L, Agnoletto L, Dolci S (2010) Extraordinary blowing snow transport events in East Antarctica. Clim Dyn 34(7):1195–1206. https://doi.org/10.1007/s00382-009-0601-0

Shuman CA, Scambos TA, Fahnestock MA, Suchdeo VP (2011) Antarctic megadunes-assessment of topographic form and windward progradation from ICESat Altimetry, 2003–2009. In American Geophysical Union, Fall Meeting 2011, Abstract Id. C32B-07

Shurr GW (1982) Geometric hierarchy of sandstone bodies in the Shannon sandstone member near the northern Black Hills, Montana and South Dakota

Slack PB (1981) Paleotectonics and hydrocarbon accumulation, Powder River basin, Wyoming. Am Assoc Pet Geol Bull 65(4):730–743. https://doi.org/10.1306/2f9199bd-16ce-11d7-8645000102c1865d

Swithinbank C (1988) Antarctica. US Government Printing Office

Wise DU, Cianfarra P, Salvini F (2007) Megadunes and geologic maps of snow/firn of East Antarctica: implications for major climatic change, accumulation rates, ice flowage, and bedrock structures. In AGU Fall Meeting Abstracts C51A-0073

Investigating the Effect of Environmental Variables on the Isotopic Composition of Transpiration: Implications to Study the Monsoon Processes

Amey Datye, Charuta Murkute, S. Chakraborty, Pramit K. Deb Burman, M. N. Patil, and T. Dharmaraj

Abstract The isotopic composition of plant sap water and leaf water has been studied extensively to investigate the plant physiological processes and their interaction with the environmental parameters. Yet, the knowledge of ecosystem water pools, such as soil water and transpiration, remain poorly constrained. We propose an alternative method to study the effect of environmental variables on the isotopic composition of plant transpiration.

A small-scale field experiment was carried out to collect transpiration water samples along with rainwater throughout the summer monsoon season of 2017 and 2018. It was observed that the isotopic records of the transpired water very well captured the variabilities observed in the isotopic values of the rainwater. The difference between $\delta^{18}O$ of rainfall and $\delta^{18}O$ of transpired water is more during low rainfall events and small during high rainfall events. Isotopic values of the transpired water showed strong sensitivity to the soil water content. It means that the soil water uptake by plants is strongly controlled by the soil water dynamics, which modulate the isotopic values of the transpired water. The break phases of monsoon result in characteristic isotopic values of precipitation and the transpired water. These characteristics features could be useful in delineating the low rainfall regime during the monsoon season.

Keywords Transpiration · Precipitation · Water isotopes · Soil water · Monsoon

1 Introduction

Trees and other vegetation influence the atmospheric water cycle in various ways (Sheil 2018). Plants uptake water from the soil via roots, use it for metabolic and physiologic functions and ultimately release it to the atmosphere. Transpiration, the

A. Datye · C. Murkute · S. Chakraborty (✉) · P. K. Deb Burman · M. N. Patil · T. Dharmaraj
Indian Institute of Tropical Meteorology, Ministry of Earth Sciences (MoES), Pune, India
e-mail: supriyo@tropmet.res.in

B. Phartiyal et al. (eds.), *Climate Change and Environmental Impacts: Past, Present and Future Perspective*, Society of Earth Scientists Series,
https://doi.org/10.1007/978-3-031-13119-6_13

loss of water vapor from plants, is a physical process controlled both by external physical and plant physiological processes (Pallardy 2008). Transpiration accounts for 39% of terrestrial precipitation and contributes up to 61% of the vapor generated through evapotranspiration on a global scale (Schlesinger and Scott 2014). It also plays a significant role in influencing the regional and global climate (LeMone et al. 2007). Hence the study of transpiration is a crucial component of ecohydrological research. Atmospheric parameters, such as temperature, relative humidity, rainfall, wind speed, etc., control the transpiration process (Hanami 1974). Solar radiation is essential for photosynthesis, and transpiration and photosynthesis are strongly coupled (Tuzet 2011). Similarly, soil parameters, such as soil water (SW) content, and soil temperature (ST), also regulate the process of transpiration (Wang et al. 2011).

Evapotranspiration (ET) is a gross measure of surface-atmosphere water flux comprising of physical and biological components, namely evaporation and transpiration, respectively. Several methods exist to measure ET, such as flux-gradient technique (Singh et al. 2014), lysimeter (Bala et al. 2016), Bowen ratio energy balance (Unlu et al. 2010). On the other hand, the eddy-covariance (EC) technique is widely used to study the biosphere-atmosphere interaction, including the estimation of the ET (Baldocchi et al. 2001). In the Indian context, the EC technique has been used to measure the ET and to characterize the energy and water exchanges of several ecosystems with the atmosphere in central India (Jha et al. 2013; Rodda et al. 2021), northeast India (Deb Burman et al. 2017; Chatterjee et al. 2018; Sarma et al. 2018; Sarma et al. 2022; Deb Burman et al. 2019), North India (Watham et al. 2014), south India (Patil et al. 2014), west India (Dharmaraj et al. 2012) and also the mangrove environments (Rodda et al. 2016; Gnanamoorthy et al. 2019; Chakraborty et al. 2020).

Apart from the meteorological observations, isotopic studies of plant materials have been carried out to investigate the transpiration process. Plants ingest soil water through the roots, and the soil is mostly recharged by the precipitation process. Hence the isotopic signature of the rainwater is transferred to the plant. Yu et al. (2014) performed the isotopic analysis of rainwater and leaf water at a south-eastern site of the Tibetan plateau. They argued that hydrogen isotopic values of leaf water reliably inherited the isotopic characteristics of Indian monsoon variability.

Water sources used by plants have been studied by several investigators. Xu et al. (2011) observed that dominant canopy trees used groundwater primarily, but midstory and understory trees used rainwater but switched to groundwater during the dry season. Schwinning et al. (2005) performed the isotopic analysis of stem water and observed that deeper levels waters were used by different species during drought conditions. However, the same species sourced their water from shallower levels after the heavy rainfall events.

The use of multi-level soil water by the plants plays a vital role in determining the isotopic composition of the transpired water. The reason being that soil waters at different levels experience different extents of evaporation and hence isotopic fractionation. Near-surface waters usually undergo rapid evaporation, especially during low humidity and windy condition. This makes the remaining water isotopically enriched (Tang and Feng 2001) subsequently used by trees. It is known that the plant physiological processes do not alter the isotopic composition of the water that they ingest from the soil (Ehleringer and Dawson 1992). Hence, the isotopic composition of transpired water usually represents that of the soil water (Cernusak et al. 2016). Since soil water is primarily recharged by rainwater, it is believed that the isotopic values of the transpired water mimic that of the rainwater (Gibson et al. 2008). However, Chakraborty et al. (2018) demonstrated that this assumption is valid only during heavy rainfall conditions. These authors showed that during the active phases of monsoon, soil water content remained high and hence suffered little fractionation. But during the break phases of monsoon, soil water undergoes relatively rapid evaporation, causing isotopic enrichment in the remaining water. This isotopically enriched water, when ingested by plants, produces transpired water enriched in heavier isotopes. The work done by Chakraborty et al. (2018) mostly addressed the issue of surface soil water. But the effect of deeper level waters on the isotopic characteristics of the transpired water was not taken into account. Soil water dynamics are greatly influenced by the rainfall processes and groundwater movement. This is more prevalent, especially during the rainy season when the recharge process is strongly influenced by the active and break phases of the monsoon.

Hence to understand variation in the isotopic composition of transpired water during high and low rainfall events and their response to soil water dynamics, a field-scale experiment was conducted in a semi-arid environment of western peninsular India. Transpiration and rainwater samples were collected throughout the monsoon season of 2017 and 2018, and its isotopic analysis was carried out. Based on two years of observations (2016 and 2017) Chakraborty et al. (2018) also showed that the isotopic values of the transpired water could be used to delineate the active and break phases of monsoon. We also like to test this hypothesis for the year 2018.

The objectives of this study are i) how the isotopic values of transpired water are modulated by the soil water dynamics during the monsoon season, and ii) whether the isotopic records of plant transpiration provide useful information about the active and break phases of monsoon in 2018.

2 Study Area

The experiment was conducted at the Indian Institute of Tropical Meteorology, Pune Campus. Pune (18°43′N, 73°51′E, 559 m AMSL) is an urban city in the state of Maharashtra. Geographically it is situated in the western peninsular region of India, having a semi-arid climate. The area receives rain during the summer monsoon, which extends from June to September. Mean annual rainfall is about 740 mm (India Meteorological Department; IMD). More than 90% of precipitation is received during the summer monsoon season. The IITM campus has abundant vegetation and is surrounded by several hillocks on its western side.

3 Materials and Methods

3.1 Collection of Transpired Water

Plant transpiration samples were collected during the monsoon season of 2017 and 2018. Samples were collected from multiple plants. In 2017 samples were collected from two plants, namely, *Cassia fistula* and *Tabebuia* sp. In 2018, the same *Cassia fistula* was used, additionally with different plant; *Mahogany* sp. Cassia plant is smaller in size (~ 10 ft) while the other two plants are mature. To collect the transpired water, the 'chamber method' of Menchaca et al. (2007) was followed.

A branch having 4–5 leaves was placed inside a pre-weighed transparent plastic bag. It is ensured that the leaves did not have any intercepted water. The bag was sealed by cable ties to ensure that it maintained a closed system. Sampling was started during the morning hours (typically 08:00) when transpiration peaks up. It continued until about 17:00 in the afternoon. At the end of the sampling period, the plastic bag was taken to the laboratory, weighed, and the collected water was taken out using a micro-pipette. Water samples were transferred to leak-proof plastic bottles and labelled. Precautions were taken to avoid any evaporation after sample collection.

Apart from the transpired water, rainwater was collected using an ordinary rain gauge in the same vicinity on a daily scale. The time of rainwater collection was fixed at 08:30 (local time) to commensurate with the daily collection of meteorological data by the IMD. It is to be noted that collected rainwater practically represents the previous day's rainfall, whereas the transpired water means the same day process. So, there may be an offset of about 24 hours, but a finite time lag between the percolation of rainwater through the soil and ingestion by the plant reduces this offset by a few hours.

Finally, the Isotopic analysis (both $\delta^{18}O$ and δD) of the samples was carried out by using an LGR water isotope analyzer (Model: TIWA-45-EP). The analytical precision of the instrument was $<1‰$ for Hydrogen and $< 0.1‰$ for oxygen, respectively (Sinha et al. 2019). The deuterium excess, defined as $d = \delta D - 8\delta^{18}O$, is a secondary isotope parameter often used to study the moisture source conditions (Johnsen et al. 1989; Pfahl and Wernli 2008). For example, in an oceanic environment, d-excess is strongly anti-correlated to the relative humidity and positively correlated with the sea surface temperature. On the other hand, rainwater or dews derived from secondary moisture sources, such as from the re-evaporated raindrops or the evapo-transpirated vapor, usually show high d-excess values. In this study, we will examine the d-excess characteristics both for rainwater and the transpired water.

3.2 Environmental Parameters

Radiation and Wind parameters were collected from an eddy-covariance micrometeorological tower located about 100 m off the sampling site. Radiation parameters were recorded using a Kipp and Zonen radiation sensor (Model-CNR4); Wind Parameters were recorded using a Sonic Anemometer (Model- Gill WindPro). The reanalysis data product of ERA5 (Hersbach et al. 2020) was used to get soil temperature and soil moisture information. Monsoon active and break period record was obtained from the Monsoon Report 2017 and 2018 published by IMD.

4 Results

Figure 1 shows the isotopic time series of rainfall (open circle) and transpired water (closed circle). Bars represent the amount of rain (mm/day). Panel A presents the variation in $\delta^{18}O$ of transpired water of *Tabebuia* sp. and rainwater (Year 2017). Panel B shows the same for Cassia *fistula* and rainwater (Year 2017); similarly, Panel C depicts the variation in $\delta^{18}O$ of transpired water of the *Mahogany* plant and rainwater (Year 2018), and Panel D shows the variation in $\delta^{18}O$ of transpired water of *Cassia fistula* and rainwater (Year 2018).

Figure 2(a) shows the Local meteoric water line (LMWL) for the year 2017. Figure 2(b) shows the relationship between d-excess and $\delta^{18}O$ of rainfall. Figure 2(c, d) represent the same for the year 2018. The open circles indicate data points when rainfall was low (<5 mm/day), and filled circles indicate high rain events (>5 mm/day).

5 Discussion

5.1 Isotopic Characteristics of Rain and Transpired Water

For the monsoon season of 2017, the minimum and maximum values of $\delta^{18}O$ are approximately −15 and + 1‰, while the same for δD are −114 and +1‰ respectively. The values are reported on the Vienna Standard Mean Ocean Water (VSMOW) scale. The 2018 monsoon season minimum and maximum values of $\delta^{18}O$ are approximately −2, +4‰ and for δD are 4, +26‰, (relative to VSMOW) respectively. Total seasonal rainfall during 2017 and 2018 was 785 and 601 mm, respectively.

Higher depletion in $\delta^{18}O$ was observed in 2017, especially during September and October. The month of September 2017 is marked with high rainfall events. This, followed by the formation of Deep Depression over Gangetic West Bengal and adjoining the north Bay of Bengal took place during 9–10 October (IMD Report

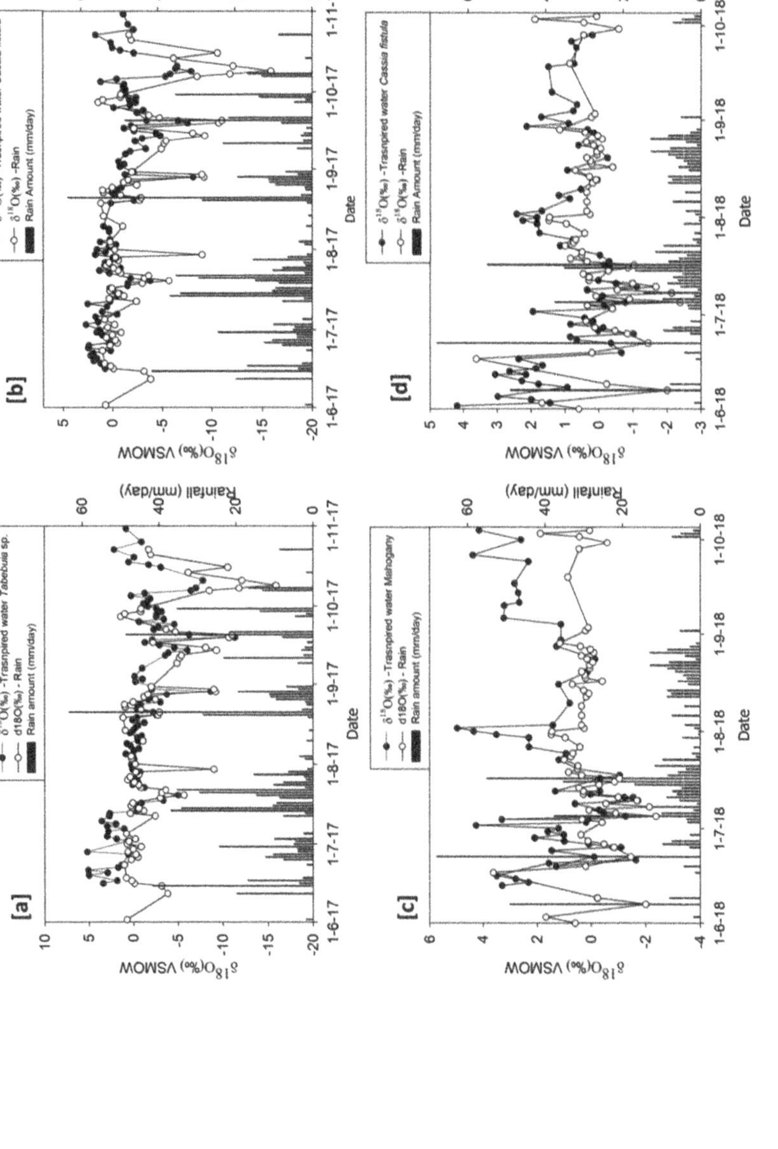

Fig. 1 Oxygen isotopic records of rainwater and transpired water. Open and closed circles represent the rainfall, and the transpired water, respectively. Bars represent daily scale rainfall variability. **(a)** Tabebuia **(b)** Cassia fistula **(c)** Mahogany **(d)** Cassia fistula

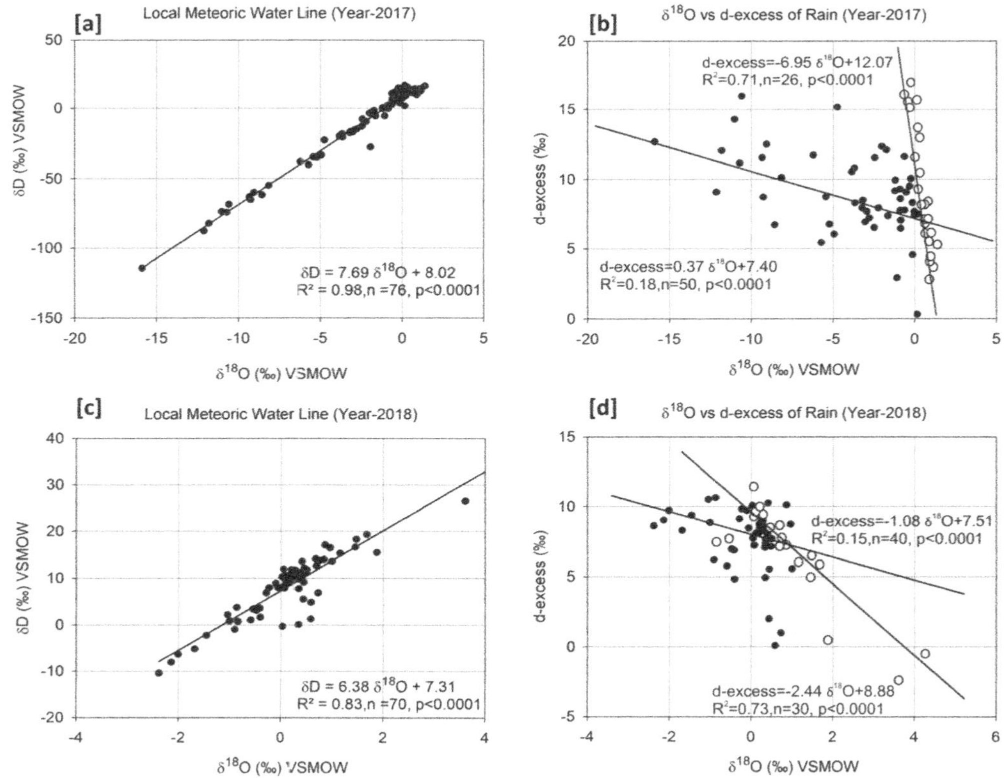

Fig. 2 Local Meteoric Water Line. (**a**) LMWL for the year 2017, (**b**) the relationship between $\delta^{18}O$ and d-excess for two rainfall regimes (see text for details). (**c**) same as (**a**) but for the year 2018. (**d**) same as (**b**) but for the year 2018. Filled and open circles in (**b**) and (**d**) represent high and low rainfall events respectively

October 2017). Since low-pressure systems are known to produce depleted isotopic values in precipitation (Lawrence and Gedzelman 1996; Chakraborty et al. 2016), the higher depletion in the late phase of 2017 was attributed to the deep depression over the Gangetic West Bengal. It is observed that the isotopic pattern of the transpired water had a close resemblance with that of the rainwater (Fig. 1). This implies that the source of soil water is mainly rainwater and its isotopic composition is more or less conserved. However, careful observation of these two time series reveals that the transpired water$\delta^{18}O$ maintains a small positive offset concerning the rain samples. During periods of high rainfall, this offset reduces. On the other hand, a significant difference between the isotopic values of rainwater and transpired water was noted during the weak rainfall period.

The local meteoric water line (LMWLs) represents the site-specific variation of hydrogen and oxygen stable isotope ratios. LMWL of 2017 and 2018 show different characteristics.

The slope of LMWL for the year 2017 (7.69 $\pm$ 0.11; Fig. 2a) is not significantly different from that of the Global Meteoric Water Line (GMWL slope = 8; Dansgaard 1964), while the intercept (8.02 $\pm$ 0.55) is somewhat smaller (GMWL intercept = 10).

Figure 2(b) shows the relationship between d-excess and $\delta^{18}O$ of rainfall for the year 2017. This kind of plot is generally used to study the sub-cloud raindrop evaporation. A significant amount of raindrop evaporation increases the isotopic values of rain but decreases its d-excess values (Gat 1996; Crawford et al. 2017). Two distinct trends are visible in the plot. Data points during high rain events are marked with filled circles, while data points associated with low rainfall are marked with open circles. A higher slope (−6.95) during low rain events indicates that raindrops experienced sub-cloud evaporation. These low rain events, which mostly occurred during early June and mid-August, coincided with the break phases of the monsoon. But during other rain events (marked with lower slope, i.e., 0.37), raindrops did not undergo significant sub-cloud evaporation. The rainfall events which produced a lower slope in the $\delta^{18}O$-d-excess line mostly belonged to the periods which witnessed relatively heavy rainfall (towards the end of July and August). This aspect has also been discussed later.

Similarly, Fig. 2(c) represents LMWL for the year 2018. Slope (6.38 $\pm$ 0.34) and intercept (7.31 $\pm$ 0.36) are significantly lower compared to that of GMWL. This indicates rainwater during this year experienced a significant amount of evaporation.

The d-excess and $\delta^{18}O$ plot (Fig. 2d) for the year 2018 also show two trends.

During the low rainfall events, d-excess and $\delta^{18}O$ show a strong negative correlation, suggesting evaporative enrichment of raindrops.

These low rainfall events occurred during the end of July and the first week of August. On the other hand, a lower slope (−1.8) was observed during high rain events. High rain events occurred during mid-July and towards the end of August.

As stated earlier, soil water plays an important role in modulating the isotopic composition of the transpired water. Since the water content in soil is significantly controlled by rainfall, we examine the characteristic behaviour of the isotopic composition of the transpired water for high and low rainfall regimes. To characterise the rainfall variability (the Pune city), the climatological mean of daily rainfall was calculated based on the last 22-year (1996–2017) rainfall data obtained from the IMD, which is 4.85 $\pm$ 9.76 mm. Accordingly, daily rainfall within the range of 0–5 mm was considered a low rainfall regime, and $>$ 5 mm was considered a high rainfall regime. To gain a comprehensive understanding, rain and transpiration samples collected during the years 2016, 2017, and 2018 have been considered for this purpose (the data for 2016 and 2017 have been obtained from Chakraborty et al. 2018). The $\delta^{18}O$ and δD of the transpired water are plotted, and the regression lines are generated, known as Local Evaporation Line or LEL (*op cit.*). The corresponding LMWL and the LEL were plotted for the two rainfall regimes and are shown in Fig. 3 (upper panel for $>$5 and lower panel for $<$5 mm/day rainfall).

It is observed that the LMWL representing the high rainfall events yielded a slope of 7.77, and the corresponding slope of LEL was 6.23. Since the slope is linked to evaporation, the LMWL slope 7.77, which is not significantly different from that of the GMWL (slope = 8), implies that raindrops during the heavy rainfall events did not undergo sizeable evaporation. On the other hand, the LMWL slope for the low rainfall events registered a slope of 6.93, implying a significant number of raindrops underwent evaporation. This also means that relative humidity was relatively higher during the high rainfall events than the low rainfall events. This is also reflected in the slope value of the LEL, which produced slope values of 6.23 and 4.36 for the high and low rainfall regimes, respectively. High humidity suppressed soil evaporation, which did not allow significant isotopic fractionation and helped produce a relatively higher slope in the corresponding LEL.

6 Effect of Environmental Parameters on Transpiration Isotopes

In this section, we examine the effect of Soil Water (SW) content and Soil Temperature (ST) on the transpiration isotopes. Towards this, we consider these two parameters at multiple depths, such as Level-1(0–7 cm), Level-2 (7–28 cm), Level-3 (28–100 cm), Level-4 (100–289 cm) for both years.

Tables 1 and 2 present the correlation values between $\delta^{18}O$ of transpired water vs. SW content, and $\delta^{18}O$ transpired water vs. ST respectively.

In 2017, $\delta^{18}O$ of transpired water showed a good negative correlation with soil water content for all levels, though the correlation is stronger for the upper-level water content. But during the low rainfall regime, the correlation between $\delta^{18}O$ of transpired water and soil water content increased with depth.

Figure 4 shows the variation of soil water content at different levels. Rainfall variation is also displayed. Level-1 and Level-2 soil water content showed near synchronous variation with rainfall. While the soil water content of Level 4 showed a progressively increasing trend. Due to this increased soil water availability at lower levels, plants tend to uptake water from this stratum during the low rainfall periods.

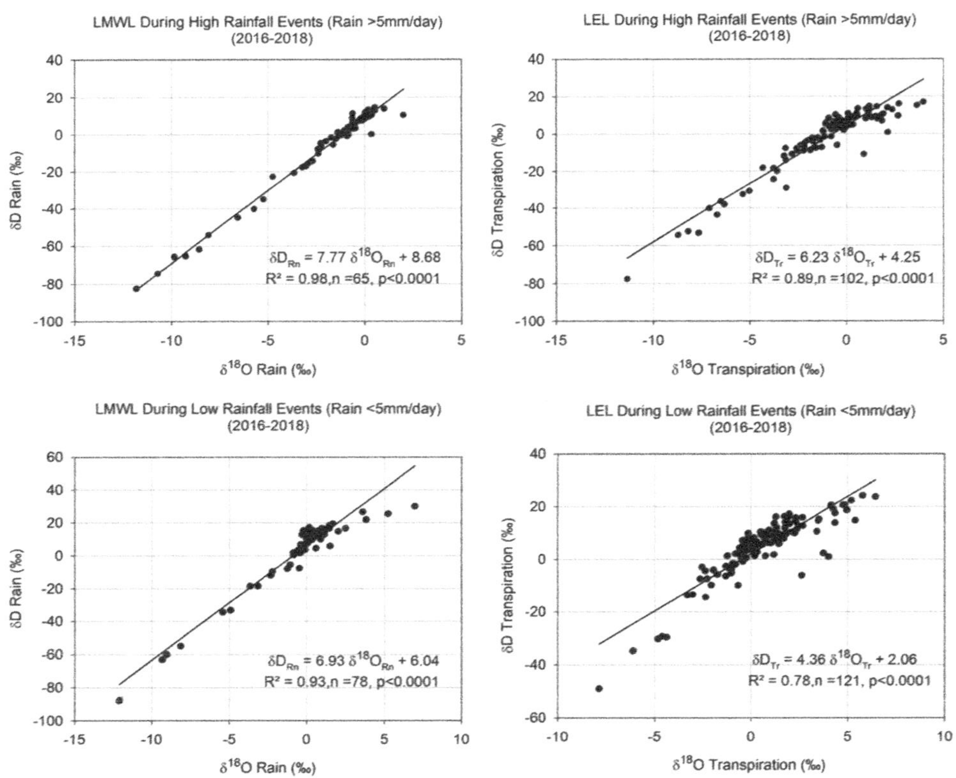

Fig. 3 Local meteoric water line and evaporation line for the high rainfall (>5 mm/day) and the low rainfall (<5 mm/day) events

Table 1 Correlation coefficient between $\delta^{18}O$ of transpiration and soil water content at various depths

Soil Water Content	Plant—*Tabebuia sp.*		Plant—*Cassia fistula*		Plant—*Mahogany sp.*		Plant—*Cassia fistula*	
Year	2017	2017	2017	2017	2018	2018	2018	2018
Soil Depth	Rain (>5 mm)	Rain (<5 mm)	Rain (>5 mm)	Rain (<5 mm)	Rain (>5 mm)	Rain (<5 mm)	Rain (>5 mm)	Rain (<5 mm)
1(0–7 cm)	−0.64	−0.52	−0.61	−0.42	−0.32	−0.58	−0.71	−0.52
2(7–28 cm)	−0.63	−0.72	−0.60	−0.55	−0.33	−0.50	−0.66	−0.58
3(28–100 cm)	−0.53	−0.78	−0.56	−0.60	0	0	0	0.48
4(100–289 cm)	−0.51	−0.77	−0.57	−0.77	0.32	0.35	0	0.36

Table 2 Correlation coefficient between $\delta^{18}O$ of transpiration and soil temperature at various depths

Soil Temperature	Plant—*Tabebuia sp.*		Plant—*Cassia fistula*		Plant—*Mahogany sp.*		Plant—*Cassia fistula*	
Year	2017	2017	2017	2017	2018	2018	2018	2018
Soil Depth	Rain (>5 mm)	Rain (<5 mm)	Rain (>5 mm)	Rain (<5 mm)	Rain (>5 mm)	Rain (<5 mm)	Rain (>5 mm)	Rain (<5 mm)
1(0–7 cm)	−0	0	0	0	0.35	0	0.57	0.51
2(7–28 cm)	0	0	0	0	0.30	0	0.63	0.50
3 (28–100 cm)	0	0.59	0.36	0.45	0	0	0.56	0.36
4(100–289 cm)	0.58	0.83	0.65	0.79	0	−0.35	0	0.42

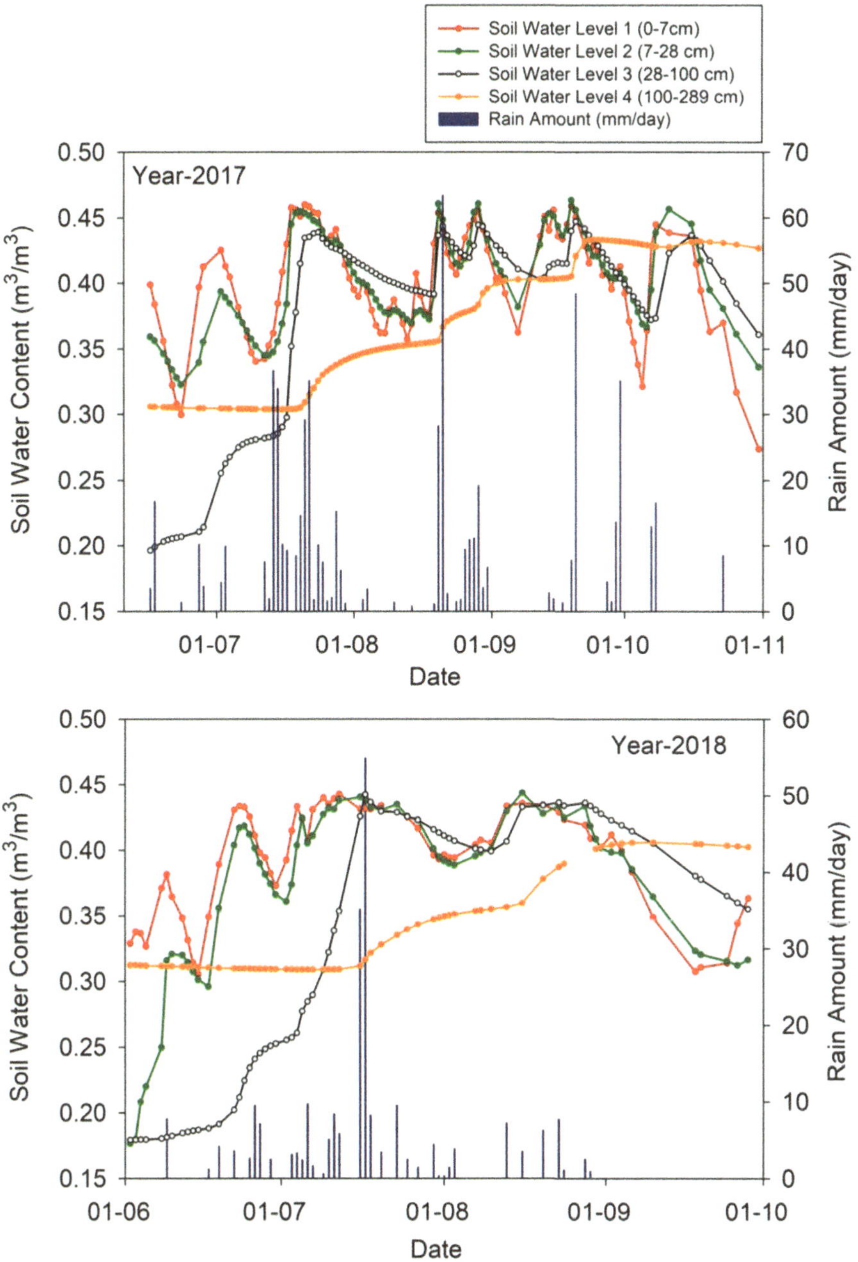

Fig. 4 Variation of soil water content with rainfall at different depths for 2017 (upper panel) and 2018 (lower panel) respectively

Since the low rainfall events enhance evaporation from the upper soil layers, the plants switch to lower levels to meet their water demand during this time. Hence the transpired isotopic values are expected to be more influenced by the deeper level soil water. This is indeed observed in the case of the year 2017 for both the plants when daily rainfall was less than 5 mm (Table 1). Similar behaviour was observed in the case of $\delta^{18}O$ and soil temperature as well. As evident in Table 1, during the low rainfall events, $\delta^{18}O$ of the transpired water had a stronger relationship with the deeper level of soil temperature.

In 2018, $\delta^{18}O$ of transpired water shows a better correlation with the upper levels soil water content. In contrast to the Year 2017, even during the weak rainfall regime, the correlation between $\delta^{18}O$ of transpired water and soil water content did not change with depth.

Similar to 2017, upper soil levels (Level-1 and Level-2) show variability in soil water content with the amount of rain received. Even soil water content of lower-level (Level-4) showed an increasing trend. But this trend is not prominent as seen in 2017 (Fig. 4). Low values in soil water content of Level-4 are observed during Jun and July. Higher values in soil water content of Level-4 are observed towards the end of the monsoon season. Due to this reason, plants were unable to uptake water from lower soil levels even during low rain events.

The *Mahogany* plants are known to have large roots that spread laterally or stay near the surface (Gilman and Watson 2011). This characteristics feature could explain the higher correlation with SW at upper soil levels throughout the season.

Transpiration is strongly influenced by several meteorological and soil parameters. So, it is expected that isotopic values of the transpired water would also be controlled by them, which has been discussed earlier. However, these environmental parameters are often interconnected; as a result, a simple bivariate analysis may lead to over or underestimation of the correlation values (Le Duy et al. 2018). To circumvent this problem, principal component analysis (PCA) is performed in which an equal number of synthetic variables are generated which are orthogonal to each other. To examine the inter-dependency, the following parameters are considered. These are, $\delta^{18}O$ of rainfall ($\delta^{18}O_{rain}$) and transpired water ($\delta^{18}O_{cassia}$), d-excess of rainwater (d_{rain}) and transpired water (d_{cassia}), soil water content at four levels (SW-L1, SW-L2, SW-L3, SW-L4), soil temperature at four levels (ST-L1, ST-L2, ST-L3, ST-L4), net radiation (R_{net}), relative humidity (RH), and wind speed (WS). In a two-dimensional space, two leading principal components are represented by the x and y-axis; then, the biplot technique (Gabriel 1971) is used to describe the PCA result. The arrow vectors describe the variables in this space. The length of an arrow represents the variance, and the cosine between two arrows represents the linear correlation between the two variables. The variables that are better explained by the two principal components will be longer. Acute and obtuse angle represents positive and negative correlation, respectively, while a right angle implies a lack of correlation. Such kind of analysis is often performed to study atmospheric processes (Lone et al. 2020; Metya et al. 2021a, 2021b). We have done this analysis only for 2018 since all the essential parameters are available for this year. Figure 5 shows the result, in which we see that $\delta^{18}O$ of rain and transpired water are closely related.

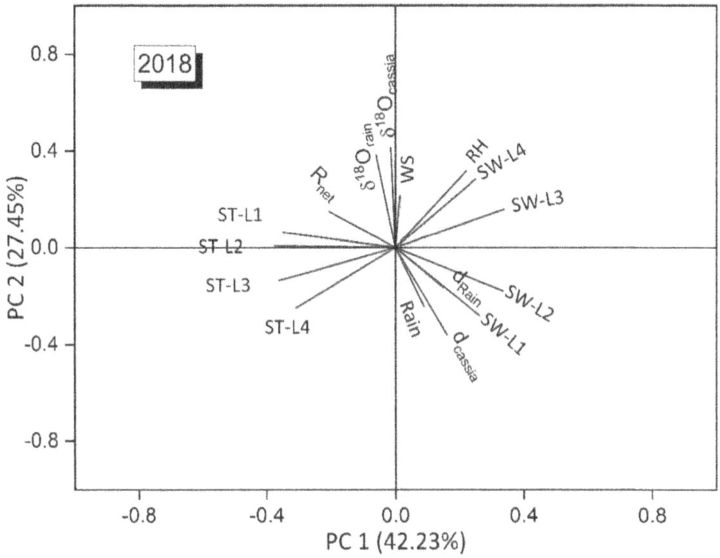

Fig. 5 Result of bi-plot analysis. The relation between different parameters is shown

Both of them show an inverse correlation with the rainfall amount, the manifestation of the amount effect. The isotopic values show a reasonably strong negative correlation with the soil water content at Level-1. But this is slightly reduced for Level-2 and becomes weakly positive for the lower levels SW content. $\delta^{18}O$ of transpired water show practically no correlation with the soil temperature at the upper two levels, but with a negative correlation with the lower-level soil temperature.

SW content strongly depends on radiation; hence a strong inverse correlation was found between net radiation and upper-level SW. However, this dependency decreases with depth, as revealed in this figure, i.e., reduced correlation between net radiation and soil water at levels 3 and 4. Since increased radiation promotes evaporation and hence enriches the isotopic composition of remaining soil water, a positive correlation is expected between $\delta^{18}O$ of transpired water and the radiation parameter, which is indeed shown in this figure.

7 Active and Break Phases of Monsoon

Active and Break phase is an important attribute of the Indian Summer Monsoon. Active and break events are defined as periods during the peak monsoon months of July and August, in which the normalized anomaly of the rainfall over a specified area, called the monsoon core zone, exceeds +1.0 or is less than−1.0 respectively, provided the criteria are satisfied for at least three consecutive days (Rajeevan et al. 2010). The monsoon core zone is denoted approximately by an area covering from

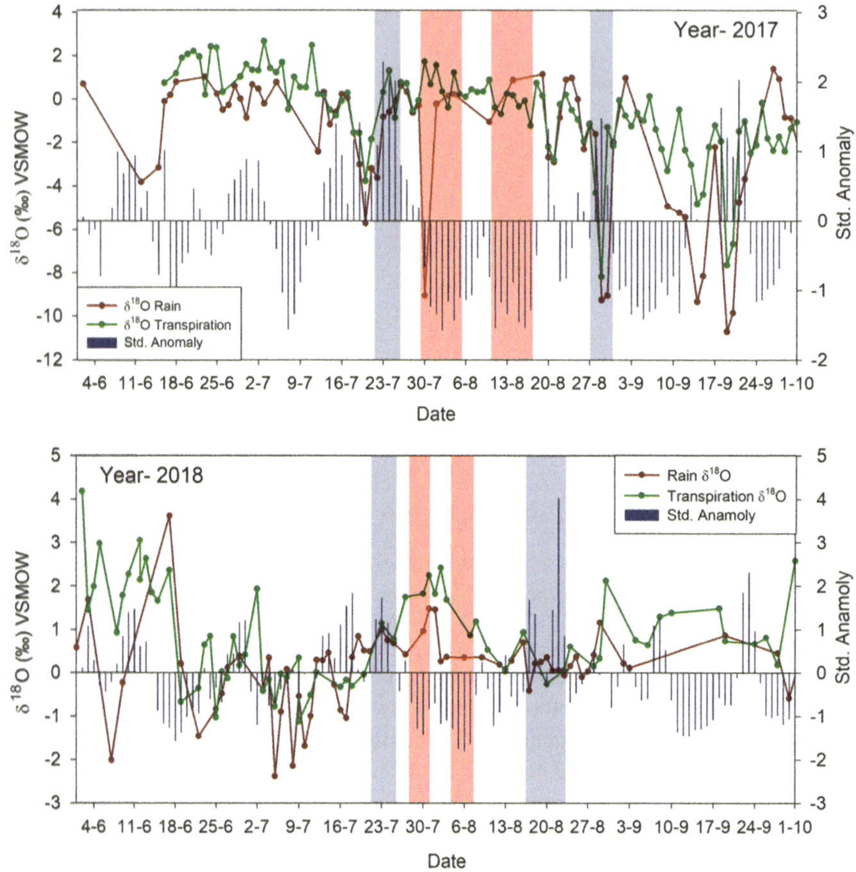

Fig. 6 Isotopic time series of rain and transpired water and their association with the active-break phases of monsoon for 2017 and 2018. The stippled bars show the standardised rainfall anomaly over the core monsoon zone of India. The thick blue (pink) bars depict the active (break) phases of rainfall over the core monsoon zone of India

18°N to 28°N and 65°E to 88°E. Sampling site Pune is situated at the periphery of this zone.

To study the relationship between the isotopic composition of rain and transpiration during active and break phases, the normalized anomaly of the rainfall over the core monsoon zone was considered (the year 2017 and Year 2018) (Monsoon Report 2017 and 2018, India Meteorological Department).

In Fig. 6 red line indicates $\delta^{18}O$ of rain; the green line indicates $\delta^{18}O$ of transpired water (*Cassia fistula*). Bars represent standardised anomalies of rainfall over the core monsoon zone. Red shading shows monsoon break periods, while blue shading shows active periods of rain. The upper panel represents variability for the year 2017, while the lower panel indicates variability for 2018.

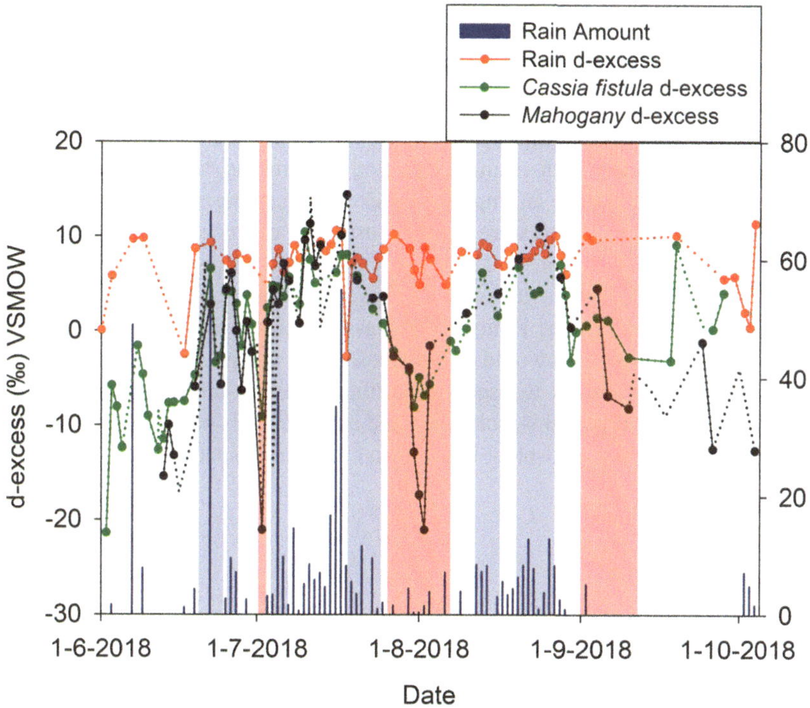

Fig. 7 d-excess time series and their sensitivity with the active and break phases of monsoon is illustrated. The blue bars represent the active, while the red bars show the break phases of the monsoon. As shown here, during the active phases, the difference between rain and transpired water d-excess is small. But during the break phases, this difference is relatively large

The isotopic composition of rain is governed by the amount of rainfall (Dansgaard 1964); it generally reduces with an increase in rainfall amount. It is observed that during the events of positive rainfall anomaly, the proportion of ^{18}O is reduced, both in the rain and transpired water. On the other hand, enrichment in ^{18}O of precipitation and transpiration is witnessed during the events of negative rainfall anomalies.

During the break phases of monsoon 2017 (31 July to 7 August and 11 August to 18 August) difference between rain $\delta^{18}O$ and transpiration $\delta^{18}O$ is high. The difference diminishes during active phases of monsoon (20 July to 24 July and 27 August to 31 August).

Similar results were obtained for the year 2018. During break phases of monsoon (28 July to 31 July and 4 August to 7 August) difference between rain $\delta^{18}O$ and transpiration $\delta^{18}O$ is more with the difference between $\delta^{18}O$ of rain and $\delta^{18}O$ of transpiration diminishing during active phases of monsoon (21 July to 24 July and 18 August to 25 August). It was observed that d-excess of transpired water is generally lower than that of rain (Fig. 7) owing to evaporative enrichment of soil water. During the events of deficient rainfall, this difference is increased (marked with red shading in Fig. 7). It was also noted that the difference between d-excess of rain and that of transpired water reduced drastically during high rain events (marked with blue shading in Fig. 7). High humidity and soil moisture availability are responsible for lowering the soil evaporation, leading to a reduction in the difference between d-excess of rain and transpired water.

d-excess is associated with changes in relative humidity and temperature, so active and break periods have an impact on d-excess characteristics of rainwater as well as transpired water. This feature could be useful to get more information about active and break phases of monsoon at a regional scale by using isotopic records of transpiration. The study shows that a dry environment would cause isotopic enrichment of the transpired vapour. Hence, if dryness increases, the isotopic values of transpiration and precipitation would also increase. This characteristic feature would provide an alternative means to assess the climate change-related scenario for a specific region. In addition to this, collection and isotopic analysis of soil water could provide information about soil-plant interaction of different species in the changing climate.

8 Conclusions

Isotopic analysis of plant transpired water and rainwater were carried out during the monsoon season of 2017 and 2018. Different plants were used for this purpose. The isotopic values of the transpired water were found to be strongly controlled by the soil water dynamics, that is, soil water influx through precipitation and soil water evaporation modulated by environmental conditions during the monsoon seasons of the studied years. Active period increases while break periods help reduce the soil water content. This, in turn, affects soil moisture availability and hence the isotopic composition of transpiration. Investigation of soil moisture content from multiple depths and corresponding transpiration isotopes suggest that plants switch from upper soil levels to deeper soil levels to mitigate the water need during periods of low rainfall.

It was observed that the break periods strongly affect the isotopic composition of both rain and the transpired water. This signal is stronger in the secondary isotope parameter, the d-excess. Hence, the combined analysis of the isotopic characteristics of rainfall and transpired water could provide an alternative means to delineate the break phases of monsoon.

Acknowledgments IITM is fully supported by the Ministry of Earth Sciences, Govt. of India. The National Data Centre and Weather Service Section, India Meteorological Department, Pune are gratefully acknowledged for providing the meteorological data.

References

Bala A, Rawat KS, Misra AK, Srivastava A (2016) Assessment and validation of evapotranspiration using SEBAL algorithm and Lysimeter data of IARI agricultural farm, India. Geocarto Int 31(7):739–764. https://doi.org/10.1080/10106049.2015.1076062

Baldocchi DD, Falge E, Gu L, Olson R, Hollinger DY, Running SW et al (2001) FLUXNET: a new tool to study the temporal and spatial variability of ecosystem-scale carbon dioxide, water vapor, and energy flux densities. Bull Am Meteorol Soc 82:2415–2434

Cernusak LA, Barbour MM, Arndt SK, Cheesman AW, English NB, Feild T et al (2016) Stable isotopes in leaf water of terrestrial plants. Plant Cell Environ 39:1087–1102. https://doi.org/10.1111/pce.12703

Chakraborty S, Sinha N, Chattopadhyay R, Sengupta S, Mohan PM, Datye A (2016) Atmospheric controls on the precipitation isotopes over the Andaman Islands, Bay of Bengal. Sci Rep 6(1): 19555. https://doi.org/10.1038/srep19555

Chakraborty S, Belekar AR, Datye A, Sinha N (2018) Isotopic study of intraseasonal variations of plant transpiration: an alternative means to characterise the dry phases of monsoon. Sci Rep 8: 8647. https://doi.org/10.1038/s41598-018-26965-6

Chakraborty S, Tiwari YK, Deb Burman PK, Baidya RS, Valsala V (2020) Observations and modeling of GHG concentrations and fluxes over India. In: Krishnan R, Sanjay J, Gnanaseelan C, Mujumdar M, Kulkarni A, Chakraborty S (eds) Assessment of climate change over the Indian region. Springer, Singapore. https://doi.org/10.1007/978-981-15-4327-2

Chatterjee A, Roy A, Chakraborty S, Karipot AK, Sarkar C, Singh S, Ghosh SK, Mitra A, Raha S (2018) Biosphere atmosphere exchange of CO2, H2O vapour and energy during spring over a high altitude Himalayan Forest in eastern India. Aerosol Air Qual Res 18:2704–2719. https://doi.org/10.4209/aaqr.2017.12.0605

Crawford J, Hollins SE, Meredith KT, Hughes CE (2017) Precipitation stable isotope variability and subcloud evaporation processes in a semi-arid region. Hydrol Process 31:20–34. https://doi.org/10.1002/hyp.10885

Dansgaard W (1964) Stable isotopes in precipitation. Tellus 16:436–468

Deb Burman PK, Sarma D, Williams M et al (2017) Estimating gross primary productivity of a tropical forest ecosystem over north-east India using LAI and meteorological variables. J Earth Syst Sci 126:99

Deb Burman PK, Sarma D, Morrison R, Karipot A, Chakraborty S (2019) Seasonal variation of evapotranspiration and its effect on the surface energy budget closure at a tropical forest over north-east India. J Earth Syst Sci 128:127. https://doi.org/10.1007/s12040-019-1158-x

Deep Depression over Gangetic West Bengal and adjoining North Bay of Bengal (09–10 October, 2017): a report (cyclone warning division-India Meteorological Department, New Delhi, October 2017). http://www.rsmcnewdelhi.imd.gov.in/images/pdf/publications/preliminary-report/dd09-10oct.pdf

Dharmaraj T, Patil MN, Waghmare RT, Ernest Raj P (2012) Carbon dioxide and water vapour characteristicson the west coast of Arabian Sea during Indian summer monsoon. J Earth Syst Sci 121(4):903–910

Ehleringer JR, Dawson TE (1992) Water uptake by plants: perspectives from stable isotope composition. Plant Cell Environ 15:1073–1082. https://doi.org/10.1111/j.1365-3040.1992.tb01657.x

Gabriel KR (1971) The biplot graphic display of matrices with application to principal component analysis. Biometrika 58:453–467

Gat JR (1996) Oxygen and hydrogen isotopes in the hydrologic cycle. Annu Rev Earth Planet Sci 24(1):225–262

Gibson JJ, Birks SJ, Edwards TWD (2008) Global prediction of δA and δ2H-δ18O evaporation slopes for lakes and soil water accounting for seasonality. Global Biogeochem Cycles 22: GB2031. https://doi.org/10.1029/2007GB002997

Gilman EF, Watson DG (2011) ENH-766. UF/IFAS, University of Florida, Gainesville. https://edis.ifas.ufl.edu/st608

Gnanamoorthy P, Selvam V, Ramasubramanian R et al (2019) Diurnal and seasonal patterns of soil CO_2 efflux from the Pichavaram mangroves, India. Environ Monit Assess 191:258. https://doi.org/10.1007/s10661-019-7407-2

Hanami Y (1974) Effect of environmental factors upon transpiration. Acta Hortic 39:183–196. https://doi.org/10.17660/ActaHortic.1974.39.18

Hersbach H, Bell B, Berrisford P et al (2020) The ERA5 global reanalysis. Q J R Meteorol Soc 146:1999–2049. https://doi.org/10.1002/qj.3803

Jha CS, Thumaty KC, Rodda SR et al (2013) Analysis of carbon dioxide, water vapour and energy fluxes over an Indian teak mixed deciduous forest for winter and summer months using eddy covariance technique. J Earth Syst Sci 122:1259–1268. https://doi.org/10.1007/s12040-013-0350-7

Johnsen SJ, Dansgaard W, White JWC (1989) The origin of Arctic precipitation under present and glacial conditions. Tellus B 41:452–468

Lawrence JL, Gedzelman SD (1996) Low stable isotope ratios of tropical cyclone rains. Geophys Res Lett 23:527–530

Le Duy N, Heidbüchel I, Meyer H, Merz B, Apel H (2018) What controls the stable isotope composition of precipitation in the Mekong Delta? A model-based statistical approach. Hydrol Earth Syst Sci 22:1239–1262. https://doi.org/10.5194/hess-22-1239-2018

LeMone MA, Chen F, Alfieri JG, Tewari M, Geerts B, Miao Q, Coulter RL (2007) Influence of land cover and soil moisture on the horizontal distribution of sensible and latent heat fluxes in Southeast Kansas during IHOP_2002 and CASES-97. J Hydrometeorol 8(1):68–87. https://doi.org/10.1175/JHM554.1

Lone AM, Achyuthan H, Chakraborty S, Metya A, Datye A, Kripalani RH, Fousiya AA (2020) Controls on the isotopic composition of daily precipitation characterized by dual moisture transport pathways at the monsoonal margin region of North-Western India. J Hydrol 588:125106. https://doi.org/10.1016/j.jhydrol.2020.125106

Menchaca LB, Smith BM, Connolly J, Conrad M, Emmett B (2007) A method to determine plant water source using transpired water. Hydrol Earth Syst Sci Discuss 4(2):863–880

Metya A, Datye A, Chakraborty S, Tiwari YK, Sarma D, Bora A, Gogoi N (2021a) Diurnal and seasonal variability of CO2 and CH4 concentrations in a semi-urban environment of western India. Sci Rep 11:2931. https://doi.org/10.1038/s41598-021-82321-1

Metya AL, Chakraborty S, Datye A, Bhattacharya SK, Deb Burman PK, Dasgupta P, Sarma D, Gogoi N, Bora A (2021b) Isotopic and concentration analyses of CO2 and CH4 in association with the eddy-covariance based measurements in a tropical forest of north-east India. Earth Space Sci 8(6):e2020EA001504. https://doi.org/10.1029/2020EA001504

Pallardy S (2008) Physiology of Woody Plants, 3rd edn. Academic, ISBN 9780120887651. https://doi.org/10.1016/B978-012088765-1.50001-4

Patil MN, Dharmaraj T, Waghmare RT, Prabha TV, Kulkarni JR (2014) Measurements of carbon dioxide and heat fluxes during monsoon-2011 season over rural site of India by eddy covariance technique. J Earth Syst Sci 123(1):177–185

Pfahl S, Wernli H (2008) Air parcel trajectory analysis of stable iso-topes in water vapor in the eastern Mediterranean. J Geophys Res 113:D20104. https://doi.org/10.1029/2008JD009839

Rajeevan M, Gadgil S, Bhate J (2010) Active and break spells of the Indian summer monsoon. J Earth Syst Sci 119:229–247. https://doi.org/10.1007/s12040-010-0019-4

Rodda SR, Thumaty KC, Jha CS, Dadhwal VK (2016) Seasonal variations of carbon dioxide, water vapor and energy fluxes in tropical Indian Mangroves. Forests 7:35. https://doi.org/10.3390/f7020035

Rodda SR, Thumaty KC, Praveen M, Jha CS, Dadhwal VK (2021) Multi-year eddy covariance measurements of net ecosystem exchange in tropical dry deciduous forest of India. Agric For Meteorol 301–302:108351

Sarma D, Baruah KK, Baruah R, Gogoi N, Bora A, Chakrabotry S, Karipot A (2018) Carbon dioxide, water vapour and energy fluxes over a semi-evergreen forest in Assam, Northeast India. J Earth Syst Sci 127:94. https://doi.org/10.1007/s12040-018-0993-5

Sarma D, Deb Burman PK, Chakraborty S, Gogoi N, Bora A, Metya A, Datye A, Murkute C, Karipot K (2022) Quantifying the net ecosystem exchange at a semi-deciduous forest in northeast India from intra-seasonal to seasonal time scale. Agric For Meteorol 314:108786. https://doi.org/10.1016/j.agrformet.2021.108786

Schlesinger WH, Scott J (2014) Transpiration in the global water cycle. Agric For Meteorol 189–190:115–117. https://doi.org/10.1016/j.agrformet.2014.01.011

Schwinning S, Starr BI, Ehleringer J (2005) Summer and winter drought in a cold desert ecosystem (Colorado Plateau) part I: effects on soil water and plant water uptake. J Arid Environ 60:547–566. https://doi.org/10.1016/j.jaridenv.2004.07.003

Sheil D (2018) Forests, atmospheric water and an uncertain future: the new biology of the global water cycle. For Ecosyst 5:19. https://doi.org/10.1186/s40663-018-0138-y

Singh N, Patel NR, Bhattacharya BK, Soni P, Parida BR, Parihar JS (2014) Analyzing the dynamics and inter-linkages of carbon and water fluxes in subtropical pine (Pinus roxburghii) ecosystem. Agric For Meteorol 197, ISSN 0168-1923. https://doi.org/10.1016/j.agrformet.2014.07.004

Sinha N, Chakraborty S, Chattopadhyay R, Goswami BN, Mohan PM, Parua DK, Sarma D, Datye A, Sengupta S, Bera S, Baruah KK (2019) Isotopic investigation of the moisture transport processes over the Bay of Bengal. J Hydrol X 2, 100021, ISSN 2589–9155. https://doi.org/10.1016/j.hydroa.2019.100021

Tang K, Feng X (2001) The effect of soil hydrology on the oxygen and hydrogen isotopic compositions of plants' source water. Earth Planet Sci Lett 185(3–4) ISSN 0012-821X. https://doi.org/10.1016/S0012-821X(00)00385-X

Tuzet AJ (2011) Stomatal conductance, photosynthesis, and transpiration, modeling. In: Gliński J, Horabik J, Lipiec J (eds) Encyclopedia of agrophysics. Encyclopedia of Earth Sciences Series. Springer, Dordrecht. https://doi.org/10.1007/978-90-481-3585-1-213

Unlu M, Kanber R, Kapur B (2010) Comparison of soybean evapotranspiration's measured by weighing lysimeter and Bowen ratio-energy balance methods. Afr J Biotechnol 9:4700–4713

Wang H, Ouyang Z, Chen W, Wang X, Zheng H, Ren Y (2011) Water, heat, and airborne pollutants effects on transpiration of urban trees. Environ Pollut 159(8–9):2127–2137, ISSN 0269–7491. https://doi.org/10.1016/j.envpol.2011.02.031

Watham T, Kushwaha S, Patel N, Dadhwal V (2014) Monitoring of carbon dioxide and water vapour exchange over a young mixed forest plantation using eddy covariance technique. Current Science 107(5):858–867. Retrieved January 20, 2021, from http://www.jstor.org/stable/2410559

Xu Q, Li H, Chen J, Cheng X, Liu S, An S (2011) Water use patterns of three species in subalpine forest, Southwest China: the deuterium isotope approach. Ecohydrology 4:236–244. https://doi.org/10.1002/eco.179

Yu W, Xu B, Lai C-T, Ma Y, Tian L, Qu D, Zhu Z (2014) Influences of relative humidity and Indian monsoon precipitation on leaf water stable isotopes from the Southeastern Tibetan Plateau. Geophys Res Lett 41:7746–7753. https://doi.org/10.1002/2014GL062004

Investigating the Effect of Air-sea Carbon Dynamics and Water Quality Parameters on the Coral Reef Ecosystem of Lakshadweep Sea

Ravi Ranjithkumar, Saravanan Kumaresan, Supriyo Chakraborty, Amey Datye, Nowfer Kuly, and Kumar Balachandar

Abstract The integral parameters which assist in sustaining the coral reef ecosystem are being influenced by ocean warming and any small variations in these environmental parameters are known to cause substantial damage to this susceptible ecosystem. Lakshadweep Sea encompasses several atoll-based islands in the Arabian Sea and is known for its coral wealth. We studied the carbon dioxide concentration (pCO_2) at the air-sea interface along with important water quality parameters at the coral reef ecosystem of the Lakshadweep Sea covering different islands over a period of four years (2014–2018) to understand their influence on this ecosystem. The annual range and mean values were calculated from the monthly average values for all the parameters and presented.

The sea surface temperature (SST) and salinity varied between 26.68–30.36 °C and 34.06–36.15 psu respectively. On the other hand, pH showed variability of 8.17–8.22 during the observational period of 2014–2018. The dissolved oxygen (DO) was found with a normal range of 3.96–4.51 ppm indicating the Lakshadweep water is well mixed and aerated by localized currents. The primary productivity (PP) was found with a lower range of 0.68–1.96 mgC/m^3/hr. whereas an elevated partial pressure of air-sea pCO_2 was recorded in the range of 388–402 ppm. Increased levels of SST could be noticed during the study period which might be responsible for the prevailing coral bleaching in the island ecosystem. Average air-sea pCO_2 in the Lakshadweep Sea was found to be 400 ppm in the year 2017 that favored active dissolution of CO_2 into the sea and trigger the acidification process by decreasing the pH below 8.5. This process might have reduced the CO_3^- ion available for coral building and also rising SST have synergistically been affecting this precious ecosystem and denude its biodiversity. CO_2

R. Ranjithkumar · S. Kumaresan (✉) · N. Kuly · K. Balachandar
CAS in Marine Biology, Faculty of Marine Sciences, Annamalai University, Chidambaram, Tamil Nadu, India

S. Chakraborty · A. Datye
Indian Institute of Tropical Meteorology, Ministry of Earth Sciences, Pune, India

B. Phartiyal et al. (eds.), *Climate Change and Environmental Impacts: Past, Present and Future Perspective*, Society of Earth Scientists Series,
https://doi.org/10.1007/978-3-031-13119-6_14

sequestration was found to be insignificant since the primary productivity showed lower values. The present investigation reaffirms that the variations of these vital parameters in the Lakshadweep Sea have a notable impact on this sensitive ecosystem.

Keywords Air-sea pCO_2 · Sea surface temperature · pH · Coral reefs · Lakshadweep Sea

1 Introduction

The Indian Tropical Ocean represents only 13% of the global ocean surface. However, its role in increasing the global ocean heat constant is significant and accounts for about one-quarter of the heat elevation over the last two decades (Beal et al. 2019). Among the oceans, the Indian Ocean stands out as one of the most rapidly warming ocean basins due to marine heat waves and as a result, the wealth of corals, phytoplankton, fishes and shellfishes are being severely impacted (Gnanaseelan et al. 2017; Beal et al. 2019; Collins et al. 2019). The SST of the tropical Indian Ocean has risen by 1 °C during 1951–2015 on an average of 0.15 °C/decade, while the global average has risen by 0.7 °C (0.11 °C/decade) (Krishnan et al. 2020). Therefore, the decadal observations show that the Indian Ocean especially its northern part becomes more vulnerable due to the rise in SST.

Approximately 30% of CO_2 emitted each year is being absorbed by the surface ocean, causing a shift in the chemistry of the seawater carbonates (Sabine et al. 2004; Canadell et al. 2007) and it influences the reduction in pH and Carbonate (CO_3) ion. The reduction in 0.1 unit of pH has been reported to have increased the acidity by 30% and the CO_3 ion decreased simultaneously by 11-15% (Orr et al. 2005; Feely et al. 2009; Hofmann et al. 2010; IPCC 2014). Further, it has been projected that there will be a reduction in pH by 03-0.5units at the end of this century if the ocean continues to absorb the emitted CO_2 at the same level (IPCC 2007). Precipitation of calcium carbonate in the ocean also plays an important role in the carbon cycle and ~40% of the $CaCO_3$ in the ocean is being precipitated in the coastal zones of tropical seas, where coral reefs are a major component (Guttuso et al. 1998). The decrease in surface ocean aragonite (carbonate mineral) may affect the calcification of coral reefs (Guttuso et al. 1999; Kleypas et al. 1999) and hence reduce this CO_2 source to the atmosphere. In general, the air-sea exchange of CO_2 is balanced in the coral reef ecosystem when flourished with phytoplankton diversity and the net primary productivity and respiration are more or less equal. At the same time, when the partial pressure of CO_2 in the air is higher, it would tend to dissolute in the water and as a result, the free hydrogen ion binds with CO_3^- ion which is available for reef-building. Consequently, the aqueous carbon dioxide CO_2 and bicarbonate (HCO_3) levels increase while the concentrations of carbonate (CO_3^-) and the pH decrease (Sabine et al. 2004) which lead to ocean acidification. It is evident that ocean acidification affects phytoplankton diversity and lessens the primary productivity in the coral reef ecosystem. Therefore, when the sequestration of CO_2 becomes insignificant which leads to degradation of the coral reef ecosystem. It is considered a complicated ecosystem because when there are effective $CaCO_3$ precipitation and lesser phytoplankton diversity, this ecosystem acts as a source for CO_2 and if the phytoplankton community is rich in the ecosystem, this will act as a sink for CO_2. The integral parameters (physico-chemical) which assist in sustaining the coral reef ecosystem are also affected by ocean warming and ocean acidification. Any small variation in these environmental parameters causes huge physical damage to this susceptible ecosystem.

The Lakshadweep is a Union Territory of India and is about 200-400 km off the Kerala coast. The Lakshadweep Sea is bestowed with atoll based islands with fringing reefs and submerged banks, having a lagoon area of about 4200 km^2, the territorial waters of 20,000 km^2, the exclusive economic zone (EEZ) of 400,000 km^2 and the total shelf area of around 7,770 km^2 (Varghese 1990). The Lakshadweep Sea is an attractive place for oceanographers since it is encompassed with all ideal aspects of oceanographic features (Mukhopathyay et al. 2007). Studies were made in the previous decades to understand the hydro-biology of the Lakshadweep Sea (Jayaraman et al. 1959, 1960; Patil and Ramamirtham 1963; Uda 1973; Pillai et al. 2001) in terms of environmental variables and their role in this ecosystem. These studies also described the role of ridges and atolls which influence the movement of water in the islands. The shoreline dynamics of the Lakshadweep islands were studied by Chandramohan et al. (1993). A review on the distribution of SST by using satellite-derived data in the Lakshadweep Sea was made by Abhiya et al. (2015) and Shaji et al. (2017). In Lakshadweep islands, localized currents are playing a major role in determining the physico-chemical parameters. Several surge channels are present in the outer reef oriented towards NW-SE and NE-SW directions. The lagoon current occurred towards the southeast during February-August followed by south-southwest movement in the subsequent periods. The current speeds vary between 0 and 0.7 m/s with an average value of around 0.2 m/s inside the Lagoon (Prakash et al. 2015). This localized current facilitates the normal range of DO at the near-shore waters of the Lakshadweep Islands. Sedimentation load and nutrient enrichments due to terrestrial run off would severely impact the coral reefs but the coral reefs of Lakshadweep Sea are away from the mainland would not have been affected. However, a drastic depletion of coral reefs in these remote places might be because of the common phenomena of rising in SST and acidity, as the Lakshadweep provides a favorable geographic setting for the rapid increase of SST causing coral bleaching (Shenoi et al. 1999) (see supplementary photos). However, considerable importance is being given worldwide for the mitigation of climate change impacts (Furnas 2003; Hutchings et al. 2005) as it offers some management options that will enhance the reef resilience, prior to the critical threshold levels being reached due to climate change. The present study was carried out with a view to investigate the impact of air-sea pCO_2 movement along with other physico-chemical parameters on the current status of the coral reef ecosystem of the Lakshadweep Sea.

2 Materials and Methods

2.1 Study Area

The present study was carried out in the lagoons of near-shore shelf region of the Kavaratti, Agati, Bangaram and Thinnakara islands of the Lakshadweep Sea (8°-12° 3' N latitude and 71° E-74° E longitude) during 2014–2018 (Fig. 1).

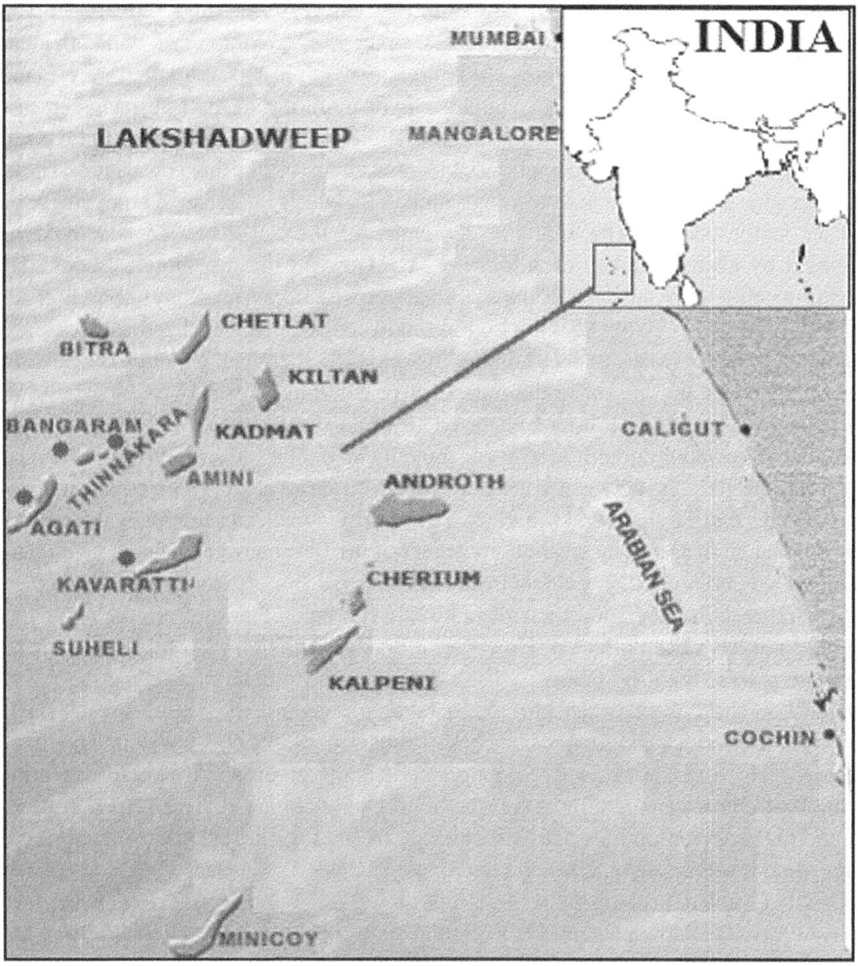

Fig. 1 Map showing the study area in the selected Lakshadweep islands (Agati, Kavaratti, Bangaram and Thinnakara)

2.2 *Sample Collection and Analysis*

SST, pH, and Salinity were measured *in situ* by a calibrated water quality instrument (Model: Quanta-Qd04193). Primary Productivity (PP) measurement was carried out by incubating the light and dark bottles *in situ* as described by Strickland and Parsons (1972) for ~12 hrs (day). Niskin water sampler was used to collect the sea surface water samples at 0.5 to 1.0 meter depth. The collected water samples in duplicate were transferred gently to the DO (Dissolved Oxygen) bottles without making air bubbles. Both PP and DO samples were fixed by adding magnesium sulphate and alkaline iodide in the field and brought to the laboratory for the estimation of dissolved oxygen by following Winkler's titration method. The primary productivity was calculated based on the variations in the oxygen levels both in the dark and light bottles (Grasshoff et al. 1999).

2.3 *Air-Sea* pCO_2

The pCO_2 measurements were done the at air-sea interface (~1meter above the sea level) at the study sites by using LI-COR 820-CO_2 gas analyzer and it was periodically calibrated using NOAA gas cylinders. Regressions between the parameters were carried out and also Analysis of Variance (ANOVA) was done for the parameters on the temporal basis (year wise) using SPSS software (Version 17.0).

3 Results

Inter-annual range and mean values for sea surface temperature (SST) (Fig. 2), Salinity (Fig. 3), pH (Fig. 4), dissolved oxygen (DO) (Fig. 5), primary productivity (PP) (Fig. 6), and air-sea pCO_2 (Fig. 7) are obtained from the monthly average values

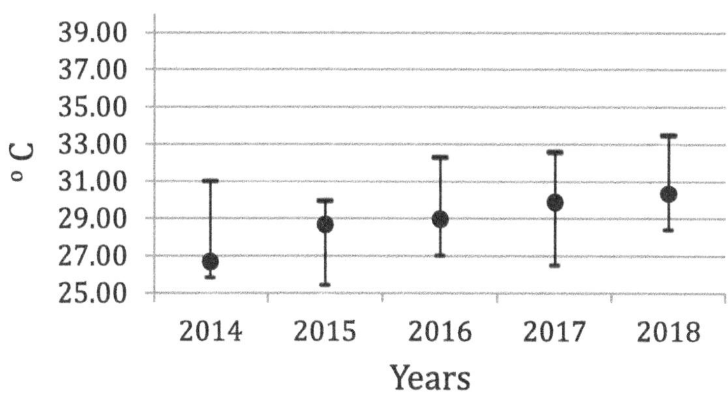

Fig. 2 Inter-annual range and mean values of sea surface temperature (SST) during 2014–2018 at Lakshadweep Sea

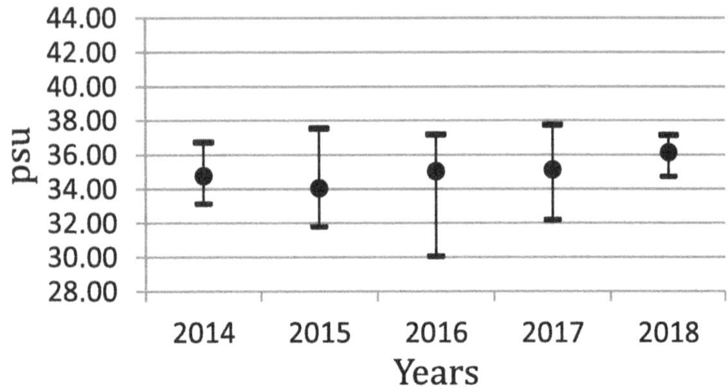

Fig. 3 Inter-annual range and mean values of salinity during 2014–2018 at Lakshadweep Sea

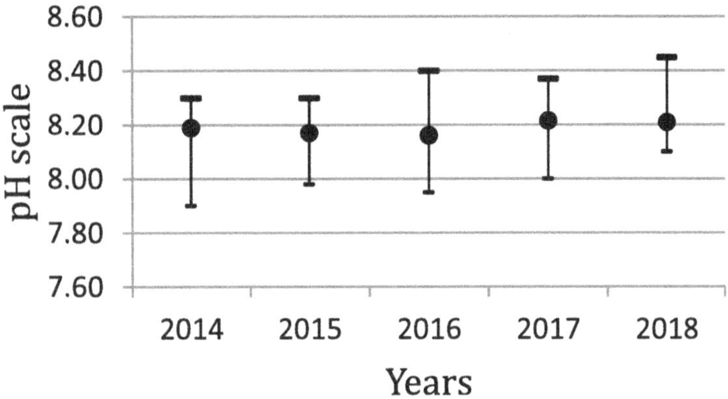

Fig. 4 Inter-annual range and mean values of pH during 2014–2018 at Lakshadweep Sea

during 2014–2018 at Lakshadweep Sea are shown in the respective figures. The overall annual variations of SST, Salinity, pH, DO, PP, and pCO_2 were recorded in the range of 25.42 to 33.50 °C, 30.07 to 37.75 psu, 7.90 to 8.45, 2.87 to 6.21 ppm, 0.44 to 3.88 mgC/m^3/hr. and 360 to 418 ppm respectively.

The range within the average values obtained annually is also shown in Figs. 2, 3, 4, 5, 6, and 7 and the values of SST (26.68–30.36 °C) and salinity (34.06–36.15 psu) were noticed with higher side whereas the average values of pH ranged between 8.17–8.22 with varying trends. A normal range of average levels of DO (3.96–4.51 ppm) was noticed whereas the lower range of average values was found for the PP (0.68–1.96 mgC/m^3/hr). The range of annual average levels of air-sea pCO_2 (388-402 ppm) showed a slightly increasing trend during the study period.

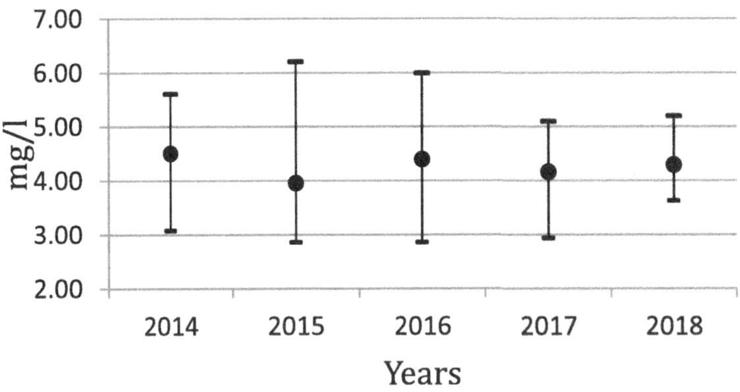

Fig. 5 Inter-annual range and mean values of dissolved oxygen (DO) during 2014–2018 at Lakshadweep Sea

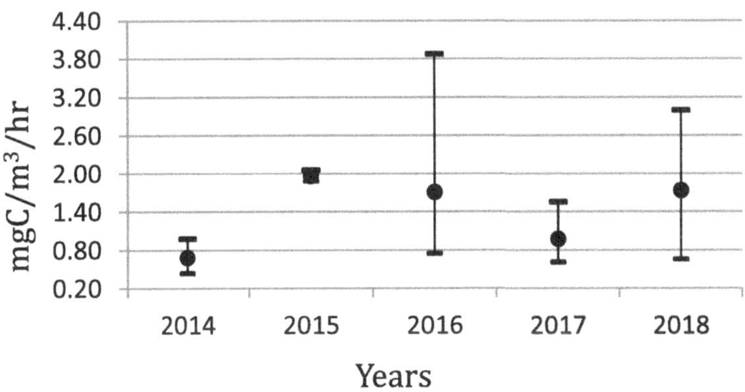

Fig. 6 Inter-annual range and mean values of primary productivity (PP) during 2014–2018 at Lakshadweep Sea

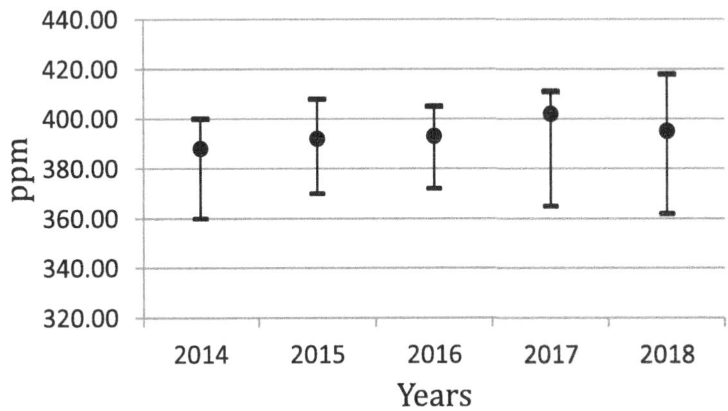

Fig. 7 Inter-annual range and mean values of air-sea pCO_2 during 2014–2018 at Lakshadweep Sea

4 Discussion

The Sea of Lakshadweep is known for its coral reef wealth and associated fauna and flora. However, it was observed that in recent years, due to climate change impact the vital environmental parameters essential for the health of the coral reefs are dwindling. Inter annual variations of these parameters were found to be vulnerable for this ecosystem in the Lakshadweep Sea. The overall range and their monthly average values of the parameters such as SST, salinity, pH, DO and PP was reported by Saravanan Kumaresan et al. (2018) for the Lakshadweep Sea for the period from 2014 to 2015. Those data have been used in this investigation to study the interannual variations for the subsequent years (2016, 2017 and 2018).

The year-wise variations of the parameters were statistically assessed using ANOVA (Table 1). The results showed that the PP and pCO_2 were noticed with significant variations as their values changed inter annually with higher side and PP was found on the lower side. The parameters such as SST, pH, DO and salinity did not show any significant variation as their values were maintained with a similar trend in the Lakshadweep Sea during the period of study (Table 1).

ANOVA was also made for the same parameters between the initial year (2014) and the final year (2018) to understand the variation over a period of four years (Table 2). Highly significant variation for SST and pCO_2 was noticed between these years indicating a significant change occurred during the study period. Likewise, pH and PP also showed slightly significant variation between these years showing its declining phase in the Lakshadweep Sea. However, pH showed little elevated level during 2018 but the values were found well below the normal level for the sea waters (pH -8.5). The parameters of DO and salinity were not found with significant variations, showing a similar trend.

The linear regression was made between the parameters of sea surface temperature (SST) and salinity (Fig. 8), salinity and pH (Fig. 9), air-sea pCO_2 and pH

Table 1 ANOVA for all the parameters between the years from 2014 to 2018

Parameters	2014–2015	2015–2016	2016–2017	2017–2018
SST-mean	29.043 (0.121)	28.780 (0.567)	29.339 (0.114)	30.108 (0.230)
SST-Variance	0.540	0.840	1.789	1.607
PP-mean	1.346 (0.000)^a	1.820 (0.043)^c	1.328 (0.000)^a	1.353 (0.000)^a
PP-Variance	0.012	0.120	0.146	0.154
DO -mean	4.294 (0.126)	4.213 (0.314)	4.294 (0.568)	4.254 (0.774)
DO-Variance	0.491	0.487	0.313	0.195
pH-mean	8.181 (0.551)	8.164 (0.961)	8.189 (0.176)	8.268 (0.002)^b
pH-Variance	0.016	0.015	0.008	0.005
SAL-mean	35.779 (0.202)	35.561 (0.101)	34.823 (0.325)	34.606 (0.737)
SAL-Variance	0.738	1.647	1.993	1.126
pCO_2-mean	377.311 (0.000)^a	382.380 (0.367)	387.237 (0.009)^c	389.267 (0.138)
pCO_2-Variance	5.126	18.038	47.911	41.620

Note: ^{c}0.05 level significant; ^{b}0.001 level significant; ^{a}0.0001 level significant

Table 2 ANOVA for all the parameters between years of 2014 and 2018

Parameters	2014–2018
SST-mean	29.921
SST-Variance	1.048 (0.024)[b]
PP-mean	1.219
PP-Variance	0.137 (0.020)[b]
DO -mean	4.401
DO-Variance	0.259 (0.258)
pH-mean	8.259
pH-Variance	0.011(0.007)[b]
SAL-mean	35.114
SAL-Variance	0.660 (0.016)
pCO_2-mean	380.139
pCO_2-Variance	12.267 (0.000)[a]

Note: [c]0.05 level significant; [b]0.001 level significant; [a]0.0001 level significant

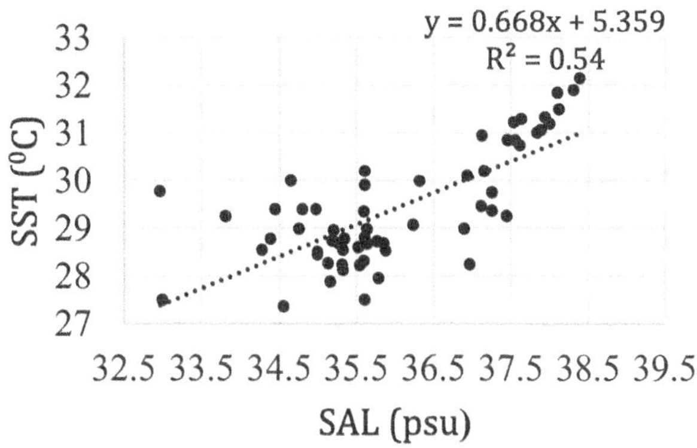

Fig. 8 Linear regression between sea surface temperature (SST) and salinity

(Fig. 10) and for sea surface temperature (SST) and dissolved oxygen (DO) (Fig. 11) using the average values obtained annually. The results showed a highly significant correlation between SST and salinity (R^2 0.54) and also between SST and pH (R^2 0.158) whereas a less significant correlation was found between air-sea pCO_2 and pH (R^2 0.010) and also between SST and DO (R^2 0.021). The correlation analysis shows that the SST is strongly coupled with the salinity and pH but weakly related to DO. The possible reason for a weak correlation between SST and DO is that the SST has no influence over the DO but the localized current favours for its normal range in the near-shore waters of the Lakshadweep islands. The pH was influenced by CO_2 dissolution in the sea but the less significant correlation between air-sea pCO_2 and pH might be due to the regression made between the water pH and air pCO_2 (Fig. 10).

SST of the Lakshadweep Sea could also be noticed with the increasing trend as compared to the studies carried out previously (Jayaraman et al. 1960). Mohit Arora et al. (2019) reported significantly high SST ca. 30.71 °C, 30.77 °C and 30.86 °C

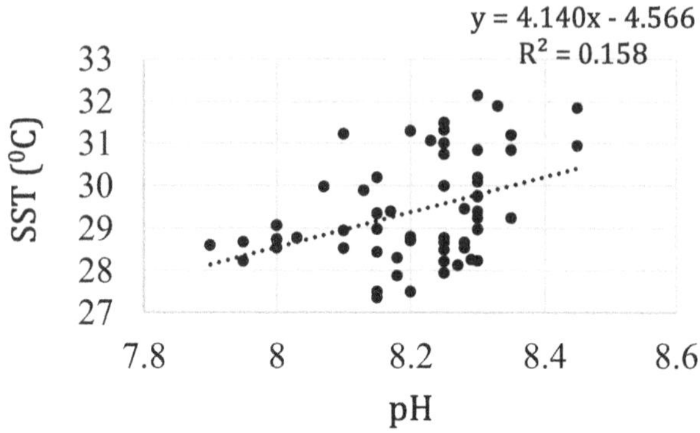

Fig. 9 Linear regression between salinity and pH

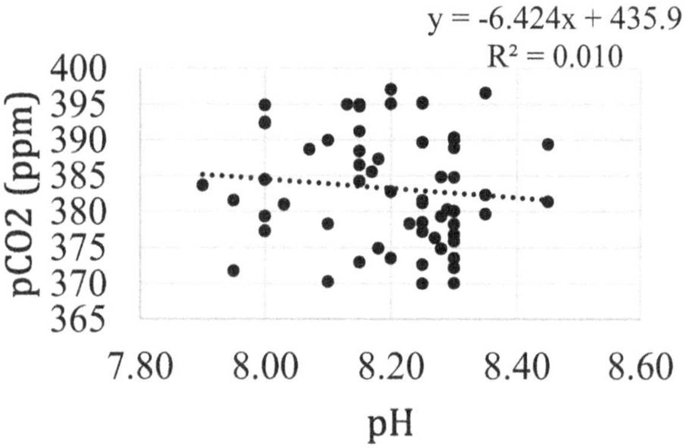

Fig. 10 Linear regression between air-sea pCO_2 and pH

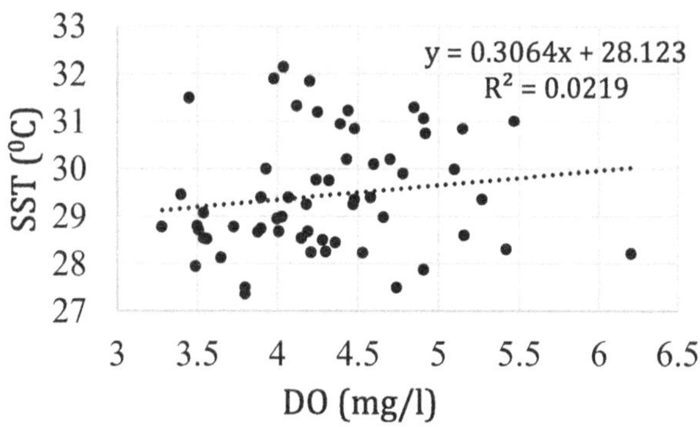

Fig. 11 Linear regression between sea surface temperature (SST) and dissolved oxygen (DO)

during the warmest months (April–May) in 1998, 2010 and 2016 respectively. They reported that the year 2016 recorded highest level and it was 0.15 °C and 0.09 °C more than that of 1998 and 2010 respectively. They also reported that the thermal threshold as 29.99 °C and coral bleaching threshold as 30.40 °C for the Lakshadweep Sea. Our study also indicates an increasing level of SST and exhibiting the similar threshold limit for coral bleaching as the average level is crossing 30.36 °C during the study period. Abhiya et al. (2015) reported that the SST of the Lakshadweep Sea has been increasing at a rate of 0.2 °C per decade. Shaji et al. (2017) analyzed the satellite-derived data which seem to be corroborating our observations. A similar annual SST trend of 28.50 °C in 1985 to 28.92 °C in 2005 was reported in the Lakshadweep Sea (Vivekanandan et al. 2008). It was also reported that most serious threat to Lakshadweep coral reef happened in 1998 with massive bleaching of corals and more than 80% of the coral cover was severely affected (Srivastava and Syedismailkoya 1998; Arthur 2000; Vivekanandan et al. 2008). Recovery of corals was witnessed by Idrees Babu and Suresh Kumar (2016) in some locations and also specified that the growth rate was slowed down due to the stress exerted by increased SST, higher rate of bleaching with maximum coral mortality was also reported during 2015–2016 coupled with El Nino event. Salinity values in the present study were found to be higher and coincided with the previous studies in the Lakshadweep Sea (Jayaraman and Gogate 1957; Beal et al. 2000). It was reported that the higher values were due to the sinking of Red Sea water of high salinity from the northern part to the southern part of the Arabian Sea.

The level of pH recorded in the present study showed that the acidification is set off and reduces the process of formation of carbonate ions which are vital for the calcifying organisms (corals) and it is substantiated through the reports made by Sabine et al. (2004) and IPCC (2007) that the reductions in pH of 0.3–0.5 units are projected by the end of this century as the oceans continue to absorb anthropogenic CO_2. A similar level of pH (7.55–8.26) in the Arabian Sea was reported by Raj and Rajashekhar (2014) during the contemporary period. As far as DO level is concerned, it was found to be independent in the Lakshadweep Sea as it is being influenced by the local currents irrespective of temperature. Shirodkar et al. (2009) reported more or less similar DO values from the Mangalore coast in the Arabian Sea and by Iyer et al. (2003) in the coastal waters of Cochin. The PP was noticed below the normal level in the Lakshadweep Sea (0.68–1.96 $mgC/m^3/hr$) as the availability of nutrients is limited in the lagoon area (Saravanan Kumaresan et al. 2018). Imbalance in the nutrients due to poor loading might be the reason for lesser productivity as reported by Arrigo (2005) and Bhavya et al. (2016a; b).

The average global atmospheric CO_2 was 409.8 $\pm$ 0.1 ppm in 2019. Currently, it is increasing at a rate of approximately 2.5 $\pm$ 0.1 ppm per year (https://gml.noaa.gov/ccgg/trends/graph.html). This is significantly higher than the rate observed during the 1960s, ca. 0.6 $\pm$ 0.1 ppm. Chakraborty et al. (2020) reported surface CO_2 concentration at a western Indian site is increasing over the last few years. However, the CO_2 variability at this terrestrial environment is controlled both by the biosphere activity and the monsoon circulation that carries pristine air from the southern ocean. Hence the amplitude of CO_2 variability in the western Indian region is quite higher than that observed in the marine environment of the Lakshadweep Islands. Nevertheless, the increasing trends of CO_2 observed at the western Indian site and in the nearby Arabian Sea environment seem to be in tandem with each other.

The higher partial pressure of CO_2 at the air-sea interface tends to dissolute the CO_2 into the sea and accelerates the acidification process. Kleypas et al. (1999) have described that the rising CO_2 in the atmosphere could lower the saturation state of the

CO_3^- mineral (i.e. coral-aragonite) in the surface waters of the oceans. The present study also indicates that a high CO_2 concentration scenario likely to adversely affect the coral-mineralisation process leading to their progressive degradation.

Water reacts with CO_2 that dissolute from the atmosphere to form HCO_3^- ions and in the process, depletes the CO_3^- ions. CO_3^- and HCO_3^- ions are in equilibrium with each other in the oceans; so an increase in the abundance of one causes a decrease in the abundance of the other. At a typical pH (8.5), most of the inorganic carbon is stored in the form of HCO_3^- ions but there is still enough CO_3^- available for the formation of $CaCO_3$. The continuous addition of CO_2 brings the free H^+ ions in the sea which in turn binds with CO_3^- ions, reducing the availability of the important CO_3^- ions for the shell-forming marine organisms. This could be another reason for the destruction of corals in the Lakshadweep Sea in addition to a rise in SST.

As more carbon dioxide enters the oceans, the saturation horizons (the boundary between saturated and under-saturated waters) for both aragonite and calcite move closer to the surface, thereby shrinking the area in which calcification can take place (Gehlen et al. 2007). Reports are indicating that the reefs would continue to degrade and get threatened with extinction when the atmospheric CO_2 levels reach 450 ppm (Ove et al. 2007). pH of the ocean surface has already fallen by 0.1 unit-representing a 30% increase in acidity which not only slows down the coral growth but also weakens the coral skeletons (Ove et al. 2007).

5 Conclusions

We have carried out atmospheric and geochemical investigations in the Lakshadweep Sea from 2014 to 2018. We found the average air-sea pCO_2 levels varied between 388 and 402 ppm during the study period. These levels surpassed the threshold limit to the coral reef ecosystem in the Lakshadweep Sea. This might have affected the reef-building capacity of the ecosystem since the current level of air-sea pCO_2 is believed to cause substantial damage to coral reefs. Various investigations including our study show that SST is increasing in the Lakshadweep Island ecosystem. Our study also reaffirms that lowering pH and reducing the availability of CO_3^- ions would synergistically affect this precious ecosystem and denudes its biodiversity. Hence, an integrated management strategy has to be implemented to safeguard the coral reef ecosystem since they are serving as an early warning sign to the entire biodiversity on the earth, especially for the marine environment.

Acknowledgements The authors are thankful to the Director, IITM-MoES, Govt. of India, Pune (Grant No. IITM/Metflux/SK/2018–19) for the financial support through the Metflux India project and the authority of Annamalai University for providing necessary facilities for carrying out the work. The authors also sincerely thank Dr. K. J. Ramesh, former Director General of Meteorology, IMD-MoES, Govt. of India for his scientific advice and encouragement.

References

Abhiya AM, Minu S, Ramachandrankizhur (2015) Salient long-term observations of SST along Kerala coast and its comparative variation with Lakshadweep coast. Aquatic Procedia 4:556–562

Arora M, Chaudhury NR, Gujarati A, Patel RC (2019) Bleaching stress on Indian coral reef regions during mass coral bleaching years using NOAA OISST data. Curr Sci 117(2):242–250

Arrigo KR (2005) Erratum: marine microorganisms and global nutrient cycles. Nature 438:122–122

Arthur R (2000) Coral bleaching and mortality in three Indian reef regions during an El Nino southern oscillation event. Curr Sci 12:1723–1729

Beal LM, Field A, Gordon AL (2000) Spreading of red sea overflow waters in the Indian Ocean. J Geophys Res 105:8549–8564

Beal LM, Vialard J, Roxy MK et al (2019) IndOOS-2: a roadmap to sustained observations of the Indian Ocean for 2030. CLIVAR-4/2019. https://doi.org/10.36071/clivar.rp.4-1.2019

Bhavya PS, Kumar S, Gupta GVM, Sudharma KV, Sudheesh V, Dhanya KR (2016a) Carbon isotopic composition of suspended particulate matter and dissolved organic carbon in the Cochin estuary during post-monsoon. Curr Sci 110:1539–1543. https://doi.org/10.18520/cs/v110/18/1539-1543

Bhavya PS, Kumar S, Gupta GVM, Sudheesh V, Sudharma KV, Varrier DS, Saravanane N (2016b) Nitrogen uptake dynamics in a tropical eutrophic estuary (Cochin, India) and adjacent coastal waters. Estuar Coasts 39:54–67. https://doi.org/10.1007/s1223-015-9982-y

Canadell, J.G., Corinne Le Quere, Michael R. Raupach, Christopher B. Field, Erik T. Buitenhuis, Philippe Ciasis, Thomas J. Conway, Nathan P. Gillet, R.A. Houghton. And Gregg Marland (2007) Contribution to accelerating atmospheric CO2 growth from economic activity, carbon intensity and efficiency of natural sinks. Proc Natl Acad Sci U S A, 104, 18866–18870

Chakraborty S, Tiwari YK, Deb Burman PK, Baidya RS, Valsala V (2020) Observations and modeling of GHG concentrations and fluxes over India. In: Krishnan R, Sanjay J, Gnanaseelan C, Mujumdar M, Kulkarni A, Chakraborty S (eds) Assessment of climate change over the Indian region. Springer, Singapore. https://doi.org/10.1007/978-981-15-4327-2

Chandramohan P, Anand NM, Nayak BU (1993) Shoreline dynamics of the Lakshadweep islands. Indian. J Mar Sci 22:198–202

Collins, M., Sutherland, M., Bouwer, L., Cheong, S.M., Frolicher, T., Jacotes Combes, H., M. K., Losada, I., McInnes, K., Ratter, B., Rivera-Arriga, E., Susanto, R.D., Swingedouw, D. and Tibig, L. (2019) Extremes, abrupt changes and managing risks. In: Portner et al. (edu) IPCC special report on ocean and cryosphere in a changing climate. Cambridge University Press, Cambridge

Feely RA, Doney SC, Cooley SR (2009) Ocean acidification: present conditions and future changes in a highCO2 world. Oceanography 22:36–47

Furnas MM (2003) Catchments and corals: terrestrial runoff to the great barrier reef. Australian Institute of Marine Science & CRC Reef Research Centre, Townsville

Gehlen M, Gangsto R, Schneider B, Bopp L, Amount O, Ethe C (2007) The fate of pelagic $CaCO_3$ production in a high CO_2 ocean: a model study. Biogeosciences 4:505–519

Gnanaseelan C, Roxy MK, Deshpande A (2017) Variability and trends of sea surface temperature and circulation in the Indian Ocean. In: Observed climate variability and change over the Indian Region. Springer, Singapore, pp 165–179

Grasshoff K, Kremling K, Ehrhardt M (1999) Methods of seawater analysis, vol 600, 3rd edn. Wiley-VCH. Completely revised and extended version. Seawater Analysis

Guttuso JP, Frankignoulle M, Bourge I, Romaine S, Buddemeier RW (1998) Effect of calcium carbonate saturation of seawater on coral calcification. Global Planet Change 18:37–46

Guttuso JP, Allemand D, Frankignoulle M (1999) Photosynthesis and calcification at cellular organisms and community levels in coral reefs: a review on interactions and control by carbonate chemistry. Am Zool 39:160–183

Hofmann GE, Barry JP, Edmunds PJ, Gates RD, Hutchins DA, Klinger T, Sewell MA (2010) The effect of ocean acidification on calcifying organisms in marine ecosystems: an organism-to-ecosystem perspective. Annu Rev Ecol Evol Syst 41:127–147

Hutchings, P., Haynes,D., Goudkamp, K. and McCook, L. (2005) Catchment to reef: water quality issues in the great barrier reef region-an overview of papers. Marine Poll Bull 51(1), 3–8

Idrees Babu KK, Suresh Kumar S (2016) Status and changing trends of coral reefs in Lakshadweep archipelago after 1998 mass bleaching event-long term monitoring survey. Int J Appl Pure Sci Agric 02:2394–5532

IPCC (2007) In: Solomon S, Qin D, Manning M, Chen Z, Marquis M, Averyt KB et al (eds) Climate change 2007: the physical science basis. Contribution of working group I to the fourth assessment report of the Intergovernmental Panel on Climate Change. Cambridge University Press, Cambridge, p 996

IPCC (2014) Climate change 2014: synthesis report. In: Core Writing Team, Pachauri RK, Meyer LA (eds) Contribution of working groups I, II and III to the fifth assessment report of the intergovernmental panel on climate change. IPCC, Geneva, 151 pp

Iyer CP, Sindhu M, Kulkarni SG, Tambe SS, Kulkarni BD (2003) Statistical analysis of the physic-chemical data on the coastal waters of Cochin. J Environ Monit 5:324–327

Jayaraman R, Gogate SS (1957) Salinity and temperature variations in the surface waters of the Arabian Sea off the Bombay and Saurashtra coasts. Proc Plant Sci 45:151–164

Jayaraman R, Ramamirtham CP, Sundaraman KV (1959) The vertical distribution of dissolved oxygen in the deeper waters of the Arabian Sea in the neighbourhood of the Laccadives during the summer of 1959. J Mar Biol Assoc India 1:206–211

Jayaraman R, Ramamirtham CP, Sundaraman KV, Nair CP (1960) Hydrography of the Laccadives offshore waters. J. Mar Biol Assoc India 2:24–34

Kleypas JA, Buddemeier RW, Archer D, Guttuso JP, Langdon C, Opdyke BN (1999) Increased atmospheric carbon dioxide on coral reefs. Science 284:118–120

Krishnan R, Sanjay J, Gnanaseelan C, Mujumdar M, Kulkarni A, Chakraborty S (eds) (2020) Assessment of climate change over the Indian region. A report of the Ministry of Earth Sciences (MoES), government of India. Springer, Singapore, p 226. https://doi.org/10.1007/978-981-15-4327-2

Kumaresan S, Shekhar S, Chakraborthy S, Sundaramanickam A, Kuly N (2018) Environmental variables and nutrients in selected islands of Lakshadweep Sea; Addressing Coral Bleaching. Reg Stud Mar Sci 22(2018):38–48

Mukhopathyay R, Ghosh AK, Iyer SD (2007) The Indian Ocean nodule Field: geology and resource potential, vol 10. Elsevier, p 306

Orr JC, Fabry VJ, Aumont O, Bopp L, Doney SC, Feely RA, Gnanadesikan A, Gruber N, Ishida A, Joos F, Key RM, Linday K, Maier-Reimer E, Matear R, Monfray P, Mouchet A, Najjar RG, Plattner GK, Rodgers KB, Sabine CL, Sarmiento JL, Schlitzer R, Slater RD, Totterdell IJ, Weirig MF, Yamanaka T, Yool A (2005) Anthropogenic ocean acidification over the twenty-first century and its impact on calcifying organisms. Nature 437:681–686

Ove H-G, Mumby P, Hooten AJ, Steneck RS (2007) Coral reefs under rapid climate change and ocean acidification. Science 318:1737–1742

Patil MR, Ramamirtham CP (1963) Hydrography of the Laccadives offshore water-a study of the winter conditions. J Mar Biol Assoc India 5:61–169

Pillai PP, Naser AKV, Gopakumar G, Yohannan TM, Koya KP, Sivadas M, Livingston P (2001) Problems and prospects of marine fisheries at Lakshadweep. Geol Surv India 56:95–101. Special Publication

Prakash TN, Sheela Nair L, Shahul Hameed TS (2015) Geomorphology and physical oceanography of the Lakshadweep coral Islands in the Indian Ocean. Springer Briefs Earth Sci. https://doi.org/10.1007/978-3-319-12367-7217-18

Raj SV, Rajashekhar M (2014) Seasonal assessment of hydrographic variables and phytoplankton community in the Arabian Sea waters of Kerala, southwest coast of India. Braz J Oceaogr 62: 279–289

Sabine CL, Freely RA, Gruber N, Key RM, Lee K, Bullister JL, Wanninkhof R, Wong CS, Wallace DWR, Tilbrook B, Millero FJ, Peng T-H, Kozyr A, Ono T, Rios AF (2004) The oceanic sink for anthropogenic CO_2. Science 305:367–371

Shaji C, Sreejith KS, Reba Mary R, Sundaresan J (2017) On the seasonal variability of sea surface temperature and air-sea fluxes in the Lakshadweep Sea. Proc Natl Acad Sci India 87(4):781–795

Shenoi SSC, Shankar D, Shetye SR (1999) On the sea surface temperature high in the Lakshadweep Sea before the onset of the southwest monsoon. J Geophys Res Oceans 104:15703–15712

Shirodkar PV, Mesquita A, Pradhan UK, Verlekar XN, Babu MT, Vetahmony P (2009) Factors controlling physic-chemical characteristics in the coastal waters off Mangalore-a multivariate approach. Environ Res 109:245–257

Srivastava G, Syedismailkoya MS (1998) Current status and impacts on coral reefs in Lakshadweep: management actions and future scenario. In: STAPCOR-98, Seminar on status and protection of coral reefs. NIO and U.T. of Lakshadweep, pp 13–14

Strickland JD, Parsons TR (1972) A practical handbook of seawater analysis. J Fish Res Board Can 167:310

Uda M (1973) Hydrography in relation to tuna fisheries in the Indian Ocean. Mar Biol Assoc India 1:276–292

Varghese (1990) Thirty years of fisheries development in Lakshadweep. Dept of Fisheries publication, UT of Lakshadweep

Vivekanandan E, Ali MH, Jasper B, Rajagopalan M (2008) Thermal thresholds for coral bleaching in the Indian seas. J Mar Biol Assoc India 2:209–214

Trend Analysis and Change Point Detection of Annual and Seasonal Precipitation Timeseries Over Varanasi District, Uttar Pradesh

Antara Gupta and Ajai Mishra

Abstract Urbanization has burdened cities with many problems associated with growth and the physical environment. Some urban areas are becoming increasingly vulnerable to natural disasters such as floods and drought. As a result, researching the features of these events and their physical explanations becomes increasingly necessary. This work studies rainfall trends in Varanasi district, one of the biggest urban cities in Uttar Pradesh, India, from 1901 to 2019. Monthly, seasonal, and annual rainfall patterns have been studied using daily data series of 119 years. The total duration was divided into six different periods of 1901–1920; 1921–1940; 1941–1960; 1961–1980; 1981–2000, and 2001–2019 for a detailed analysis. Two non-parametric statistical tests were used, including the Mann-Kendall test (MK) and the Sequential Mann-Kendall test (SQ-MK). Analysis of rainfall data for 1901–2019 did not show a clear trend for the region. Sen's slope estimator tests describe a statistically significant positive trend is observed in the monsoon season (0.02 mm/year) from 1901 to 1920. However, during the last 40 years, the rainfall trend appears to be increasing, though the result is not statistically significant. The Sequential Mann-Kendall test (SQ-MK test) was also applied for change point detection in annual and seasonal precipitation time series. SQMK results show the mid-1960s is the most probable change point year in the Varanasi time series.

Keywords Climate change · Precipitation · MK and SQMK test analysis · Change-point detection

1 Introduction

Investigations of changing the rainfall pattern are of great significance for India because of its dependency on rainfall for food security and the country's economy (Gadgil and Gadgil 2006). Urbanization has established a network of competitive

A. Gupta (✉) · A. Mishra
Department of Geology, University of Lucknow, Lucknow, India

B. Phartiyal et al. (eds.), *Climate Change and Environmental Impacts: Past, Present and Future Perspective*, Society of Earth Scientists Series,
https://doi.org/10.1007/978-3-031-13119-6_15

urban centers that set the physical reference points for today's globalization. Some of the urban locations in India, i.e., Delhi, Mumbai, Kolkata, Lucknow, and Varanasi, are becoming increasingly vulnerable to natural hazards related to weather and climate (De and Dandekar 2001). The globally averaged precipitation is expected to rise, but regional and continental scales precipitation patterns are unlikely to have any systematic pattern (IPCC 2007). A recent Ministry of Earth Science (MoES) report highlighted that the summer monsoon precipitation (June to September) over India has declined by around 6% from 1951 to 2015, with notable decreases over the Indo-Gangetic Plains and the Western Ghats (Krishnan et al. 2020). The report further emphasizes an emerging consensus, based on multiple datasets and climate model simulations, that the radiative effects of the anthropogenic aerosol forcing over the Northern Hemisphere have considerably offset the expected precipitation increase from GHG warming to the observed decline in summer monsoon precipitation trends. Intergovernmental Panel on Climate Change (IPCC) report has clearly said that climate change will affect agriculture, and increase the risk of human hunger and water scarcity (IPCC 2007). Worldwide, many attempts have been made to study rainfall trends' spatial and temporal variability. It is expected that average precipitation may increase in some areas, but over some areas, it may decrease as well. Both intensification and weakening are expected at continental and regional levels (Dore 2005). The summer monsoon accounts for nearly 80% of the annual rainfall and primarily regulates the hydrology of the Indian subcontinent. A small change in the monsoon rainfall can have profound economic and environmental impacts since a large population in India depends on agriculture and allied sectors that are climate-dependent. The frequent occurrence of climatic extremes in recent years has given rise to questions and concerns on whether the characteristics of hydro-climatic variables are changing or not. Precipitation is one of the major indicators in climate change impact studies that influence the Varanasi region's water management system and agriculture sector. Therefore, this study examined the historical variations in monthly, seasonal, and annual rainfall patterns in the Varanasi district to observe the effect of climate change on a regional scale. Thus, it is pertinent to identify and deliberate on the changes in key hydro-climatic variables that will help in the judicious management of water resources in the Varanasi region.

2 Literature Review: Rainfall Trends in India

Several studies have reported the important issue of rainfall trends in India and at regional scales (e.g., Parthasarathy et al. 1987; Srivastava et al. 1992; De 2001; Prakasa Rao et al. 2004; Kothawale and Rupa Kumar 2005; Dash et al. 2007; Jain et al. 2017; Kulkarni et al. 2020) using the observed daily data as well as model/satellite data sets. Subbaramayya and Naidu (1992) analyzed the monsoon rainfall for various sub-divisions of India using the 1871–1988 period data. They reported a decreasing trend until the end of the nineteenth century; an increasing trend followed this until the middle of the twentieth century, and then the trend reversed again until

1970. After that, an increasing trend was observed. Over the last six decades of their study, they noted a significant increasing trend in southeast peninsular India, which is in the opposite phase of the rest of the country. Prakasa Rao et al. (2004) studied the impacts of urbanization on meteorological parameters in fifteen Indian cities. They found that the radiation levels, bright sunshine hours, wind speeds, and total cloud amounts have decreased in some cities over the last 40–50 years, while relative humidity and rainfall amounts have been increasing in other areas. Rainfall variability and its trend highly depend on the temporal scale. Pramanik and Jagannathan (1954) analyzed rainfall records consisting of 80–100 years of annual records until 1950. These authors discovered systematic variations over certain regions and random fluctuations in long-term annual rainfall trends over India as a whole. Kumar et al. (2010) analyzed a long-term data set (1871–2005) and found no significant trend for rainfall (i.e., annual, seasonal, and monthly scale) in India as a whole. However, in the last 60 years, India's rainfall trend is mostly decreasing (Kulkarni et al. 2020). No trends were reported over the western Himalayas on seasonal and annual rainfall patterns from 1893–1990 (Pant et al. 1999). Rajeevan et al. (2006) and Guhathakurta and Rajeevan (2008) analyzed rainfall time series from 1901 to 2003, consisting of 1476 rain-gauge stations across India. They observed a significant decreasing trend during the monsoon period for three sub-meteorological divisions (Chhattisgarh, Jharkhand, and Kerala). A significant increasing trend for eight sub-divisions (Jammu and Kashmir, west Uttar Pradesh, Gangetic West Bengal, Rayalseema, Madhya Maharashtra, Coastal Andhra Pradesh, Konkan and Goa, and north interior Karnataka) Jain et al. (2017) analyzed precipitation and discharge data for the last 61 years in the Indian River basin. They concluded that annual peak rainfall decreases in upstream areas of the basins over the same period, while annual peak rainfall is rising in plain areas of the basins. The number of different magnitudes of daily rainfall events is higher in the head regions and at the mouth of the basins than in the whole basins. The Indian summer monsoon (ISM) is one of the most energetic and vigorous regional monsoon systems and exhibits highly complex Spatio-temporal variability from June to September (Goswami 2011). Since the 1950s, several observational studies indicated a substantial rise in the rate of intense precipitation events (Goswami et al. 2006; Rajeevan et al. 2006) and a decreasing pattern in low to moderate rainfall events and low-level circulation correlated with ISM over the Indian subcontinent (Dash et al. 2009). The intensity of summer monsoon rainfall is modulated by the intra-seasonal variability, which is characterized by active/inactive spells of increased/decreased precipitation over India. Active spells and weak spells/breaks of the ISM have been extensively studied since Ramamurthy's (1969) study, particularly in the last decade (e.g., Singh et al. 1992; Krishnamurthy and Shukla 2000; Krishnan et al. 2000; Goswami et al. 2003; Kripalani and Kumar 2004; Goswami 2005; Rajeevan et al. 2006). The frequency of heavy rainfall events during the monsoon season was found to be rising over the Andaman and Nicobar Islands, Lakshadweep, the west coast (i.e., 30 cm in coastal Karnataka followed by 25 cm in Konkan and Goa coast), and some pockets in central and northwest India (i.e., 12 cm in Punjab). It was, however, found to be declining over most of India during the winter, pre-monsoon, and post-monsoon seasons (Dash et al. 2007). They further show that Seven subdivisions regions such as Tamil Nadu, Madhya Maharashtra, West Madhya Pradesh, East Rajasthan, Bihar Plateau, East Uttar Pradesh, and North Assam, do not show any trend in ISM. Mall et al. (2007) observed a westward shift in rainfall activity over the Indo-Gangetic Plain region. Goswami et al. (2006) reported that despite considerable year-to-year variability, there has been a significant increase in the frequency and the intensity of extreme monsoon rain events in central India over the past 50 years; the observed trends predicted enhanced risks associated with extreme rainfall over India in the coming decades. However, some pockets of significant long-term rainfall trends and change point detection in precipitation series have been investigated by many researchers throughout the world (Raghavendra 1974; Chaudhary and Abhyankar 1979; Serra et al. 2001; Turkes and Sumer 2004; Partal and Kahya 2006; Dash et al. 2007; Kumar and Jain 2009; Zarenistanak et al. 2014; Mukhopadhyay et al. 2016; Zarenistanak 2018). In the Varanasi district and adjacent region, along with rapid urban development, population growth, excessive use of groundwater, and the main river Ganga have increased significantly day by day. Several studies have examined the changes in water quality, Land use, Land cover under climate change, Ganga River hydrological, and sedimentological assessment. Still, few have considered the long-term rainfall trend pattern and the exact year when those changes started. This type of research is extremely important for groundwater, irrigation, and future water resource management in the Varanasi district and other parts of South Asia. The main objective of the present study is to detect trends and change points in annual and seasonal precipitation series from 1901 to 2019 over the Varanasi region, Eastern India.

3 Study Area

Varanasi, also earlier known as Benaras or Kasi, is a famous Hindu holy city situated on the banks of the river Ganga in the Indian state of Uttar Pradesh. In Rigveda, the city is known as Kasi, which means "*the luminous*." It is often referred to as a "city of temples and learning." Varanasi district is located between the latitude 25°9′30" N to 25°34′22" N and longitude 82°40′4″ E to 83°11′10″ E (Fig. 1b). The region is about 50 Km North of the Vindhyan ranges, which form the northern flank of the peripheral bulge and mark the southern limit of the Ganga basin (Craton). Varanasi district and its environs are located at an average height of about ~76 to 80 m above the mean sea level (MSL) and, by and large, have largely even topography with a low relief in km scale undulations. The Ganga is the principal river of Varanasi, flowing incised into its narrow valley from south to north direction (Shukla and Raju 2008). The climate is mostly humid subtropical, with hot, humid summers, severe monsoons, and mild winters. The area receives about 80% of its annual rainfall of 80–120 mm from the southwest monsoon from July to August (Singh 1994; Fig. 1b). In summer, the maximum temperature rises to 47 °C, and in winter, the minimum temperature drops to 4 °C, with the average annual temperature being 24 °C. The

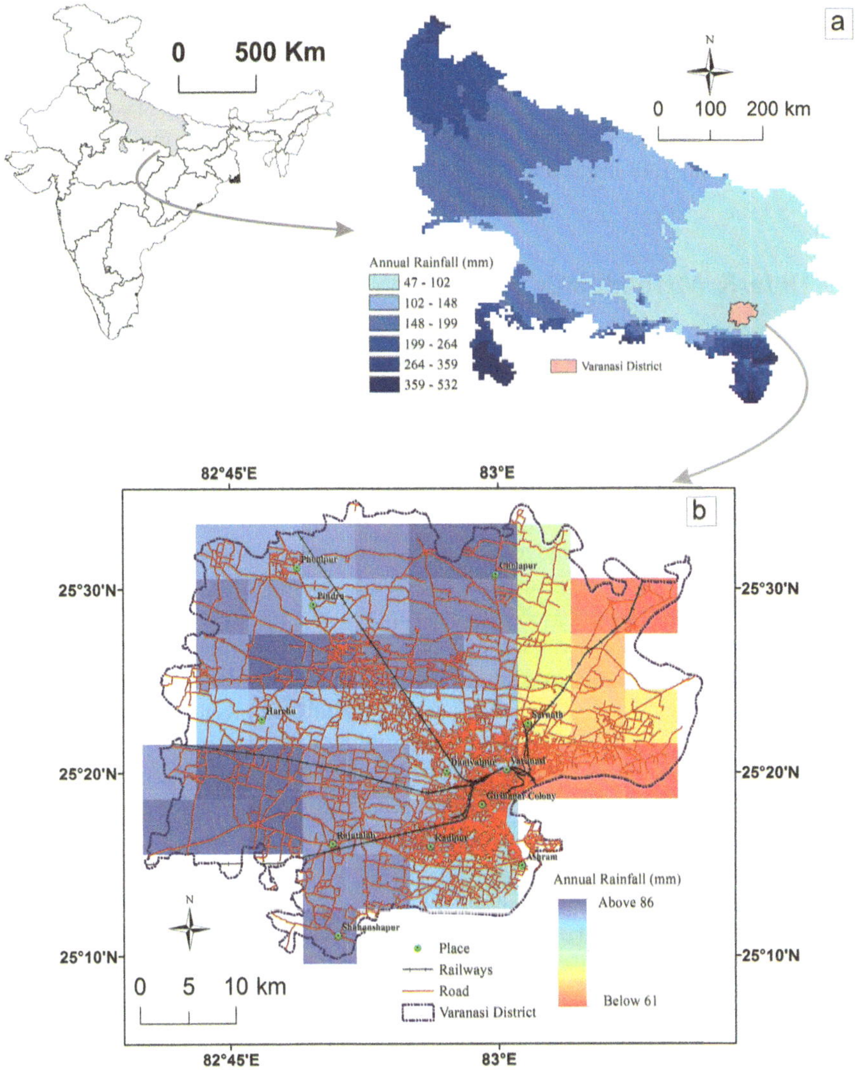

Fig. 1 Study area map, (**a**) Rainfall distribution pattern (mm/day) derived from TRMM satellite data for the Uttar Pradesh region, presented on an annual scale (**b**) showing the annual rainfall distribution pattern (mm/day) map derived from TRMM satellite data for the Varanasi district region

groundwater elevation contour map of the Varanasi area reveals that the general slope of the water table is from north to south. The water table contours indicate that Ganga and Varuna rivers are affluent rivers that receive water from the groundwater table (Shukla and Raju 2008). The water table elevation varies from 60 to 70 m above MSL. Intensive pumping of water due to population increase and urbanization in Varanasi shows a lowering trend of the groundwater level in some parts of the area. Gupta et al. (2018) show that the water quality parameters like $CaCO_3$, HCO_3, Electrical Conductivity (EC), and Total hardness are increasing in the last few years, and the average decline of groundwater level in the last 15 years in the Varanasi city environs is about 1.43 m (Sinha 2003).

4 Data & Methodology

Daily rainfall data obtained were converted to monthly, seasonal, and annual averages for Varanasi district, Uttar Pradesh. Seasons were defined as follows: Winter (January–March), Pre-Monsoon (April–June), Monsoon (July–October), and Post-Monsoon (November–December). Observed daily precipitation of 119 years' data from 1901 to 2019 obtained from the Indian Meteorological Organization (IMD), Pune. The data were carefully analyzed for homogeneity and missing data. To test the homogeneity of the data, a standard normal homogeneity test (Alexandersson 1986) and a standard normal homogeneity test by Alexandersson and Moberg (1997) were applied. For the trend analysis, monthly rainfall data were used to form seasonal and annual series of these variables. Basic statistics such as minimum, maximum, mean, standard deviation (SD), and coefficient of variation (CV) of annual and seasonal rainfall are given in Table 2.

Trends in rainfall data can be identified either by using parametric methods such as those described by Kashyap and Rao (1976) or by using non-parametric methods such as those followed by Fox et al. (1990). The parametric methods depend mainly on prior knowledge of the frequency distribution followed by the data. Dooge (1986) stated that no amount of statistical refinement can overcome the disadvantage of not knowing the frequency distribution involved. Parametric methods are also unsuitable if the data contain outliers. On the contrary, non-parametric methods are not influenced by outliers in the data, and these can be used without the imposition of any frequency distribution on the data. Hence, non-parametric methods were selected for analysis. These nonparametric methods are preferable for detecting trends because they are less affected by the presence of outliers in the data (for more details, see Kothyari and Singh 1996). In the present study, the magnitude of the trend in a rainfall time series was determined using a non-parametric method known as Sen's estimator (Sen 1968), and the statistical significance of the trend in the time series was analyzed using the Mann-Kendall (MK) test (Mann 1945; Kendall 1975) and the Sequential-Mann-Kendall (SQ-MK) test (Sneyers 1990) used for determining the approximate year of the change or beginning of a significant trend. The following sections give details of these techniques.

4.1 Mann-Kendall (MK) Test

The non-parametric Mann-Kendall (MK) test (Mann 1945; Kendall 1975) checks the null hypothesis of no trend versus the alternative hypothesis of the existence of an increasing or decreasing trend. The s (S) is defined as follows (Salas 1993), and Kumar et al. (2010) follow these steps:

$$S = \sum_{i=1}^{N-1} \sum_{j=i+1}^{N} sgn\,(Xj - Xi), \quad (1)$$

Where N is the number of data points; Xj and Xi are sequential data for the ith and jth terms; Assuming (Xj-Xi) = θ, the value of Sgn (θ) is computed as follows:

$$Sgn(\theta)\begin{cases} 1\ if\ \theta > 0 \\ 0\ if\ \theta = 0 \\ -1\ if\ \theta < 0 \end{cases} \quad (2)$$

This statistic shows the number of positive differences minus the number of negative differences for all differences analysed. The test is performed on a large sample (N > 10) using a normal distribution, with the mean and variance as follows:

$$\mathrm{var}(S) = \frac{N(N-1)(2N+5) - \sum_{k=1}^{n} tk(tk-1)(2tk+5)}{18} \quad (3)$$

The number of trends (zero difference between comparison values) groups is n, and the number of data points in the kth linked group is tk. For the one-tailed test of the statistic (S), the standard normal deviate (Z-Statistics) is calculated as follows (Hirsch et al. 1993):

$$Z = \begin{cases} \dfrac{S-1}{\sqrt{\mathrm{var}\ (S)}} \\ 0 \\ \dfrac{S+1}{\sqrt{\mathrm{var}\ (S)}} \end{cases} \quad (4)$$

In a two-sided test, if the computed value of [Z] > $z_{\alpha/2}$, the null hypothesis (H0) is rejected at the level of significance, and the alternative hypothesis is accepted. The null hypothesis was evaluated at a 95% confidence level in this analysis.

4.2 Sen's Slopes Estimation

Sen's (1968) method assumes a linear trend in the time series and has been frequently used to determine the degree of trend in hydro-meteorological time series (Lettenmaier et al. 1994; Jain et al. 2012). The Sen's slopes (Ti) of all data pairs are initially calculated in this manner,

$$Ti = \frac{\mathrm{Xj} - \mathrm{Xk}}{j - k} \text{For i} = 1{,}2 \ldots \mathrm{N} \tag{5}$$

Where, Xj and Xk denote data values at time j and k (j > k), respectively. Sen's estimator of the slope is obtained as the median of these N values of Ti.

$$\beta = \begin{cases} T\dfrac{N+1}{2} & N \text{ is odd,} \\ \dfrac{1}{2}\left(T\dfrac{N}{2} + T\dfrac{N+2}{2}\right) & N \text{ is even.} \end{cases} \tag{6}$$

A positive value of β shows an upward (increasing) trend in the time series, while a negative value suggests a downward (decreasing) trend.

4.3 Sequential Mann- Kendall (SQ-MK) Test

Sneyers (1990) suggested the Sequential Mann-Kendall test find the probable year of starting a major trend. This SQ-MK test creates two series, one progressive u(t) and one backward *u'(t)* (details see Sneyers 1990). If they cross each other, there is a statistically significant trend, then diverge and acquire specific threshold values. The point at which they cross indicates the approximate year when the trend begins. In this case, u(t) is a standardized variable with a zero mean and a unit standard deviation. As a result, its sequential behaviour oscillates about zero. u(t) is the sum of the z values found from the first to the last data point. Similarly, u'(t) values are calculated backward, beginning at the end of the series. This test considers the relative values of all terms in the time series (x_1, x_2,. . ., x_n). The following steps are taken from Zarenistanak et al. (2014), which are in sequence:

I. The magnitudes of x_j the annual mean time series (j = 1. . . n) are compared with x_k (k = 1,. . ., j-1). The number of cases $x_j > x_k$ is counted and denoted by n_j each comparison.
II. The equation then gives the test statistic t:

$$t_j = \sum\nolimits_1^j n_j \tag{7}$$

III. The mean and variance of the statistic are:

$$e(t) = \frac{n(n-1)}{4} \tag{8}$$

And

$$\operatorname{var} t_j = \frac{j(j-1)(2j+5)}{72} \tag{9}$$

IV. The sequential vales of statistic u are then calculated as:

$$u(t) = \frac{t_j - e(t)}{\sqrt{\operatorname{var}(t_j)}} \tag{10}$$

Similarly, the values u'(t) are computed backward, starting from the end of the series.

V. The magnitudes of $x_{j,}$ the annual mean time series (j = n,. . .1) are compared with x_k (k = j-1,. . .,1). At each comparison, the number of cases $x_j > x_k$ is counted and denoted by n_j

VI. The equation then gives the test statistic t:

$$t_j = \sum_1^j n_j \tag{11}$$

VII. The mean and variance of the statistic are:

$$e'(t) = \frac{(N-j+1)(N-j)}{4} \tag{12}$$

And

$$\operatorname{var}'(t_j) = \frac{(N-j+1)(N-j)(2(N-j+1)+5)}{72} \tag{13}$$

VIII. The sequential values of statistics $u'(t)$ is then calculated as follows:

$$u'(t) = \frac{\sum t_j - e't}{\sqrt{var'(t_j)}} \tag{14}$$

The sequential Mann-Kendall could be considered an effective method of determining a trend's beginning year(s). The point at which the curves representing the forward and backward u connect signifies the beginning of a trend or a change in value. The critical value for a 95% confidence level is ± 1.96.

5 Result & Discussion

5.1 Statistical Characteristics of Annual, Seasonal, and Monthly Rainfall

Basic statistical attributes of annual and seasonal (monsoon, post-monsoon, pre-monsoon, and winter) rainfall events for the whole period of 119 years (1901–2019) of Varanasi district, such as mean, standard deviation (SD), and coefficient of variations (CV), were analyzed and annual, seasonal results are given in Table 1. The mean and SD of the annual rainfall data of Varanasi districts are 83.33 mm and 18.34 mm, respectively. In the case of seasonal rainfall, these values are 221.74 mm and 50.77 mm (monsoon), 17.07 mm and 18.47 mm (post-monsoon), 9.05 mm and 7.86 mm (pre-monsoon), and 17.33 mm and 15.45 mm (winter) throughout 1901–2019. These values indicate that the regions with medium rainfall have less variability than the regions with relatively lower rainfall. Box plot diagram showing the monthly rainfall characteristics of Varanasi district (Fig. 2). The annual mean rainfall over Varanasi from 1901 to 2019 is 83.33 mm with a standard deviation of 18.34 mm, respectively. For Varanasi, rainfall during July is the highest (772.96 mm) and contributes to 29.32% of annual rainfall, followed by June (10.18%), August (29.1%), and September (20.2%) of the annual rainfall. Rainfall in December is the least (4.49 mm) and contributes only 0.45% to the annual rainfall. During the southwest monsoon (July–September), Rainfall contributes 71% of the annual rainfall. The contribution of Pre-Monsoon (April–June), Post-Monsoon (November–December), and winter rainfall (January–March) to annual rainfall is 19, 8, and 2%, respectively.

The simple regression analysis of mean annual and monsoonal rainfall over Varanasi showed a long-term insignificant negative trend (Fig. 3). One of the most significant consequences of global warming would be an increase in the magnitude

Table 1 Summary statistics of annual and seasonal rainfall (mm) series of Varanasi district (1901 to 2019)

Statistics summary	Annual	Winter	Pre-Monsoon	Monsoon	Post-Monsoon
Min	40.12	0	0.07	114.42	0
Max	130.65	85.63	34.36	344.56	99.65
Mean	83.33	17.33	9.05	221.74	17.07
Std. error	1.68	1.42	0.72	4.65	1.69
Variance	336.31	238.84	61.91	2577.55	341.28
SD	18.34	15.45	7.86	50.77	18.47
CV (%)	22.01	89.16	86.96	22.89	108.21

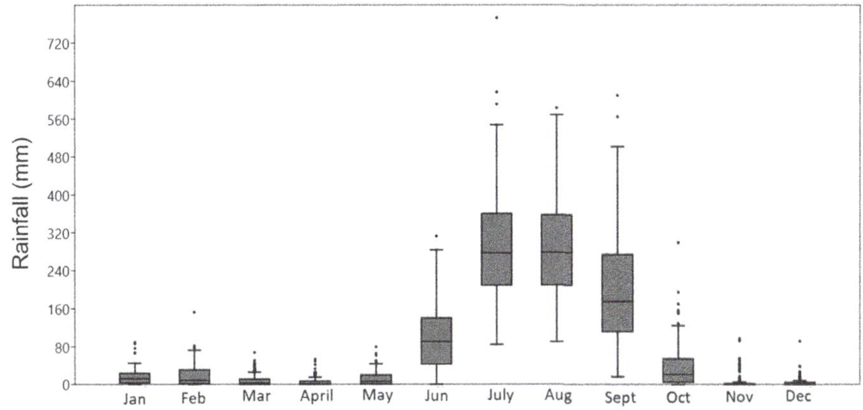

Fig. 2 Box-plot shows the climatological variation of monthly rainfall in Varanasi district during the last 119 years (grey dots represent the outlier's rainfall value; black horizontal lines in the box are median; upper box line upper quartile and lower line lower quartile)

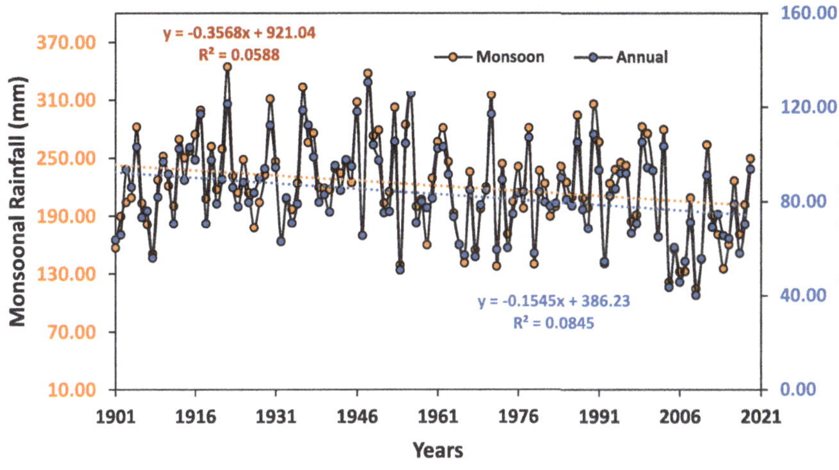

Fig. 3 Long term pattern of rainfall trend both on annual (line with blue dots) and seasonal (summer monsoonal; line with red dots) timescale for the Varanasi district (1901–2019)

and frequency of extreme precipitation events (Goswami et al. 2006). We have characterized the time series into six-time intervals (1901–20; 1921–40; 1941–60; 1961–80; 1981–00, and 2001–19). The Mann-Kendall and Sen slope estimator has been used to determine the rainfall trend and rate of change. The 119-year trend analyses for the Varanasi district were carried out using monthly, average annual, and seasonal (monsoon, pre-monsoon, post-monsoon, and winter) rainfall data.

5.2 Quantification of the Trend

To understand the magnitude or rate of change in the precipitation time series, we have used the Sen slope estimator, which results are given in Fig. 4, Table 2. The analysis of the trends of rainfall variations shows large variability in the magnitude and direction of the trend. The Monthly data series indicated that the majority of months have very little change throughout the months, but they are not statistically significant. Only in February and March during 1921–40 the rainfall amount shows a gentle increasing trend, which is statistically significant at a 95% confidence level (Fig. 5). Also, the same kind of positive and statistically significant trend was noticed in July after 2000. The maximum reduction was observed in April (−0.97 mm/year) during 1901–20, and the maximum increase in December (1921–40; 1 mm/year) and February (1981–00; 1 mm/year) is not statistically significant. Seasonal analysis showed that in the last 20 years, most of the seasons were increasing their rainfall slope except winter and post-monsoon seasons. In the case of monsoon rainfall from 1901–1920 years, the rainfall trend was positive by 0.02 mm/year and statistically significant at 95% (Fig. 6). The maximum increase in monsoon rainfall was of 0.38 mm/year from 1981-to 2000; after 2000, the rainfall magnitude was reduced by half (0.16 mm/year) of the previous duration. Box plot (Fig. 4) provides the Sen slope of precipitation time series at an annual and seasonal time scale over the Varanasi district from 1901 to 2019. According to this box plot diagram, median annual rainfall, the series slope is positive (0.17 mm/year), but the monsoonal rainfall slope is negative (−0.22 mm/year). The highest median negative slopes

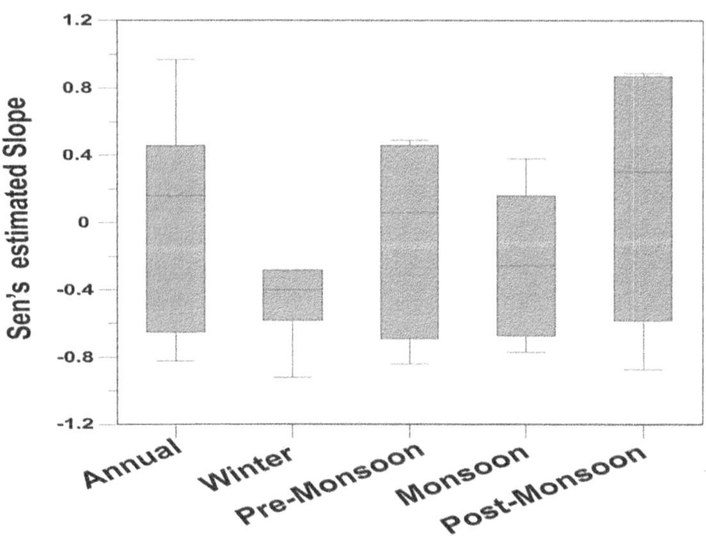

Fig. 4 Box Plot presents the Sen slope of Varanasi Annual and seasonal rainfall pattern (Horizontal line in the box present median value; box upper limit horizontal lines are upper quartile and lower lines are lower quartile)

Table 2 Sen's estimator of slope (mm/year) for monthly rainfall in Varanasi district, Uttar Pradesh (*Bold values indicate statistical significance at 95% confidence level as per the Mann-Kendall test ("+" for increasing and "−" for decreasing)

Year	Jan	Feb	Mar	April	May	Jun	Jul	Aug	Sep	Oct	Nov	Dec
1901–20	−0.05	−0.97	0.79	−0.97	0.72	0.07	0.18	0.16	−0.92	−0.89	−0.73	0.17
1921–40	−0.41	**0.04**	**0.02**	0.97	−0.92	0.77	−0.23	0.35	−0.42	0.97	−0.33	1
1941–60	0.67	−0.01	0.2	−0.69	−0.31	−0.58	0.23	−0.09	−0.18	0.54	−0.86	−0.48
1961–80	0.62	−0.92	0.42	−0.13	−0.72	−0.92	0.58	−0.16	−0.82	0.97	−0.95	−0.76
1981–00	−0.11	1	−0.74	0.72	0.97	0.28	−0.82	0.42	0.87	−0.79	0.32	−0.51
2000–19	0.47	−0.6	0.67	0.51	−0.85	−0.63	**0.04**	−0.77	0.31	−0.13	−0.85	−0.68

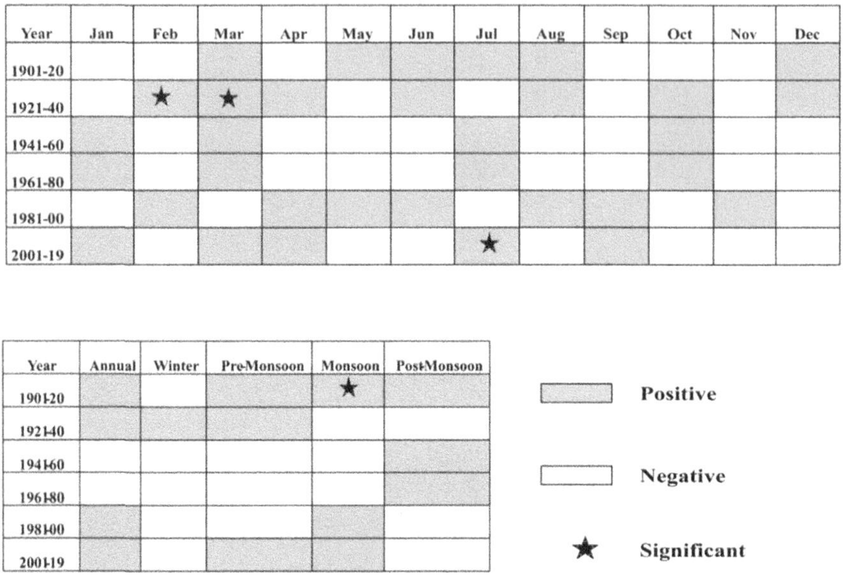

Fig. 5 Trend's detection diagram presents the annual, seasonal, and monthly rainfall for the Varanasi study area. The grey color indicates increasing, and the white color indicates a decreasing trend. The Star symbol indicates a significant trend at a 95% confidence level

are observed in winter (−0.4 mm/year). This implies that the trend line has the steepest negative slope in the winter month, followed by the monsoonal month. Dash et al. (2007), analyzing the rainfall data for the 1871–2003 period, found the same three sub-divisions showing the maximum increase in monsoon rainfall, while they found a maximum decrease in Nagaland, Manipur, Mizoram & Tripura, followed by East Madhya Pradesh, and Orissa.

On the other hand, Kumar et al. (2010) analyzed the all-India rainfall data period of 1870 to 2008, and the results highlighted on an all-India basis February, April, August, October, and November experienced increasing rainfall, whereas June, July, and September showed decreasing rainfall pattern. The months of January, March, May, and December showed little or no change in rainfall trend. The monsoon rainfall showed a decreasing trend, and all other seasons' rainfall showed an increasing trend. The maximum magnitude of the trend was found in the monsoon season (negative), and the minimum in the winter season (positive). Roxy et al. (2015) observed a significant weakening trend in Indian summer monsoonal (ISM) rainfall from 1901 to 2012 over the northern regions of India, along the Ganga-Brahmaputra-Meghna basins. They further show that increased Indian Ocean warming could reduce rainfall over parts of South Asia by weakening the land-sea thermal contrast, dampening the summer monsoon Hadley circulation in the upper atmosphere, and dampening the land-sea thermal contrast. Our studies also suggest that monsoonal rainfall has decreased but not significantly in the last 119 years data (Fig. 3). Some studies have suggested the role of anthropogenic aerosols in

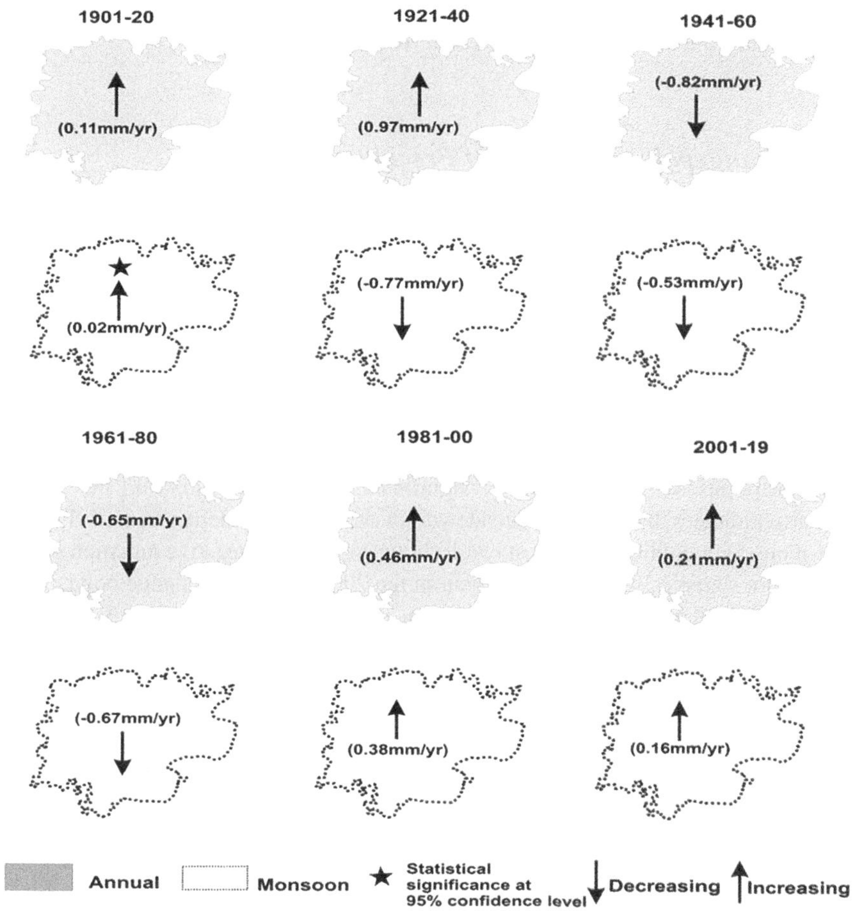

Fig. 6 Trends and magnitude of changes in annual and monsoonal rainfall for the Varanasi district. The up-arrow indicates increasing and down-arrow decreasing trends, respectively

weakening the monsoon circulation during recent decades (Sanap et al. 2015; Bollasina et al. 2011; Meehl et al. 2008). But the observed window of 20 years of data in the last 39 years (1980–2000 & 2001–2019), the monsoon rainfall appears to be increasing though it may not be statically significant. The regional variation in summer monsoon rainfall in the northern Indian region, including the Varanasi district, maybe remotely influenced by the sea surface temperature (SST) of the Indian Ocean (Roxy et al. 2015). So, we hypothesize that this regional variation in the Varanasi region SST factor could control the summer Monsoon, although this hypothesis needs to be tested. In comparison, new research indicates that anthropogenic aerosols have a cooling effect and interact with greenhouse warming over the Indian Ocean (Dong et al. 2014; Ning et al. 2014). Considering the strong seasonal variability in the northern Indian region and short-term persistence in atmospheric changes, it is evident that, on longer time scales, the warmer ocean plays a larger role in weakening the monsoon and vice-versa (Roxy et al. 2015).

5.3 Changepoint Detection of Precipitation

Graphical representations of the results obtained by applying the SQ-MK test to annual rainfall and seasonal series are shown in Fig. 7. Figure 7 shows some examples of warming trends observed in annual and seasonal series. The intersection points of *u(t)* and *u'(t)* time-series are the mutation points that indicate the year when the trend begins (Mosmann et al. 2004). Here we have used the SQ-MK test for change point detection to find out climate trends in the climate of the Varanasi district. The result shows that the annual and monsoon precipitation pattern remained almost the same, most of the (annual and monsoon rainfall) negative mutation points that are statistically significant occurred in 1963 and the 2000s. This also indicates that the 2000 decade was characterized by deficit rainfall (Fig. 7). Post-monsoon rainfall SQ-MK test result shows most of the negative mutation points that are not statistically significant began in the 2000s. Multiple change points have been noticed; it may not be recognized as a significant turning point due to a probability value much higher than the accepted level of significance ($\rho \leq 0.05$).

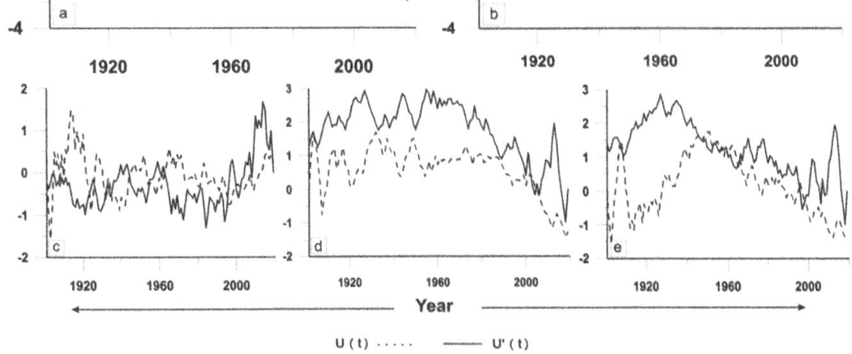

Fig. 7 Graph showing the sequential values of the forward series *u(t)* (dashed line) and backward series *u'(t)* (solid line) obtained by SQ-MK test for (**a**) annual and seasonal series: (**b**) Monsoon; (**c**) Pre-Monsoon; (**d**) Post-Monsoon and (**e**) Winter

All mutation points (either positive or negative) in the winter series occurred after the 1960s. Similarly, all negative mutation points in the seasonal period occurred in the ~1963s. Following the 1960s, most of the mutation points lie in the 2000s. Mukhopadhyay et al. (2016) found a significant mutation point around1950–66 s in annual precipitation over the Lower Ganga plain, West Bengal region. Another study by Tiwari and Pandey (2019) applied the SQMK test for the long-term (1851–2006) rainfall time series under seven meteorological regions of India. These authors noted that a mutation point in North Central India (NCI), the Gangetic region, and West Peninsular India (WPI) was around 1955. Consistent with this finding, precipitation data from the northern parts of India (mostly in the middle and lower Ganga plain region, including the Varanasi district region) have also shown significant change years in or around the mid-1955 to 1963s.

6 Conclusions

The present study has examined trends and change point detection in the monthly, seasonal, and annual rainfall time series in the Varanasi district region. The analysis revealed that the annual and monsoon rainfall had similar trends during the observation period of the last 119 years (1901–2019). There is a gentle increase in southwest monsoon rainfall while a small decreasing trend in winter and the post-monsoon season has been observed over Varanasi for the last 40 years. This suggests that the annual and seasonal rainfall trend in the Varanasi district has not changed significantly in the last 40 years. The annual precipitation rate varied from 1.59 mm/year (1901–1920) to 1.27 mm/year (2001–2019), whereas monsoonal precipitation varied from 2.37 mm/year to 1.39 mm/year. No noticeable decreasing trends were detected by statistical tests in annual and seasonal precipitation time series. Therefore, it is inferred that no significant change in overall precipitation was observed in the last century (1921–2019). However, an increasing trend in rainfall was observed during the last 40 years, though the result is not statistically significant.

On the other hand, the SQMK results conclude that most of the significant mutation points in the annual and seasonal precipitation series began in the 1963s. The long-term trend of annual and seasonal rainfall in pre and post-change point periods (1963, 2000) in the Varanasi district does not indicate the role of the ENSO-ISMR inverse relationship but is remotely influenced by the sea surface temperature of the Indian Ocean. These trend analysis results and change point detection in precipitation series are important for policymakers, water resources management, and agriculture. In this paper, we discuss the climate change issues in the Varanasi region in respect of long-term rainfall variability; however, for a comprehensive assessment, other factors such as changes in temperature, heatwaves, etc. should also be taken into account in the future study.

Acknowledgments The authors thankfully acknowledge the IMD, Pune, for providing the rainfall data of the Varanasi district. AG thanks her brother Dr. Yugantak Raj Gupta, for helping with some of the statistical analysis. We also thank the anonymous reviewers and the editor Supriyo Chakraborty for their constructive suggestions and critical review that helped us to improve the manuscript significantly.

References

Alexandersson HA (1986) A homogeneity test applied to precipitation data. Int J Climatol 6:661–675

Alexandersson H, Moberg A (1997) Homogenization of Swedish temperature data. Part I: homogeneity test for linear trends. Int J Climatol 17:25–34

Bollasina MA, Ming Y, Ramaswamy V (2011) Anthropogenic aerosols and the weakening of the South Asian summer monsoon. Science 334:502–505

Chaudhary A, Abhyankar VP (1979) Does precipitation pattern foretell Gujarat climate became arid. Mausam 30:85–90

Dash SK, Jenamani RK, Kalsi SR, Panda SK (2007) Some evidence of climate change in twentieth-century India. Clim Chang 85:299–321

Dash SK, Kulkarni MA, Mohanty UC, Prasad K (2009) Changes in the characteristics of rain events in India. J Geophys Res 114:D10109. https://doi.org/10.1029/2008JD010572

De US (2001) Climate change impact: Regional scenario. Mausam 52(1):201–212

De US, Dandekar MM (2001) Natural disasters in urban areas. Deccan Geographer 39:1–12

Dong L, Zhou T, Wu B (2014) Indian Ocean warming during 1958–2004 simulated by a climate system model and its mechanism. Clim Dyn 42:203–217

Dooge JE (1986) Looking for hydrologic laws. War Resour Res 22:465–585

Dore MHI (2005) Climate change and changes in global precipitation patterns: what do we know? Environ Int 31:1167–1181

Fox JP, Morgan TR, Miller WJ (1990) Trends in freshwater inflow to San Francisco Bay from the Sacramento-San Joaquin delta. Wat Resour Bull 26:101–106

Gadgil S, Gadgil S (2006) The Indian monsoon, GDP and agriculture. Econ Polit Wkly 41(47): 4887–4895

Goswami BN (2005) Intra-seasonal variability (ISV) of south Asian summer monsoon. In: Lau K, Waliser D (eds) Intra-seasonal variability of the atmosphere-ocean climate system. Springer, Chichester

Goswami BN (2011) South Asian summer monsoon. In: Lau WK-M, Waliser DE (eds) Intra-seasonal variability of the atmosphere-ocean climate system, 2nd edn. Springer, Berlin, pp 21–72

Goswami BN, Ajayamohan RS, Xavier PK, Sengupta D (2003) Clustering of low-pressure systems during the Indian summer monsoon by intra-seasonal oscillations. Geophys Res Lett 30(8). https://doi.org/10.1029/2002GL016734

Goswami BN, Venugopal V, Sengupta D, Madhusoodanam MS, Xavier PK (2006) Increasing trends of extreme rain events over India in a warming environment. Science 314:1442–1445

Guhathakurta P, Rajeevan M (2008) Trends in the rainfall pattern over India. Int J Climatol 28(11): 1453–1469

Gupta A, Mishra A, Singh DS (2018) Environmental impact assessment of water resources and urbanization in Varanasi District, UP, India. J Environ Sci Toxicol Technol 12(1):62–66

Hirsch RM, Helsel DR, Cohn TA, Gilroy EJ (1993) Statistical treatment of hydrologic data. In: Maidment DR (ed) Handbook of Hydrology. McGraw-Hill, New York, pp 17.1–17.52

IPCC (2007) Summary for policymakers. In: Solomon S, Qin D, Manning M, Chen Z, Marquis M, Averyt KB, Tignor M, Miller HL (eds) Climate change: the physical science basis. Cambridge University Press, New YorkContribution of Working Group I to the Fourth Assessment Report of the Intergovernmental Panel on Climate Change

Jain SK, Kumar V, Saharia M (2012) Analysis of rainfall and temperature trends in Northeast India. Int J Climatol 33(4):968–978. https://doi.org/10.1002/joc.3483
Jain SK, Nayak PC, Singh Y, Chandniha SK (2017) Trends in rainfall and peak flows for some river basins in India. Curr Sci 112(8):1712–1726
Kashyap RL, Rao AR (1976) Dynamic and stochastic models from empirical Duta. Academic, New York, pp 315–323
Kendall MG (1975) Rank correlation methods. London, Charles Griffin
Kothawale DR, Rupa Kumar K (2005) On the recent changes in surface temperature trends over India. Geophys Res Lett 32:L18714. https://doi.org/10.1029/2005GL023528
Kothyari UC, Singh VP (1996) Rainfall and temperature trends in India. Hydr Processes 10:357–372
Kripalani RH, Kumar P (2004) Northeast monsoon rainfall variability over south peninsular India vis-à-vis Indian Ocean dipole mode. Int J Climatol 24:1267–1282
Krishnamurthy V, Shukla J (2000) Intra-seasonal and inter-annual variability of rainfall over India. J Clim 13:4366–4377
Krishnan R, Zhang C, Sugi M (2000) Dynamics of breaks in the Indian summer monsoon. J Atmos Sci 57(9):1354–1372
Krishnan R, Gnanaseelan C, Sanjay J, Swapna P, Dhara C, Sabin TP, Jadhav J, Sandeep N, Dey Choudhury A, Singh M, Mujumdar M, Parekh A, Tewari A, Mehajan R (2020) Introduction to climate over Indian region. In: Krishnan R, Sanjay J, Gnanaseelan C, Mujumdar M, Kulkarni A, Chakraborty S (eds) Assessment of climate change over the Indian region. A report of the Ministry of Earth Sciences (MoES), Government of India. Springer, Singapore, pp 1–20. https://doi.org/10.1007/978-981-15-4327-2
Kulkarni A, Sabin TP, Chowdary JS, Rao KK, Priya P, Gandhi N, Bhaskar P, Buri VK, Sabade SS (2020) Precipitation changes in India. In: Krishnan R, Sanjay J, Gnanaseelan C, Mujumdar M, Kulkarni A, Chakraborty S (eds) Assessment of climate change over the Indian region. A report of the Ministry of Earth Sciences (MoES), Government of India. Springer, Singapore, pp 48–72. https://doi.org/10.1007/978-981-15-4327-2
Kumar V, Jain SK (2009) Trends in seasonal and annual rainfall and rainy days in Kashmir valley in the last century. Quaternary Int 212(1):64–69. https://doi.org/10.1016/j.quaint.2009.08.006
Kumar V, Jain SK, Singh Y (2010) Analysis of long-term rainfall trends in India. Hydrol Sci J 55: 484–496
Lettenmaier DP, Wood EF, Wallis JR (1994) Hydro-climato-logical trends in the continental United States, 1948–88. J Clim 7:586–607
Mall RK, Bhatia R, Pandey SN (2007) Water resources in India and impact of climate change. Hydrol Rev 22:157–176
Mann HB (1945) Non-parametric tests against trend. Econometrica 13:245–259
Meehl GA, Arblaster JM, Collins WD (2008) Effects of black carbon aerosols on the Indian monsoon. J Clim 21:2869–2882
Mosmann V, Castro A, Fraile R, Dessens J, Sanches JL (2004) Detection of statistically significant trends in the summer precipitation of mainland Spain. Atmos Res 70:43–53
Mukhopadhyay S, Kulkarni S, Kulkarni P, Dutta S (2016) Rainfall statistics change in West Bengal (India) from period 1901-2000. In: Raju N (ed) Geostatistical and geospatial approaches for the characterization of natural resources in the environment. Springer, Cham. https://doi.org/10.1007/978-3-319-18663-4_29
Ning H, Li-Juan L, Bin W (2014) The role of the aerosol indirect effect in the northern Indian Ocean warming simulated by CMIP5 models. Atmos Oceanic Sci Lett 7:411–416
Pant GB, Rupa Kumar K, Borgaonkar HP (1999) Climate and its long-term variability over the Western Himalaya during the past two centuries. In: Dash SK, Bahadur J (eds) The Himalayan environment. New Age International (P) Limited Publishers, New Delhi, pp 171–184
Partal T, Kahya E (2006) Trend analysis in Turkish precipitation data. Hydrol Process 20:2011–2026

Parthasarathy B, Sontakke NA, Munot AA, Kothawale DR (1987) Droughts/floods in the summer monsoon season over different meteorological subdivisions of India for period 1871–1984. J Climatol 7:57–70

Prakasa Rao GS, Jaswal AK, Kumar MS (2004) Effects of urbanization on meteorological parameters. Mausam 55(3):429–440

Pramanik SK, Jagannathan P (1954) Climate change in India-1: rainfall. Indian J Meteorol Geophys 4:291–309

Raghavendra VK (1974) Trends and periodicities of rainfall in sub-divisions of Maharashtra state. Indian J Met Geophys 25:197–210

Rajeevan M, Bhate J, Kale JD, Lal B (2006) High resolution daily gridded rainfall data for the Indian region: analysis of break and active monsoon spells. Curr Sci 91(3):296–306

Roxy MK et al (2015) Drying of Indian subcontinent by rapid Indian Ocean warming and a weakening land-sea thermal gradient. Nat Commun 6:7423. https://doi.org/10.1038/ncomms8423

Salas JD (1993) Analysis and modeling of hydrologic time series. In: Maidment DR (ed) Handbook of hydrology. McGraw-Hill, New York, pp 19.1–19.72

Sanap S, Pandithurai G, Manoj M (2015) On the response of Indian summer monsoon to aerosol forcing in CMIP5 model simulations. Clim Dyn 2:1–13

Sen PK (1968) Estimates of the regression coefficient based on Kendall's tau. J Am Stat Assoc 63: 1379–1389

Serra C, Burgueno A, Lana X (2001) Analysis of maximum and minimum daily temperatures recorded at Fabra observatory (Barcelona, NE Spain) in the period 1917–1998. Int J Climatol 21:617–636

Shukla UK, Raju NJ (2008) Migration of Ganga River and its implication on hydrogeological potential of Varanasi area UP. J Earth Syst Sci 117(4):489–498

Singh RL (1994) India: a regional geography. National Geographical Society of India, Varanasi

Singh SV, Kripalani RH, Sikka DR (1992) Interannual variability of the Madden–Julian oscillations in the Indian summer monsoon rainfall. J Clim 5:973–978

Sinha TK (2003) Groundwater condition and its quality in Varanasi city. Ind J Geomorphol 8:153–154

Sneyers S (1990) On the statistical analysis of series of observations; Technical note no. 143, WMO No. 725 415, Secretariat of the World Meteorological Organization, Geneva, 192p

Srivastava HN, Dewan BN, Dikshit SK, Rao PGS, Singh SS, Rao KR (1992) Decadal trends in climate over India. Mausam 43:7–20

Subbaramayya I, Naidu CV (1992) Spatial variations and trends in the Indian monsoon rainfall. Int J Climatol 12:597–609

Tiwari H, Pandey BK (2019) Non-parametric characterization of long-term rainfall time series. Meteorog Atmos Phys 131(3):627–637. https://doi.org/10.1007/s00703-018-0592-7

Turkes M, Sumer UM (2004) Spatial and temporal patterns of trends and variability in diurnal temperature ranges of Turkey. Theor Appl Climatol 77:195–227

Zarenistanak M (2018) Historical trend analysis and future projections of precipitation from CMIP5 models in the Alborz Mountain area, Iran. Meteorog Atmos Phys 131(5):1259–1280. https://doi.org/10.1007/s00703-018-0636-z

Zarenistanak M, Dhorde AG, Kripalani RH (2014) Trend analysis and change point detection of annual and seasonal precipitation and temperature series over Southwest Iran. J Earth Syst Sci 123(2):281–295

Assessment of Snow Cover Changes Over the Tons River Basin During Last Two Decades (2000–2019)

Aradhana Thakur, Anupma Sharma, and L. N. Thakural

Abstract Tons river originating from Bandar punch mountain of Uttarakhand, is a major tributary of Yamuna river. Covering a stretch of 148 km, the river contributes runoff from an area of 5146 km^2. The water availability in the river depends on the precipitation received in different forms. The present study is principally concerned with snow cover changes over the Tons river basin on monthly, annual, and seasonal basis over two decades (2000–2019). In the present study, MODIS data was used with a non-parametric Mann-Kendall test (MK test) and Sen's slope estimator. The monthly zone wise snow cover area analysis reveals spatial and temporal variations in snow cover. A notable variation is detected in higher elevation area (>3000 m elevation) of the basin. Annual snow cover variation is also found in the middle of the basin. The average snow cover for the area varies from 38 to 50%.

Keywords Tons · Snow cover · Trend · MK test · Sen's slope

1 Introduction

Snow is a key component of the hydrological cycle in hilly regions and a vital resource in many parts of the world, especially in mountainous regions. Mountain regions hold a substantial amount of freshwater in the form of snow and glaciers and thawing occurs during the summer season, which offers critical flow of water in the pro-glacial rivers supporting downstream ecology, agricultural activities, power generation, and domestic water supplies. In India, snow and glacier melt plays an important role in streamflow contribution to Himalayan rivers (Bhambri et al. 2011). The snow cover area is influenced by many factors like elevation, temperature, aspects, sunshine hours and wind velocity (Shukla et al. 2017). Several studies have been carried out for mapping the snow cover over the Himalayas and the estimation of snowmelt runoff contribution on stream network. Kumar et al. (2007)

A. Thakur (✉) · A. Sharma · L. N. Thakural
National Institute of Hydrology, Roorkee, Uttarakhand, India

B. Phartiyal et al. (eds.), *Climate Change and Environmental Impacts: Past, Present and Future Perspective*, Society of Earth Scientists Series,
https://doi.org/10.1007/978-3-031-13119-6_16

have analysed the contribution of runoff from snow and glacier melt in the Beas river at Pandoh dam and found that about 35% runoff was received from the snow and glacier melt during the summer. Singh and Jain (2002) have evaluated the snow and rainfall contribution over the Satluj river at Bhakra dam location and found that 59% of runoff is contributed from thawing of snow and glacier. Singh et al. (2018) assessed the zone wise snow cover variability for the North-West Himalayan basin (NWH), Lesser Himalaya elevation zone (LHZ), Middle Himalaya elevation zone (MHZ) and Higher Himalaya elevation zone (UHZ) during 2001–2016. They found an insignificant negative trend of snow cover over NWH, LHZ and MHZ before shift (2001–2010) while UHZ followed an increasing trend. Similarly, Sood et al. (2020) analysed the temporal variation in snow cover over Himalayas for the decade 2008–2019 and found a shift in snow cover. In LHZ, snow cover followed the increasing trend in first half of decade while negative trend in second half of decade. A negative trend in first half of decade and positive trend in second half of decade is seen in MHZ and UHZ. The Tons river flows through Garhwal region of Himalyas. The runoff over the Tons river basin depends on snow cover and rainfall, and approximately 50% area of the basin is layered with snow during the winter season. Climate dissimilarities have an intense effect on snow cover changes that unfavourably influences the snowmelt runoff, glacier mass balances and river discharge. The present study analyses the snow cover changes over the Tons river basin on monthly, annual, and seasonal basis over the last two decades (2000–2019).

2 Material and Methodology

2.1 Study Area

Tons river originating from Bandar punch mountain of Uttarakhand in India, lies between 30°30″ to 31°30″N and 77°25″ to 78°45″E. The elevation of Tons river basin ranges from 462 to 6280 m and a major part of the basin lies in Uttarakhand while some part of basin is in Himachal Pradesh (Fig. 1). Tons river is a major tributary of the Yamuna river in the Himalayan reaches and contributes the highest amount of water to the Yamuna. Precipitation and snowmelt in the Tons river basin are the sources of water to the river.

2.2 Data Sources

To fulfil the objective of the study, MODIS data and SRTM digital elevation model (DEM) were employed to study the snow cover changes. The extent of snow enveloped area was extracted using 912 MODIS Terra (MOD10A2) imageries at 8 days interval for the period February 2000 to December 2019.To delineate the elevation zones, SRTM DEM was used, which was procured from the USGS Earth Explorer.

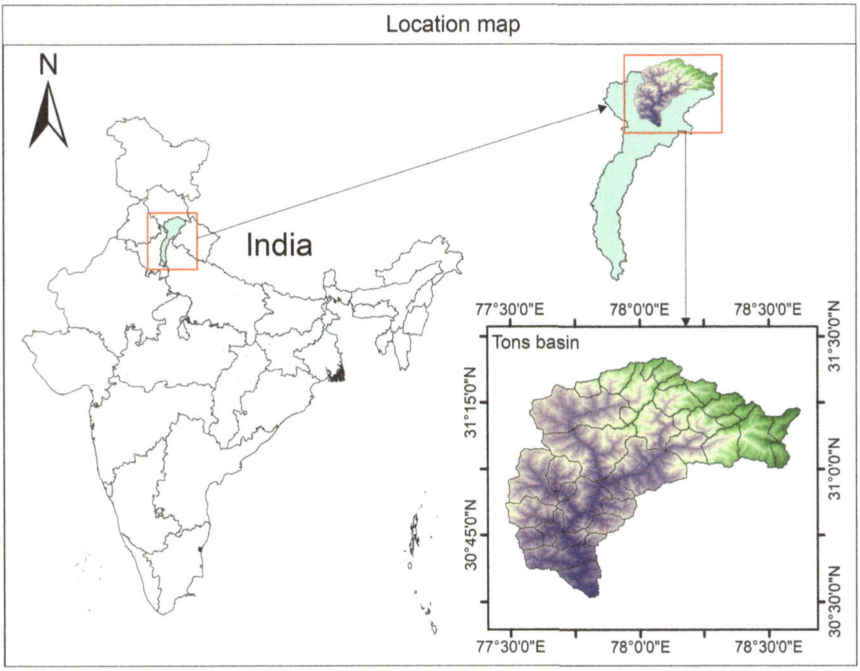

Fig. 1 Location map of Tons basin

2.3 Methodology

In the present study, zone wise snow cover trends on monthly, seasonal and annual basis were investigated. To accomplish the study, the Tons basin was divided into seven elevation zones (Fig. 2, Table 1). Fig. 3 shows the step by step procedure to extract the snow cover area and trend estimation. Mann Kendall and Sen's slope test were used to determine the trend and its magnitude, respectively (Mann 1945; Sen 1968; Kendall 1975; Thakur et al. 2020).

3 Results and Discussion

In the present study, zone wise snow cover trends were investigated. The first zone (zone1) is a non-snow zone while the last zone (zone 7) shows constant snow cover area over the study period. The middle five zones are considered for the snow cover change detection analysis on monthly, seasonal and annual basis. Figure 4 shows the month wise snow cover area over the Tons basin for the year 2019 and the zone wise results are discussed below:

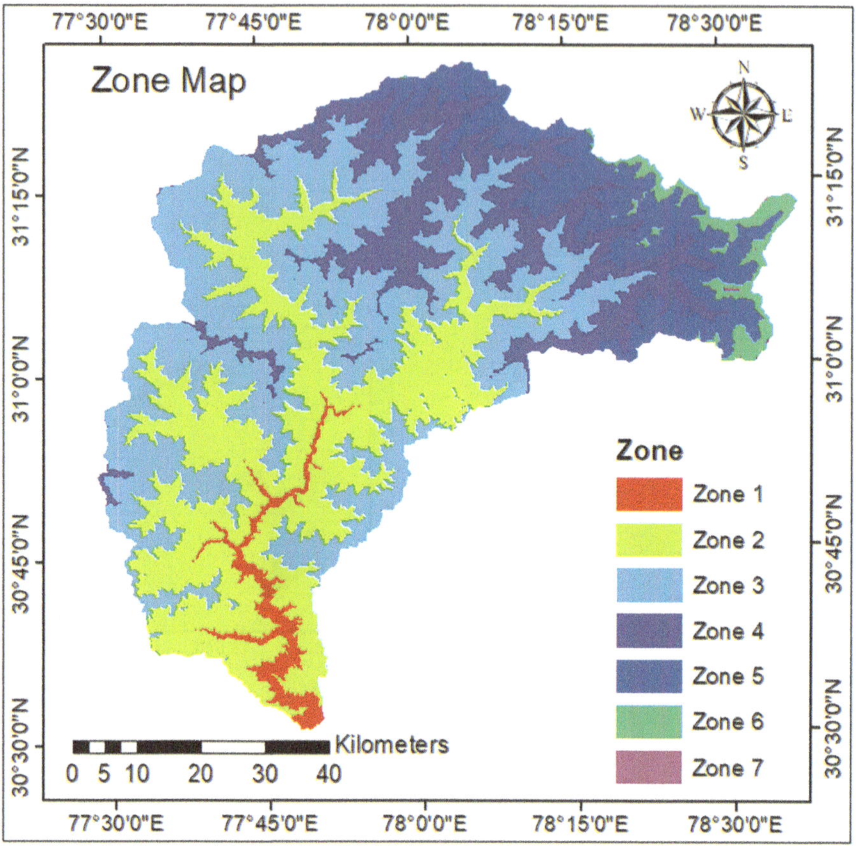

Fig. 2 Zone map of Tons river basin

Table 1 Elevation zone classification

Zone	Area (km^2)	Elevation (in meters above msl)
Zone 1	130.73	<1000
Zone 2	1535.02	1000–2000
Zone 3	1908.53	2000–3000
Zone 4	826.01	3000–4000
Zone 5	612.42	4000–5000
Zone 6	131.16	5000–6000
Zone 7	1.93	>6000

Zone 2 The zone 2 was enveloped with snow during the winter season only i.e. January to March and covers a small fraction of the area. In zone 2, snow cover followed the insignificant rising trend during January to March. In January, the snow cover is increasing with 0.98 z-statistics and 0.002 Sen's slope while February and March reveal zero Sen's slope (Table 2).

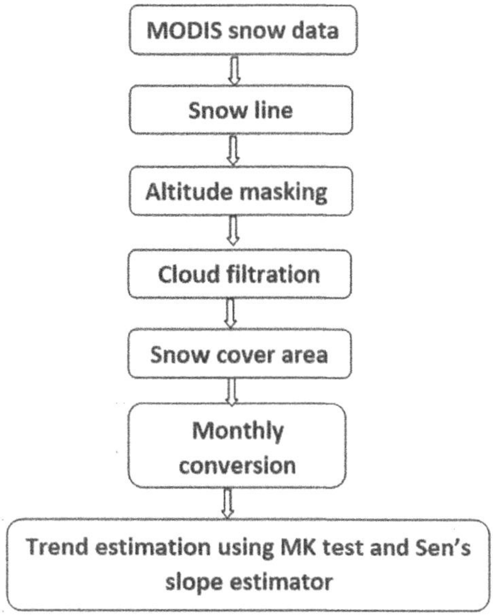

Fig. 3 Snow cover extraction and trend analysis flowchart

Zone 3 This zone covers an area of 1908.5 km^2 which is the highest area covered by an individual zone. In this zone, snow cover changed during monsoon season but it is not significant. The January month is showing upward insignificant changes with z-statistics 1.4 and 0.009 magnitude of trend. In February and July to October, snow cover over the area is following a negative insignificant trend with z-statistics −0.10, −0.48, −1.46, −1.13, and −0.50,respectively. There were no changes in snow cover during the month of March and April while June, November and December showing the insignificant positive trend (Table 2).

Zone 4 Trend analysis of snow cover over zone 4 shows some significant trend. Trend analysis on monthly basis reflected a significant positive trend during the month of June and July with z statistics 1.65 and 2.88 followed by 10% and 1% level of significance, respectively. February, April, September and October months show insignificant negative trend of snow cover while in the remaining months snow cover has increased insignificantly. Snow cover is increasing in all the seasons however, a significant trend with 10% level of significance was observed during the monsoon season. The average annual snow cover is also increasing with 1.78 Z-statistics and 0.003 magnitude of slope (Table 2).

Zone 5 It covers an area of 612.4 km^2, having upward significant trend of snow cover in June and July with Z-statistics 2.50 and 1.78 at 5% and 10% respective level of significance. In zone 5, the snow cover is declining in January, February and September while in remaining months, there was a rising trend. The seasonal analysis indicates, during pre-monsoon to post-monsoon season there is upward change while in pre-monsoon season snow cover is increasing notably with 10% level of significance. In the winter season, snow cover area followed an insignificant

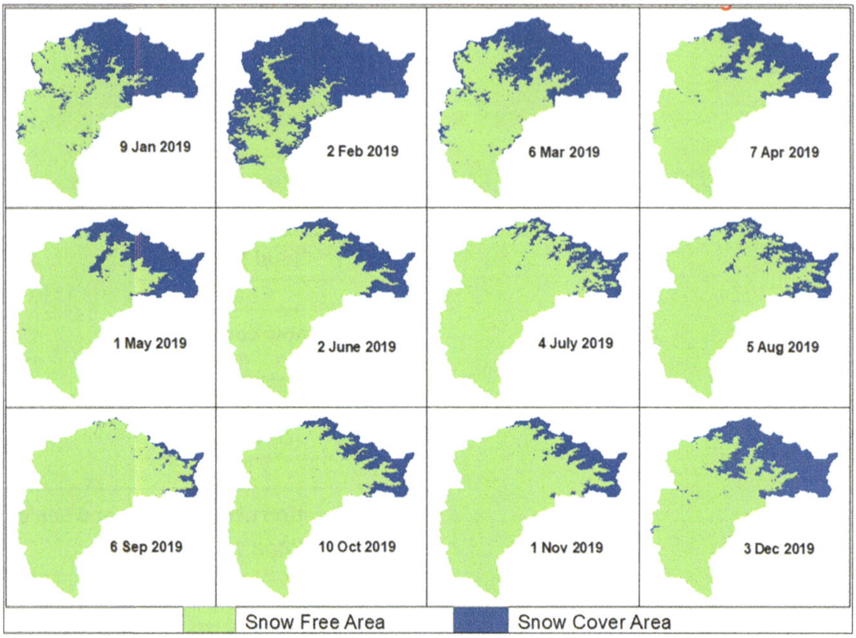

Fig. 4 Snow cover area maps of study area based on MODIS terra snow cover products

downward trend with z-statistics −0.03. The annual snow cover shows a rising trend with z-statistics 1.59 and Sen's slope 0.003 (Table 2).

Zone 6 In this zone, monthly snow cover shows both upward and downward trend. In the month of January, May, August and December, there was negative trend while remaining months followed an upward trend. In the month of June, a significant trend of snow cover with z-statistics 2.27 at 5% level of significance was found (Table 2).

4 Summary and Conclusion

Topography and weather have a very notable connection, especially in higher altitude regions. The present study highlights the spatiotemporal trend of snow cover over the Tons river basin, which occupies a small portion of the Himalayan region. From the analysis, it was observed that the average annual snow cover with maximum area (50%) was present in year 2019 while minimum area (42%) was present in the year 2000. In case of winter season, the average winter snow cover was maximum (74%) in 2019 while minimum (53%) was present in the years 2013 and 2016. It was also noticed that the snow cover during the month of June is increasing with time in upper part of the basin. At higher elevations, snow cover area is

Table 2 Statistics of zone wise snow cover area

	Zone 2			Zone 3			Zone 4			Zone 5			Zone 6		
Time series	Test Z	Sign.	Q	Test Z	Sign.	Q	Test Z	Sign.	Q	Test Z	Sign.	Q	Test Z	Sign.	Q
Jan	0.98	NS	0.00	1.40	NS	0.01	0.91	NS	0.00	−0.42	NS	0.00	−0.63	NS	0.00
Feb	0.13	NS	0.00	−0.10	NS	0.00	−0.62	NS	0.00	−0.75	NS	0.00	0.42	NS	0.00
Mar	0.80	NS	0.00	0.55	NS	0.00	1.20	NS	0.00	0.66	NS	0.00	0.69	NS	0.00
Apr		NS		0.00	NS	0.00	−0.03	NS	0.00	0.00	NS	0.00	0.69	NS	0.00
May		NS		0.00	NS	0.00	1.40	NS	0.01	1.27	NS	0.00	−0.07	NS	0.00
Jun		NS		0.61	NS	0.00	1.65	#	0.00	2.50	*	0.02	2.27	*	0.00
Jul		NS		−0.48	NS	0.00	2.89	**	0.00	1.78	#	0.01	0.42	NS	0.00
Aug		NS		−1.46	NS	0.00	0.19	NS	0.00	0.03	NS	0.00	−0.23	NS	0.00
Sep		NS		−1.13	NS	0.00	−0.84	NS	0.00	−0.81	NS	0.00	0.49	NS	0.00
Oct		NS		−0.50	NS	0.00	−0.36	NS	0.00	0.00	NS	0.00	0.29	NS	0.00
Nov		NS		0.88	NS	0.00	0.55	NS	0.01	0.10	NS	0.00	0.36	NS	0.00
Dec		NS		0.10	NS	0.00	0.81	NS	0.01	0.94	NS	0.00	−1.27	NS	0.00
Monsoon		NS		−1.06	NS	0.00	1.85	#	0.00	1.40	NS	0.01	0.36	NS	0.00
Pre-Monsoon		NS		0.75	NS	0.00	0.75	NS	0.01	1.72	#	0.00	1.40	NS	0.00
Post-Monsoon		NS		0.72	NS	0.00	0.36	NS	0.00	0.29	NS	0.00	1.49	NS	0.00
Winter		NS		0.68	NS	0.00	0.55	NS	0.00	−0.03	NS	0.00	−0.23	NS	0.00
Average Annual		NS		0.42	NS	0.00	1.78	#	0.00	1.59	NS	0.00	1.78	#	0.00

Where ***, **, * and # show the level of significance at 0.1%, 1%, 5%, and 10% respectively; negative (−) and positive (+) values indicate the decreasing and increasing trends, respectively; Sign: significance; NS: Non-significant increasing in the month of June and maximum changes were detected in the middle part (3000 to 5000 m elevation) of the basin.

References

Bhambri R, Bolch T, Chaujar RK, Kulshreshtha SC (2011) Glacier changes in the Garhwal Himalaya, India, from 1968 to 2006 based on remote sensing. J Glaciol 57(203):543–556
Kendall MG (1975) Rank correlation methods, 4th edn. Charles Griffin, London
Kumar V, Singh P, Singh V (2007) Snow and glacier melt contribution in the Beas River at Pandoh Dam, Himachal Pradesh, India. Hydrol Sci J 52(2):376–388
Mann HB (1945) Nonparametric tests against trend. Econometrica:245–259
Shukla S, Kansal ML, Jain SK (2017) Snow cover area variability assessment in the upper part of the Satluj River basin in India. Geocarto Int 32(11):1285–1306
Singh P, Jain SK (2002) Snow and glacier melt in the Satluj River at Bhakra Dam in the western Himalayan region. Hydrol Sci J 47(1):93–106
Singh DK, Gusain HS, Mishra V, Gupta N (2018) Snow cover variability in north-west Himalaya during last decade. Arab J Geosci 11(19):579
Sood V, Singh S, Taloor AK, Prashar S, Kaur R (2020) Monitoring and mapping of snow cover variability using topographically derived NDSI model over north Indian Himalayas during the period 2008–19. Appl Comput Geosci 8:100040
Thakur A, Mishra PK, Nema AK, Sahoo SP (2020) Spatio-temporal trends and shift analysis of temperature for Wainganga sub-basin, India. Int J Environ Stud 77(3):464–479

Extreme Rainfall Trends and their Statistical Significance

Sanjay Kumar, L. N. Thakural, Sunil Gurrapu, and J. P. Patra

Abstract Atmospheric temperature is likely to increase due to increase in greenhouse gases concentration (Global Warming), which in turn is likely to impact evapotranspiration and atmospheric water content there by significantly changing the rainfall magnitude, frequencies and its intensities. Global warming may also affect the seasonal, inter-annual variabilities and spatial distribution of rainfall. The implication of these changes are significant for climatically and ecological sensitive Himalayan regions as small changes in the climate can relatively produce large changes in the probabilities of extreme events, which might have severe consequences in this region. Therefore, an understanding of climate related extremes in the region is important to mitigate the negative impacts of climate change. This study investigates the extreme rainfall events observed over a period of 25 years in a small town in the foothills of Himalaya. Consecutive day extreme rainfall i.e. annual one day, two days and three days maximum rainfall values are extracted from daily rainfall data. These extreme rainfall values are investigated for trend analysis using non-parametric (Man-Kendall and Sen Slope) tests. Results show an increasing trend in the extreme rainfall events over the time. However, inferences from significance test show that these trends are not significant at 95% level of confidence. The study estimates, Mann Kendall, and "Sen Slope Estimator" statistics to infer the above results.

Keywords Extreme rainfall · Trends · Non-parametric · Mann-Kendall

1 Introduction

Global warming intensifying the occurrence of unprecedented hot spells and downpours faster than predicted by historical trends (Diffenbaugh 2020; Seneviratne et al. 2012). Diffenbaugh (2020) mentioned that predictions that relied only on historical

S. Kumar · L. N. Thakural (✉) · S. Gurrapu · J. P. Patra
National Institute of Hydrology, Roorkee, India

B. Phartiyal et al. (eds.), *Climate Change and Environmental Impacts: Past, Present and Future Perspective*, Society of Earth Scientists Series,
https://doi.org/10.1007/978-3-031-13119-6_17

observations underestimated by about half the actual number of extremely hot days in Europe and East Asia, and the number of extremely wet days in the U.S., Europe and East Asia. The change in climate has occurred at a global scale, still its effects often vary from region to region. Several studies in recent past have been conducted to quantity the rainfall variability and trend over India under the changing climate. Moreover, rainfall distribution and trend analysis is being targeted by the climatologists, hydrologist and agriculturist to evaluate the influence of climate globally (Ghosh et al. 2009; Kampata et al. 2008). In recent past, number of climatic studies have been conducted to examine rainfall trend and variability over Indian region (Lal 2001; Sinha Ray and De 2003), metrological sub-divisions (Kumar et al. 2010) and river basins scale (Jain and Kumar 2012; Thakural et al. 2017, 2018). It is also revealed that trend and variability at large-scale may vary from regional scale (Barsugli et al. 2009; Bisht et al. 2017; Raucher 2011; Taxak et al. 2014).

The extreme climatic events in Himalayan region have also been studies by different authors (Rimi et al. 2018; Ma et al. 2017; Wang et al. 2015). Rimi et al. (2018) mentioned that anthropogenic climate change doubled the likelihood of the 2017 pre-monsoon extreme 6-day rainfall event at northeast Bangladesh and also showed that the magnitude of this contribution is sensitive to the climatological period in use. Wang et al. (2015) studied the Himalayan snowstorm of October 2014 resulted from the unusual merger of a tropical cyclone with an upper trough, and their collective changes and mentioned that climate warming have increased the odds for similar events.

Understanding of rainfall trends and their distribution in Himalayan region is needed to formulate the adaptation strategies and optimal management of water resources management under climate change conditions. Several parametric and non-parametric statistical techniques are accessible for the investigation of trend in hydro-meteorological variables. However, non-parametric tests are considered to be more robust owing to their ability to handle both ordinal and ranked data along with the outliers. The present study aims to quantity extreme rainfall trends and their significance for Roorkee city using non-parametric statistical approach and inferences are made.

2 The Study Area

Roorkee city lies in the Haridwar district of Uttarakhand state in North India. It is spread over a flat terrain under Shivalik hills having Himalayas in east & north-east direction and situated on the banks of Northern Ganga canal. The city is located at 29 52 N 53 E with an average elevation of 268 m above msl. An absolute unpredictable continental climate exists over Roorkee due to its closeness to gigantic Himalaya experiencing a warm and temperate climate. There is lot of rain in driest month. The average annual temperature and rainfall of the city is 23.7 °C and 1170 mm respectively. Figure 1 shows the index map of study area.

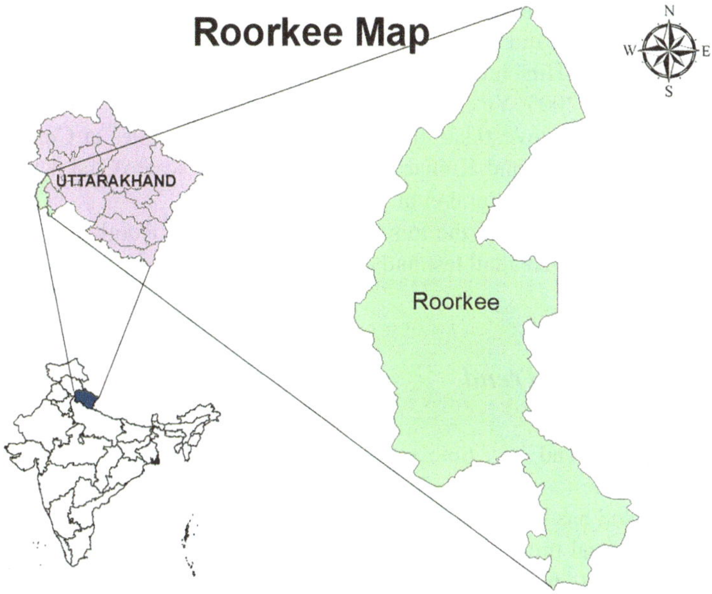

Fig. 1 Location map of study area

3 Materials & Methods

3.1 Data Used

The daily rainfall for 25 years (1995–2019) recorded at the metrological observatory in National Institute of Hydrology campus has been used for the analysis. For the detection and quantification of trends, consecutive day extreme rainfall i.e. annual one day, two days and three days maximum rainfall values are extracted from daily rainfall data. These extreme rainfall values are investigated for trend analysis using non-parametric (Man-Kendall and Sen Slope) tests, however, a linear trend line (with slope 'm') is also fitted for preliminary investigation of trends and their significance.

3.2 Trend Analysis

Spatio-temporal variability and investigation of rainfall trends are essential input for climatic studies in a region. Trends in data can be identified by using either parametric or non-parametric methods, and both the methods are widely used. The non-parametric methods do not require normality of time series and also are less sensitive to outliers and missing values. The non-parametric methods are extensively used for analyzing the trends in several hydrologic series namely rainfall, temperature, pan evaporation, wind speed etc. (Chattopadhyay et al. 2011; Dinpashoh et al. 2011; Fu et al. 2004; Hirsch et al. 1982; Jhajharia and Singh 2011; Jhajharia et al. 2009; Tebakari et al. 2005; Yu et al. 1993). A number of studies have been attempted using both methods to investigate the trend of climatic variables at Country scale (Kumar et al. 2010; Jain and Kumar 2012) and regional scale (Chakraborty et al. 2013; Patra et al. 2012; Thakural et al. 2017, 2018).

The present study analyzes the trend of rainfall series of Roorkee station using non-parametric (Mann-Kendall test and Sens's estimator of slope).

3.3 Magnitude of Trend

The magnitude of trend in a time series was determined using a non-parametric method known as Sen's estimator (Sen 1968). This method assumes a linear trend in the time series and has been widely used for determining the magnitude of trend in hydro-meteorological time series (Lettenmaier et al. 1994; Yue and Hashino 2003; Partal and Kahya 2006). In this method, the slopes (T_i) of all data pairs are first calculated by

$$T_i = \frac{x_j - x_k}{j - k} \quad for\ i = 1,2,\ldots,N \tag{1}$$

where x_j and x_k are data values at time j and k ($j > k$) respectively. The median of these N values of T_i is Sen's estimator of slope which is calculated as

$$\beta = \begin{cases} T_{\frac{N+1}{2}} & \text{if } N \text{ is odd,} \\ \frac{1}{2}\left(T_{\frac{N}{2}} + T_{\frac{N+2}{2}}\right) & \text{if } N \text{ is even.} \end{cases} \tag{2}$$

A positive value of β indicates an upwards (increasing) trend and a negative value indicates a downwards (decreasing) trend in the time series.

3.4 Significance of Trend

To ascertain the presence of a statistically significant trend in hydrologic climatic variables such as temperature, precipitation and streamflow with reference to climate change, the non-parametric Mann–Kendall (MK) test has been employed by a number of researchers (Yu et al. 1993; Douglas et al. 2000; Burn et al. 2004; Singh et al. 2008a, b). The MK method searches for a trend in a time series without specifying whether the trend is linear or non-linear. The MK test checks the null hypothesis of no trend versus the alternative hypothesis of the existence of an increasing or decreasing trend. Following Bayazit and Onoz (2007), no pre-whitening of the data series was carried out as the sample size is large ($n \geq 50$) and slope of the trend was high (> 0.01).

The statistic S is defined as (Salas 1993):

$$S = \sum_{i=1}^{N-1} \sum_{j=i+1}^{N} \operatorname{sgn}\left(x_j - x_i\right) \tag{3}$$

where N is the number of data points. Assuming $(x_j - x_i) = \theta$, the value of sgn(θ) is computed as follows:

$$\operatorname{sgn}(\theta) = \begin{cases} 1 & if \quad \theta > 0, \\ 0 & if \quad \theta = 0, \\ -1 & if \quad \theta < 0. \end{cases} \tag{4}$$

This statistic represents the number of positive differences minus the number of negative differences for all the differences considered. For large samples ($N > 10$), the test is conducted using a normal distribution (Helsel and Hirsch 1992) with the mean and the variance as follows:

$$\mathrm{E[S]} = 0 \tag{5}$$

$$Var(S) = \frac{N(N-1)(2N+5) - \sum_{k=1}^{n} t_k(t_k - 1)(2t_k + 5)}{18} \tag{6}$$

where n is the number of tied (zero difference between compared values) groups and t_k is the number of data points in the kth tied group. The standard normal deviate (Z-statistics) is then computed as (Hirsch et al. 1993):

$$Z = \begin{cases} \dfrac{S-1}{\sqrt{Var(S)}} & if \quad S > 0 \\ 0 & if \quad S = 0 \\ \dfrac{S+1}{\sqrt{Var(S)}} & if \quad S < 0. \end{cases} \tag{7}$$

If the computed value of $|Z| > z_{\alpha/2}$, the null hypothesis H_0 is rejected at the α level of significance in a two-sided test. In this analysis, the null hypothesis was tested at 95% confidence level.

4 Results and Discussions

Prior to non-parametric test, preliminary trend lines are fitted to the extreme rainfall values as shown in Figs. 2, 3 & 4 and their results (slope and their significance) are shown in Table 1. Subsequently non-parametric statistical methods, Sen's estimator

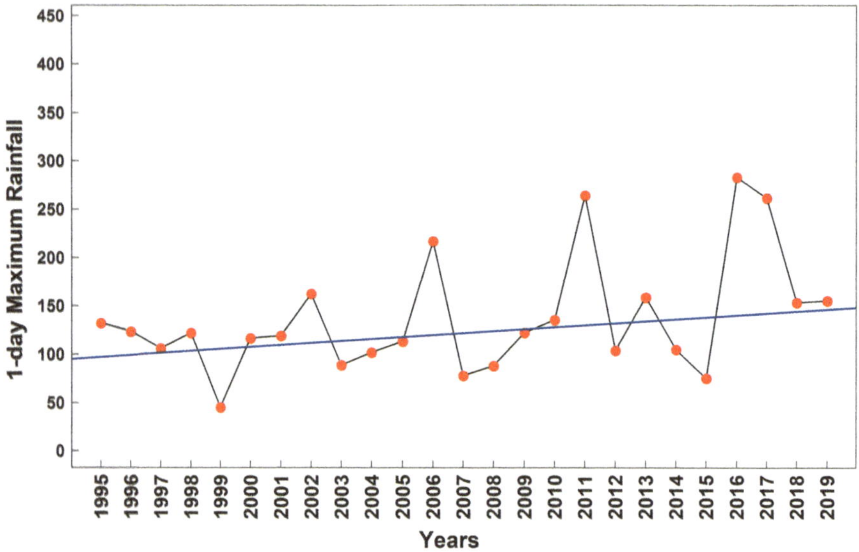

Fig. 2 One-day maximum rainfall and its trend

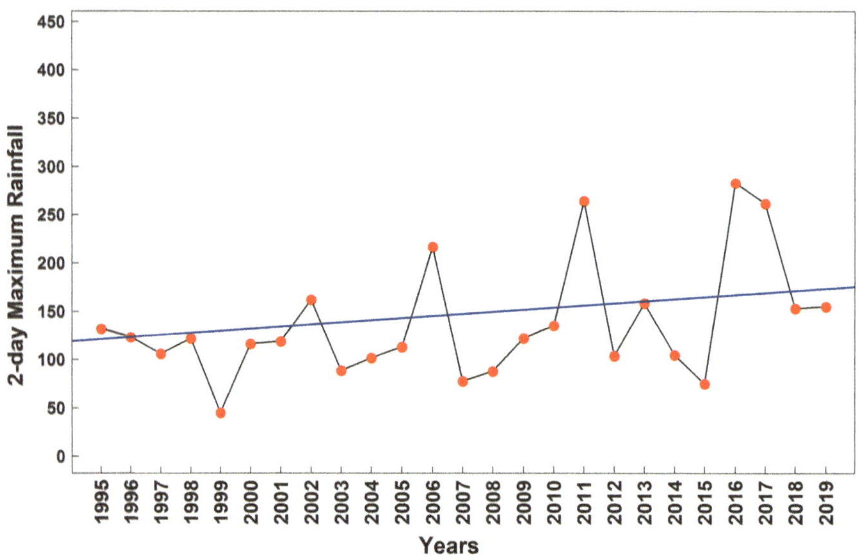

Fig. 3 Consecutive two-day maximum rainfall and its trend

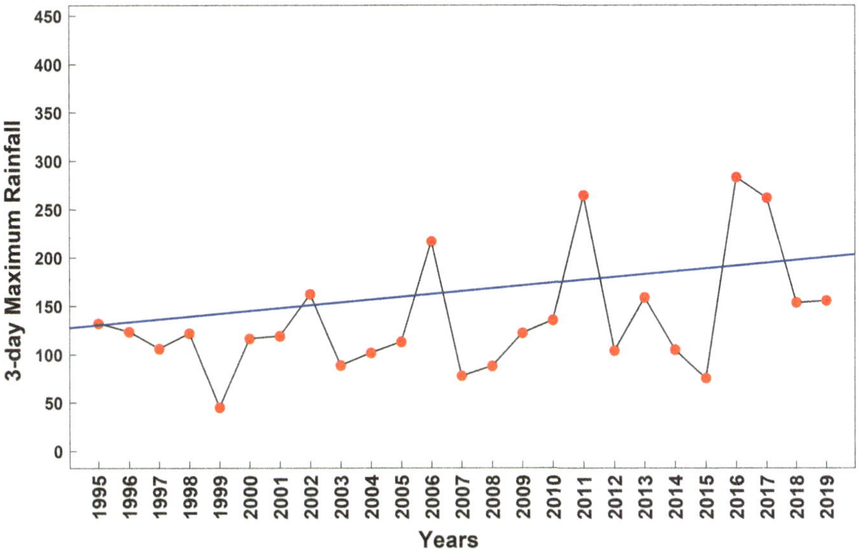

Fig. 4 Consecutive three-day maximum rainfall and its trend

Table 1 Preliminary trend line and its slope

Trend line (y = mx + c)						
S. N.	Data Series	Trend (m)	P values	t-test	Lower limit (95%)	Upper limit (95%)
1	One-day max. rainfall	3.39	0.04	2.17	0.16	6.62
2	Consecutive two-day max	3.59	0.12	1.63	−0.96	8.15
3	Consecutive three-day max	4.10	0.07	1.91	−0.33	8.54

Table 2 Sen slope (mm/year) for consecutive 1 day, 2-day, 3-day rainfall at Roorkee

Rainfall	Z Statics	Sen Slope	significance at 95%
1-day maximum rain	1.19	2.03	Non-significant
2-day maximum rain	1.00	2.17	Non-significant
3-day maximum rain	1.33	2.90	Non-significant

of slope (SE) and Mann–Kendall (MK) test have been carried out for estimating the magnitude and testing the statistical significance of trend (at confidence interval of 95%) respectively. However, prior to non-parametric test, data series was tested for presence of any auto-correlation. The outcomes of the analysis are shown in the form of Table 2.

It is revealed from Table 2 that the consecutive 1 day, 2-day and 3-day rainfall at Roorkee station is having the increasing trend. Moreover, 3-day rainfall is showing the maximum slope (2.902 mm/yr) during the study period. However, the trend is found to be in-significant at the 95% confidence level.

5 Conclusion

The impact of climate change on extreme rainfall events has been an area of interest for researcher because of the severe consequences these heavy rainfall events have on human life. It has been experienced that extreme rainfall events and their frequencies are increasing because of global warming and associated climate changes. These extreme rainfall events resulted in disastrous floods in different regions of the world including Indian sub-continent causing loss of property and life. The impact of these changes in climatically and ecological sensitive Himalayan regions would be significant. Therefore, an understanding of these extreme rainfall events in this region would be important for mitigating the adverse impacts of climate change and associated phenomenon.

The present study attempted to find empirical evidences of increasing trends in extreme rainfall events in a small town in the foothills of Himalaya. Twenty-five years of daily rainfall data is used for extracting the one-day annual maximum rainfall and consecutive two (and three) day maximum rainfall. Non-parametric statistical tests (Man-Kendall Tests) (Mann-Kendall and Sen's Slope) have been used to detect and quantify the trend in extreme rainfall events. The results show that there is an increasing trend in the extreme rainfall values at 1-day, 2-day and 3-day maximum rainfall given by the slopes of the trend lines fitted in the observed extreme rainfall events. The non-parametric significance tests, however indicate that these trends are not significant at 95% confidence level. The study signifies the importance of empirical evidences over climatic projections within statistical level of significance of these trends.

References

Barsugli J, Anderson C, Smith JB, Vogel JM (2009) Options for improving climate

Bayazit M, Onoz B (2007) To prewritten or not to prewritten in trends analysis. Hydrol Sci J 52(4): 611–624

Bisht DS, Chatterjee C, Raghuwanshi NS, Sridhar V (2017) Spatio-temporal trends of rainfall across Indian river basins. Theor Appl Climatol 132:419–436. https://doi.org/10.1007/s00704-017-2095-8

Burn DH, Cunderlick JM, Pietroniero A (2004) Hydrological trends and variability in the Liard river basins. Hydrol Sci J 49(1):53–67

Chakraborty S, Pandey RP, Chaube UC, Mishra SK (2013) Trend and variability analysis of rainfall series at Seonath River Basin, Chhattisgarh (India). Int J Appl Sci Eng Res 2(4):2013

Chattopadhyay S, Jhajharia D, Chatopadhyay G (2011) Univariate modeling of monthly maximum temperature time series over north East India: neural network versus Yule-walker equation based approach. Meteorol Appl 18:70–82. https://doi.org/10.1002/met.2011

Diffenbaugh NS (2020) Verification of extreme event attribution: using out-of-sample observations to assess changes in probabilities of unprecedented events. Sci Adv 6(12):eaay2368. https://doi.org/10.1126/sciadv.aay2368

Dinpashoh Y, Jhajharia D, Fakheri-Fard A, Singh VP, Kahya E (2011) Trends in reference evapotranspiration over Iran. J Hydrol 399:422–433. https://doi.org/10.1016/j.jhydrol.2011.01.021

Douglas EM, Vogel RM, Kroll CN et al (2000) Trends in floods and low flows in the United States: impact of spatial correlation. J Hydrol 240:90–105p

Fu G, Chen S, Liu C, Shepard D (2004) Hydr0-climatic trends of the yellow river basin for the last 50 years. Climate Change 65:149–178

Ghosh S, Luniya V, Gupta A (2009) Trend analysis of Indian summer monsoon rainfall at different spatial scales. Atmos Sci Lett 10:285–290. https://doi.org/10.1002/asl.235

Helsel DR, Hirsch RM (1992) Statistical methods in water resources. Elsevier, Amsterdam, p 522

Hirsch RM, Slack JR, Slack RA (1982) Techniques of trend analysis for monthly water quality data. Water Resour Res 18(1):107–121

Jain SK, Kumar V (2012) Trend analysis of rainfall and temperature data for India. Curr Sci 102: 37–49

Jhajharia D, Shrivastava SK, Sarkar D, Sarkar S (2009) Temporal characteristics of pan evaporation trends under the humid conditions of northeast India. Agric For Meteorol 149:763–770

Jhajharia D, Singh VP (2011) Trends in temperature, diurnal temperature range and sunshine duration in northern India. Int J Climatol 31(9):1353–1367. https://doi.org/10.1002/joc.2164/

Kampata JM, Parida BP, Moalafhi DB (2008) Trend analysis of rainfall in the headstreams of the Zambezi River basin in Zambia. Phys Chem Earth Parts A/B/C 33:621–625. https://doi.org/10.1016/j.pce.2008.06.012

Kumar V, Jain SK, Singh Y (2010) Analysis of long-term rainfall trends in India. Hydrol Sci J 55: 484–496. https://doi.org/10.1080/02626667.2010.481373

Lal M (2001) Climatic change — implications for India' s water resources key sources of water in India. J Soc Econ Dev 1:57–87

Lettenmaier DP, Wood EF, Wallis JR (1994) Hydro-climatological trends in the continental United States, 1948-88. J Clim 7:586–607

Ma S, Zhou T, Angélil O, Shiogama H (2017) Increased chances of drought in Southeastern periphery of the Tibetan plateau induced by anthropogenic warming. J Clim 30:6543–6560. https://doi.org/10.1175/JCLI-D-16-0636.1. Modeling to Assist Water Utility Planning for Climate Change. Water Util. Clim Alliance, pp. 144

Partal T, Kahya E (2006) Trend analysis in Turkish precipitation data. Hydrol Process 20:2011–2026

Patra JP et al (2012) Detecting rainfall trends in twentieth century (1871-2006) over Odisha State, India. Clim Change 111:801–817

Raucher RS (2011) The future of research on climate change impacts on water: a workshop focusing on adaptation strategies. Water Research Foundation records. WMO Bulletin no, 52

Rimi R, Haustein K, Barbour EJ, Allen MR (2018) Risks of pre-monsoon extreme rainfall events of Bangladesh: is anthropogenic climate change playing a role? Bull Am. Meteor Soc 100:S1–S5. https://doi.org/10.1175/BAMS-D-18-0097.1

Salas JD (1993) Analysis and modeling of hydrological time series. In: Maidment DR (ed) Handbook of Hydrology. McGraw-Hill, New York, pp 19.1–19.72

Sen PK (1968) Estimates of the regression coefficient based on Kendall's tau. J Am Stat Assoc 63: 1379–1389

Seneviratne SI, Nicholls N, Easterling D, Goodess CM, Kanae S, Kossin J, Luo Y, Marengo J, McInnes K, Rahimi M, Reichstein M, Sorteberg A, Vera C, Zhang X (2012) Changes in climate extremes and their impacts on the natural physical environment. In: Field CB, Barros V, Stocker

TF, Qin D, Dokken DJ, Ebi KL, Mastrandrea MD, Mach KJ, Plattner G-K, Allen SK, Tignor M, Midgley PM (eds) Managing the risks of extreme events and disasters to advance climate change adaptation. A Special Report of Working Groups I and II of the Intergovernmental Panel on Climate Change (IPCC), Cambridge University Press, Cambridge, pp 109–230
Singh P, Kumar V, Thomas T, Arora M (2008a) Change in rainfall and relative humidity in different river basins in the northwest and Central India. Hydrol Process 22:2982–2992
Singh P, Kumar V, Thomas T, Arora M (2008b) Basin-wide assessment of temperature trends in the northwest and Central India. Hydrol Sci J 53(2):421–433
Sinha Ray KC, De US (2003) Climate change in India as evidenced from instrumental. 52(1):53
Taxak AK, Murumkar AR, Arya DS (2014) Long term spatial and temporal rainfall trends and homogeneity analysis in Wainganga basin, Central India. Weather Clim Extrem 4:50–61. https://doi.org/10.1016/j.wace.2014.04.005
Tebakari T, Yoshitani J, Suvanpiomol C (2005) Time-space trend analysis in pan evaporation Kingdom of Thailand. J Hydrol Eng 10(3):205–215
Thakural LN, Kumar S, Jain SK, Ahmad T (2017) The impact of climate change on rainfall variability: a study in Central Himalayas. Water Science and Technology Library, pp 181–192. https://doi.org/10.1007/978-981-10-5714-4_15
Thakural LN, Anbu Kumar S, Ansari MI (2018) Trend analysis of rainfall for the Chaliyar Basin. South India. Int J Res Appl Sci Eng Technol 6(7):91–100
Wang SYS, Fosu B, Gillies RR, Singh PM (2015) The deadly Himalayan snowstorm of October 2014: synoptic conditions and associated trends, [in "Explaining extremes of 2014 from a climate perspective"]. Bull Amer Meteor Soc 96(12):S89–S94
Yu YS, Zou S, Whittemore D (1993) Non-parametric trend analysis of water quality data of rivers in Kansas. J Hydrol 150:61–80
Yue S, Hashino M (2003) Temperature trends in Japan: 1900-1990. Theoret Appl Climatol 75:15–27

Examination of Historical Trends and Future Projections for Climate and Land-use Variables and its Impacts on Kalna River Flow in Goa, India

Ashwini Pai Panandiker, B. Venkatesh, Shubham Gude, K. Mahender, and A. G. Chachadi

Abstract There is always pressure on water availability due to increasing levels of societal demand and from economic activities. Hence, understanding the effect of climate change on various components of the water cycle is crucial in management of this resource. To devise sustainable water resource strategies, seeing how much change in climate and land-use/land cover (LULC) affects hydrologic regimes can help decision-makers to incorporate necessary measures in the policy instruments. The objective of this study was to analyze the impact of climate coupled with land-use change on Kalna river flow situated in North Goa. The assessment involved temporal rainfall analysis to understand the historical trends. Further, future climate and land-use change projections were evaluated to comprehend the impact on the river flow. An ensemble of models was used for future predictions. For climate modeling, the Norwegian Earth System Model (NorESM) was used under two scenarios that included RCP 4.5 and RCP 8.5. The land-use change was simulated using the Land Change Modeler (LCM). Finally, hydrological modeling was done using the Soil and Water Assessment Tool (SWAT) model. The results from NorESM and LCM were used as an input to SWAT model to predict future flow for Kalna River.

A. P. Panandiker (✉)
School of Earth, Ocean and Atmospheric Science, Goa University, Taleigao, Goa, India

The Energy and Resources Institute, St Cruz, Goa, India
e-mail: ashwini.panandiker@teri.res.in

B. Venkatesh
National Institute of Hydrology, Belagavi, Karnataka, India

S. Gude
Water Resources Department, Government of Goa, Goa, India

K. Mahender · A. G. Chachadi
School of Earth, Ocean and Atmospheric Science, Goa University, Taleigao, Goa, India

B. Phartiyal et al. (eds.), *Climate Change and Environmental Impacts: Past, Present and Future Perspective*, Society of Earth Scientists Series,
https://doi.org/10.1007/978-3-031-13119-6_18

The historical trend in rainfall was statistically scrutinized using Mann-Kendall Test and Sen's Slope method. Average annual rainfall data from India Meteorological Department rain gauge station at Mapusa for the period between 1980 and 2018 was used. An increasing linear trend was observed which was supported by Kendall's tau and Q (Sen's slope) indicating strong positive correlation between rainfall and duration. The land-use change analysis was done using satellite images of 1993, 2014 and 2019 map for validation. A Kappa co-efficient of 0.73 indicated acceptable accuracy. Multi-Layer Perceptron neural network was used for prediction of land-use for 2030 and 2040. These two future land-use maps were used as an input and SWAT model was calibrated for the years 2011 to 2015 and validated for 2016 to 2018. Two statistical measures, Nash Sutcliffe Efficiency (NSE) and R^2 with value of 0.7 showed goodness of calibration. It was then used to predict the future streamflow till 2050. As compared to baseline average monsoon rainfall data (26.87 mm), the future projections under both RCPs (4.5 and 8.5) scenarios indicate an increase in rainfall and streamflow. This increase in average streamflow is more pronounced in RCP 4.5 as compared to RCP 8.5. As per the LCM projections, the forest area is likely to decrease by 2040 with a distinct increase of 14% in barren land owing to quarrying and mining activities. The decrease in the forest cover along with changing climate decreases the streamflow clearly demonstrating the importance of the green cover. Currently, around 10% of the water required by the water treatment plant at Chandel is extracted from Kalna River. Based on the simulations, site-specific recommendations are given to aid in the strategic planning of this watershed.

Keywords Climate change · Land-use change · Hydrological model · River flow

1 Introduction

Land cover and climate change and their impact on hydrological processes are widespread concern and great challenge to researchers and policy makers. The current understanding of the link between land cover change and watershed functions at larger scales and their possible variations with vegetation, geology and rainfall pattern are still limited (Vaidyanathan and Venkatesh 2011). The LULC change especially the alteration in the forest cover can alter the rainfall process into runoff by modifying hydrological parameters like surface runoff, percolation, lateral flow, and evapotranspiration (Lele and Krisnaswamy 2019; Sajikumar and Remya 2015). On the contrary, climate change alters precipitation, the water content in the atmosphere and soil moisture (Wang et al. 2008). Surface temperature is projected to rise over the twenty-first century under all assessed emission scenarios. Extreme precipitation events will likely become more intense and frequent in many regions (IPCC's *Fifth Assessment Report* (AR5) 2014). Climate change is likely to affect global water availability through compounding changes in seasonal precipitation and evaporation (Konapala et al. 2020; Moors et al. 2011; Asokan and Dutta 2008). Climate change studies in India predict intense impacts such as higher annual average rainfall and increased drought on water resources. This can have negative impacts on water supply (Krishnan et al. 2020).

In general, assessing the impacts of climate change on watershed hydrology requires the use of watershed models and General Circulation Models (GCMs) or Regional Climate Models (RCMs). Several studies have recently been carried out on the impacts of climate change on streamflow or runoff in (Sharma et al. 2019; Narsimlu et al. 2013; Givati et al. 2019).

Further, land use is one of the foremost drivers of hydrologic processes, influencing the available water resources and flow regimes in a river basin worldwide (Gashaw et al. 2018). Assessing land-use/land cover (LULC) impacts on hydrology is essential for watershed management and ecological restoration (Panandiker et al. 2019). Investigations on land-use change impacts on water resources in India have been done by Wagner et al. (2019), Anand et al. (2018), Babar and Ramesh (2015), and Wagner et al. (2013). However, to devise sustainable water resource strategies, it is also crucial to establish interaction between climate coupled with land-use changes and local hydrology through proper assessment. To be precise, seeing how much change in climate and LULC affects hydrologic regimes or which land use shall be appropriate for the local hydrological regime can help decision-makers to incorporate necessary measures in the policy instruments (Anand et al. 2018). A few old studies, such as Wilk and Hughes (2002), examined the impacts of land-use and climate change on India's water resource availability using empirical models. There are very few studies (Sinha et al. 2020; Woldesenbet et al. 2018) that have tried to examine the impact of both land use and climate change together. More studies at a regional or local level in India are required to closely examine the impact of future land-use coupled with climate change on the river flow using distributed numerical models.

The overall objective of this study include: (i) LULC change detection analysis; (ii) impact assessment of climate change on surface runoff under RCP 4.5 and 8.5 emission scenarios; and (iii) combined impact assessment of LULC and climate change on the surface runoff for the Kalna river watershed in North Goa, India.

2 Study Area

Kalna river originates from a mountain in the State of Maharashtra from a village known as Kalane (Karmali wada). It enters Goa near village Assapur (Pernem taluka, North Goa district) from where it flows southwards and joins Colvale river at Ozorim village and finally uniting with Chapora river at Taramas. Tidal influence is up to Bailpur village. This is 9.5 km long and passes through Assapur, Kristanvadi, Bailpur, Kutwal, Kessarwane, Muryavaddi, and Tamaras. The location of the river is shown in Fig. 1. Kalna is a perennial river. The depth of the river is around 300–500 cm during monsoons; 70 cm between November to February and later drops down in summer. The discharge for the river is monitored by WRD during the months of June to October at Hasapur station. The watershed area with Hasapur as

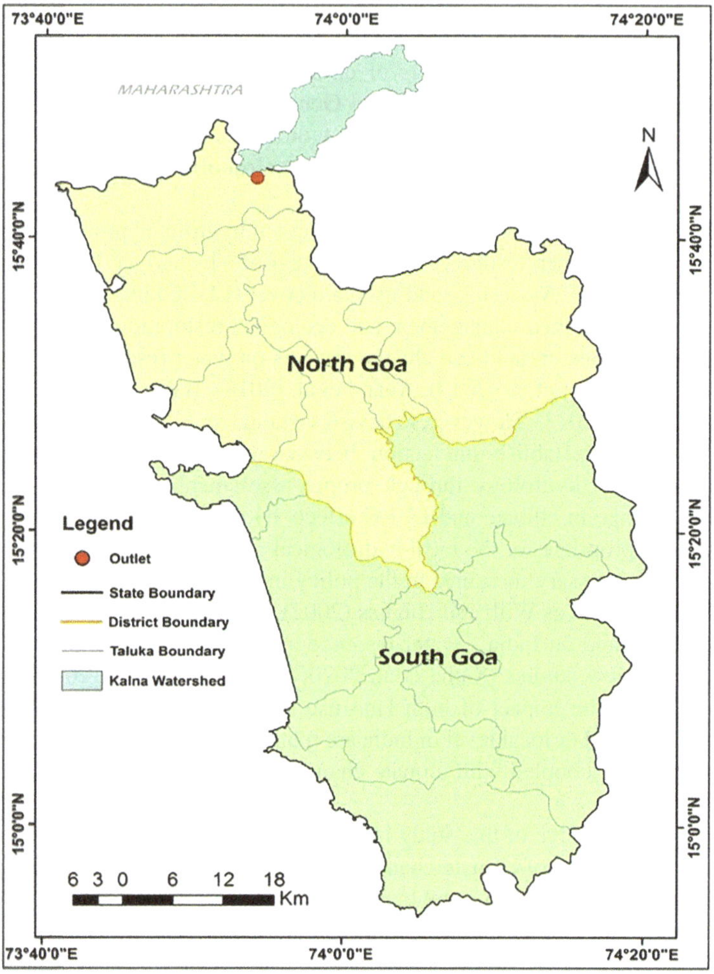

Fig. 1 Location of the Kalna river watershed

an outlet is 118.51sq.km. The Water Treatment Plant (WTP) at Chandel which is operated by the Public Works Department (PWD), Government of Goa is situated close to the Hasapur monitoring station. This WTP pumps 15% of its raw water directly from the Kalna river while the rest is from Tillari canal. Treated water from this plant is supplied to Dhargal, Pernem, Mandrem and Morjim areas. Agriculture is the major occupation of the local population in the study area. Coconut, bananas, areca-nuts, chillies, vegetables are the major cash crops of the area and paddy forms the conventional food crop. Direct pumping from the river for irrigation is also being practiced by the kulaghars, orchards and agricultural fields. The climate of the area is warm and humid. The area receives an annual rainfall of about 3400–3600 mm mainly from Southwest (SW) monsoon during June to September.

3 Methodology

The overall objective was to understand the impact of land-use coupled with climate change on the near surface hydrology. To do so various activities were undertaken which were: (1) trend analysis of past precipitation data; (2) predictions for future climatic parameters; (3) estimation of probable land-use change; (4) forecasting the future river flow using a hydrological model with inputs from the climate and land-use change model. The methodology that was adopted is described below:

3.1 *Trend Analysis*

Rainfall is a crucial parameter in a hydrology study. The historical trend in rainfall was statistically scrutinized using Mann-Kendall Test and Sen's Slope method. The precipitation data was collected from India Meteorological Department (IMD) for 39-year period (1981–2019) for Mapusa station ($15^0 36' 0.21''$ N $73^0 48' 43.39''$ E) that is located close to the Kalna river watershed. Significance of trends was determined using Mann Kendall statistical test and magnitude by Sen's slope.

3.1.1 Mann-Kendall Test

The non-parametric Mann–Kendall criterion was originally developed by Mann (1945) and rephrased by Kendall (1975). Detection of trend is a complex subject because of characteristics of data, and the main idea of trend analysis is to detect whether values of data are increasing, decreasing or trendless over time (Kisi and Ay 2014). The Mann–Kendall statistical test has been frequently used to quantify the significance of trends in hydro-meteorological time series (Duhan and Pandey 2013; Silva et al. 2013). It is calculated as given in Eq. (1):

$$S = \sum_{i=1}^{n-1} \sum_{j=i+1}^{n} \mathrm{sgn}\left(x_j - x_i\right) \tag{1}$$

Where, n is the number of data points, xi and xj are the data values in the time series i and j (j > i), respectively, and sgn(xj − xi) is the sign function as shown in Eq. (2):

$$\mathrm{Sgn}\,(xj - xi) = \begin{cases} +1 & \text{if } x_j - x_i > 0 \\ 0 & \text{if } x_j - x_i = 0 \\ -1 & \text{if } x_j - x_i < 0 \end{cases} \tag{2}$$

The variance is computed using Eq. (3):

$$\mathrm{Var(S)} = \frac{n(n-1)(2n+5)}{18} \quad (3)$$

Where, n is the number of data points. The standard normal test statistic ZS is computed using Eq. (4):

$$ZS = \begin{cases} \frac{S-1}{\sqrt{\mathrm{Var(S)}}} \text{ ie3waf } S > 0 \\ 0 \text{ if } S = 0 \\ \frac{S+1}{\sqrt{\mathrm{Var(S)}}} \text{ if } S < 0 \end{cases} \quad (4)$$

If the calculated value of |Zsk| > Zα/2, the null hypothesis is rejected at significance level α.

3.1.2 Sen's Slope

The magnitude of trend in time series can be determined using Sen's slope. Sen (1968) developed a non-parametric procedure for estimating the slope of trend in a sample of n pairs of data. The Sen's method uses a linear model to estimate the slope of the trend, and the variance of the residuals should be constant in time, calculated using Eq. (5):

$$Q_i = \frac{x_j - x_k}{j - k} \text{ from } i = 1,2,3,\ldots.n \quad (5)$$

Where, xj and xk are the data values at times j and k (j > k), respectively. The median of these N values of Qi Sen's slope estimator is calculated as shown in Eq. (6):

$$Q = \begin{cases} Q_{\left(\frac{n}{2}\right)} \text{ if n is odd} \\ \frac{1}{2}\left(Q_{\left(\frac{n}{2}\right)} + Q_{\left(\frac{n+2}{2}\right)} \text{ if n is even}\right. \end{cases} \quad (6)$$

A positive value of Q indicates an increasing trend and a negative value indicates a decreasing trend in time series.

3.2 Climate Modeling

The projected climate scenarios RCP 4.5 and 8.5 were used. For climate simulation (precipitation and rainfall), a 30-year baseline period was generated using the Norwegian Earth System Model (NorESM). The resolution used was a horizontal grid of 25 km × 25 km. The following steps were followed to get the data from the NorESM model for a historical period (1950–2005) and projected period (2006–2100) from NASA's Climate Data Services (CDS) portal.

1. Python script was used to download the NorESM data from NASA Earth Exchange Global Daily Downscaled Projections (NEX-GDDP)
2. Climate Data Operator (CDO) was used for selecting the study area and converting the files into the required format
3. R software was used for data analysis and plotting.

The precipitation (mm/day) output from the NorESM was compared with the gridded data from the India Meteorological Department (IMD). The historical data (1950–2005) was used for this verification. Statistical methods of linear scaling and quantile mapping were used for bias correction of the model outputs. The bias-corrected data was used for future rainfall and temperature predictions until 2060.

3.3 Land-use Change Analysis

The satellite images were organized and classified for analysis and understanding. Landsat images are among the widely used satellite remote sensing data and their spectral, spatial and temporal resolution make them useful input for mapping and planning projects (Panandiker et al. 2019; Mishra et al. 2014; Sadidy et al. 2009). Landsat TM 5 images for 1993 and Landsat 8 images for the years 2014 and 2019 were used in developing the future land-use/cover maps for Kalna river watershed. These images were mostly for the month of April/May. The images were projected to WGS-1984 and UTM Zone—43 N co-ordinate system. ERDAS Imagine software was used to perform land-use/cover classification in multi-temporal approach. Each image was separately classified using the supervised classification maximum likelihood algorithm in ERDAS Imagine. Based on the NRSA land-use/land cover classification system, level 1 has been used for this study. Five categories of land-use were that were used for classification were agricultural land, built-up or urban area, barren land, forests (dense vegetation, shrubs, and marshes) and water bodies (rivers, ponds, lakes and open mine pits filled with water). The built-up category includes settlements, mining areas, and man-made construction activities such as roads.

The TerrSet Land Change Modeler (LCM) was used to compare and identify the land-use maps generated using the Maximum Likelihood method. Use of such model gives a better understanding of the functions of the land-use systems and the support for planning and policy making. Such models can also predict the possible future change and use of the land cover under different scenario (Anand et al. 2018; Mishra et al. 2014; Ahmed and Ahmed 2012). The change analysis panel of LCM provides a rapid assessment of quantitative change by graphing gains and losses by land cover categories. A second option, net change, shows the result of taking the earlier land cover areas, adding the gains and then subtracting the losses. The third option is to examine the contribution of changes experienced by single land cover (Clark Labs 2009). The change analysis between 1993 and 2014 were computed and studied using LCM. Accordingly, the transitions and exchanges that took place between the various land-use/cover categories during the years were obtained in a graphical form. All the land cover categories were used in sq. km unit. The transition sub-models (Forest to Agriculture and Forest to Barren) along with DEM and Slope images were used to predict future land-use for 2019. Amongst the modeling algorithms, Multi-layer Perceptron (MLP) neural network was selected since it uses minimal parameters and has been extensively enhanced to offer an automatic mode that requires no user intervention. Simulated map was validated with the actual land-use map of 2019. The validation was statistically tested using Kappa co-efficient. Further, future maps for 2030 and 2040 were generated.

3.4 SWAT: Hydrological Modeling

The Soil and Water Assessment Tool (SWAT2012) version with the QSWAT 1.9 interface was used for this research. It is a physically based semi-continuous time model developed by the United States Department of Agriculture (USDA) at the Grassland, Soil, and Water Research Laboratory in Temple, Texas. SWAT is designed to predict the effects of land management practices on hydrology, sediment, and contaminant transport in agricultural watersheds under varying soils, land use, and management conditions. It can operate on a large basin and can simulate several processes such as flows in rivers and sediment transport on a daily/sub-daily time step. SWAT is based on the concept of hydrologic response units (HRUs), which are portions of a sub-basin that possess unique land-use, management, and soil attributes. The runoff, sediment, and nutrient loadings from each HRU are calculated separately based on weather, soil properties, topography, vegetation, and land management and then summed to determine the total loading from the sub-basin (Arnold et al. 2013; Park et al. 2011; Neitsch et al. 2012; Kiniry et al. 2000). SWAT simulates various hydrologic processes; including surface runoff generation using either SCS curve number method or Green and Ampt infiltration equation. For the estimation of evapotranspiration, Penman-Monteith method was employed. Groundwater flow, lateral flow and percolation are assessed through mass balance of the underlying system.

4 Results and Discussion

4.1 *Simulations from Climate Model*

The bias correction of the NorESM output concerning IMD observations was done using the statistical linear scaling method. The reduction in the root mean square error (RSME) value after bias correction of historical precipitation data is shown in Table 1. Using the bias-corrected historical data, future rainfall, and temperature (minimum and maximum) were projected under RCP 4.5 and RCP 8.5 for the period from 2006 to 2100. These outputs were used in the hydrological model to predict future streamflow.

4.2 *Simulations from Land Change Modeler*

An evaluation of resemblance between the simulated LULC map for 2019 and the actual 2019 map was done to examine LCM's applicability to predict the change. Kappa statistics (K) for similarity was estimated to understand the similarity between the projected and the observed land-use maps for 2019. The Kappa coefficient range is between 0 and 1, with 0 representing poor and 1 indicating excellent accuracy. A kappa coefficient of 0.73 was obtained, indicating good accuracy. Using this calibrated model, future land-use maps for the years 2030 and 2040 were prepared. The future land-use projections for the Kalna river watershed are depicted in Fig. 2. The area under different land-use categories for the years 1993, 2014, 2030, and 2040 is shown in Fig. 3. As compared to the baseline in 2019, the forest area is likely to decrease by 1.5% in 2030 and 3.1% in 2040 respectively. Most of the north-eastern region of this watershed is located in the State of Maharashtra. Being an undulating terrain, only a marginal increase in the settlement area is expected. The settlements may increase by 3.3% by 2030 and 7.4% by 2040. Around 7.1% rise in agricultural activities is expected by 2030 and 12.7% by 2040. There is presence of basalt (minor mineral) in this watershed. Further, the Dodamarg tehsil of Maharashtra (North-eastern region in the watershed) had open cast activities for mining of iron/manganese. This along with quarrying of basalt may lead to increase in the barren area by 7.8% in 2030 and 14.5% in 2040.

Table 1 Statistical analysis of bias correction of precipitation data

		Before bias correction			After bias correction		
Latitude	Longitude	Mean model	Mean observed	RSME	Mean model	Mean observed	RSME
15.375	73.875	6.26	9.46	3.20	9.44	9.46	0.02

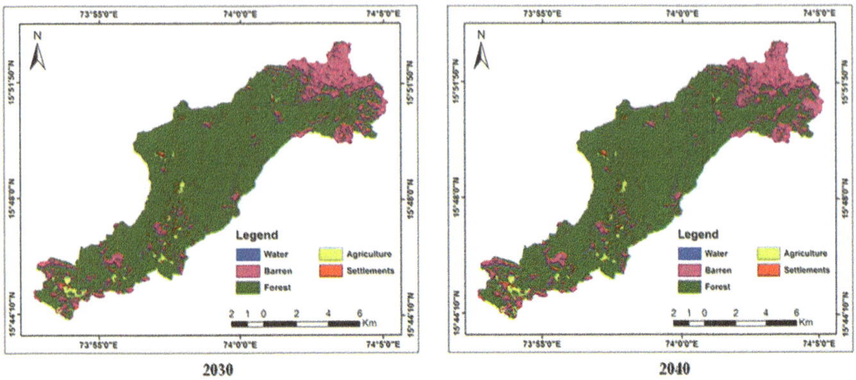

Fig. 2 Kalna river watershed: Land-use change projections for the year 2030 and 2040

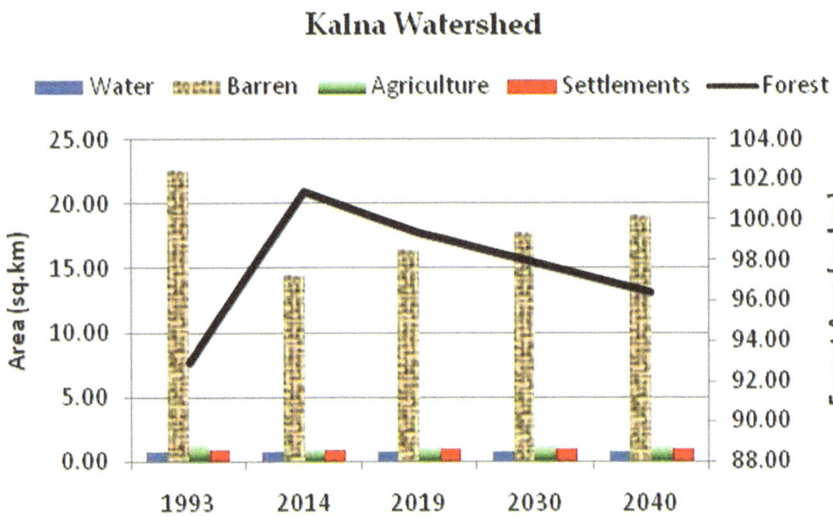

Fig. 3 Time-based variation in the area under various land-use categories

4.3 Simulations from SWAT and Future Projections

The future LULC maps for the years 2030 and 2040 were used as an input and QSWAT model was calibrated and validated. The 2019 LULC map was used to assess the current or baseline situation. The calibration efficiency was evaluated using statistical measures such as NSE and R^2. The calibrated and validated models for Kalna watershed is presented in Fig. 4. Since the NSE and R^2 values were above 0.6, they were acceptable and, hence, used for future streamflow predictions. Further, using the rainfall and temperature projections from the NorESM model under the RCP 4.5 and 8.5 scenarios, the streamflow was simulated till 2050. The variation

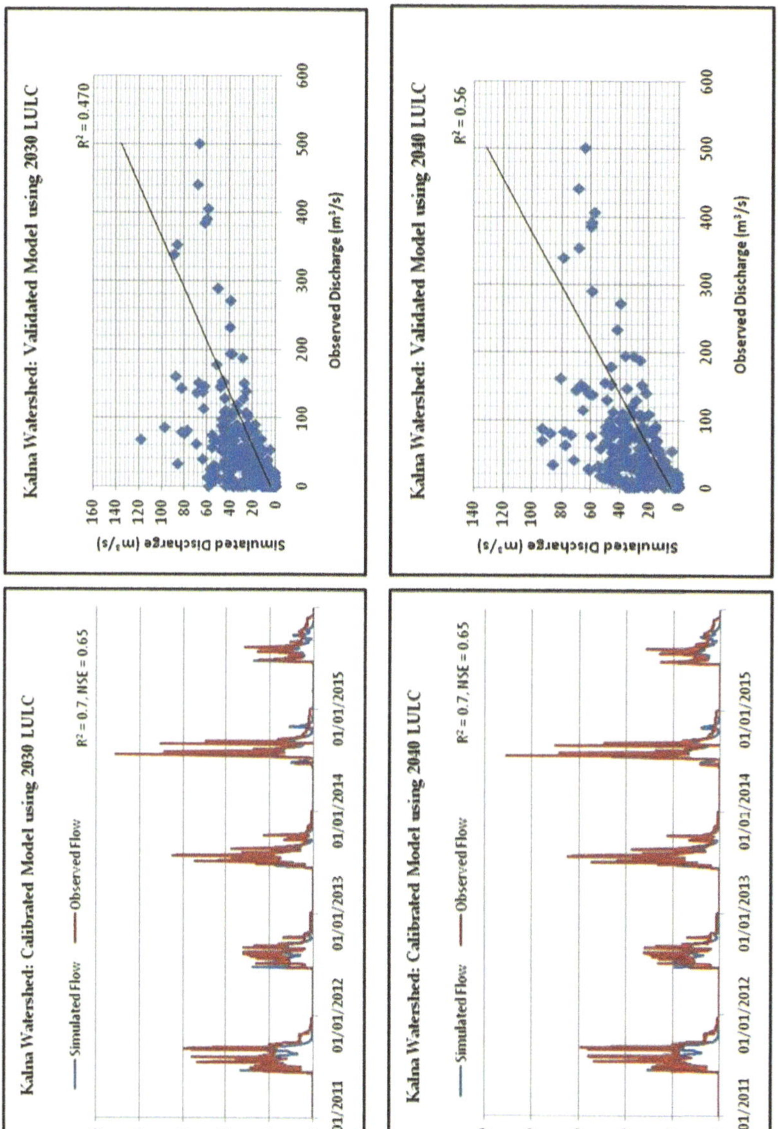

Fig. 4 Calibration and validation of QSWAT model using 2030 and 2040 LULC maps

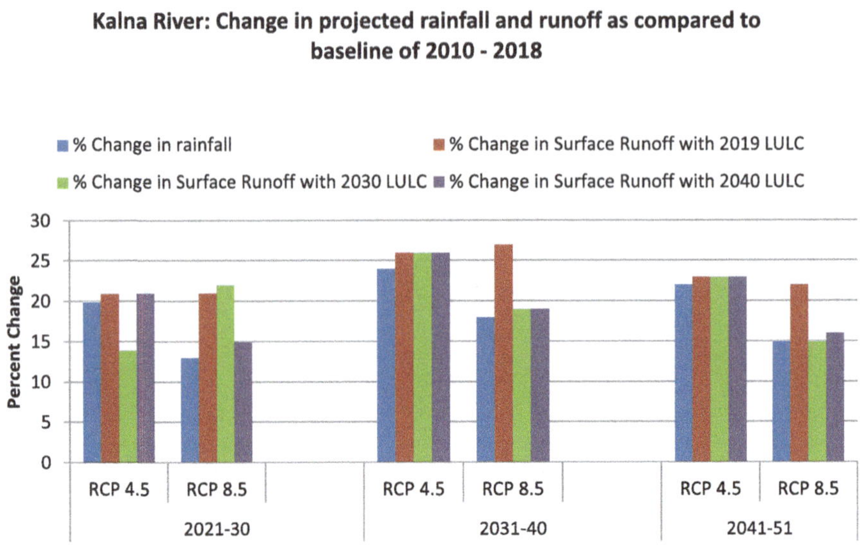

Fig. 5 Variation in mean rainfall and runoff in JJAS (monsoon) during future time period

in the mean rainfall and runoff (flow) of JJAS (monsoon) during the Kalna river's future time is presented in Fig. 5.

The Kalna river basin (up to gauging point) is largely dependent on monsoon rainfall for its flows. As compared to this baseline data, the future projections under both RCP (4.5 and 8.5) scenarios indicate an increase in rainfall and streamflow. Under the changing climate conditions, this increase in average streamflow is more pronounced in RCP 4.5 as compared to RCP 8.5. The runoff is likely to be more during the decade of 2031–2040. The scenarios that combine climate and land-use change indicate a further decrease in the streamflow quantity. The forest cover is likely to decrease by 3% up to 2040s and an increase in the agricultural and barren areas (14%) is expected. A decrease in the forest cover coupled with changing climate is likely to decrease the run-off, clearly demonstrating the importance of a green cover.

5 Conclusion and Recommendations

While analysing the climate change scenarios exclusively, it was observed that the streamflow or runoff is likely to increase under both the RCP scenarios (4.5 and 8.5) when compared with the baseline of 2010–18. The flow is likely to be less under RCP 8.5 as compared to RCP 4.5. It must be noted that under RCP 8.5, the high levels of greenhouse concentrations are assumed until the end of the twenty-first century. Further, climate combined with land-use change reveals that the streamflow is likely to decrease with the decrease in forest cover. As a general tendency in land cover change, open forest usually gets converted into agricultural/urban settlement. In such a case, a slight increase in runoff and a corresponding decrease in evapotranspiration is predicted (Aggarwal et al. 2012). Around 10% of the water required by the water treatment plant at Chandel is extracted from Kalna river. Hence, new water intensive industries should be restricted. In case of water guzzling activities/ industries, rain water harvesting should be made mandatory and compliance should be checked regularly. Recharging of groundwater could enhance the base-flow, thereby improving the water availability during non-monsoon periods. Considering the delivery, conveyance, and application losses, actual irrigation will be considerably more, depending upon the type of irrigation method. Hence, agriculture and horticulture activities in these sub-basins need attention. Water conservation measures such as drip irrigation; system of rice intensification should be encouraged. Regulating the use of canal water to avoid wastage and over-use should be looked into. Developing irrigation design and scheduling guides for local crops needs to be done to enhance water-use efficiency of agriculture sector. The functioning of the existing Water User's association (WUAs) needs to be scrutinized and regularised.

Acknowledgements Authors gratefully acknowledge Water Resources Department, Government of Goa, for providing the financial support and data to carry out this study. We thank India Meteorological Department, the National Aeronautics and Space Administration (NASA) for sharing the data. We sincerely acknowledge the timely suggestions extended by Dr. Nandagiri (Professor, National Institute of Technology Karnataka, Surathkal).

References

Aggarwal SP, Vaibhav G, Prasun KG, Bhaskar RN, Praveen KT (2012) Climate and LULC change scenarios to study its impact on hydrological regime. International Archives of the Photogrammetry, remote sensing and spatial information sciences, Volume XXXIX-B8, 2012 XXII ISPRS Congress, 25 August–01 September 2012, Melbourne, Australia

Ahmed B, Ahmed R (2012) Modeling urban land cover growth dynamics using multi-temporal satellite images: a case study of Dhaka, Bangladesh. ISPRS Int J Geoinform 1:3–31. https://doi.org/10.3390/ijgi1010003

Anand J, Gosain AK, Khosa R (2018) Prediction of land use changes based on land change modeler and attribution of changes in the water balance of Ganga basin to land use change using the SWAT model. Science of Total Environment 644:503–519. https://doi.org/10.1016/j.scitotenv.2018.07.017

Arnold JG, Kiniry JR, Srinivasan R, Williams JR, Haney EB, Neitsch SL (2013) Soil and water assessment tool: input/output documentation. Version 2012. Texas Water Resource Institute (TR-439 650)

Asokan S, Dutta D (2008) Analysis of water resources in the Mahanadi River Basin, India, under projected climate conditions. Hydrol Process. https://doi.org/10.1002/hyp.6962

Babar S, Ramesh H (2015) Streamflow response to land-use land cover change over the Nethravathi River Basin, India. J Hydrol Eng 20(10). https://doi.org/10.1061/(ASCE)HE.1943-5584.0001177

Clark Labs (2009) The land change modeler for ecological sustainability. IDRISI Focus Paper, Worchester: Clark University. Available on http://www.clarkslabs.org/applications/uploads/Land-Change-Modeler-IDRISI-Focus-Paper-pdf

Duhan D, Pandey A (2013) Statistical analysis of long term spatial and temporal trends of precipitationduring 1901–2002 at Madhya Pradesh, India. Atmos Res 122:136–149

Gashaw T, Tulu T, Argaw M, Worqlul AW (2018) Modeling the hydrological impacts of land use/land cover changes in the Andassa watershed, Blue Nile Basin, Ethiopia. Sci Total Environ 619–620:1394–1408

Givati A, Thirel G, Rosenfeld D, Paz D (2019) Climate change impacts on streamflow at the upper Jordan River based on an ensemble of regional climate models. Journal of Hydrology: Regional Studies 21:92–109

IPCC (Intergovernmental Panel on Climate Change) (2014) *Climate change 2014 synthesis report, Summary for policymakers*. https://www.ipcc.ch/site/assets/uploads/2018/02/AR5_SYR_FINAL_SPM.pdf

Kiniry J, Williams J, Srinivasan R (2000) Soil and water assessment tool user's manual. Texas Water Resources Institute, College Station, Texas TWRI Report TR-191. https://swat.tamu.edu/media/1290/swat2000theory.pdf

Kisi O, Ay M (2014) Comparison of Mann–Kendall and innovative trend method for water quality parameters of the Kizilirmak River, Turkey J Hydrol 513(26):362–375

Konapala G, Mishra AK, Wada Y, Michael EM (2020) Climate change will affect global water availability through compounding changes in seasonal precipitation and evaporation. Nat Commun 11:3044. https://doi.org/10.1038/s41467-020-16757-w

Krishnan R, Sanjay J, Chellappan G, Mujumdar M, Kulkarni A, Chakraborty S (2020) Assessment of climate change over the Indian region, Ministry of Earth Sciences (MoES), Government of India. Springer Open, Singapore. https://doi.org/10.1007/978-981-15-4327-2

Lele S, Krisnaswamy J (2019) Climate change and India's forests. In: Dubash NK (ed) India in a warming world, integrating climate change and development. Oxford University Press, New Delhi

Mishra V, Narayan R, Praveen K, Mohan K (2014) Prediction of land use changes based on land change Modeler (LCM) using remote sensing: a case study of Muzaffarpur (Bihar), India. J Geogr Inst Cvijic 64(1):111–127

Moors EJ, Groot A, Biemans H, van Scheltinga CT, Siderius C, Stoffel M, Huggel C, Wiltshire A, Mathison C, Ridley J, Jacob D, Kumar P, Bhadwal S, Gosain A, Collins DN (2011) Adaptation to changing water resources in the Ganges basin, northern India. Environ Sci Pol 14(7):758–769

Narsimlu B, Gosain AK, Chahar BR (2013) Assessment of future climate change impacts on water resources of Upper Sind River Basin, India using SWAT model. Water Resources Management 27:3647–3662. https://doi.org/10.1007/s11269-013-0371-7

Neitsch SL, Arnold JG, Kiniry JR, Srinivasan R, Williams JR (2012) Soil and water assessment tool user's manual. GSWRL Report

Panandiker AP, Gude S, Venkatesh B, Kotha M, Chachadi AG (2019) Examining temporal change and prediction of future land use using geospatial approach: A case study of Talpona river watershed in Goa, India. Journal of Indian Associations for Environmental Management 30(1–4):25–29. http://op.niscair.res.in/index.php/JIAEM

Park JY, Park MJ, Joh HK, Shin HJ, Kwon HJ, Srinivasan R, Kim SJ (2011) Assessment of MIROC3.2 HIRES climate and CLUE-s land use change impacts on watershed hydrology using SWAT. Transactions of the American Society of Agricultural and Biological Engineers (ASABE) 54(5):1713–1724

Sadidy J, Firouzabadi PZ, Entezari A (2009) The use of Radarsat and Landsat image fusion algorithms and different supervised classification methods to use map accuracy – case study: sari plain – Iran. Available at http://www.isprs.org/proceeding/XXXVI/5-C55/papers/sadidy_javad.pdf

Sajikumar N, Remya RS (2015) Impact of land cover and land use change on runoff characteristics. J Environ Manag 161:460–468

Sen PK (1968) Estimates of the regression coefficient based on Kendall's tau. J Am Stat As 63:1379–1389

Sharma T, Gusain A, Karmakar S (2019) Future hydrologic scenarios in India under climate change. In: Venkataraman C, Mishra T, Ghosh S, Karmakar S (eds) Climate change signals and response. Springer, Singapore

Sinha RK, Eldo TI, Subimal G (2020) Assessing the impacts of land use/land cover and climate change on surface runoff of a humid tropical river basin in Western Ghats, India. Int J River Basin Manag. https://www.tandfonline.com/doi/full/10.1080/15715124.2020.1809434

Silva RM, Santos CAG, Macedo MLA, Silva L, Freire PKMM (2013) Space–time variability of rainfall and hydrological trends in the Alto São Francisco River basin. IAHS-AISH Publ 359:48–54

Vaidyanathan J, Venkatesh B (2011) Evaluating the hydrological response to land cover change – Dasanakatte Catchment of Varahi river, Western Ghats, Karnataka. IJWREM 3:23–32

Wagner PD, Kumar S, Schneider K (2013) An assessment of land-use change impacts on the water resources of the Mula and Mutha Rivers catchment upstream of Pune, India. Hydrol Earth Syst Sci 10:1943–1985. https://doi.org/10.5194/hessd-10-1943-2013

Wagner PD, Bhallamudi SM, Narasimhan B, Kumar S, Fohrer N, Fiener P (2019) Comparing the effects of dynamic versus static representations of land-use change in hydrologic impact assessments. Environ Model Softw 122. https://doi.org/10.1016/j.envsoft.2017.06.023

Wang S, Kang S, Zhang L, Li F (2008) Modelling hydrological response to different land-use and climate change scenarios in the Zamu River basin of Northwest China. Hydrol Process 22(14): 2502–2510

Wilk J, Hughes DA (2002) Simulating the impacts of land-use and climate change on water resource availability for a large south Indian catchment. Hydrol Sci J 47(1):19–30. https://doi.org/10.1080/02626660209492904

Woldesenbet TA, Elagib NA, Ribber L, Heinrich J (2018) Catchment response to climate and land use changes in the Upper Blue Nile sub-basins, Ethopia. Sci Total Environ 644:193–206

Temporal Trends in Water Discharge Characteristics of the Large Peninsular Rivers: Assessing the Role of Climatic and Anthropogenic Factors

Harish Gupta, S. Kiran Kumar Reddy, and Vamshi Krishna Gandla

Abstract In recent years, the growing water scarcity at the regional and global level has become a topic of increased discussion, particularly concerning its response to changing societal use and climate vagaries. For the Indian peninsula, the southwest monsoon accounts for >80% of annual precipitation, and rain-fed peninsular rivers serve as an important source for domestic and agricultural uses. Therefore, this study examines the trends of peninsular rivers' annual water discharge (WQ). We computed location-specific trends of more than 300 locations covering ten large and several coastal river basins using daily WQ data provided by the Central Water Commission, India. The results are based on 50 years' annual discharge data for 7 locations, more than 30 years of data for around 150 locations, and more than 20 years of data for additional 100 locations spread across the Indian peninsula. The discharge trends for the peninsular rivers were statically calculated. On-parametric Mann-Kendall test and linear regression analysis were applied to identify peninsular rivers with remarkable discharge variations. In the case of large rivers, the change varies between −79.1% and 29.5%. Large peninsular rivers except Mahi and Sabarmati experience a regular reduction in their annual water flux. Annual discharge chart of the rivers such as the Brahmani (−14%), Mahanadi (−12%), Godavari (−11%), and Sabarmati (−20%) illustrated a relatively small reduction, whereas a significant reduction of the annual water discharge of the Krishna river (−79%; $p < 0.001$), Pennar river (−78%; $p < 0.05$), Cauvery river (−54%; $p < 0.01$), Tapti river (−28%; $p < 0.1$) and the Narmada river (−62%; $p < 0.01$) observed. However, Mahi exhibited a minor increase in the water discharge. Overall, large peninsular rivers showed a 30% reduction in annual water discharge. Studies

H. Gupta (✉)
Dept. of Civil Engineering, Osmania University, Hyderabad, India
e-mail: harishgupta78@osmania.ac.in

S. K. K. Reddy
CSIR-National Geophysical Research Institute, Hyderabad, India

V. K. Gandla
Dept. of Applied Geochemistry, Osmania University, Hyderabad, India

B. Phartiyal et al. (eds.), *Climate Change and Environmental Impacts: Past, Present and Future Perspective*, Society of Earth Scientists Series,
https://doi.org/10.1007/978-3-031-13119-6_19

suggest that despite a significant increase in the frequency and the intensity of extreme monsoon rain events, there is no appreciable upward trend in average annual rainfall in most peninsular basins. The peninsular region hosts 45 mega and more than 3800 large dams and several thousand minor structures. Therefore, diverting water from dams for various purposes could be one of the major reasons fora decline in annual WQ in these basins. We anticipate that local or sub-basin-scale changes in rainfall patterns might be another important factor for observed water discharge trends based on location-specific discharge trends. Factor analysis, a data reduction technique, revealed that in the case of large rivers, storage capacity, rainfall, and evapotranspiration rates (3 factors) have eigenvalue>1 and explain about 87% of the variance. The statistical analysis endorses that (i) the construction of dams has significantly impacted the water annual and seasonal discharge patterns across large basins; (ii) although the precipitation is the major water source to the peninsular basins, it becomes a subordinate factor in the case of large peninsular rivers, following evapotranspiration losses.

Keywords Water discharge · Temporal trends · Indian Peninsula · Large rivers · Rainfall · Dam

1 Introduction

Rivers, lakes, and wetlands as surface water resources account for about 70% of the water used by humans globally. The surface water supply is vulnerable to fluctuation in annual rainfall, climatic conditions, and anthropogenic pressure. The rising demand for water for societal use, myopic policy-making, unregulated urbanization, and uncertainties associated with freshwater supply due to variations in annual rainfall and global warming are challenging ecological balance and affecting society's social and economic well-being. Several global, regional, and local scale studies have already evaluated water resources' vulnerability to global warming and human activities (Vörösmarty et al. 2000). It is projected that under the enhanced greenhouse effect, the hydrological cycle will strengthen leading to a substantial alteration in annual/seasonal precipitation, its intensity, and the frequency of rainy days throughout the world (Pachauri et al. 2014). According to Kummu et al. (2016), while over the past century, global water use has grown four times, the population facing water scarcity increased from 14% of the world population (0.24 billion) in the 1900s to 58%(3.8 billion) in the 2000s. Mekonnen and Hoekstra (2016) submitted that 4.0 billion people (making up two-thirds of the global population) face severe water scarcity at least a month each year; and pointed out that nearly 50% of those people reside in China and India.

India, which houses 17.7% of the global population, occupies about 2.4% of the global land area with 4% of freshwater resources. India's population has grown multifold, from 330 million in 1947 to over 1330 million in 2011. Consequently, the per capita surface water availability declined from over 5400 m^3in 1951 to 1900 m^3in 2001 (Kumar et al. 2005). The present surface water accessibility is less than

1500 m^3 per capita (CWC 2017). According to Kumar et al. (2005), the water requirement in the year 1997–98 for different uses in India was about 629 km^3. In 2010, this requirement was 712 km^3 and is projected to increase to 833 km^3 by 2025 and 899 km^3 by 2050 (UNICEF 2013). Thus, for a diverse and highly populated country like India, intermittent river discharge assessments are vital for proper planning and management of water resources.

Indian rivers are broadly divided into the peninsula and the Himalayan rivers (Jain et al. 2007). Despite draining 45% of the geographical area, the Himalayan rivers contribute about 65% of the annual water discharge. The peninsula basins occupy the remaining area yet contribute only 35% of the annual flux. However, 50 urban cities are located in peninsular India, which creates enormous pressure on available water resources for domestic, agriculture, and industrial activities. Hence, periodically monitoring of the water resources is critical to this region. Earlier attempts on water discharge trends from Peninsular India rivers are limited to either a single large river (Gupta and Chakrapani 2005; Bastia and Equeenuddin 2016) or a group of small-scale rivers. Recently, Gupta et al. (2021) provided a comprehensive account of Peninsular India's water discharge and temporal trends. Continuation with the previous efforts, this study's key objective is to offer accurate and updated information on the water discharge of the Indian peninsula using annual water discharge data from 10 large rivers, covering a considerable length of time (5–50 yrs). A subordinate objective was to understand the key factors determining temporal changes in large peninsular rivers' water discharge.

2 Study Area and Source of Data

Godavari, Krishna, Mahanadi, Cauvery, Pennar, and Brahmani (flowing to the Bay of Bengal), and Narmada, Tapti, Mahi, and Sabarmati (flowing into the Arabian Sea) are the ten large rivers that drain the Indian peninsular region (Fig. 1). The catchments of these ten large rivers constitute an area of 1.11×10^6 km^2. Most of these rivers are well covered with a gauge-discharge (GD) network. Relatively smaller rivers and streams mostly drain from either side of these areas, covering Western and Eastern Ghats. Kale (2003) pointed out that the peninsular rivers are generally incised in rock or alluvium and have stable channels. Hence, it is uncommon to have bank-line changes and shifts in channel position due to large floods.

Mountain chains confine the triangular-shaped Indian peninsulato the north, east, and the west. Indian peninsula experiences diverse climatic regimes. The elevation of this region ranges between the mean sea level (MSL) and 2674 m. Except for the rivers draining the north-western parts of the peninsula (20% of total area), all other large rivers debouch into the Bay of Bengal, indicating the east-ward general slope of the Indian peninsula. The Aravalli and Vindhyan mountains mark the northern boundary, whereas the Eastern Ghats and the Western Ghats form the eastern and western margins. The Western Ghats (50 to 80 km wide) run parallels the Arabian Sea coast for almost 1600 km from the southern tip of Tamil Nadu to the

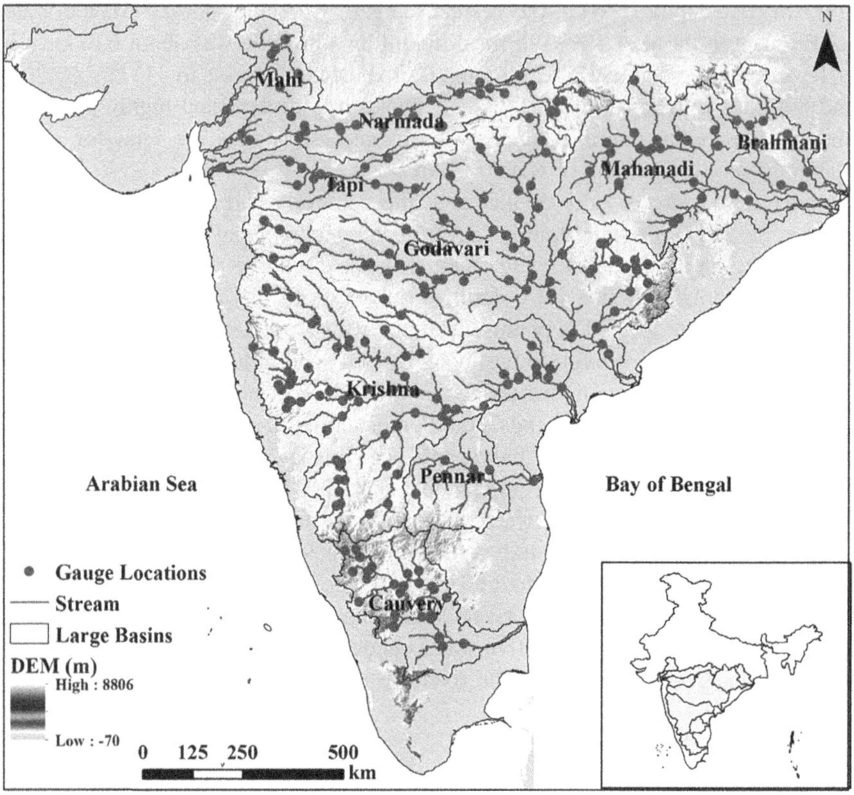

Fig. 1 Map showing the network of gauge stations on the ten large river basins

Maharashtra-Gujarat border (Reddy et al. 2019). About 1400 km long, northeast-southwest trending (along the Bay of Bengal),the Eastern Ghats are a group of discontinuous and dissimilar ranges of small relict mountains. The Eastern Ghats region is 100 to 200 km wide with an average height of 600 m. The Eastern Ghats lies parallel to the Bay of Bengal branch of the southwest monsoon resulting in low precipitation. Hence, the Pennar is the only large river originating from the Eastern Ghats. The Arabian Sea branch of monsoon winds brings the rainfall to the Western Ghats. The orographic effect of the Western Ghats(average elevation of about 1200 m) favors high precipitation on the windward side of the mountain chain, resulting in a remarkably high runoff in a relatively smaller west-flowing coastal rivers (Reddy et al. 2019). However, the leeward side of the Western Ghats remains relatively dry, as most of the moisture of these rain-bearing winds has already been lost. Therefore, the large rivers (i.e., Cauvery, Krishna, and the Godavari) originating from the east side of the Western Ghats are known for low surface runoff. Thus, the Western Ghats have an "unfair" advantage over the Eastern Ghats. Only Mahanadi and Brahmani originate from the Satpura range among the east-flowing large rivers.

These basins receive plentiful rainfall from the Bay of Bengal branch of the southwest monsoon.

The responsibility to monitor and develop surface water resources in India lies with the Central Water Commission (CWC), an apex organization of Govt. of India. The Water Year Book (available on www.cwc.nic.in) is a periodical publication of the CWC detailing the annual discharge, and data can also be retrieved through the WRIS portal (https://india-wris.nrsc.gov.in). The methods and procedures followed while making gauge and discharge measurements are detailed in CWC's annual handbook for hydrometeorological observations and analytically explained by Bisoyi et al. (2018). We have obtained data on daily water discharge of 244gauge and discharge stations of the CWC, spread across the basins (upstream of tidal influence) to calculate the area-weighted seaward water discharge from the Indian peninsula covering ten large rivers (Fig. 1). The 90 m spatial resolution Shuttle Radar Topography Mission-Digital Elevation Model (SRTM-DEM) (http://srtm.csi.cgiar.org/) was used to extract basin properties such as basins boundaries and channel gradient. The basin-wise population density was extracted using the block-level population data from the 2011 Census of India (www.censusindia.gov.in). The non-parametric Mann-Kendall test was applied for the long-term trend analysis for annual water discharge. Climatological normal of district-wise rainfall, obtained from the Indian Meteorological Department's (available at https://data.gov.in/keywords/annual-rainfall) data for 1951–2000, used for obtaining the basin-wise mean annual rainfall. The National Register of Large Dams (cwc.gov.in/main/downloads/NRLD_06042018.pdf) lists the basin-wise storage capacity of dams.

3 Results and Discussions

3.1 The Hydrological Regime of Large Peninsular Rivers

The runoff derived from the gauge data shows remarkable variability in the ten large peninsular rivers. For instance, runoff produced by the Krishna river (81 mm) and Pennar river (23 mm) is remarkably lower than the Brahmani river (489 mm), Mahanadi river (372 mm), Narmada river (308 mm), and the Godavari river (277 mm). Hence, the yearly mean runoff variability amid the large rivers differs by orders of magnitude (Fig. 2). Among these rivers, the Krishna and the Pennar, with mean annual runoff <100 mm, are categorized as arid rivers. The Tapti, Cauvery, Mahi, and the Sabarmati rivers are categorized as semi-arid rivers (100 to 250 mm). The Brahmani, Mahanadi, Godavari, and the Narmada river come under the humid category (250 to 750 mm). The parts of the Cauvery, Krishna, and the Godavari covering the leeward side of the Western Ghats are only categorized as wet or high-runoff (>750 mm) regions of the large peninsular rivers (Fig. 2). It is interesting to note that in spite of draining through a relatively small proportion of the global land, the peninsular rivers witness all four runoff classes (Fig. 2). A wide range of climatic conditions acting in conjunction with a range of topographic

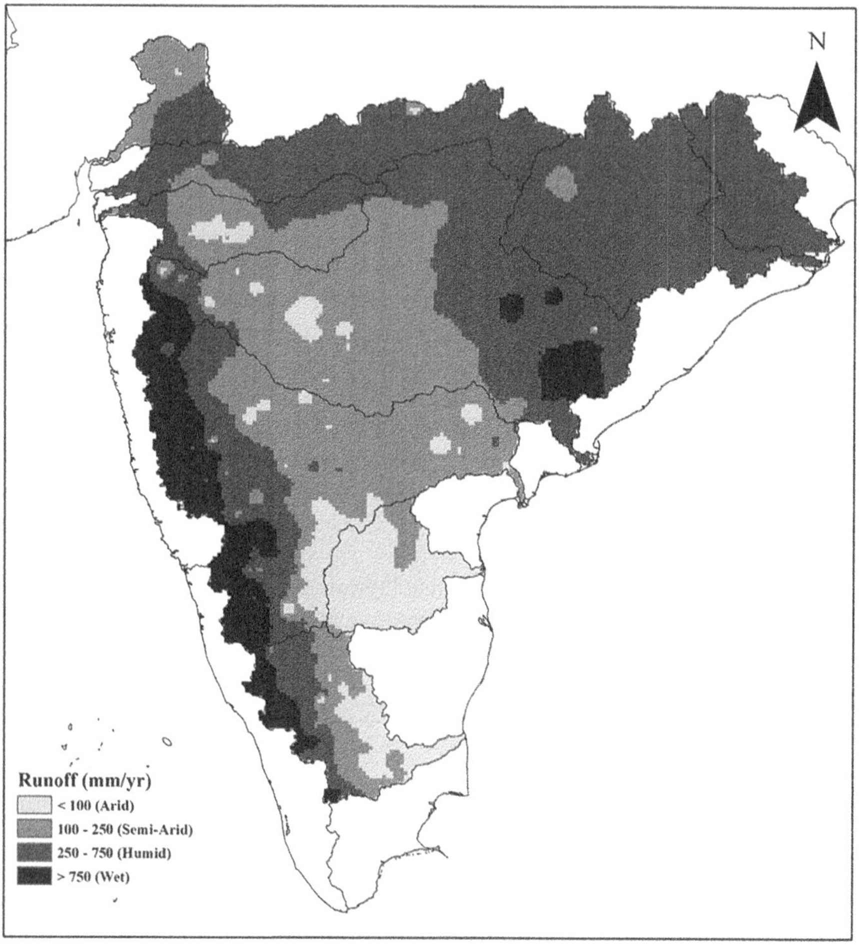

Fig. 2 Runoff distribution map of the ten large rivers flowing in peninsular India

and geological variations produce this intricate and attention-grabbing pattern of runoff over the peninsular region. Overall, the large rivers average runoff (212 mm) is analogous to the global mean (360 mm; Milliman and Farnsworth 2011). The long-term combined seaward mean annual discharge of the large rivers calculated at respective terminal gauge stations is 219 km^3. Among them, the Godavari river alone contributes about 39% of water flux (85.4 km^3), followed by 21% from the Mahanadi, 12% from the Narmada, 9% from the Krishna, and 8% from the Brahmani river. Thus, the remaining 11% of annual water flux comes from five large rivers (Pennar, Cauvery, Tapti, Mahi, and Sabarmati).

3.2 Temporal Variability of Water Discharge from the Large Peninsular Rivers

The long-term temporal variation in the annual water discharge for the ten large rivers is depicted in Fig. 3. To identify rivers with significant discharge changes, the discharge trends for these rivers suggest that among the large peninsular rivers, the Mahi and the Sabarmati show a growing trend. However, the remaining large peninsular rivers witness a regular decline in their mean annual flux (Fig. 3). This decline in annual water discharge is smaller for the rivers such as Brahmani (−13.7%), Mahanadi (−12.2%), Godavari (−10.6%), and Sabarmati (−20%). However, this decline become considerably large for the Krishna river (−79.1%; $p < 0.001$), Pennar river (−77.9%; $p < 0.05$), Cauvery river (−53.7%; $p < 0.01$),

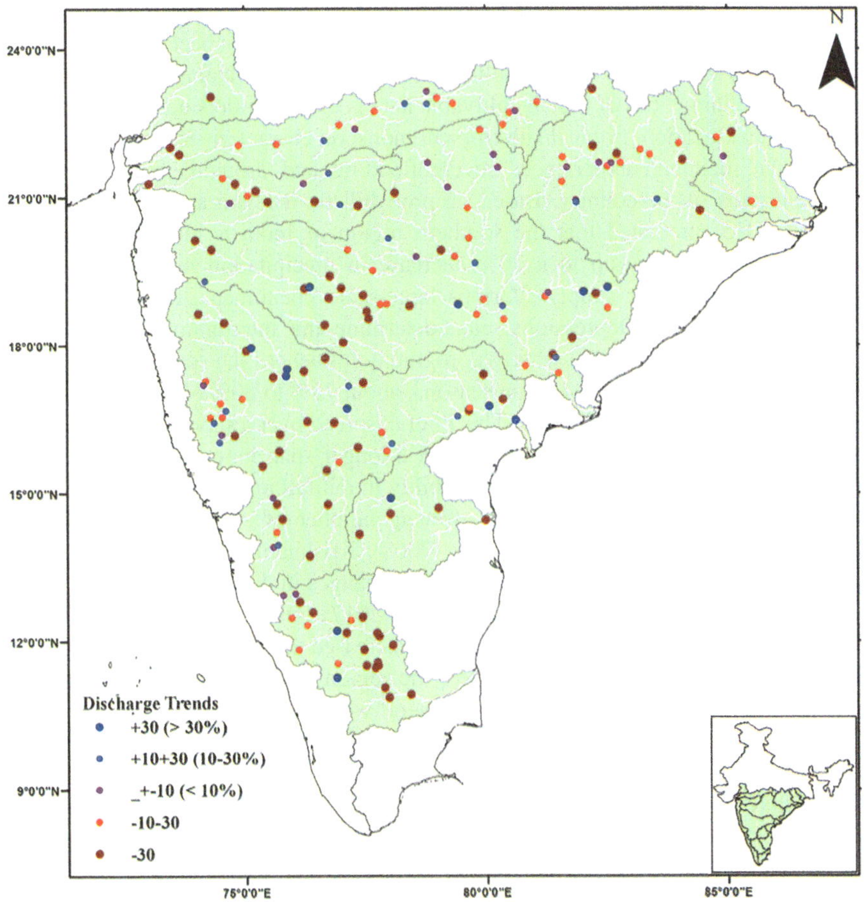

Fig. 3 The *long-term* variability in the mean annual water discharge trends of the ten large peninsular rivers

Tapti river (−27.8%; $p < 0.1$) and the Narmada river (−62%; $p < 0.01$) . On the other hand, the Mahi river exhibits an insignificant increase in discharge.

3.3 Controls to Variations in Long Term River Discharge

The impact of different natural and anthropogenic controls on the hydrological regime is evident from the observed sizeable variability in runoff across the peninsula. Basin parameters such as area, length, mean basin elevation, and lithology are not directly related to the long-term water discharge. India holds the largest arable land globally and ranks second in the agricultural output. Besides, the Indian agricultural sector remains the largest employer. In the year 2000, Indian agriculture accounted for 85.3% of water use, which is going to be 83.3% by 2025 (CWC 2007). It implies that the agricultural sector will remain the largest water user. Out of 140 million hectares of land under agriculture, about 22 million hectares are irrigated by the networks of canals and 39 million hectares by groundwater pumping. It means that the remaining 56% of Indian cultivable land is monsoon-dependent. Due unique geographical location, tropical/sub-tropical monsoon characteristics, and landscape, the Indian mainland experiences severe flooding and droughts concurrently in different parts of the country. Based on probabilistic Budyko analysis, Singh and Kumar (2015) predicted that the southern region of India is most vulnerable to vagaries of climate as up to a 10% decrease in precipitation may cause a 25% reduction in the water supply. Various large-scale phenomena, like El Niño, La Niña, Indian Ocean Depressions, etc., determine the inter-annual variability in monsoon rainfall over India, whereas orography influences the spatial variabilities.

During four months of monsoon seasons, about 75% to 80% of annual rainfall is received and flows in rivers peaks. This rainfall spread over a short time may cause a flood, but the less/non-rainfall spread over a longer span may result in a drought. Climate change and meteorological variability in rainfall in India have already been analyzed and discussed. Besides, a growing number of studies by Rao (1993), Goswami et al. (2006), Rajeevan et al. (2008), Krishnamurthy et al. (2009), Jain (2017), and Bisht et al. (2018) have reported, changes in daily and seasonal rainfall patterns across Indian mainland. Roxy et al. (2017) found that between 1950 and 2015, extreme rain events over central India increased three-fold. This study notes a 10–30% increase in rainfall events over the region where more than 150 mm of rainfall is registered in a day despite a general weakening of monsoon circulation. Jain (2017) examined trends in mean annual rainfall, number of rainy days, rainfall intensity, and riverine flow of a few selected Indian peninsular rivers. They observed that despite the increase in yearly peak rainfall in most basins, the number of flood peaks of smaller magnitude showed a slight falling trend in different decades, whereas there was no trend for severe floods (Fig. 4). According to Jain (2017), river regulation through storage reservoirs has reduced peak flows in the past 50 years.

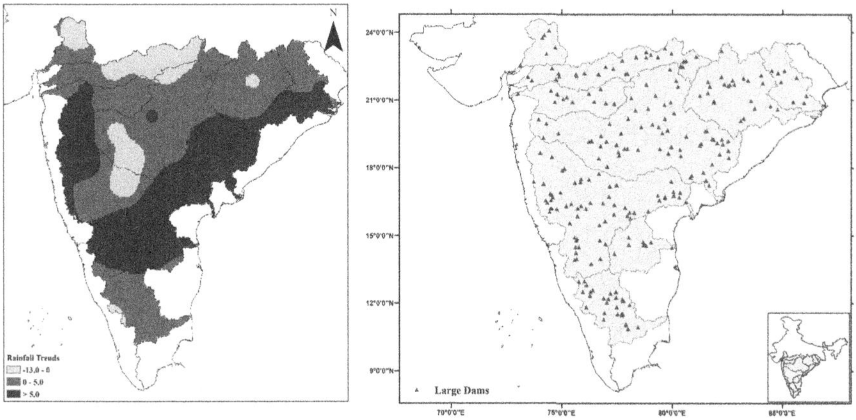

Fig. 4 Map showing the rainfall trends and the distribution of large dams over the ten large peninsular rivers

The peninsular region supports about 46% (559 million) of the Indian population (Population Census 2011). Also, out of 51 million-plus Indian cities, 27 lie in this region. Besides, about 0.91 million km^2 (>60% of total) land is under agriculture. The peninsular rivers act as a key freshwater source for this relatively more urbanized part of the mainland. Over the last few decades, agriculture and population pressure have resulted in a notable alteration in land-use and land-cover (LULC) patterns and witnessed a growing number of water flow regulating structures and inter-basin transfers for societal uses (Gupta et al. 2012). Water conflicts, disproportional distribution of water resources, and water stress Considering the increasing water requirement, the National Water Development Agency (NWDA), India, has suggested its rivers' interlinking. Inter-River Linking (IRL) is a mega project that will ease water scarcities in southern and western India. It will also mitigate the recurrence of floods in eastern India. The project aims to link 30 major rivers and transfer 185 km^3 of water through an about 15,000-km network of canals and about 3000 small and large reservoirs. The Godavari and Krishna rivers are already linked through a 174 km canal. Currently, the Ministry of Water Resources and River Development is commencing projects to interlink four rivers in the first phase of this initiative, covering the Ken-Betwa, Par-Tapi-Narmada, and Damanganga-Pinjal projects.

Direct water consumption for India's domestic, industrial, and agricultural sectors has increased significantly over the past decades. The mounting demand for water forced the administrators to devise different ways to utilize the surface runoff. As a result, 41 mega-dams, over 3800 large dams, and several thousands of minor irrigation structures have been constructed. Several more are planned across the peninsula to cope with the growing demand for food and energy (CWC 2007). These dams have an adequate storage capacity of about 216 km^3. Among the large dams, over 3200 are constructed in the large basins, and the remaining dams are on the coastal rivers (NRLD). The Krishna basin alone has 935 large dams, followed by the

Godavari (837), Tapti (445), Narmada (353), and Mahanadi (266; Fig. 4) basins. The cumulative storage capacity (49 km^3) of all dams in the Krishna basin far exceeds its current seaward flux (6.24 km^3). The cumulative storage capacity of dams in the Godavari basin is about 37 km^3, followed by Narmada (29 km^3), Mahanadi (18 km^3), and Tapti (11.8 km^3) basins.

4 Conclusion(s)

Altogether, ten large Indian peninsular rivers discharge237 km^3 of water annually. Rainfall and water loss through natural processes and human consumption are identified as major factors influencing riverine discharge. Despite low runoff, it was found that large rivers deliver higher water flux by virtue of basin size. Spatial variance in precipitation exerts direct control on the surface runoff. Water loss from the low rainfall basins is greater than from the higher rainfall basins, and large rivers lose more water than the minor rivers. Only 30% of total rainfall is transformed into riverine discharge in the peninsula. In the absence of any noteworthy decline in annual precipitation across the peninsular basins, the significant decline in the annual discharge of the Krishna river, Pennar river, Cauvery river, Tapti river, and the Narmada river can be attributed to human influence. The presence of several mega-dams and over 3800 large dams might also explain the decline in water discharge from the Indian peninsula. The inter-annual variability of rainfall might not play an important role, as discharge decline is mainly confined to heavily regulated large basins and a few minor rivers. However, the impact of climate change, which can be measured as rainfall intensity and frequency of extreme events on the peninsular rivers' annual water discharge, remains to be examined.

Acknowledgments HG is thankful to the University Grants Commission (UGC), India, for the faculty position under Faculty Recharge Programme. KKR (IF150795) and VKG (IF160993) are grateful to the DST-Inspire Fellowship Program, India, for providing financial support. Database accessible online on the Central Water Commission and WRIS India portals was greatly helpful in carrying out this study.

References

Bastia F, Equeenuddin SM (2016) Spatio-temporal variation of water flow and sediment discharge in the Mahanadi River, India. Glob Planet Chang 144:51–66

Bisht DS, Chatterjee C, Raghuwanshi NS, Sridhar V (2018) Spatio-temporal trends of rainfall across Indian river basins. Theor Appl Climatol 132(1–2):419–436

Bisoyi N, Gupta H, Padhy NP, Chakrapani GJ (2018) Prediction of daily sediment discharge using a back propagation neural network training algorithm: a case study of the Narmada River, India. Int J Sediment Res 34(2):125–135

Census 2011 India (2011). https://www.census2011.co.in/

CWC (2007) Reassessment of water resources potential of India. Central Water Commission Ministry of Water Resources, Government of India
Goswami BN, Venugopal V, Sengupta D, Madhusoodanan MS, Xavier PK (2006) Increasing trend of extreme rain events over India in a warming environment. Science 314(5804):1442–1445
Gupta H, Chakrapani GJ (2005) Temporal and spatial variations in water flow and sediment load in Narmada River Basin, India: natural and man-made factors. Environ Geol 48(4–5):579–589
Gupta H, Kao S-J, Dai M (2012) The role of mega dams in reducing sediment fluxes: a case study of large Asian rivers. J Hydrol 464–465:447–458
Gupta H, Reddy KK, Gandla V, Paridula L, Chiluka M, Vashisth B (2021) Freshwater discharge from the large and coastal peninsular rivers of India: a reassessment for sustainable water management. Environ Sci Pollut Res 29:1–18
Jain SK (2017) Water resource management in India. Curr Sci 13:1211–1212
Jain SK, Agarwal PK, Singh VP (2007) Hydrology and water resources of India. Springer, Heidelberg
Kale VS (2003) Geomorphic effects of monsoon floods on Indian rivers. In: Flood Problem and Management in South Asia. Springer, Dordrecht, pp 65–84
Krishnamurthy CKB, Lall U, Kwon H-H (2009) Changing frequency and intensity of rainfall extremes over India from 1951 to 2003. J Clim 22(18):4737–4746
Kumar R, Singh R, Sharma K (2005) Water resources of India. Curr Sci:794–811
Kummu M, De Moel H, Salvucci G, Viviroli D, Ward PJ, Varis O (2016) Over the hills and further away from coast: global geospatial patterns of human and environment over the 20th–21st centuries. Environ Res Lett 11(3):034010
Mekonnen MM, Hoekstra AY (2016) Four billion people facing severe water scarcity. Sci Adv 2 (2):e1500323. https://doi.org/10.1126/sciadv.1500323
Milliman JD, Farnsworth KL (2011) River discharge to the coastal ocean: a global synthesis. Cambridge University Press
NRLD CWC (2017) National register of large dams. Central Water Commission. Ministry of Water Resources, Government of India. www.cwc.gov.in/main/downloads/NRLD_06042018.pdf
Pachauri RK,Allen MR, Barros VR, Broome J, Cramer W, Christ R, Church JA, Clarke L Dahe Q, Dasgupta P, Dubash NK (2014) Climate change 2014: synthesis report. Contribution of Working Groups I, II and III to the fifth assessment report of the Intergovernmental Panel on Climate Change. IPCC
Rajeevan M, Bhate J, Jaswal A (2008) Analysis of variability and trends of extreme rainfall events over India using 104 years of gridded daily rainfall data. Geophys Res Lett 35(18)
Rao PG (1993) Climatic changes and trends over a major river basin in India. Clim Res 2(3): 215–223
Reddy SKK, Gupta H, Reddy DV (2019) Dissolved inorganic carbon export by mountainous tropical rivers of the Western Ghats, India. Chem Geol 530:119316
Roxy MK, Ghosh S, Pathak A, Athulya R, Mujumdar M, Murtugudde R, Rajeevan M (2017) A threefold rise in widespread extreme rain events over Central India. Nat Commun 8(1):1–11
Singh R, Kumar R (2015) Vulnerability of water availability in India due to climate change: a bottom-up probabilistic Budyko analysis. Geophys Res Lett 42(22):9799–9807. https://doi.org/10.1002/2015GL066363
UNICEF, FAO, and SaciWATERs (2013) Water in India: situation and prospects
Vörösmarty CJ, Green P, Salisbury J, Lammers RB (2000) Global water resources: vulnerability from climate change and population growth. Science 289(5477):284–288

Groundwater Responses to Climate Variability in Punjab, India

Gopal Krishan, Nitesh Patidar, N. Sudarsan, Rajesh Vasisth, and B. S. Sidhu

Abstract Groundwater has the potential to show resilience to the climate variability due to its high residence time in the aquifer systems which makes it less vulnerable to climate change. The recently recharged water in the shallow groundwater aquifer is more susceptible to the climate variability than the very old water in the deep aquifers. However, irregular patterns of rainfall and change in its frequency and intensity over a period of time can influence the groundwater storage through variable recharge. Northwestern states of India—Punjab, Haryana and Rajasthan fall in arid to semi-arid regions with mean annual rainfall ranging from 500–600 mm/year and the region is underlain by Indus river aquifer systems. The present paper highlights groundwater responses to rainfall variations during 1901 to 2019 using a simplistic empirical relationship of annual rainfall and recharge in Punjab. It has been found that 20% decrease in the annual normal rainfall volume of 600 mm would result in 65% decreased recharge. From this it is inferred that there are possible impacts of climate change in respect of projected change in mean annual precipitation on groundwater resources.

Keywords Groundwater recharge · Rainfall · Punjab · Climate variability · 1901 · 2019

G. Krishan (✉) · N. Patidar · N. Sudarsan
National Institute of Hydrology, Roorkee, Uttarakhand, India

R. Vasisth
Dept of Agriculture and Farmers' Welfare, Sangrur, Punjab, India

B. S. Sidhu
PSFC, Mohali, Punjab, India

B. Phartiyal et al. (eds.), *Climate Change and Environmental Impacts: Past, Present and Future Perspective*, Society of Earth Scientists Series,
https://doi.org/10.1007/978-3-031-13119-6_20

1 Introduction

In North-West region of India viz. Punjab, Haryana, Rajasthan and Delhi; being an agrarian belt water demand for irrigation is very high and population is profoundly dependent on groundwater and as a result groundwater draft is leading to falling water table and permanent decrease in groundwater storage (Rodell et al. 2009; MacDonald et al. 2016; CGWB 2017). Depletion of groundwater levels have become a global concern due its high negative consequences (Aeschbach-Hertig and Gleeson 2012; Famiglietti 2014). However, globally 24% of the recharge water is extracted (Aeschbach-Hertig and Gleeson 2012) but rapid declines are observed in local wells measurements and generally not shown by large scale gravity experiments data (Rodell et al. 2009) making high resolution monitoring data indispensable (Scanlon et al. 2012; MacDonald et al. 2016). In North-West, India the high resolution data is lacking and whatever data existed have highlighted high depletion (Lapworth et al. 2014, 2015, 2017) which indicated depletion in groundwater in the Indo-Gangetic plan caused by pumping and, to some extent, by climate variability (MacDonald et al. 2016). These variations are discussed in detail at a regional scale by Bonsor et al. (2017).

There are several issues related to the groundwater depletion such as lowering of water table, reduced storage (Inflow -means recharge<outflow means-discharge/ extraction) and base flow, saline water ingression and increased pollution (Sophocleous 2000; Foster and Chilton 2003; Konikow and Kendy 2005).

Extraction of groundwater is often followed by the capture (Bredehoeft and Durbin 2009) and this process helps in the water storage and the renewal capacity of groundwater depends on residence time defined as the time taken from recharge to discharge while groundwater age is the time taken from infiltration to reaching the sub surface (Kazemi et al. 2006). For efficient groundwater management more emphasis needs to be given on supply component which is specifically groundwater recharge. Renewal of groundwater takes place as a result of recharge and this depends on several factors amongst which rainfall is the most important. However, groundwater has responded to climate variability (Scanlon et al. 2006) but any long term irregularities in rainfall will affect the recharge.

In the present study, rainfall pattern of period 1901–2019 was studied to find groundwater responses for the period 2000–2019 in Punjab state of India. Punjab state has relatively variable geology and geomorphology consisting of shivaliks, alluvial plains and arid regions (Gopal et al. 2015). Groundwater systems are either single aquifer in shivaliks or multi aquifer in other parts of the state (CGWB 2017). More pressure on groundwater resources have been found due to the unprecedented demographic growth rates (Census 2011) and increasing demands for irrigation in entire state. Alarming groundwater depletion rates have been observed in central parts of Punjab due to higher dependency on groundwater in urban and rural areas (Gopal et al. 2015). Irregular and unpredicted rains in the state have compelled the people to opt for digging more wells for meeting the irrigation demands.

The urbanization and deforestation have increased during the past decades in the area leading to decreased recharge to groundwater. Increasing impervious area associated with urbanization has led to decreased infiltration and consequently affected the groundwater recharge. Moreover, the changes in land cover have also affected the groundwater recharge. The climate change is another driver which altering the groundwater balance. Therefore, the assessment of impacts of these changes on groundwater resources in this area is important. It is observed that the groundwater abstraction is relatively high during kharif periods as compared to other periods.

The research presented here investigates the multi-decadal impacts of increasing groundwater abstraction in the context of possible impacts of irregularities in rainfall pattern. Its underlying aquifer system can be considered typical of many other rapidly growing cities and the wider developing world.

This study investigates the impacts of groundwater extraction during the past decades considering the spatial-temporal variability in precipitation. The aquifer system of the study area is an important source of water which satisfies a considerable fraction of irrigation and domestic needs. The key objective of the study was to assess the impact of recent groundwater development and its probable future evolution under various management scenarios. The trend in precipitation and groundwater levels were analyzed. In addition, the recharge estimation and modelling were performed.

2 Materials and Methods

2.1 Study Area

Present study is carried out for the Punjab state which lies in latitude 29°32' to 32°28' N and longitude 73°50' to 77°00' E having a total area of 50,362 km^{2}which is home to 27 thousand people (2011 Census). The entire state is divided into 3 socio-cultural regions (Fig. 1): Malwa (comprising of districts Roopnagar, SAS Nagar, Patiala, Fatehgarh Sahib, Sangrur, Ludhiana, Moga, Barnal, Mansa, Bhatinda, Firozepur, Faridkot, Muktsar and Fazilka), Majha (comprising of districts Pathankot, Gurdaspur, Amritsar and Tarn taran), Bist Doab (comprising of districts Nawnashahr, Hoshiarpur, Jandhar and Kapurthala). Punjab is an agrarian state contributing grains enormously to the central pool. There are 3 perennial rivers in the state Ravi, Beas, Satluj, and one seasonal river. Ghaggar. Water from these rivers has been diverted into network of canals but the water from these rivers and canals are not able to fulfil the water demand leading to development of a large number of groundwater structures.

The deterioration of the human health and environment due to land use/land cover (LULC) changes has been a principal concern in an urban environment.

Landsat satellite of NASA has been providing a continuous data since 1970s and have been one of the most commonly used satellite data for land cover analysis.

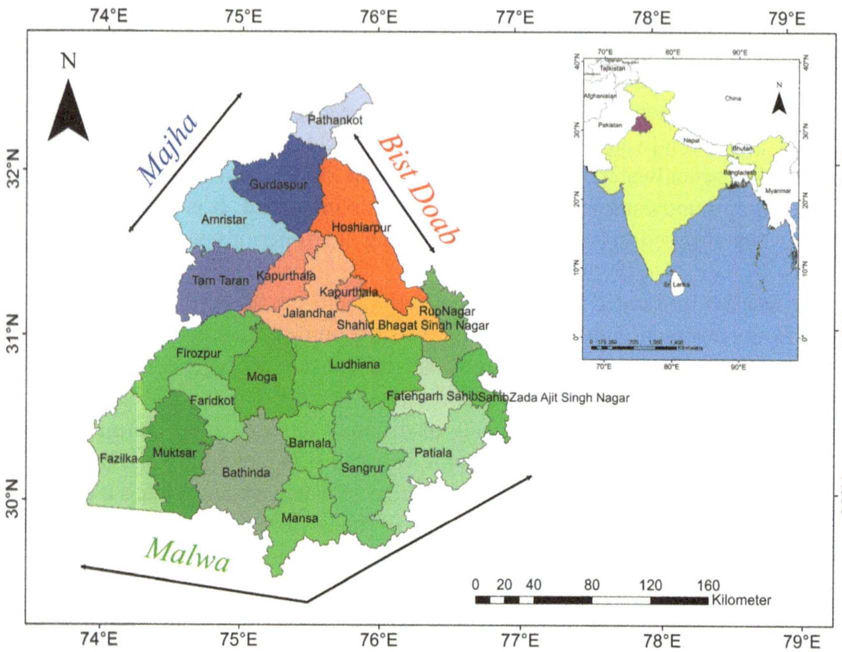

Fig. 1 Study area

Table 1 Landsat 8 OLI tiles used for LULC

Sl. No	Season	Date of Sensing (dd/mm/yyyy)	Tiles (Path-Row)
1	Rabi 2018–2019	26/11/2018	147–038
2		28/12/2018	147–039
3		19/12/2018	148–038
4		19/12/2018	148–039
5		26/12/2018	149–038
6		26/12/2018	149–039

Landsat data was acquired for this study. The bands record the radiations between visible and infrared regions of Electro-magnetic spectrum. Landsat satellite images (Table 1) were freely obtained from the data centre of USGS (www.earthexplore.usgs.gov.in). The Operational Land Imager (OLI) of Landsat-8 was acquired which is a multi-spectral data at 30 m spatial resolution. The visible, NIR and SWIR bands were selected for the LULC mapping for Rabi season (Dec-Mar) of 2018.

The SRTM-DEM (Shuttle Radar Topography Mission—Digital Elevation Model, 1-arc sec spatial resolution) was used. The downloaded tiles were stacked together and cropped out with the study area shapefile. The elevation map is prepared by dividing the range of altitudes into six ranges. The contour lines were generated in the 10 m interval by using the 'Contour' Spatial analysis tool in Arcmap 10.5.

Good quality images with less than 10% cloud cover were obtained. Processed Level 2 data product was used in the study where top of atmospheric reflectance converted to Surface reflectance. Image pre-processing involves geometric correction being applied to the images. Since datasets used were multi-temporal, they needed to be analyzed and registered uniformly to compare the change detection. After stacking the bands together, the tiles were mosaicked and the study area was clipped out. The supervised classification technique was adopted for the area already have ground truth data, drawn signature polygons were grouped into respective classes. The image was classified into five different classes such as, Water body, Agricultural land, Fallow land, Settlements and Hilly area. The area under each land cover is estimated.

2.2 *Groundwater Level Data*

Groundwater level data for the years 2000 and 2016 was collected from State Department of Agriculture and Farmers Welfare, Punjab. Aquifer data was taken from Bonsor et al. (2017). Groundwater availability and draft data was taken from CGWB (2017) for finding stage of groundwater development. Groundwater data for the years 2000 and 2016 was used to assess the spatiotemporal variations. Inverse distance Interpolation (IDW)interpolation technique was used to generate spatial maps of groundwater levels for the year 2000 and 2016.

2.3 *Rainfall Data*

Rainfall data for the period 1901–2019 was acquired from India Meteorological Department (IMD). The data is 0.25 degree gridded daily data which was generated using a dense network of point stations.

2.4 *Groundwater Recharge*

Groundwater recharge was estimated using a simplistic empirical equation given by Cave et al. (2003) for semi-arid regions:

$$Y = 148\ \ln(x) - 880 \tag{1}$$

2.5 *Trend Analysis*

The magnitude and significance of trend in the estimated 119-year recharge was determined using modified Mann-Kendall (MK) (Hamed and Rao 1998). The used modified MK test can handle autocorrelated series in the trend analysis. The MK test is a non-parametric trend analysis and thus can be applied on the data of any distribution. It has widely been used to ascertain trend magnitude and significance in the meteorological and hydrological variables, such as precipitation, temperature and river discharge. The MK test determines whether the 'null-hypothesis' is correct and provides level of significance (p-value) and trend (increasing/decreasing). If the 'null-hypothesis' exists, there is no-trend in the data series, otherwise, direction of the trend (increasing or decreasing) is identified. In this study, the modified MK test was performed on estimated recharge. The data series is divided into seven different periods considering three different intervals, i.e. 117-year, 60-year, 30-year. The created data series include 1901–2019, 1901–1960, 1961–2019, 1901–1930, 1931–1960, 1961–1990, 1991–2019.

3 Results and Discussion

3.1 *Land use, Elevation and Contour Maps*

Land use, elevation and contour maps of Punjab state are shown in Fig. 2. 55% of the total area is cultivated and comes under the category of agriculture, 22% area is fallow; 17% area comes under settlement and remaining 6% area is either water body or hills. From the elevation and contour maps it is found that Punjab state is a plain with gentle slope and an approximate elevation of 650 m AMSL in the northeast. In the north adjoining Himachal Pradesh and Jammu and Kashmir, the elevation is approximately 180 m AMSL. The south-western area is desertic and undulating.

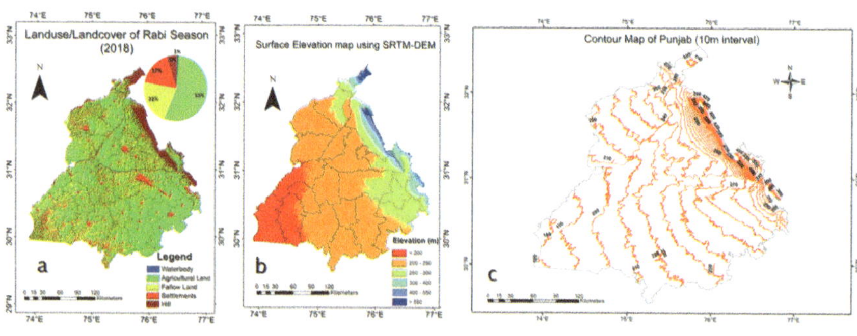

Fig. 2 Land use, elevation and contour maps of Punjab state

3.2 Groundwater Level Variations

Groundwater level variation map (Fig. 3) was prepared for the years 2000 and 2016 and it was observed that groundwater level is decreasing only in the central parts of the Punjab and it is seen there is no water level fall in the south-western districts of Punjab. The entire Fazilka districts and parts of Muktsar and Ferozepur district were supposed to have water table of less than 5 m below ground level (bgl). The most exploited districts were Hoshiarpur, Patiala, Sangrur, Barnala and parts of Ludhiana and Jalandhar. In the year of 2000, water level varied from 0.44 to 92.32 m with an average of 9.4 m and in the year of 2016, water level varied from 0.94 to 93.28 m with an average of 12.96 m. There was increase in average depth to water level by 3.56 m indicating a decline of 0.22 m/yr.

Groundwater decline effects can be clearly seen in the stage of groundwater development (Fig. 4) where out of 22 districts 18 districts have more than 100% stage of groundwater development (CGWB 2017). It has been found that key driver of groundwater depletion in Punjab is growing demands for various purposes, the same was reported by Lapworth et al. (2015) for Bist doab area; and reported for entire Punjab by Gopal et al. (2015).

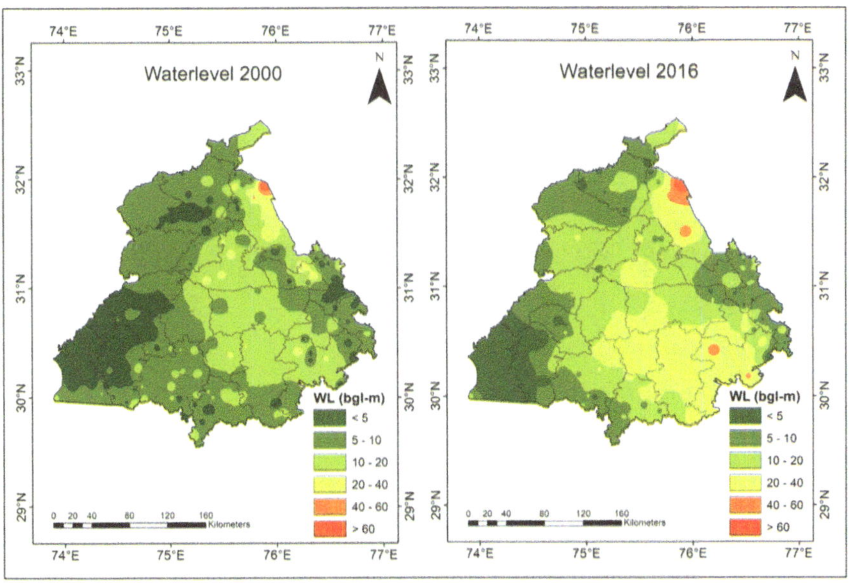

Fig. 3 Groundwater level difference of year 2000 and 2016

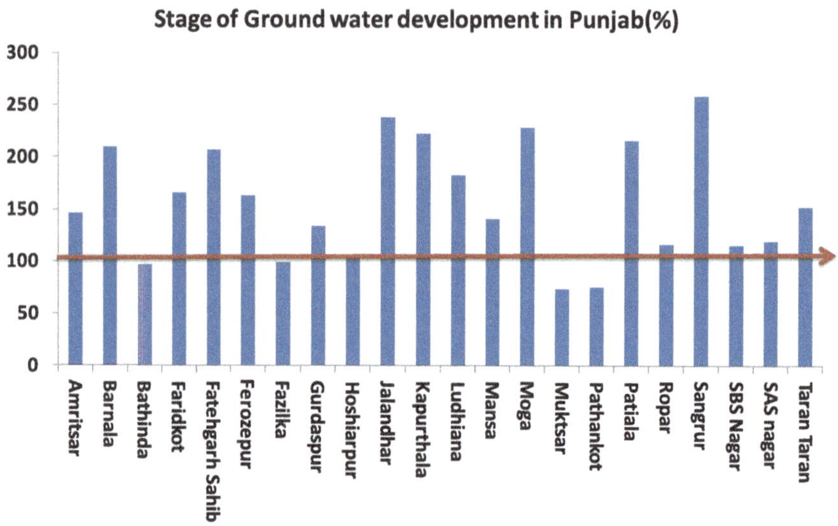

Fig. 4 Stage of groundwater development

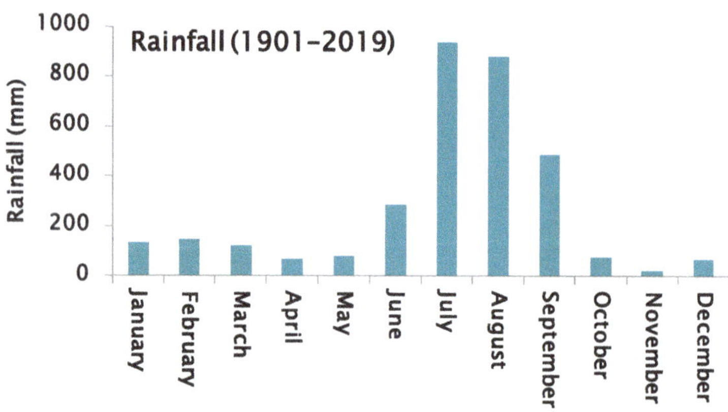

Fig. 5 Month wise rainfall pattern (1900–2019)

3.3 *Rainfall Analysis*

From the rain data (1900–2019), this can be clearly seen that 80% of the rainfall occur in monsoon period (July to September) (Fig. 5).

3.4 *Rainfall Recharge*

The trend analysis was performed on annual recharge from 1901 to 2019 using modified MK test. The test was performed considering seven different periods. The results obtained are shown in Table 2. As expected, the value of slope, intercept, p-value and trend direction is different for each period. The test suggests that there is a significant increasing trend in recharge for the period of 1901–1960 with p-value of 0.0009. However, no significant trend was found when the entire period between 1901–2019 was analyzed. Similarly, all other periods exhibit no-significant trend as indicated by higher p-values, similar trends were observed by Gopal et al. (2015).

This study provided an overview of the impact of long term rainfall on groundwater recharge. Rain is the recharge source and recharge is found varying with the rainfall pattern and variations. It has been found 20% decrease in the annual normal rainfall volume of 600 mm would result in 65% decreased recharge (Fig. 6).

4 Conclusions

The study shows that the annual rainfall variations will affect the recharge. The trend analysis was based on daily precipitation data, therefore an analysis using sub-daily data is required to understand the impacts of short events on recharge. Moreover, scanty precipitation will affect the replenishable groundwater and will ultimately affect groundwater availability in the area. The spatio-temporal analysis of groundwater indicates that the central part of the study is the most affected due high groundwater extractions. It can be inferred that if the groundwater extractions continuous at current pace, the aquifer is likely to deplete below safe limits. Such overexploitation might expand laterally to neighbouring areas and therefore the impact may expand from local to regional scale. These impacts might reduce the baseflow contribution to river, lead to drying of wetlands and land subsidence (Fig. 7).

Table 2 Summary of modified MK test on recharge for different periods between 1901 and 2017

Sr. No.	Period	Trend	P-value	Slope	Intercept
1	1901–2017	No-trend	0.0905	0.14	57.43
2	**1901–1960**	**Increasing**	**0.0009**	**0.56**	**46.10**
3	1961–2017	No-trend	0.4212	−0.22	80.50
4	1901–1930	No-trend	1.0000	0.00	56.95
5	1931–1960	No trend	0.0932	1.18	53.93
6	1961–1990	No trend	0.6177	0.49	72.26
7	1991–2017	No trend	0.8676	0.08	61.62

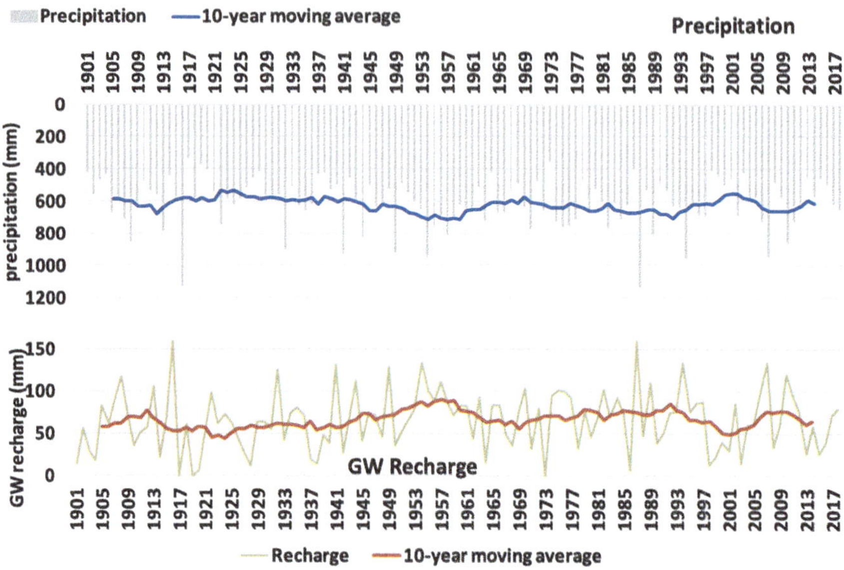

Fig. 6 Rainfall recharge

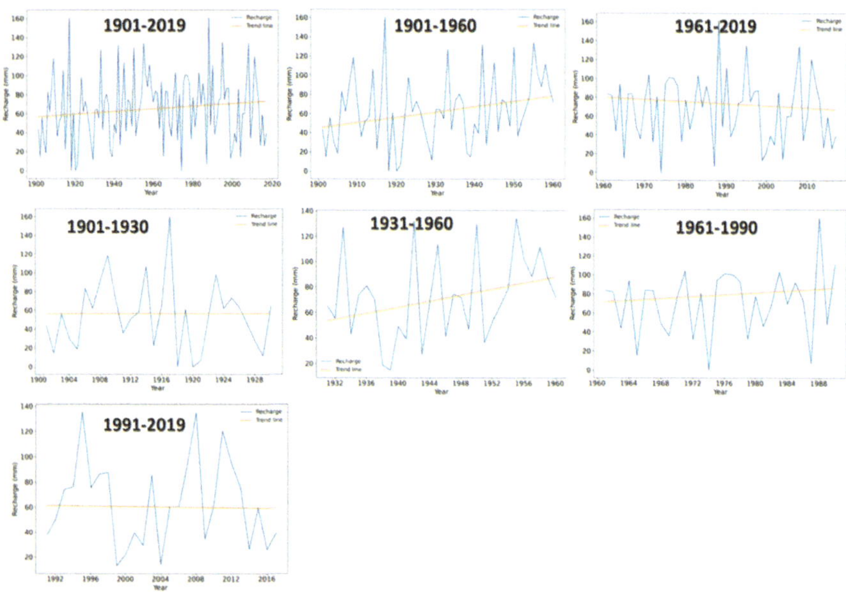

Fig. 7 Trend plots for different durations between 1901 and 2019

Acknowledgements Funding received from Punjab State farmers and Farm workers' commission is duly acknowledged.

References

Aeschbach-Hertig W, Gleeson T (2012) Regional strategies for the accelerating global problem of groundwater depletion. Nat Geosci 5:853–861. https://doi.org/10.1038/ngeo1617

Bonsor HC, MacDonald AM, Ahmed KM, Burgess WG, Basharat M, Calow RC, Dixit A, Foster SSD, Gopal K, Lapworth D, Lark RM, Moench M, Mukherjee A, Rao MS, Shamsudduha M, Smith L, Taylor R, Tucker J, van Steenbergen F, Yadav SK (2017) Hydrogeological typologies of the Indo-Gangetic basin alluvial aquifer. South Asia Hydrogeol J 25(5):1377–1406. https://doi.org/10.1007/s10040-017-1550-z

Bredehoeft JD, Durbin T (2009) Ground water development—the time to full capture problem. Ground Water 47:506–514

Cave L, Beekman HE, Weaver J (2003) Impact of climate change on groundwater recharge estimation. In: Xu Y, Beekman HE (eds) Groundwater recharge estimation in Southern Africa. UNESCO, Paris. UNESCO IHP Series No. 64, ISBN92-9220-000-3

Census (2011) https://www.censusindia.co.in/states/punjab

CGWB (2017) Groundwater resources of Punjab state

Famiglietti JS (2014) The global groundwater crisis. Nat Clim Chang 4:945–948. https://doi.org/10.1038/nclimate2425

Foster SSD, Chilton PJ (2003) Groundwater: the processes and global significance of aquifer degradation. Philos Trans R Soc Lond B 358:1957–1972

Gopal K, Chandniha S, Lohani AK (2015) Rainfall trend analysis of Punjab, India using statistical non-parametric test. Curr World Environ 10(3):792–800

Hamed KH, Rao AR (1998) A modified Mann-Kendall trend test for autocorrelated data. J Hydrol 204(1–4):182–196. https://doi.org/10.1016/S0022-1694(97)00125-X

Kazemi GA, Lehr JH, Perrochet P (2006) Groundwater age. Wiley, p 325

Konikow LF, Kendy E (2005) Groundwater depletion: a global problem. Hydrogeol J 13:317–320

Lapworth D, Krishan G, Rao MS, MacDonald A (2014) Intensive groundwater exploitation in the Punjab – an evaluation of resource and quality trends. Technical Report. NERC Open Research Archive, BGS-UK

Lapworth DJ, MacDonald AM, Krishan G, Rao MS, Gooddy DC, Darling WG (2015) Groundwater recharge and age-depth profiles of intensively exploited groundwater resources in Northwest India. Geophys Res Lett 42(18):7554–7562. https://doi.org/10.1002/2015GL065798

Lapworth DJ, Krishan G, MacDonald AM, Rao MS (2017) Groundwater quality in the alluvial aquifer system of northwest India: new evidence of the extent of anthropogenic and geogenic contamination. Sci Total Environ 599–600(2017):1433–1444

MacDonald A, Bonsor H, Ahmed K, Burgess W, Basharat M, Calow R, Dixit A, Foster S, Krishan G, Lapworth D, Lark M, Moench M, Mukherjee A, Rao MS, Shamsudduha M, Smith L, Taylor R, Tucker J, Frank S, van, Yadav, Shobha. (2016) Groundwater depletion and quality in the Indo-Gangetic Basin mapped from in situ observations. Nat Geosci 9:762–766

Rodell M, Velicogna I, Famiglietti JS (2009) Satellite-based estimates of groundwater depletion in India. Nature 460:999–1002

Scanlon BR et al (2006) Global synthesis of groundwater recharge in semiarid and arid regions. Special Issue: Emerging Issues in Rangeland Ecohydrology 20(15):3335–3370

Scanlon BR et al (2012) Groundwater depletion and sustainability of irrigation in the US High Plains and Central Valley. Proc Natl Acad Sci U S A 109:9320–9325

Sophocleous M (2000) From safe yield to sustainable development of water resources — the Kansas experience. J Hydrol 235:27–43

Reflections on Temporal Trends in Water Quality and Climate Variability at Three Degradation Hotspots of Leading Rivers in India

Apoorva Bamal, Akash Sondhi, Niharika Singh, and Priyam Saxena

Abstract River and riparian ecosystem restoration have significantly become the new approaches to reverse the past degradation activities and prevent future losses. Factors such as climate change, clearing of riparian zones, construction on wetlands and degrading river water quality are contributing to deterioration of the river and riparian ecosystem. These activities cumulatively pose challenges to the short and long-term agendas proposed by various organizations. It is important to look for solutions that will help reduce the extent of the degradation caused by the key drivers. The current restoration practices that are taken as the action plans for rejuvenation of rivers and associated riparian zones are not available to achieve the legally mandated goals, which are meant for the improvement of the structure and function of the degraded units.

While conditions of riverine ecosystem at specific location are also determined by the status of the riparian zone and vice-versa, conditions of upstream and downstream of particular riverine ecosystem are interdependent. Thus, degradation of upstream and downstream of the river is interlinked, which also connects their revival solutions. Individual rivers can be affected by similar or different key drivers of degradation; therefore, the restoration schemes differ in the magnitude of their exposure to these certain factors. Three identified leading rivers of India to assess the degradation hotspots are the Ganga, Brahmaputra and Godavari. For each identified river, degradation hotspot has been recognized based on four parameters viz. population dependence, sources of economy, river water quality, and biodiversity.

A. Bamal (✉)
The Energy and Resources Institute (TERI), New Delhi, India

A. Sondhi
Department of Energy and Environment, TERI School of Advanced Studies, New Delhi, India

N. Singh
Ministry of Environment, Forest and Climate Change, Government of India, New Delhi, India

P. Saxena
Memorial University of Newfoundland, St. Johns, Canada
e-mail: psaxena@mun.ca

B. Phartiyal et al. (eds.), *Climate Change and Environmental Impacts: Past, Present and Future Perspective*, Society of Earth Scientists Series,
https://doi.org/10.1007/978-3-031-13119-6_21

Based on their geographical location with respect to the particular river selected and the extent of secondary data available regarding the river and riparian ecosystem degradation in a particular district, rivers are studied in certain aspects to evaluate degradation issues and monitoring implementation on the same grounds.

This study is an approach to assess the degradation trends and its causal factors in the identified hotspots of the above mentioned rivers in India. The paper also reflects the impact of climatic variability on water quality of these rivers. It suggests modifications in existing conservation approaches and new action plans by taking in account their existing mortified scenario.

Keywords River ecosystem · Riparian ecosystem · Degradation hotspots · Conservation approaches · Climatic variability

1 Introduction

The Indian River system is composed of seven key rivers together with their numerous tributaries and bulk of these rivers flow into the Bay of Bengal, while the others flow into the Arabian Sea (Naiman and Décamps 1997). Rivers are of great importance to nature and humankind in one or the other possible ways. This importance of rivers is often due to the associated riparian zone that forms a transition between river and their floodplains, ultimately forming one of the most diverse and dynamic components of the landscape.

The ecosystems of these riparian zones, also known as riparian ecosystems, are the areas that occur along the water bodies and vegetation communities in these areas and play a transition role between terrestrial and aquatic ecosystems (Gregory et al. 1991). These areas exhibit distinct land properties due to unique soil and vegetation characteristics that are strongly influenced by free and unbound water in the soil. The ecological health of the river, including the relationship that the aquatic organisms have with each other and the environment, plays an important part in the maintenance of the associated riparian ecosystem (Holmes et al. 2004).

The river and riparian ecosystems are together recognized as important areas of conservation due to the services they provide to the society such as serving as corridors, connecting the otherwise disconnected landscapes through exchange of water, sediment, nutrients, pollutants, and organic materials (Tockner and Ward 2001).

The Riparian ecosystems are beneficial for the rivers as they add stability to the banks, therefore decreasing the risk of erosion. On the other hand, change in the water flow regime due to climate change, dam construction, etc. can lead to the loss of riparian ecosystem as these activities disconnect rivers from their floodplains. Therefore, there exists an ideal inter-connection between river and riparian ecosystem as the degradation of the river body affects the associated plant community and vice-versa (Nilsson and Berggren 2000).

In recent years, degradation of river and riparian ecosystem has been observed owing to multiple reasons, due to which the utilizable water resources have been limited to 28.1% of the total water available (Tonkin et al. 2018). The degradation is in terms of drop in the water quality of the river, narrowing of the river course, sedimentation, floods and vegetation clearing in the riparian zone (Karr and Chu 2000). Such conditions are correlated with each other and occur due to various key drivers such as climate change, urbanization, implementation of hydroelectric projects, and other human activities such as mass bathing in the river body and discharging industrial and domestic sewage in the same (Richardson et al. 2007). These specific drivers of degradation are not present along the whole river but at particular locations. Therefore, there are certain degradation hotspots wherein the occurring ecological disturbances (in terms of loss of function and productivity) cannot be recovered un-aided (McIver and Starr 2001).

Although majority of the rivers and their associated tributaries in India are in great need of conservation, for the current study three main rivers of the country are considered. The selection of the rivers is attributed to some of the essential parameters such as population dependence on a particular river site, occurrence of floods and water quality of the river. Considering these parameters for major rivers of the country, the identified rivers for the assessment study are Ganga, Brahmaputra and Godavari.

As the river flows across definite countries, states, districts and locations, the rivers can be segregated in terms of upstream and downstream of the same. While upstream of river is a location close to the source or origin of river, downstream is further down along the river receiving river water flowing through the upstream location (Lorenz and Feld 2013). Furthermore, factors and drivers of river and riparian ecosystem degradation vary for the upstream and downstream degradation hotspots.

Amongst the identified rivers, river Ganga has the highest population dependence and river Brahmaputra has the lowest population dependence. Population dependence is taken into consideration as there is a directly proportional relationship amongst the population and demand for resources. While for river Ganga and Brahmaputra average annual flood occurrence is 5–12, the average annual flood occurrence for Godavari is 3–5 (Dhar and Nandargi 2003). The water quality in all the three rivers is degrading. Such factors play an important role in selection of rivers and their respective hotspots for the assessment of river and riparian ecosystem degradation.

This study unveils the causal factors for degradation of three leading rivers in India to provide proximal conservation approaches for their restoration and rejuvenation.

2 Study Area

The identified rivers for the assessment study are Ganga, Brahmaputra and Godavari with one upstream and one downstream degradation hotspot identified (Fig. 1).

2.1 River Ganga

As the river Ganga originates from Gangotri glacier in the Himalayas at an elevation of 7010 m in the Uttarkashi district of Uttarakhand, the upstream district chosen is Uttarkashi (30° 43′ N and 78° 26′ E) (Singh et al. 2007). Uttarkashi district has a geographical area of 8016 km^2 and supports a population of 0.33 million people. Forest department administers 88% of the district. The forests in this district are rich in pine, deodar, spruce, kharshu, birch, and junipers and thus forestry plays an important role in economy of this district as it involves people in the business of propagation and exploitation of forest and forest produce (Table 1). Along with forestry, agriculture, animal husbandry, and the existing industries are also the sources of economy for the district (Awasthi et al. 2003).

The downstream district chosen for river Ganga is Allahabad (25° 27′ N and 81° 50′ E). Allahabad has a geographical area of 5482 km^2 and supports a population of 5.9 million people. Main sources of economy in the district are agriculture, tourism, and units of small and medium scale industries. Main agricultural crops are wheat and rice (Sharma et al. 2014). Due to the industrial and domestic wastewater, discharge in the river there has been degradation in the river ecosystem.

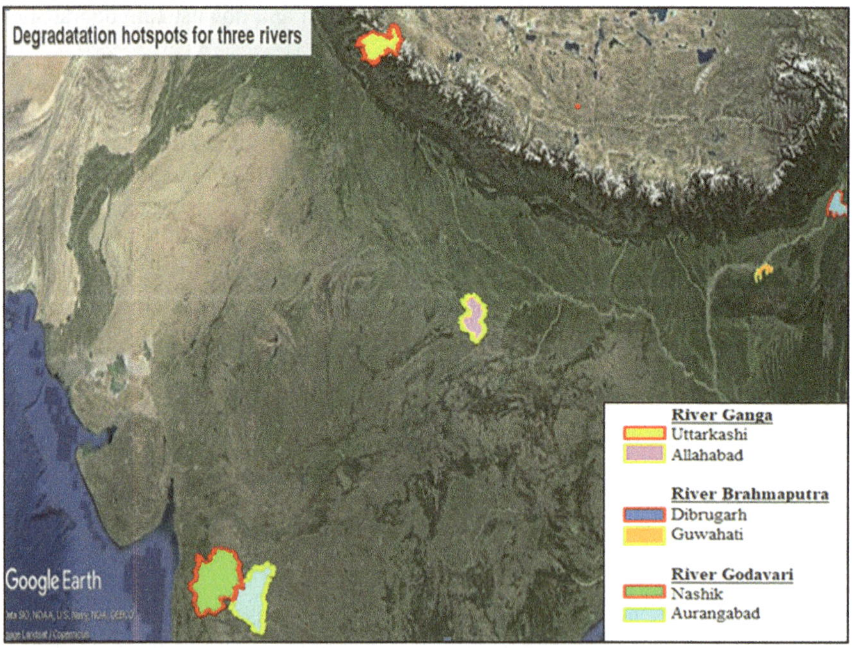

Fig. 1 Upstream and downstream degradation hotspots for three selected rivers

Table 1 Salient features of three major river basins in India

River	Ganga	Brahmaputra	Godavari
Basin extent			
Longitude Latitude	73° 2′ to 89° 5′ E 21° 6′ to 31° 21′ N	88° 11′ to 96° 57′ E 24° 44′ to 30° 3′ N	73° 24′ to 83° 4′ E 16° 19′ to 22° 34′ N
Area (km^2)	861,404	194,413	312,812
Population dependence	448 million	35.43 million	60.48 million
States covered	Uttarakhand, Uttar Pradesh, Madhya Pradesh, Chhattisgarh, Bihar, Jharkhand, West Bengal, Haryana, Himachal Pradesh, Rajasthan, Delhi	Arunachal Pradesh, Assam, West Bengal, Meghalaya, Nagaland	Maharashtra, Andhra Pradesh, Madhya Pradesh, Chhattisgarh

Sources: http://nmcg.nic.in/location.aspx
https://waterresources.assam.gov.in/portlet-innerpage/brahmaputra-river-system
http://grmb.gov.in/grmb/home
Central water commission

2.2 *River Brahmaputra*

The upstream district chosen for river Brahmaputra is Dibrugarh (27° 28′ N and 94° 54′ E). It has a geographical area of 3381 km^2 and supports a population of 1.3 million people. This district is considered as an economic hub of North East region of India. Main sources of economy in the district are oil and natural gas, tea production, tourism, power generation, fertilizer, and cottage industry (Gogoi et al. 2011). Due to increase in the resource demand by the growing population in the district, there has been rapid deforestation, encroachment on the riverbanks and increasing pollution levels in the river leading to the river and riparian zone degradation.

The downstream degradation hotspot chosen for river Brahmaputra is Guwahati (26° 8′ N and 91° 48′ E). It has geographical area of 556 km^2 and supports a population of 0.9 million people. The main sources of economy in the city are small-scale industries such as Cotton textile, etc. and large-scale industries such as Indian Oil Corporation Ltd. (Das 2008). The city being a tourist hub has increased demand for land and other resources leading to growth of industrial units in the city that are polluting the river immensely (Kotoky and Sarma 2017).

2.3 *River Godavari*

The upstream district chosen for river Godavari is Nashik (20° 00′ N and 73° 47′ E). It has a geographical area of 15,582 km^2 and supports a population of 1.48 million people. The main sources of economy in the district are agriculture and a number of large-scale industries such as Hindustan Unilever Ltd., etc. In addition to this, the district is also famous for its wine industries. Nashik district hosts the Kumbhmela that holds tremendous sacred importance (Yadav et al. 2016). It experiences great deal of deforestation that results in the degraded health of the river due to increase sediment and pollution load in the river.

For river Godavari the downstream district chosen is Aurangabad (19° 15′ N and 75° 20′ E) having a geographical area of 10,100 km^2 out of which 135.75 km^2 area is covered by forest. Aurangabad supports a population of 3.7 million people. The main sources of economy in the district are a number of industries with silk and cotton textile industries being the major ones. Since the district is a hub of large number of industrial units, the water quality of the river is generally poor due to the industrial discharge in the river (Pote et al. 2012).

3 Methodology

The methodology adopted to study the degradation hotspots of the selected rivers necessitated the use of secondary data and the unit of observation has been confined within the limits of the study area mentioned earlier:

- *Selection of rivers:* Rivers for the study were selected on the basis of population dependence, occurrence of floods and pollution levels amongst the rivers.
- *Identification of degradation hotspots:* One upstream and one downstream degradation hotspot were identified for each selected river on the basis of their population, sources of economy, river water quality and forest cover and availability of the data regarding the above-mentioned aspects.

Tools and techniques for data collection and analysis:

- The study of assessment of degradation hotspot was carried out with secondary data available for the degradation of river and riparian ecosystem of the selected degradation hotspots. Secondary data taken into account includes reports from various organizations (government, NGOs, institutes, etc.), published journals, online research articles, research papers, books, etc. that would specify the level of environmental changes that have occurred since last 10 years in the area under consideration.
- The detailed information regarding the research problem of degradation of river and riparian ecosystem degradation was collected on the basis of the following parameters:

(a) Population and its growth rate in the study area
(b) Water quality of the rivers in the identified degradation hotspots and its causes (CPCB ENVIS|Control of Pollution n.d.)
(c) Forest cover in the study area (Forest Survey of India n.d.)
(d) Occurrence and causes of floods in the degradation hotspots
(e) Other degradation factors and causes (if applicable)

- The comparative analysis was done with the help of different graphical tools and formulation of tables to assess the difference between the extent of degradation factors and causes in different hotspots to formulate conservation plans required for the revival and rejuvenation of the rivers and associated riparian zone.

4 Results

4.1 River Ganga

4.1.1 Decrease in Forest Cover (Forest Cover as Percentage of Geographical Area)

On comparing the forest cover area (in terms of forest area as percentage of Geographical area) in Uttarkashi (upstream district) and Allahabad (downstream district); it was observed that the forest cover in the upstream and downstream hotspot has changed in different ways. While in Uttarkashi, there has been a decline in the forest cover area, in Allahabad the forest cover area has increased during the years (Figs. 2 and 3).

It is a known fact that although river Ganga in its upstream is subjected to much less anthropogenic pollution but the ecosystem in this segment of the river is highly sensitive and fragile and degradation of such fragile ecosystem directly or indirectly affects the riparian ecosystem and vice-versa.

In Uttarkashi, the forest cover has decreased from 3145 km^2 (39.23%) in 2009 to 3028 km^2 (37.77%) in 2017 of the total geographical area. Such scenario is a result of the deforestation activities happening in the district since many years. Deforestation in Uttarkashi is attributed to many reasons. One of the important reasons is increase in population of the district wherein the population has increased from 295,013 in 2001 to 330,086 in 2011 (Census of India 2011). Although due to its geographical location, Uttarkashi supports only 3.48% of Uttarakhand population but the increasing population demands more resources for their survival and forest area is one the leading resources in demand. Moreover, the forest cover studied here is not solely the riparian zone but is inclusive of the associated riparian zone of the river.

Some percentage of deforestation is done to clear the land for developmental activities such as construction of residential areas, hydroelectric projects or agricultural activities; the other percentage of the deforestation is attributed to the need of

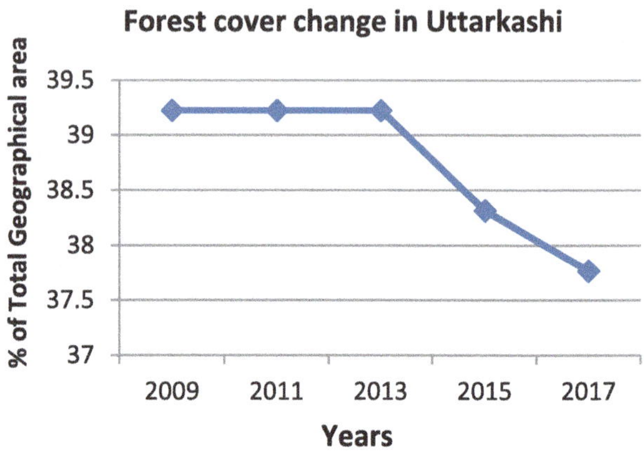

Fig. 2 Change in forest cover in Uttarkashi

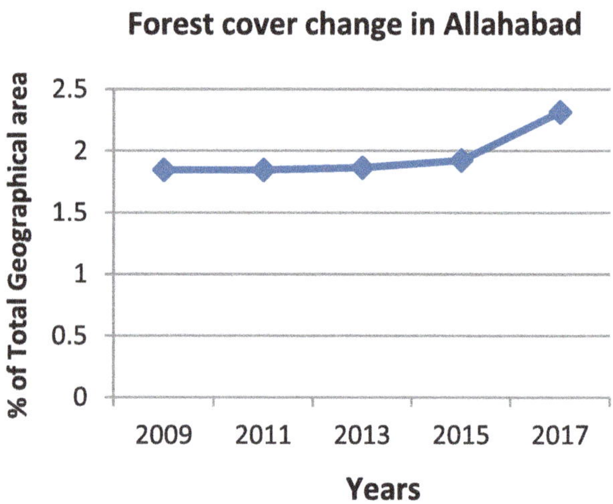

Fig. 3 Change in forest cover in Allahabad

wood material serving different purposes that can be material building, fuel, etc. Such decrease in forest cover area of the district exhibits urgency for attention of the concerned authorities as well as the people so that proper monitoring can be done in time.

In Allahabad there has been an increase in the forest cover area from 94 km^2 (1.85%) in 2009 to 127 km^2 in 2017 which is just 2.32% of the total geographical area of Allahabad. This increase in forest cover area can be a result of plantation activities done during the years.

However, the forest cover area in Allahabad is comparatively much less than that of Uttarkashi whereas the population of Allahabad is 1,112,544 in 2011 (Census of India 2011) which is larger than that of Uttarkashi. This becomes a reason for less forest cover area in Allahabad because such large population has much higher demand of land leading to less forest cover area. Also along the river, there has been observed less vegetation and tree cover, which can be attributed to the fact that due to holy importance of river Ganga in Allahabad, the demand for land in its vicinity has increased several folds that is disturbing the riparian ecosystem of river Ganga in Allahabad (Kulshrestha and Sharma 2006). Thus more plantation activities, especially in the areas, which are closely associated with the river, are required by the authorities to assure better health of the river and riparian ecosystem.

4.1.2 River Ecosystem Degradation

On comparing the different water quality parameters (pH, DO, BOD, total coliform) of river Ganga in Uttarkashi and Allahabad, it can be concluded that the water quality of river Ganga in Allahabad is much poor than that of Uttarkashi. Such scenario can be attributed to many polluting drivers present in the district. One of the key drivers is large number of industries present in Allahabad which directly discharge their effluents in the river and amongst them paper and pulp industries contribute the most pollutants by volume (Figs. 4, 5, 6, and 7).

In addition to the industrial waste, more than half of the waste generated in river Ganga is from domestic sources. This pollution happens because of the poor sewerage system of the district where more than 61% of the district area is un-sewered and there are more than 57 drains in Allahabad (National Mission for Clean Ganga n.d.). The key problems that persist here are the growing quantum of untreated sewage that is discharged in the river and the inadequate flow of water to dilute and assimilate this waste. This inadequate flow is because of the lack of vegetation cover along the river and pumping of ground water, which leads to the deepening of the ground water, and thus there occurs movement of water from river to ground water thus reducing the river flow.

Fig. 4 pH of river in river Ganga

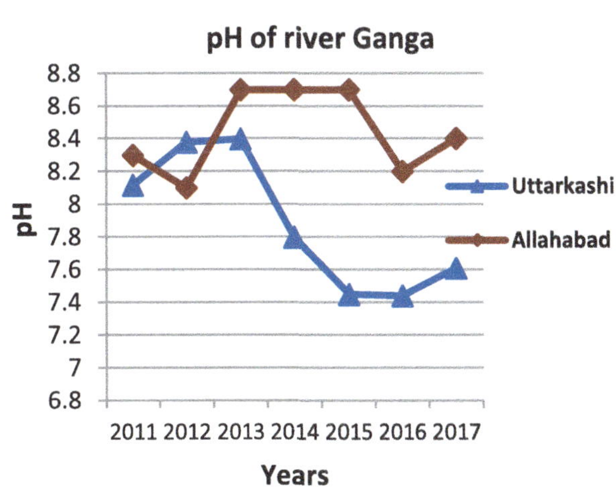

Fig. 5 Total coliform in river Ganga

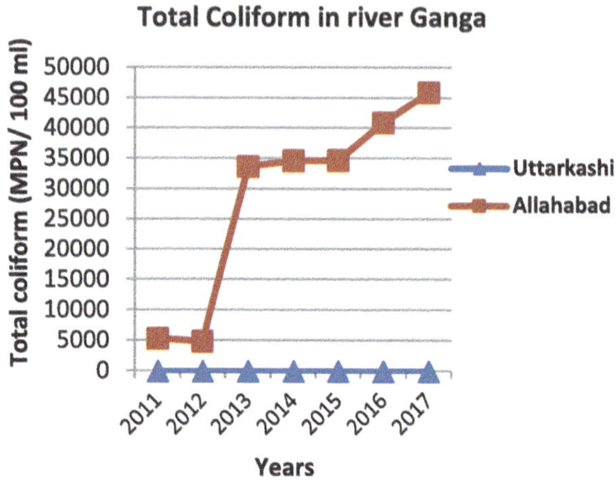

Fig. 6 Dissolved oxygen in river Ganga

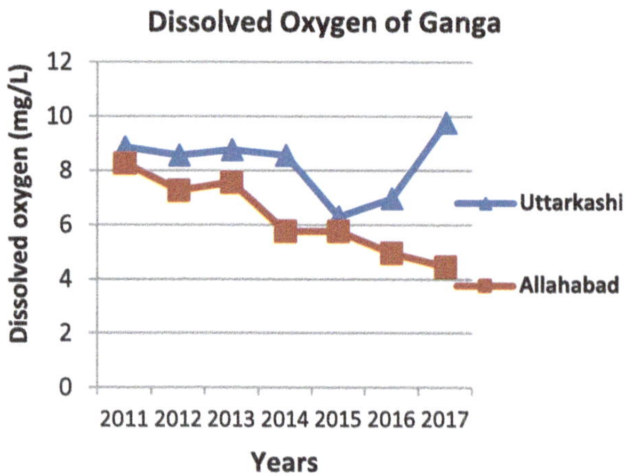

Fig. 7 BOD in river Ganga

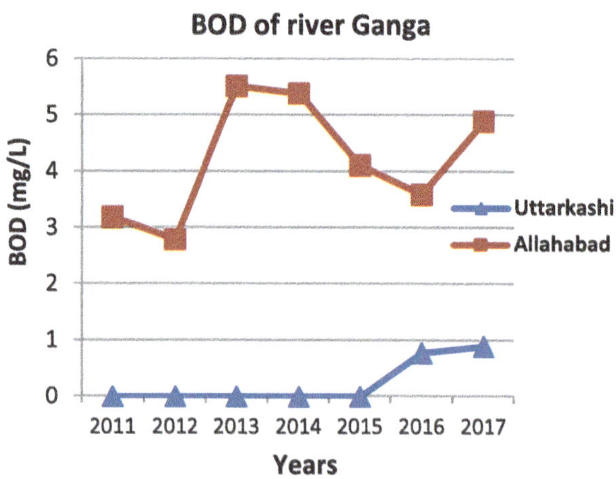

4.1.3 Impact of Dam Projects on River Water Quality

Although the anthropogenic pollution is comparatively low in Uttarkashi, there has been fragmentation of more than half of the river due to the hydroelectric projects in this district as the river in the upper segment has high potential for harnessing hydropower. Two dams in this district are ManeriBhali 1 and ManeriBhali 2. Studies have also shown that dams contribute to river water pollution (Zhang et al. 2010). Dams reduce the flow regime and water velocity in the middle and downstream of the rivers. This causes decreased water environment capacity leading to degraded river water quality (Estrats and Sl 2006).

However, upon releasing water, there is increase in sediment load in the downstream river which increases water turbidity causing anoxic conditions for inhabitants therein. Water discharged from dams into rivers also lead to sudden change in water temperature leading to fluctuations in DO level. This ultimately causes change in BOD levels impactingaquatic habitants (Sharma and Goyal 2020).

Another threat to riparian ecosystem of river Ganga that occurs due to dam construction is lowering of ground water table because of the riverbed deepening which ultimately lowers the accessibility of water tables to the plants.

4.2 River Brahmaputra

4.2.1 Decrease in Forest Cover (Forest Cover as Percentage of Geographical Area)

On comparing the forest cover of Dibrugarh and Guwahati, it can be deciphered that both the degradation hotspots are losing their forest cover at a great pace (Figs. 8 and 9). Such deforestation in both the locations can be attributed to the increasing urbanization that is taking place in both upstream as well as downstream hotspot of river Brahmaputra. This increasing urbanization is directly linked to the increasing population in Dibrugarh and Guwahati.

While in the upstream district the population growth is 11.92% from 2001 to 2011, in the downstream location the population growth is 11.24% (Census of India 2011) leading to more demand for land in the upstream district. Such increase in population is not only due to high birth rate but also due to migration of population from other states. Moreover, there has been an increase in migration of population from rural to urban areas in search of better employment opportunities, health care facilities, and education. It is evident that as the forest cover is decreasing, the built up area in the degradation hotspots has been increasing.

In addition to this, there has been a great demand of fuel wood in areas of Assam wherein there is still high dependency on fuel wood for cooking. Therefore, to carry out their daily routines people from rural as well as urban areas are involved in degrading the natural resources. Moreover, clearing of vegetation along the river has also increased due to the developmental activities taking place in the areas near the

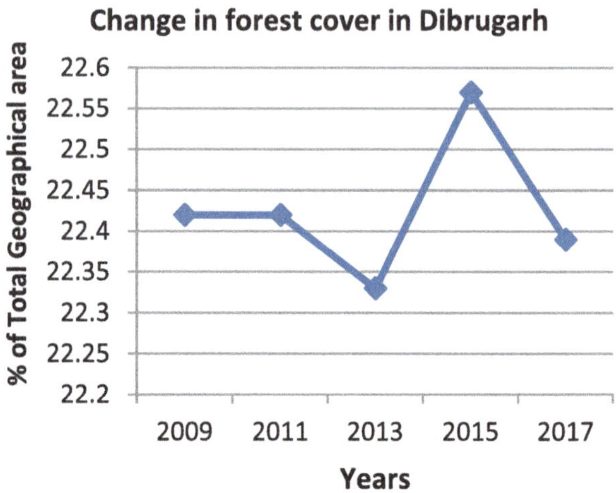

Fig. 8 Change in forest cover in Dibrugarh

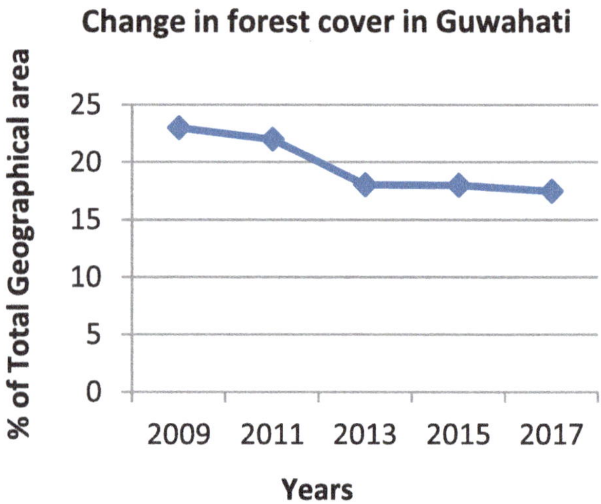

Fig. 9 Change in forest cover in Guwahati

river. This clearing leads to reduced flow of the river because of the barriers present such as constructed infrastructure, etc. and thus the sediment load keeps on increasing leading to problem of floods in the downstream of the river.

Wetlands, referred to as kidneys of nature, are also of immense importance in river and riparian ecosystem and they are shrinking due to increasing urbanization in Guwahati. The settlements in Guwahati have increased in last two decades, as there have been construction activities on its wetlands leading to fragmentation of the wetlands. Thus, the associated riparian ecosystem is getting affected due to these activities.

Deforestation activities are also major sources of degradation in the specified areas such as soil erosion, siltation, etc. These problems are also an ultimate cause of environmental hazards such as floods, landslides as the slope stability is reduced due to the settlement activities taking place. Thus, plantation activities especially along the river are required to combat the ultimate problems faced by the river and riparian ecosystem.

4.2.2 River Ecosystem Degradation

On comparing the water quality of river Brahmaputra in Dibrugarh and Guwahati, it was inferred that Guwahati has poor river water quality than that of Dibrugarh. Although the river water quality of Dibrugarh has degraded over the years, the river water quality in Guwahati is on the extremes now (Figs. 10, 11, 12, and 13).

Such poor water quality in Guwahati can be attributed to many reasons and the top most reason is the developmental activities happening in the city. The developmental activities in the city include the dynamic interplay of socio-economic, institutional, and technological activities. Along with this, the fact of Guwahati city being a tourist hub is also contributing to the poor water quality of the city. In addition, the industrial growth in the city is a major contributor of degradation factor of the river ecosystem. Many small and medium scale industries are emerging in the city amongst which cotton and textile industries are the most polluting ones.

In addition to the above-mentioned polluting drivers, there is another key driver of pollution in the river Brahmaputra. Bharalu River that is an important tributary of river Brahmaputra is the only river that is within the entry and exit of the city. There has been severe encroachment on the banks of river Bharalu that is contributing to the heavy levels of pollution. Industrial waste that is discharged into the river is highly polluting the water. On the other hand, domestic wastewater is also discharged in the river without any pre-treatment. It is thus evident that when such polluted and degraded river joins Brahmaputra in the city, it leads to high levels of pollution in the river water.

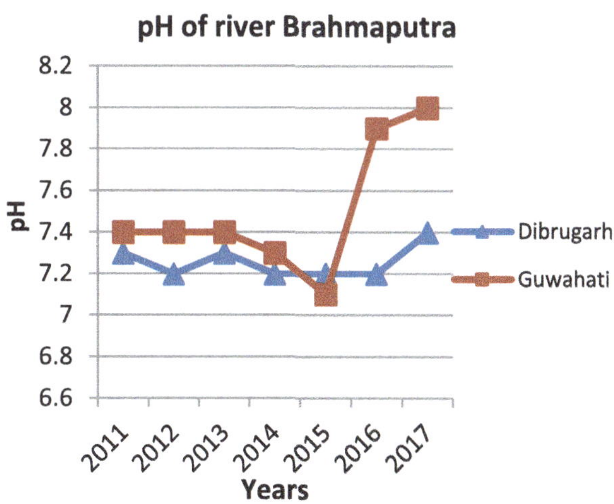

Fig. 10 pH of river Brahmaputra

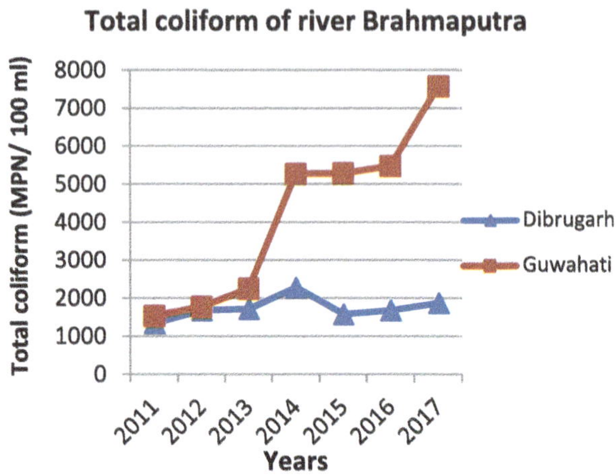

Fig. 11 Total coliform in river Brahmaputra

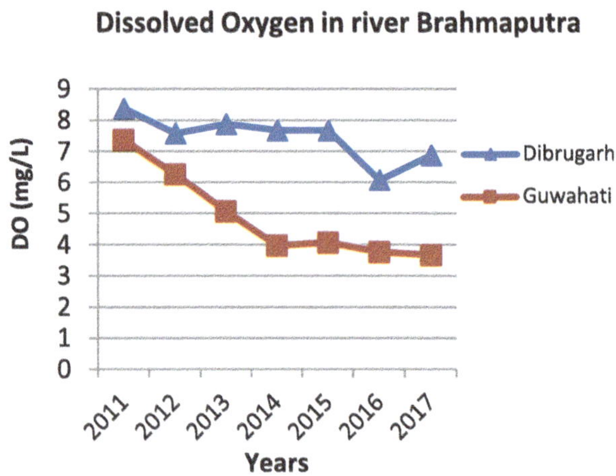

Fig. 12 DO in river Brahmaputra

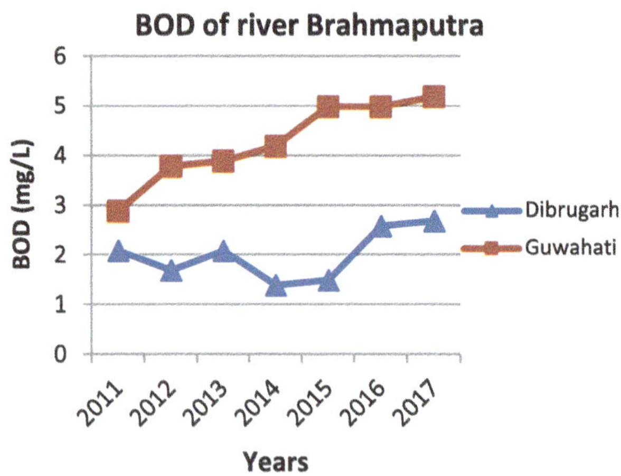

Fig. 13 BOD in river Brahmaputra

Another activity of humans such as direct dumping of waste and discarding holy material in the river is also an important driver of pollution of river Brahmaputra. Such pollution is not only making the water unsuitable for the human consumption but also making it dangerous for the aquatic life present in the river and the remaining vegetation along the riverbanks.

In addition to pollution, the urbanization in Guwahati and Dibrugarh is also an important factor leading to flash floods, which have become common in recent decades due to change in land-use pattern. Dibrugarh and Guwahati are amongst top 17 worst flood affected areas and majority of the crop area is under different flood categories (Sharma et al., 2012). Therefore, the problem of encroachment of riverbanks is not is not only affecting the river ecosystem but is also leading to the poor health of the riparian ecosystem.

4.3 River Godavari

4.3.1 Decrease in Forest Cover (Forest Cover as Percentage of Geographical Area)

On comparing the forest cover of Nashik and Aurangabad, it was ascertained that in the upstream district, i.e. Nashik, the forest cover has reduced from 1089 km^2 (7.01%) in 2015 to 1068 km^2 (6.88%) in 2017 while in the downstream district, i.e. Aurangabad, the forest cover has increased from 557 km^2 (5.51%) in 2015 to 570 km^2 (5.63%) of the total geographical area in 2017. Such scenario can be attributed to the drivers of deforestation in Nashik (Figs. 14 and 15). One of the main drivers is increase in land demand along the riverbanks due to the holy importance held by the river. The increasing population in the district accounts for the land demand that leads to clearing of vegetation and forestland along the riverbanks. This leads to conversion of natural river channel into the decorative channel thus disturbing the riparian ecosystem that is associated with the river.

It was in earlier decades that there was encroachment of flood plains in the district for the conversion of the land into agricultural fields. However, due to the increase in population and demand, the agricultural fields are now being converted to plots residential and industrial areas, thus disturbing the main source of economy in the district. Along with this, the wetlands are also threatened and constantly shrinking in the district due to the developmental activities happening in Nashik. Moreover, Nashik has been receiving high rainfall and this climatic activity leads to situation of floods in many parts of Nashik, which leads to carrying of silt and sediment in the downstream of the river thus causing problems in the lower segment of the river.

There has been an increase in the geographical area of Aurangabad from 10,107 km^2 in 2015 to 10,131 km^2 in 2017. Thus, the increase in forest cover is relative to increase in the district area. Moreover, population in Aurangabad is much larger than that of Nashik despite of the fact that geographical area of Nashik is more than that of Aurangabad. However, due to the holy importance of river in Nashik, the

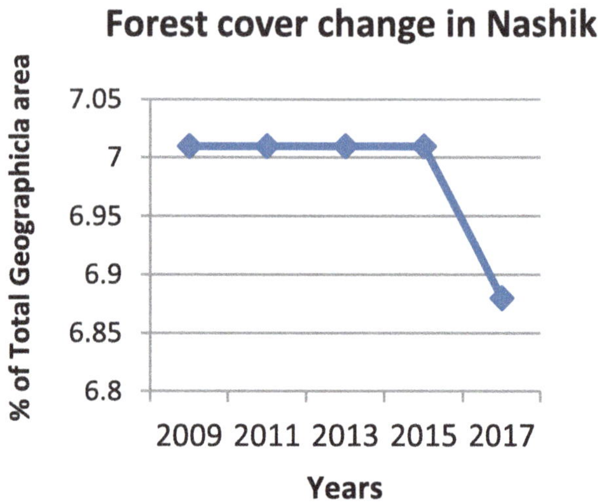

Fig. 14 Change in forest cover in Nashik

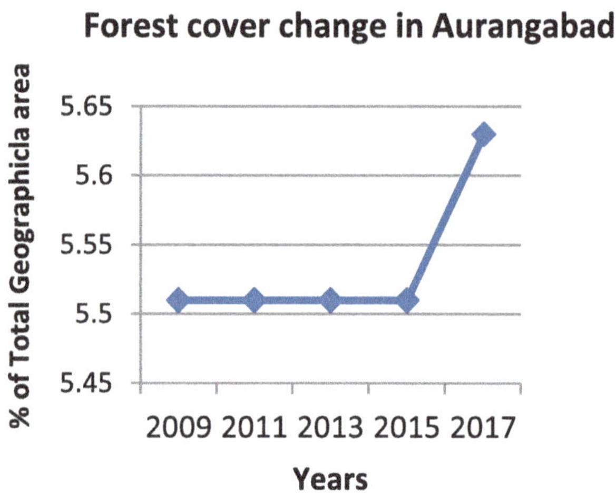

Fig. 15 Change in forest cover in Aurangabad

land demand is much higher. Thus, river Godavari close to its origin is also being degraded along with its associated lands.

4.3.2 River Ecosystem Degradation

On comparing the water quality of river Godavari in Nashik and Aurangabad, it is known that water quality in Aurangabad is much deteriorated than that of Nashik (Figs. 16, 17,18, and 19). One of the specific key drivers of this scenario is the large number of industrial units in Aurangabad. Amongst these, cotton textile industries are most polluting ones and they are quite large in number in Aurangabad. In addition to these industries, large numbers of other polluting industrial units that are directly discharging their effluents into the river water without any pre-treatment are also prevalent. The domestic waste generated by such huge population is also directly discharged into the river. The wastewater discharged in the river causes problems for the aquatic life forms present in the river by altering the water quality of river. Thus, the water becomes unsuitable not only for human consumption but also for aquatic life forms in the river.

Moreover, the water quality of Godavari River in Nashik has also started deteriorating due to different anthropogenic activities taking place in the river such as mass bathing, discarding holy material, etc. Due to high rainfall when flood conditions occur in the river, the polluted river water flows downstream, thus adding to more pollution in the downstream river segment.

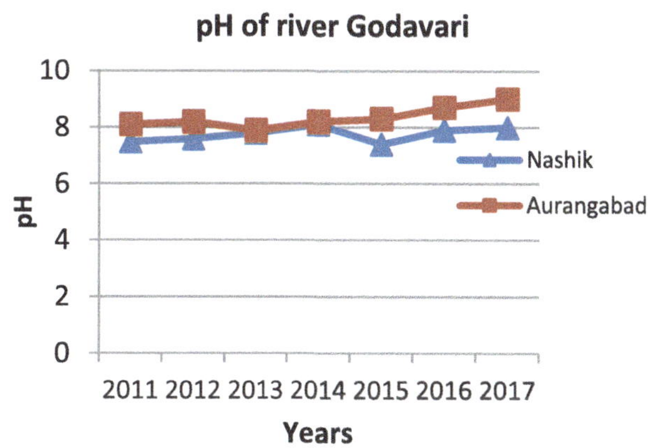

Fig. 16 pH of river Godavari

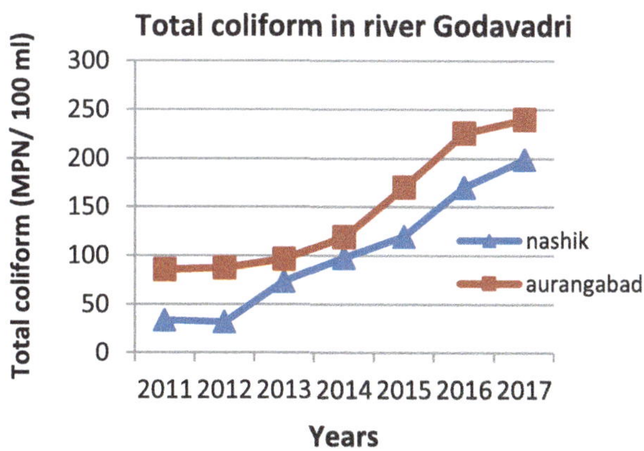

Fig. 17 Total coliform in river Godavari

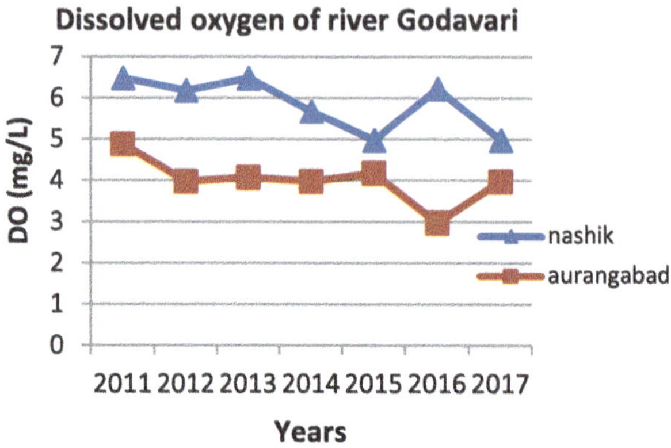

Fig. 18 DO in river Godavari

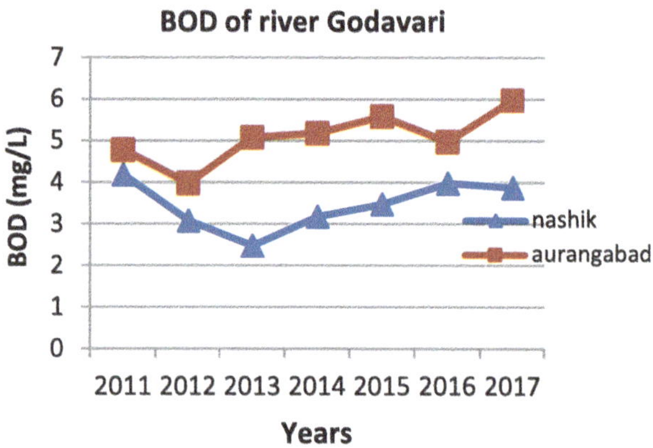

Fig. 19 BOD in river Godavari

5 Discussion

5.1 Precipitation Trend in Degradation Hotspots

Climate change and precipitation (rainfall) trend are closely associated and have been observed as a proxy for one another (Sun et al. 2018). Rainfall for both upstream and downstream hotspots in all the three rivers have been analyzed for the past decade (2009–20) to assess the impact of climate variability on river water quality (India-WRIS n.d.).

It was observed that all the upstream hotspots received higher rainfall as compared to the downstream hotspots (Fig. 20). In addition to this, except Nashik, almost all the hotspots received lesser actual rainfall than the normal rainfall with an average deviation of 50% (negative) from the normal. This is in accordance with the deteriorating water quality of rivers with due course of time. Therefore, apart from the factors mentioned in earlier sections for degrading river water quality, less rainfall is an additional agent. Due to the reduced rainfall, there is less dilution of

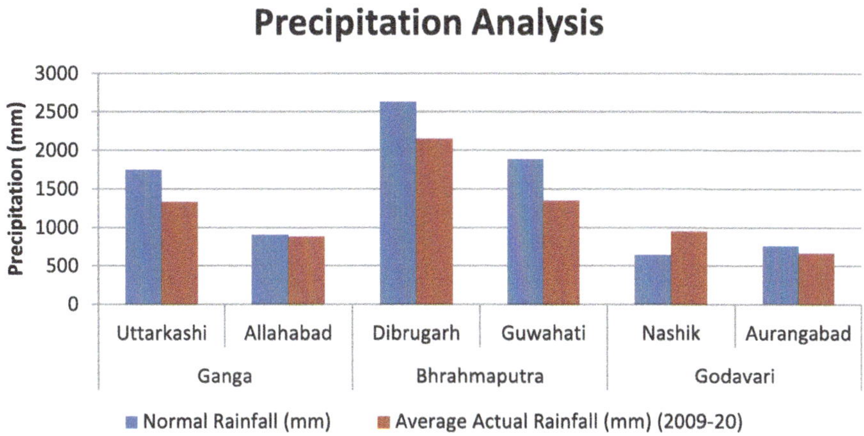

Fig. 20 Average rainfall trend in degradation hotspots

Table 2 Climate change via precipitation, temperature and flow regime alterations in the river basins

	Precipitation		Temperature		River water flow	
River	Mean decadal precipitation (mm)	Decadal change (%)	Mean decadal temperature (°C)	Decadal change (%)	Average water flow (m^3/s)	Decadal change (%)
Ganga	1051.2	−3	25.7	1.7	1.7×10^4	−6.4
Brahmaputra	2589.2	−1.5-2	17.3	1.46	2.0×10^4	−5
Godavari	1106.8	3–4	29.6	1.6	1.5×10^4	−5.2

Source: Jain and Kumar 2012; Chowdhury et al. 2016; Borse and Agnihotri 2017; Tamuly et al. 2019

the sewage discharged in the river and hence the basic water quality parameters such as TO, BOD, total coliform, etc. are not in the permissible limits.

5.2 *Climate Variability and Water Qualityin River Basins*

With due course of time, climate change has affected river basins by a number of means. This impact is assessed by analyzing temporal data of factors viz. precipitation, temperature and flow regime over a period of time (Table 2).

Upon analyzing the decadal data for precipitation, temperature and water flow for Ganga, Brahmaputra and Godavari, it is validated that apart from the anthropogenic activities mentioned in earlier sections, climate change over a period of time has led to degradation of the river ecosystem. Studies have shown that over the period of 10 years, precipitation in the three river basins has reduced up to 4% and the temperature has increased by approximately 1.46–1.7%.This disturbs the normal course of living of the river water organisms. Apart from the amount of actual rainfall, there is a huge disturbance in rainfall temporal pattern, i.e. there is untimely rainfall which impacts the agriculture of specific regions. In less or untimely rainfall conditions, there is a need to irrigate the farmlands through other sources such as groundwater (in most regions). Such situations when prevalent for a prolonged time lead to ground water depletion and disturbance in environmental sustainability.

The reduced precipitation over the years has also led to reduction in the river flow regime by 5–6.4%. International studies have highlighted that reduced water flow regime essentially impacts chemistry, hydromorphology and ecology of the rivers (Whitehead et al. 2009; Arheimer et al. 2005). Due to this, there is a serious reduction in the water environment capacity which not only affects the river ecosystem but also leads to reduced water availability of water for human beings.

Case Study of Brahmaputra River Basin: Precipitation and Temperature Trend for 1971–2000

Amongst the three selected river basins, Brahmaputra has showcased least degradation factors. However, the decadal change percentage is significant and can be validated with detailed study inclusive of precipitation and temperature trend analysis conducted by Sharma and Flügel (2015).

The trend analysis highlighted the forecasting of significant decrease in precipitation and increase in temperature (Fig. 21). Following this trend, the climate model predicted the temperature to increase by 5 °C and precipitation to decrease by 20% in next 100 years. Upon considering this pattern for other two river basins, the degradation extent can be well speculated. This is in accordance with the fact that even though narrow factors are different for different river basins, the broader factors are comparable for these basins.

Study by Sharma and Goyal (2020) has also highlighted that even though the increasing snow melt is causing increased sea water level, its stagnation state is worsening the degradation rate. As mentioned earlier, decrease in forest cover in the riparian zone is leading soil erosion into the rivers leading to increased turbidity in the river. This, in combination with reduced water flow is also impacting the important water quality parameters of river. Recent studies by Kaushal et al. (2019) and Anand et al. (2018) have provided basic indications that apart from the industrial effluent discharge in the river, climate change is also contributing an observable share to the degraded river water quality. Due to this, water quality parameters especially pH, DO, TDS and BOD are severely impacted.

6 Conclusion and Recommendations

Degradation of river and riparian ecosystem is one of the major concerns because the interplay of the river and riparian ecosystem provides the ecosystem services in association with all other ecosystems. However, there has been a disparity in the extent, factors, and key drivers of degradation of the upstream and downstream

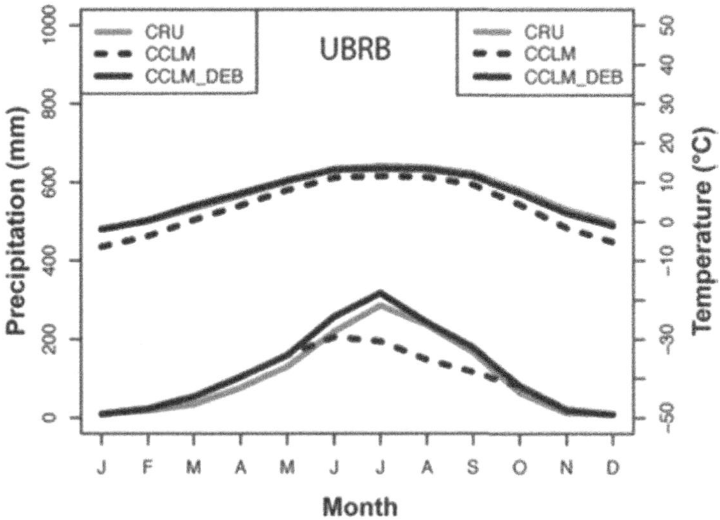

Fig. 21 Mean monthly temperature (upper lines) and precipitation of UBRB (upper Brahmaputra basin) from CCLM projections (COSMO model in climate mode) and bias corrected projections (CCLM_DEB) are compared to the observed data (CRU (climatic research unit)) for the period (1971–2000). (Adapted from Sharma and Flügel 2015)

hotspots of each of the three rivers. This is because of the difference in geographical locations in the upstream and downstream hotspots that describes the extent of anthropogenic and natural disturbance in each of the location. It was seen that while in the upstream districts of all the three rivers the disturbance was due to more demand of land for increasing population, in the downstream districts the disturbance was more related to pollution levels in the river and clearing of riverbank vegetation leading to poor health of the river and riparian ecosystem. There has also been evident reduction in precipitation in these hotspots, caused due to climate change. Over the years, while the precipitation has reduced to almost 50% of the normal rainfall, the water quality has also degraded. Moreover, degradation in the upstream of the rivers are creating problems for the downstream of river leading to increased siltation and sediment load which creates problems of flooding in the downstream of the river.

The study has also analyzed that river degradation is also driven by prolonged climate change with alterations in the precipitation, temperature and water flow regimes. Over the years, Ganga, Brahmaputra and Godavari basins, have experienced apparent variations in the aforementioned factors. This has also contributed to the water quality aspect in these rivers. Therefore, in addition to policies for anthropogenic activities, there is a need for environmental sustainability in terms of climate change.

Although the services provided by the river and riparian ecosystem are essential for human beings but encroachment of these areas to satisfy the growing is not an alternative. This is essentially visualized through a systems thinking approach wherein there is inter and intra-river structure correlation. Therefore, the ecological role played by these systems shall also be considered while retrieving the services.

To protect the river and riparian ecosystem from degradation due to several key drivers of degradation, following recommendations are suggested as conservation approaches:

- Close monitoring of the condition of the rivers and associated riparian zone should be done wherein the monitoring stations are set up at close locations and direct linkage of these monitoring stations with the government monitoring stations so that disturbances in the river and riparian zones can be immediately addressed.
- Increasing the green cover in the areas along riverbanks with species that are native to that area so that the species get sufficient weather conditions to bloom, as they are the foundation of ecosystem in their habitat and thus protect the biodiversity of the region.
- Most of the projects under STP (Sewage Treatment Plant) infrastructure are either under execution or tendering stage but only 20% of the projects are built up. Therefore, proper sewerage system in the cities that lack a proper sewerage system should be designed and connecting them with STPs so that direct discharge of domestic waste can be controlled and pollution levels of the river can be kept in check.
- As most of the degradation hotspots are tourist hub, thus, the tourism activities should be kept in check and clearing of riverbank vegetation for the purpose of developmental activities should not be allowed as such clearance leads to environmental problems such as landslides and floods.
- In addition, tourism along the rivers can be used as a medium to promote awareness regarding the conservation of rivers and associated riparian zones.
- Controlled mass bathing and holy material discarding activity in the rivers should be done when the water quality of the river is not in the permissible limits.

References

Anand J, Gosain AK, Khosa R, Srinivasan R (2018) Regional scale hydrologic modeling for prediction of water balance, analysis of trends in streamflow and variations in streamflow: the case study of the Ganga river basin. J Hydrol Reg Stud 16(March):32–53. https://doi.org/10.1016/j.ejrh.2018.02.007

Arheimer B, Andréasson J, Fogelberg S, Johnsson H, Pers CB, Persson K (2005) Climate change impact on water quality model results from southern Sweden. Ambio 34(7):559–566

Awasthi A, Uniyal SK, Rawat GS, Rajvanshi A (2003) Forest resource availability and its use by the migratory villages of Uttarkashi, Garhwal Himalaya (India). For Ecol Manag 174(1–3): 13–24

Borse K, Agnihotri PG (2017) Study of rainfall trend analysis of upper godavari basin: a case study. Hydro-2017 International April 2018:149–159

Census of India (2011) Government of India. https://censusindia.gov.in/census.website/data/census-tables
Chowdhury AI, Sayed AF, Hossain S (2016) Variation of climatic parameters (rainfall and temperature) over Ganges-Brahmaputra-Meghna river basin in Bangladesh. J Bio Environ Sci 8(6):181–189
CPCB ENVIS|Control of Pollution (n.d.). http://cpcbenvis.nic.in/water_quality_data.html. Retrieved 20 Jun 2018
Das D (2008) Urban quality of life: a case study of Guwahati. Soc Indic Res 88(2):297–310
Dhar ON, Nandargi S (2003) Hydrometeorological aspects of floods in India. Nat Hazards 28(1): 1–33
Estrats MP, Sl P (2006) The three Gorges Dam project in China: history and consequences. HMiC 4:151–188
Gogoi MM, Pathak B, Moorthy KK, Bhuyan PK, Babu SS, Bhuyan K, Kalita G (2011) Multi-year investigations of near surface and columnar aerosols over Dibrugarh, northeastern location of India: heterogeneity in source impacts. Atmos Environ 45(9):1714–1724
Gregory SV, Swanson FJ, McKee WA, Cummins KW (1991) An ecosystem perspective of riparian zones. Bioscience 41(8):540–551
Holmes TP, Bergstrom JC, Huszar E, Kask SB, Orr F (2004) Contingent valuation, net marginal benefits, and the scale of riparian ecosystem restoration. Ecol Econ 49(1):19–30
India-WRIS (n.d.). https://indiawris.gov.in/wris/#/rainfall. Retrieved 31 Mar 2021
Jain SK, Kumar V (2012) Trend analysis of rainfall and temperature data for India. Curr Sci 102(1): 37–49
Karr JR, Chu EW (2000) Sustaining living rivers. Hydrobiologia 422–423:1–14. https://doi.org/10.1023/a:1017097611303
Kaushal N, Babu S, Mishra A, Ghosh N, Tare V, Kumar R, Sinha PK, Verma RU (2019) Towards a healthy Ganga-improving river flows through understanding trade offs. Front Environ Sci 7 (JUN). https://doi.org/10.3389/fenvs.2019.00083
Kotoky P, Sarma B (2017) Assessment of water quality index of the brahmaputra river of Guwahati city of Kamrup district of Assam, India. Int J Eng Res V6(03):536–541
Kulshrestha H, Sharma S (2006) Impact of mass bathing during Ardhkumbh on water quality status of river Ganga. J Environ Biol 27(2 SUPPL):437–440
Lorenz AW, Feld CK (2013) Upstream river morphology and riparian land use overrule local restoration effects on ecological status assessment. Hydrobiologia 704(1):489–501
McIver J, Starr L (2001) Restoration of degraded lands in the interior Columbia river basin: passive vs. active approaches. For Ecol Manag 153(1–3):15–28
Naiman RJ, Décamps H (1997) The ecology of interfaces: riparian zones. Annu Rev Ecol Syst 28(102):621–658
National Mission for Clean Ganga (n.d.). https://nmcg.nic.in/pollution.aspx. Retrieved 19 Jul 2018
Nilsson C, Berggren K (2000) Alterations of riparian ecosystems caused by river regulation. Bioscience 50(9):783–792
Pote SE, Singal SK, Srivastava DK (2012) Assessment of surface water quality of Godavari river at Aurangabad. Asian J Water Environ Pollut 9(1):117–122
Richardson DM, Holmes PM, Esler KJ, Galatowitsch SM, Stromberg JC, Kirkman SP, Pyšek P, Hobbs RJ (2007) Riparian vegetation: degradation, alien plant invasions, and restoration prospects. Divers Distrib 13(1):126–139
Sharma N, Flügel WA (2015) Applied geoinformatics for sustainable integrated land and water resources management (ILWRM) in the Brahmaputra river basin. Springer, New York
Sharma A, Goyal MK (2020) Assessment of the changes in precipitation and temperature in Teesta river basin in Indian Himalayan region under climate change. Atmos Res 231(August 2019): 104670. https://doi.org/10.1016/j.atmosres.2019.104670
Sharma P, Meher PK, Kumar A, Gautam YP, Mishra KP (2014) Changes in water quality index of Ganges river at different locations in Allahabad. Sustain Water Qual Ecol 3:67–76

Sharma S, Srinivasa RG, Bhatt CM, Manjushree P, Bhanumurthy V (2012) Development of flood hazard maps for Assam State, India using historical multi-temporal satellite images. Geospatial applications and decision support systems, June, pp 87–92

Singh M, Singh IB, Müller G (2007) Sediment characteristics and transportation dynamics of the Ganga river. Geomorphology 86(1–2):144–175

Sun F, Roderick ML, Farquhar GD (2018) Rainfall statistics, stationarity, and climate change. Proc Natl Acad Sci U S A 115(10):2305–2310. https://doi.org/10.1073/pnas.1705349115

Tamuly G, Deka RL, Goswami B, Goswami J (2019) Trend of temperature regime in the Brahmaputra valley of Assam during 1986–2015. J Agrometeorol 21(1):106–112

Tockner K, Ward JV (2001) Biodiversity along riparian corridors. SIL Proceedings, 1922–2010 27(7):3981–3981

Tonkin JD, Merritt DM, Olden JD, Reynolds LV, Lytle DA (2018) Flow regime alteration degrades ecological networks in riparian ecosystems. Nat Ecol Evol 2(1):86–93

Welcome to Forest Survey of India (n.d.). https://fsi.nic.in/index.php. Retrieved 20 Jun 2018

Whitehead PG, Wilby RL, Battarbee RW, Kernan M, Wade AJ (2009) A review of the potential impacts of climate change on surface water quality. Hydrol Sci J 54(1):101–123. https://doi.org/10.1623/hysj.54.1.101

Yadav V, Karmakar S, Dikshit AK, Vanjari S (2016) A feasibility study for the locations of waste transfer stations in urban centers: a case study on the city of Nashik, India. J Clean Prod 126: 191–205

Zhang Y, Xia J, Liang T, Shao Q (2010) Impact of water projects on river flow regimes and water quality in Huai river basin. Water Resour Manag 24(5):889–908. https://doi.org/10.1007/s11269-009-9477-3

Drought Frequency Assessment and Implications of Climate Change for Maharashtra, India

Rashmi Singh, Sonal Bindal, Anil Kumar Gupta, and Madhuri Kumari

Abstract Droughts are likely to worsen in the coming decades, affecting millions of people, and ecosystem across India and the world. Agriculture employs 58% of the country's workforce and accounts for 18% of GDP. Around 55% of India's agriculture is dependent on rainfall, India will be the worst affected if drought conditions persist and are not urgently addressed. The aim of this study is to study the drought in Maharashtra, India which accounts for 15% of GDP. In Marathwada, north Maharashtra and western Vidarbha regions, water scarcity affects nearly 60% of the state. Data on soil moisture, precipitation and temperature were used to assess drought. According to the findings, Maharashtra has the highest farmer suicide rate since 2000. The results show 50 million people with 45 million hectares of crop were affected due to drought in 2001–2002. Furthermore, 4291 farmers committed suicide in 2015 due to drought. It was also found that 36.9 million people were impacted that year. In 2018–2019 Maharashtra drought,40 million people were affected in 32 districts. Hence, for assuring that government policies are efficient enough to keep countries' economy safe requires assessing drought vulnerabilities.

Keywords Drought · Vulnerability assessment · Crop failures · Climate change · Farmer suicide

1 Introduction

Drought affects an estimated 55 million people worldwide each year, and they are the most serious threat to livestock and crops in almost every part of the world. Drought threatens people's livelihoods, increases the risk of illness and death, and

R. Singh · M. Kumari
Amity School of Engineering and Technology, Amity University, Noida, Uttar Pradesh, India

S. Bindal (✉) · A. K. Gupta
National Institute of Disaster Management, Ministry of Home Affairs, Govt of India, New Delhi, India

B. Phartiyal et al. (eds.), *Climate Change and Environmental Impacts: Past, Present and Future Perspective*, Society of Earth Scientists Series,
https://doi.org/10.1007/978-3-031-13119-6_22

triggers mass migration. Water scarcity affects 40% of the world's population, and as many as 700 million people are at risk of being displaced by 2030 (WHO, Drought). The risk of drought is further burdens by impacts of climate extremes. The most serious and dreadful environmental problem that the world faces is climate change due to global warming (Shahid et al. 2016).

Climatic extremes are more likely in a warmer climate with increasing climatic variability. As a consequence, natural hazards such as flooding and drought in severity, frequency, length, and spatial scale. In all-natural disasters, in terms of the number of people directly affected, drought ranks first (Udmale et al. 2014). Drought is difficult to comprehend and describe to shifts in hydro-meteorological and socio-economic variables, and the temporal variation in water demand in various regions of the world (Scherer and Diffenbaugh 2014; Wang et al. 2015).

One of the main impacts of global climate change is more frequent and extreme hydrological catastrophes (Favre et al. 2004). Changes in the distribution of rainfall due to global climate change can, therefore, trigger regular droughts. Cook et al. (2018) and Craig et al. (2019) identified that the footprints of human-caused climate change and the warming projections were responsible for drought in the Pacific Northwest, western part of north America, California, and the Mediterranean. Projections show that warming would increase the risk and severity of drought over most of the subtropics and mid-latitudes in both hemispheres over the next century because of regional decreases in precipitation and widespread warming (Cook et al. 2018).

Globally, the drought disaster-affected area will rise with the increasing global temperature, from 15.4 to approximately 44.00% by 2100 (Yu et al. 2017). Climate forecasts are typically used on global as well as local levels to determine drought conditions for the twenty-first century (Strzepek et al. 2010; Orlowsky and Seneviratne 2013; Dai 2013). Sheffield et al. (2012) suggested that the shifts in the climatology of drought in recent times are marginal. Dai (2013) discovered, however, that droughts are ramping up due to global warming. This difference of opinion may be accredited to the inefficiency of climate models in dealing with regional level precipitation (Figs. 1 and 2).

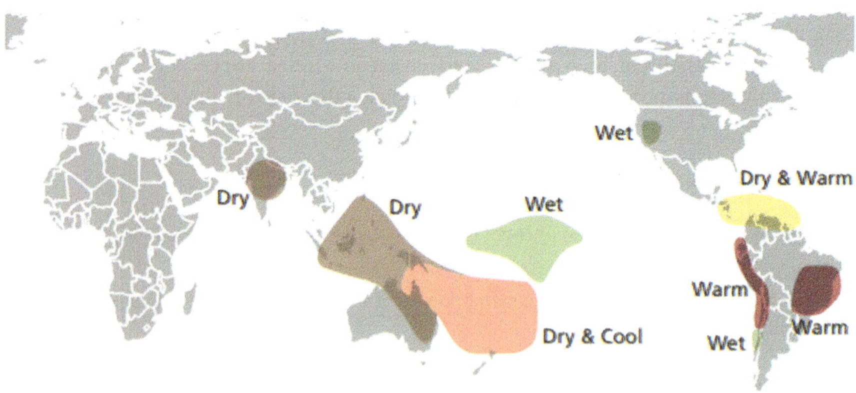

Fig. 1 Global impact of El Nino—April to September

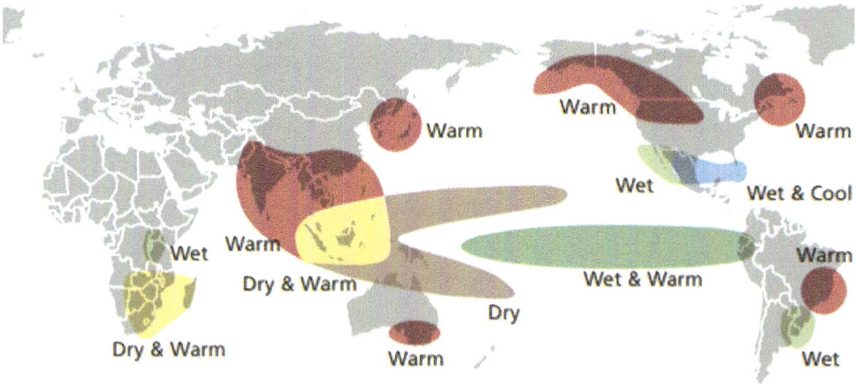

Fig. 2 Global impact of El Nino—October to March. Retrieved from http://www.srh.noaa.gov/jetstream/tropics/enso_impacts.htm

A drought is an event of prolonged shortages in the water supply, whether atmospheric (below-average precipitation), surface water or ground water. Drought is a normal feature of climate and its recurrence is inevitable (Mishra and Desai 2005). While droughts occur naturally, human activity, such as water use and management, can exacerbate dry conditions. What is considered a drought varies from region to region and is based largely on an area's specific weather patterns.

Drought is broadly classified into three types: Meteorological drought (occurs when there is a prolonged time with less than average precipitation), Agricultural drought (when the soil moisture availability to plants has dropped to such a level that it adversely affects the crop yield and hence agricultural profitability) and hydrological drought (occurs when low water supply becomes evident, especially in streams, reservoirs, and groundwater levels) (Li et al. 2009) (Fig. 3). Meteorological drought is described by the India Meteorological Department (IMD) as a situation when rainfall over an area is less than 75% of the normal precipitation (i.e., a rainfall deficiency of 25%) (MOWR, GOI).

There are four major reasons for drought in India- delay in the onset of monsoon, variability of monsoon rainfall, long break in monsoon and aerial difference in the persistence of monsoon. El Niño-related droughts have also been identified as reason for decline in Indian agricultural production. The phenomenon is that instead of the usual high-pressure air mass over the southern Indian Ocean, an ENSO-related low-pressure converging centre emerges which then draws dry air continuously from Central Asia, desiccating India during what should have been the humid summer monsoon season. Droughts in India are caused by this reversed flow of air. Figures 1 and 2 shows El Niño's global effect. It is clearly seen that the warmer and drier (drought) effect is seen in Indian region caused by El Niño.

Maharashtra produces more than a third of India's sugar cane. But droughts have become more frequent and severe in parts of Maharashtra. Since 2000, most of the state has been experiencing drought. 2013 was the worst drought in 40 years. 4000

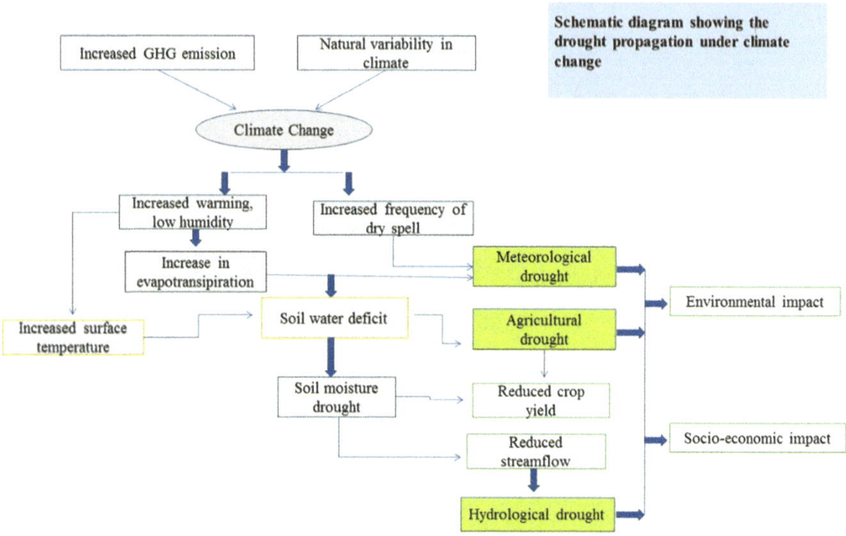

Fig. 3 Schematic diagram showing the drought propagation under climate change

farmers committed suicide after crop failures. In Maharashtra, where 64.14% of the population relies on agriculture for a living, only 16% of the total cultivated area is irrigated, less than half the national average. Rain feeds more than 80% of Maharashtra's agriculture.

1.1 Implications of Drought and Climate Change

Dai (2013) documented that during 1950–2008, the percentage of global dry areas increased by about 1.74% per decade. In Africa, East Asia, and South Asia, the largest rise in dry areas has been noticed. In many of these areas, the growing frequency and severity of droughts has seriously affected agriculture, people's livelihoods, and the national economy. The Intergovernmental Panel on Climate Change (IPCC 2007) observed that, in recent decades, increased water stress due to a combination of rising temperatures and dry spells has caused food grain production to decline in many Asian countries.

During 1901–2018, India's average temperature increased by about 0.7 ° C. This temperature increase is mainly attributable to warming caused by GHGs, counterbalanced by forcing due to anthropogenic aerosols and LULC shifts (Krishnan et al. 2020). By the end of the twenty-first century, average temperature over India is projected to rise by approximately 4.4 °C relative to the recent past (1976–2005) under the RCP8.5 scenario. In the recent 30-year period (1986–2015), temperatures of the warmest day and the coldest night of the year have risen by about 0.63 °C and 0.4 °C, respectively. By the end of the twenty-first century, these temperatures are projected to rise by approximately 4.7 °C and 5.5 °C, respectively, relative to the corresponding temperatures in the recent past (1976–2005), under the RCP8.5 scenario (Krishnan et al. 2020).

The summer monsoon precipitation (June to September) over India has declined by around 6% from 1951 to 2015, with notable decreases over the Indo-Gangetic Plains and the Western Ghats (Krishnan et al. 2020). During the last 6–7 decades, the overall decline in seasonal summer monsoon rainfall has contributed to an increased tendency for droughts in India. During 1951–2016, both the frequency and spatial scale of droughts increased significantly. During this time, areas over Central India, the southwest coast, the southern peninsula, and north-eastern India experienced on average more than 2 droughts per decade. Over the same period, the area affected by drought has also risen by 1.3% per decade (Krishnan et al. 2020). Climate model forecasts suggest a high probability of rising frequency (>2 events per decade), severity and area under drought conditions in India by the end of the twenty-first century under the RCP8.5 scenario, because of increased variability in monsoon precipitation and increased demand for water vapor in a warmer atmosphere.

In the Indian context, very little research on the study of drought scenarios incorporating climate forecasts has been published (Chaturvedi et al. 2012; Fan et al. 2012). The changing climate pattern has led to increased precipitation in some region of the country while, it has led to minimal or no precipitation in other parts. In this paper, the rising drought condition in Maharashtra is discussed along with its adverse impact on various sectors. It further assesses the Drought frequency and implications of climate change for Maharashtra.

2 Materials and Methodology

2.1 Study Area

Maharashtra State is located north of 14°N and south of 22°N in peninsular India. The Arabian Sea borders it on the west. It is India's third-largest state by area and population. It has a population of 112,372,972 and a land area of 307,713 square kilometers. The state's urban population is 45%. The state is divided into four meteorological regions: Konkan, Madhya Maharashtra, Marathwada, and Vidarbha. Maharashtra's districts are divided into six divisions: Aurangabad, Amravati, Konkan, Nagpur, Nashik, and Pune (Fig. 4).

Maharashtra has a tropical monsoon climate. Summer heat is relieved by the arrival of the monsoons in early June. Maharashtra's summers are hot, with temperatures reaching 40 degrees C in most areas. Due to its proximity to the sea, Mumbai's mercury rises Latur, Kolhapur, and Solapur, however, have higher temperatures. Winter temperatures range from 8 to 24 °C. Pune, Nashik, and Nagpur have pleasant cooling temperatures, while Mahabaleshwar, Matheran, and Khandala are typically chilly. The southwest monsoon season brings the most rain (June to September). The state receives 400–6000 mm of rain per year for 3–4 months. The wettest month is

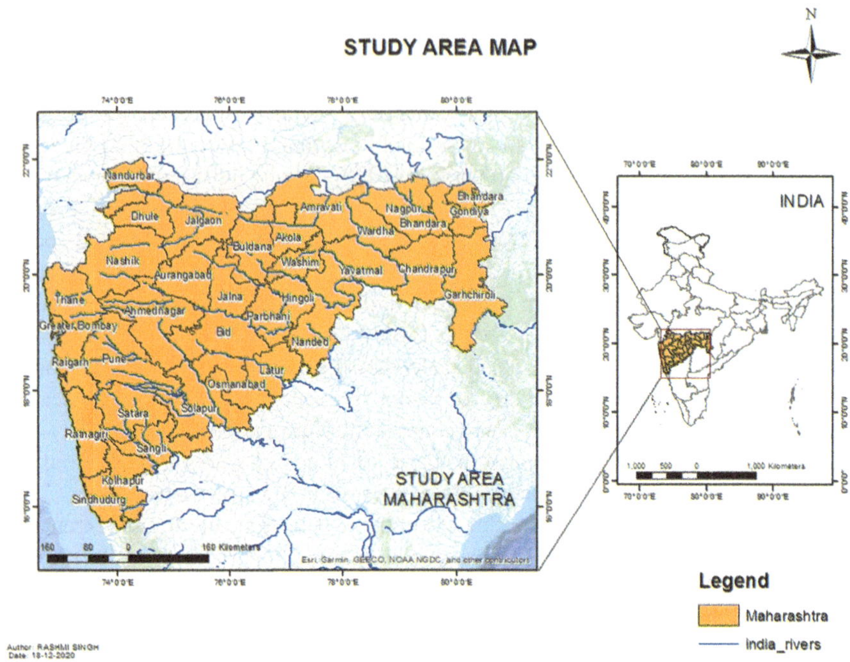

Fig. 4 Study Area Map showing the districts of Maharashtra

July, and the transition to colder weather is October. Winter arrives in November and lasts until February.

3 Methodology

The data was collected for the period of 1995 to 2020 for analyzing the occurrence of drought in India. Secondary data was collected for the affected districts along with impacts such as farmer suicide, crop loss and economic losses. Maharashtra has been considered as study area because of its severe drought frequency and severity which is increasing over the years. A thorough analysis on districts of Maharashtra was conducted as a representative for the entire India.

The secondary data was collected from various governmental agencies, research papers, policy briefs and relief documents prepared by state disaster management authorities and national disaster management authority such as reports of DAC & FW, MoA, Crisis Management plan, national, GOI, and Maharashtra govt., DAC & FW, etc. The occurrence of drought events in Maharashtra was collected, keeping in mind the authenticity and validity of data. The data was further cleaned and was linked with geospatial data, in order to assess the drought vulnerabilities.

4 Results and Discussion

The results of this study indicate that under the 3-time period chosen for data collection i.e., 1995–2005, 2005–2015 and 2015 till date, it is observed that total 32 out of 36 districts were affected due to drought along with the onset of drought from 2000–2001 in Maharashtra. Before that, it was under drought condition in the late 70s and 80s. One of the worst droughts of the century, with an average precipitation shortfall of 19%, was the 1987 drought. 59–60% of the crop area and a population of 285 million were affected globally (PACS, 2001–2008). Post 1995, Maharashtra had drought in year 1997, although it was not that significant, it affected Marathwada marginally. Drought year 2001 impacted 26 districts, thus affecting 454.99 lakh population in the state (Krishnan et al. 2020).

Another consecutive year i.e., 2002 was announced a drought year in Maharashtra. In fact, 2002 drought was one of the worst droughts in India, ranking fifth in terms of severity, had 51% precipitation shortfall. 300 million people were under the claw of this disaster spreading over 18 states of India, shrinking GDP by 3.1% and the estimated agricultural revenue loss was around Rs 38,000 crore (PACS, 2001–2008). 32 districts out of 35 were drought-hit in Maharashtra, in which 185 lakh livestock were influenced. Figure 5 shows the trend of affected districts in Maharashtra post 1995.

The 2012–2013 droughts in Maharashtra came about after the region received lower rainfall during the monsoon season June to September 2012. It is considered as the region's worst drought in 40 years affecting 34 districts. The worst-hit areas included Solapur, Beed, Pune, Nashik, Latur, Ahmednagar and Parbhani. This drought is worse drought than the one in 1972, which was termed as a 'famine' (2013, SANDRP) affected 2.5 crore people. This was also seen as a direct impact of climate change, wherein temperatures rises, soil moisture availability and rainfall declines, indicating a climate change scenario. Remote sensing approaches have

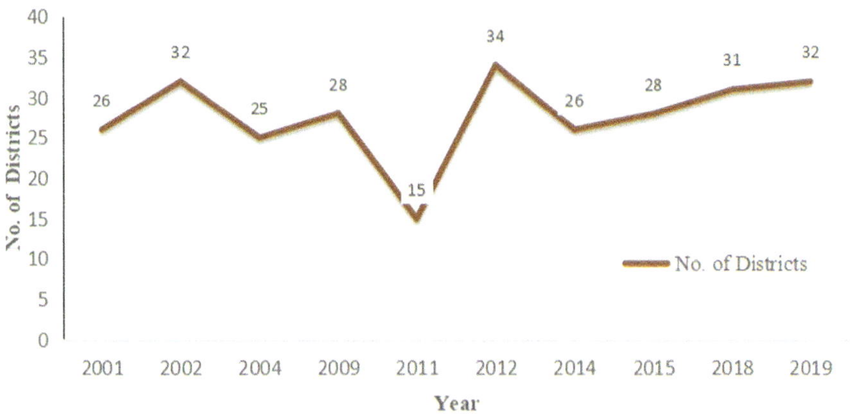

Fig. 5 Number of districts impacted by drought in Maharashtra

aided in assessing drought vulnerability by displaying the effects of precipitation loss, soil moisture, and thus vegetation due to water stress in numerical values (Berg and Sheffield 2018).

Figure 5 clearly depicts the number of districts getting affected due to drought yearly. The trend analysis also highlights the peak number of districts were impacted in the year 2012–2013. SANDRP (2013) called 2012 drought even worse than the 1972 drought for the cause that the effect of former was more intense than later. A total 34 districts were under drought in Maharashtra. Three districts which fall under Marathwada region: Aurangabad, Jalna and Osmanabad were worst impacted with more than 50% rainfall deficit. The recent 2018 drought caused suffering to 40 million people in 31 districts of Maharashtra. Similar condition was seen in 2015 as well. Figure 6 shows the frequency of drought in the districts of Maharashtra in last decade.

This data presentation is an indication of the drought scenario of the entire state. It depicts that the districts in Marathwada region i.e., Latur and Osmanabad were affected by drought every time the state was announced drought-hit. Even Beed, Aurangabad, Jalna, Nanded, Parbhani and Hingoli also were declared drought except in the year 2011. The Konkan region (Thane, Sindhudurg, Ratnagiri, Raigad,

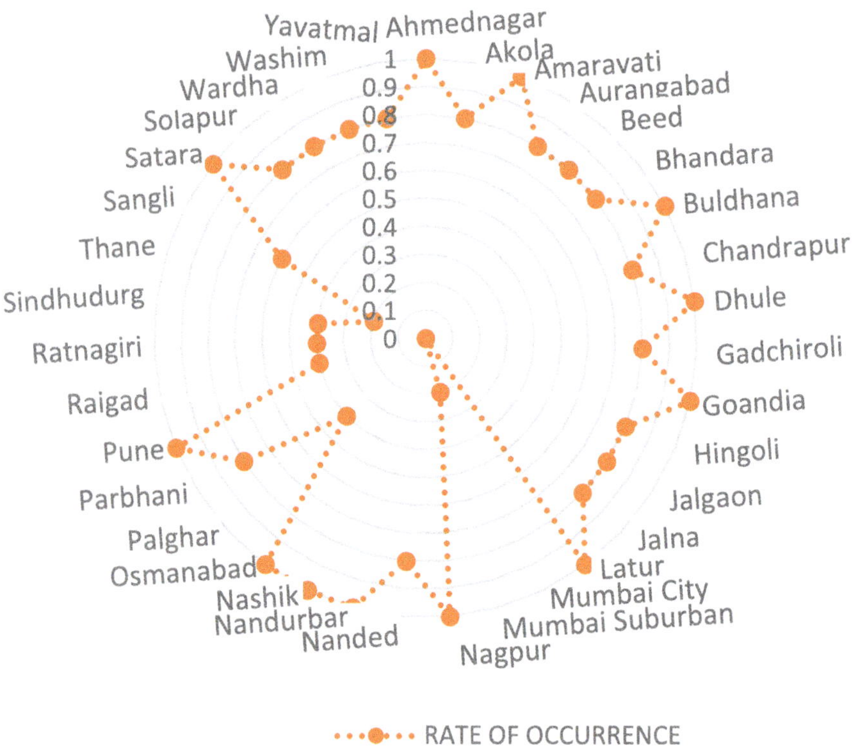

Fig. 6 Radar assessment for districts impacted by drought in Maharashtra

Palghar and Mumbai city) remained least affected by drought. Ahmednagar, Dhule, Buldhana, Satara were also constantly under drought condition. This figure thus shows that almost all the divisions in Maharashtra face drought situation frequently except for Konkan region which remains safe.

Marathwada is one of the meteorological subdivisions of Maharashtra which has seen worst drought from 2012 onwards. Although, it had no drought from 1999 to 2011, but the past decade left Marathwada dry. It has 8 districts, Latur, Aurangabad, Beed, Jalna, Parbhani, Hingoli, Osmanabad and Nanded. In 2019, Beed received 87.5 mm (32% deficit) rainfall in June, normal rainfall being 128.4 mm in this district. It was even worse in the month of July (77 mm rainfall i.e., 40% deficit, normal rainfall being 127 mm). Parbhani, having normal rainfall of 145.3 mm, received only 86 mm in June (41% deficit). The rainfall deficit in year 2015 was 90.6% for Jalna, 90.3% for Beed, 86.36% for Osmanabad, 89.3% for Parbhani, 84.75% for Latur, 83.9% for Aurangabad and 78.84% for Nanded, in the month of July (Rainfall data 2021). Similar trend also existed in the past which made Marathwada a dry region and constant hit by drought.

Another serious effect of drought is the increasing rate on farmer suicide in India. India recorded total 308,826 suicides from 1995 to 2014. Maharashtra is worst affected in this context. Compared to other drought-hit states in India, Maharashtra's farmer suicide count is maximum as per NCRB data. Figure 7 shows the farmer

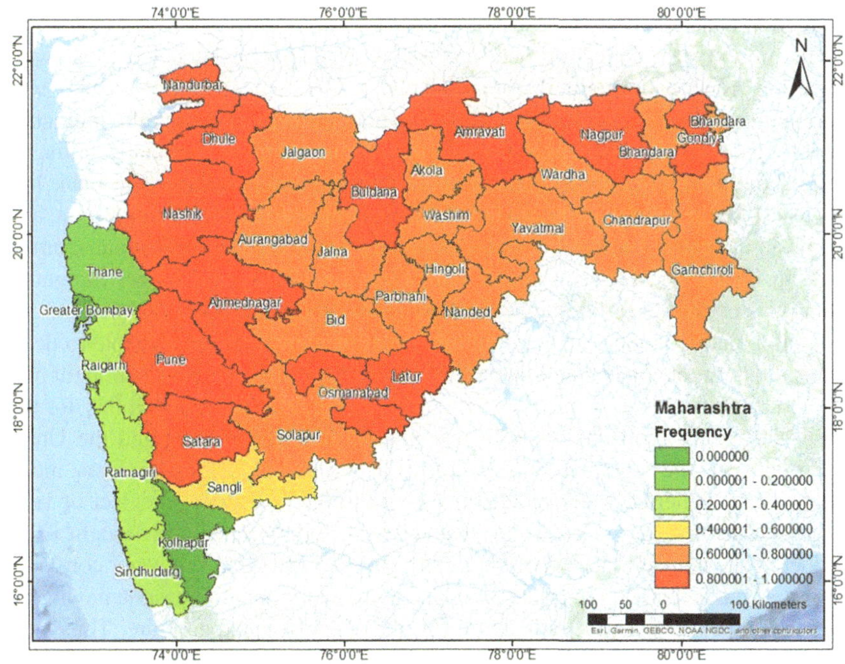

Fig. 7 Depicting the vulnerabilities od districts of Maharashtra for drought

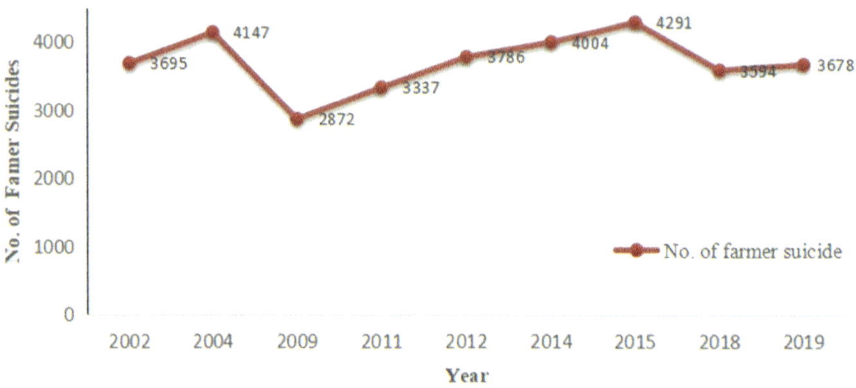

Fig. 8 Framer suicide count in Maharashtra

suicide count of four most affected states. It clearly shows that among the most affected states, Maharashtra has the maximum farmer suicide cases recorded by NCRB (Fig. 8).

Farmer suicide has also seen rising trend in Marathwada with increase in drought events and water shortage. As indicated in Fig. 8, Marathwada alone recorded 4700 farmer suicide cases in 2015–2019. 1109 farmers ended their lives in 2015. In Beed, 1681 farmer suicide was recorded till November in year 2019 and 1132 cases were recorded in Nanded (NCRB).

The present data collection and analysis showed that Maharashtra is showing increasing trend in drought disaster and the related outcome. For a state with maximum number of dams, water resource management and drought mitigation should not be a great challenge. Despite that, Maharashtra is constantly failing in tackling the drought condition and is showing poor water management plan. It is thus a need of the hour to carry out thorough drought assessment for this region, as it will be an indicative of all other states in India. Proper water resource management can help fill the gap between people and water availability and proper mitigation planning is needed to provide immediate relief to drought affected population.

From a meteorological and agricultural point of view, it is not possible to deem this year's drought worse than that of 1972. Hydrologically, this year's drought may prove to be worse for some districts than in 1972 (Fig. 9). The responsibility for this lies in the wrong decisions taken by the Maharashtra government and the Union Government on the construction of unprofitable and undesirable major dams, incorrect cropping patterns, water diversion for non-priority purposes, neglect of local water systems and unchecked water management. India's manmade drought is the product of the non-seriousness of the Indian government regarding water conservation. Question is that why the state is so heavily reliant on rainfall despite having the highest number of dams in India and the highest irrigation expenditure. This is the biggest policy failure. Why is irrigation mostly in sugarcane growing areas, leaving the rest of the crops at the mercy of the rain (Fig. 9).

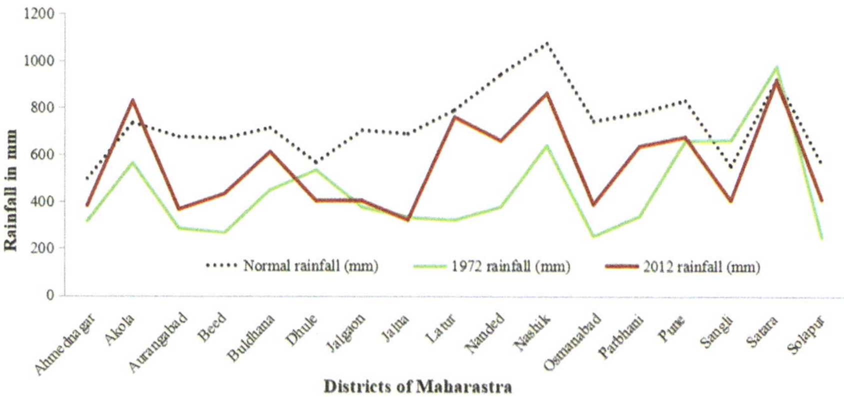

Fig. 9 Shift in rainfall patterns from the year 1972 to 2012 for the drought affected districts of Maharashtra

5 Conclusion and Recommendations

It is clear from the discussion and results that a the state of Maharastra state has varying monsoon, and the distribution of precipitation results into different climatic conditions in different districts. Thus, for a proper and more realistic drought monitoring, the study must be done from district level. Such macro level study, although, has not been conducted widely and it calls for this as a need of the hour. Drought is hitting all the possible aspect of human and environment and its now a mandatory call for including the regulations related to drought mitigation as a serious and indispensable policies (Manual for Drought Management 2009). The implementation of these policies must be checked regularly to assure proper functioning of the systems involved in the drought mitigation and management processes (Drought Management Plan 2017). Further, these policies must be implemented for increasing the capacity for local adaptation as a long term relief measures.

Detrimental effects of drought have long been observed in all regions, a condition arising from both climate change and human activity, and humanity's inability to cope with this crisis has been seen so far. In recent years, the arid and semi-arid regions have experienced several drought and below average record of precipitation, resulting in irreversible harm, particularly in the agricultural sector. It is now, a need of the hour to bring a check to this accelerating drought years. Drought monitoring, using modern indices to forecast the beginning and end of the drought season, severity and other characteristics, is therefore considered important in order to establish the necessary pre-occurrence drought management steps.

References

Berg A, Sheffield J (2018) Climate change and drought: the soil moisture perspective. Curr Clim Change Rep 4:180–191. https://doi.org/10.1007/s40641-018-0095-0

Chaturvedi RK, Joshi J, Jayaraman M, Bala G, Ravindranath NH (2012) Multi-model climate change projections for India under representative concentration pathways. Curr Sci 103(7): 791–802

Cook BI, Mankin JS, Anchukaitis KJ (2018) Climate Change and Drought: From Past to Future. Curr Clim Change Rep 4:164–179. https://doi.org/10.1007/s40641-018-0093-2

Craig CA, Feng S, Gilbertz S (2019) Water crisis, drought, and climate change in the southeast United States. Land Use Policy 88(104110) ISSN 0264-8377. https://doi.org/10.1016/j.landusepol.2019.104110

Dai (2013) Increasing drought under global warming in observations and models. Nat Clim Chang 3(1):52–58

Drought Management Plan (2017) Government of India Ministry of Agriculture And Farmers Welfare, DAC & FW. Retrieved from http://agricoop.nic.in/sites/default/files/Drought%20Management%20Plan%20.pdf

Fan K, Liu Y, Chen H (2012) Improving the prediction of the East Asian summer monsoon: new approaches. Weather Forecast 27(4):1017–1030

Favre AC, El Adlouni S, Perreault L, Thiémonge N, Bobée B (2004) Multivariate hydrological frequency analysis using copulas. Water Resour Res 40(1):W01101

Health Topic: Drought. World Health Organization. Retrieved from https://www.who.int/health-topics/drought#tab=tab_1

IPCC (2007) In: Core Writing Team, Pachauri RK, Reisinger A (eds) Climate change 2007: Synthesis report. Contribution of working Groups I, II and III to the fourth assessment report of the Intergovernmental Panel on Climate Change. IPCC, Geneva, p 104

Krishnan R, Sanjay J, Gnanaseelan C, Mujumdar M, Kulkarni A, Chakraborty S (eds) (2020) Assessment of climate change over the Indian Region. A report of the ministry of earth sciences (MoES), Government of India. Springer, Singapore, pp 226. https://doi.org/10.1007/978-981-15-4327-2

Li Y, Ye W, Wang M, Yan X (2009) Climate change and drought: a risk assessment of crop-yield impacts. Clim Res 39:31–46. https://doi.org/10.3354/cr00797

Manual for Drought Management (2009) Department of Agriculture and Cooperation, Ministry of Agriculture, Government of India, New Delhi. Retrieved from https://nidm.gov.in/PDF/manuals/Drought_Manual.pdf

Mishra AK, Desai VR (2005) Drought forecasting using stochastic models. Stoch Env Res Risk A 19(5):326–339

Orlowsky B, Seneviratne SI (2013) Elusive drought: uncertainty in observed trends and short-and long-term CMIP5 projections. Hydrol Earth Syst Sci 17(5):1765–1781

Rainfall data for Maharashtra. Department of Agriculture Maharashtra State. Accessed on 25 July 2021. Retrieved from: https://maharain.maharashtra.gov.in/

Scherer M, Diffenbaugh NS (2014) Transient twenty-first century changes in daily-scale temperature extremes in the United States. Clim Dyn 42(5–6):1383–1404

Shahid S, Wang X-J, Harun SB, Shamsudin SB, Ismail T, Minhans A (2016) Climate variability and changes in the major cities of Bangladesh: observations, possible impacts and adaptation. Reg Environ Chang 16(2):459–471

Sheffield J, Wood EF, Roderick ML (2012) Little change in global drought over the past 60 years. Nature 491(7424):435–438

Strzepek K, Yohe G, Neumann J, Boehlert B (2010) Characterizing changes in drought risk for the United States from climate change. Environ Res Lett 5(4):044012. Article 044012

Udmale P, Ichikawa Y, Manandhar S, Ishidaira H, Kiem AS (2014) Farmers' perception of drought impacts, local adaptation and administrative mitigation measures in Maharashtra State, India. Int J Disaster Risk Reduct 10:250–269

Wang X-J, Zhang J-Y, Yang Z-F, Shahid S, He R-M, Xia X-H et al (2015) Historic water consumptions and future management strategies for Haihe River basin of Northern China. Mitig Adapt Strateg Glob Chang 20(3):371–387

Yu Z, Wang J, Liu S, Rentch JS, Sun P, Lu C (2017) Global gross primary productivity and water use efficiency changes under drought stress. Environ Res Lett 12(1):014016

Multi-temporal Impact Analysis of Covid-19 Lockdown and Unlock Measures on Major Air Pollutants in Guwahati City, India

Ritwik Nigam, Kanvi Pandya, Alvarinho Luis, and Mahender Kotha

Abstract The novel coronavirus which is also known as (SARS-CoV-2) was first detected in Wuhan, China in late December 2019. It is a highly contagious and mutable disease that has become a pandemic. The majority of nations had opted for partial to total lockdown restrictions as a measure to contain the spread of COVID-19. India announced a 68-day nationwide lockdown on March 25, 2020 which was subsequently extended and succeeded by different phases (with relaxation in restriction rules) till May 31, 2020. The lockdown has dramatically impacted the economy of the country. However, it has come as the breathing space for the environment as its anthropogenic exploitation came to standstill. Drastic reduction in pollution level and increment in environment-friendly variables have been observed around the world. The present study assessed the trends of particulate matter (PM_{10}), sulphur dioxide (SO_2), and the Oxides of Nitrogen (NO_x) during the pre-lockdown, lockdown, and post-lockdown periods and similar period of 2019 for Guwahati city situated in northeast India. The meta-analysis of continuous data was performed using descriptive statistics on air pollutant data was acquired from the Central Pollution Control Board and Assam Pollution Control Board. The result revealed that as much as 63.02%, 80.28%, and 84.25% reduction in the mean concentration of PM_{10}, SO_2, and NO_x pollutants, respectively, and the fluctuation from the mean values were also drastically reduced by 93% for PM_{10}, 96.11% for SO_2 and 89.47% for NO_x in comparison to a similar period of 2019. The significance of the study is to show how the different sets of restrictions (in phases) imposed upon a city can significantly reduce pollutant concentrations. The study suggests that after

R. Nigam (✉) · M. Kotha
Earth Science, School of Earth, Ocean and Atmospheric Sciences (SEOAS), Goa University, Taleigao Plateau, Goa, India
e-mail: earthscience.ritwik@unigoa.ac.in

K. Pandya
Department of Marine Geology, Mangalore University, Mangalagangotri, Karnataka, India

A. Luis
Earth System Science Organization- National Centre of Polar and Ocean Research, Ministry of Earth Science, Government of India, Headland Sada, Goa, India

B. Phartiyal et al. (eds.), *Climate Change and Environmental Impacts: Past, Present and Future Perspective*, Society of Earth Scientists Series,
https://doi.org/10.1007/978-3-031-13119-6_23

modifications (to minimize the economic damages) the lockdown can also be one of the important measures to arrest high pollution levels in the urban and industrialized areas.

Keywords Air pollution · COVID-19 · Guwahati · Lockdown · Particulate matter

1 Introduction

The novel coronavirus, which was first detected in Wuhan, China, in late December 2019, has been commonly known as COVID-19 or SARS Cov-2. COVID-19 is a highly infectious respiratory disease. Due to its highly contagious nature, the COVID-19 had been reported in more than 210 countries around the world. Consequently, on March 11, 2020, the World Health Organization (WHO) declared it a global pandemic (WHO 2020). Humans are not naturally immune to this virus, due to which the COVID-19 is causing high mortalities. The limited availability of vaccines to control the infection and high mortality rate has consequently led to a panic across the world. Until December 4, 2020, 64,350,473 confirmed cases had been reported worldwide, out of which 1,494,668 have died due to COVID-19 (WHO 2020). As of December 2020 (at the time of editing this manuscript), an effective vaccine to build immunity against the COVID-19 was being tested.

Due to the contagious nature of COVID-19, the majority of nations were forced to introduce different measures to slow down the infection. The WHO had recommended city curfews (lockdown as a measure to build up the health infrastructure to be able to fight the COVID-19. China was the first to introduce a month-long city lockdown and social distancing and quarantine the susceptible populations. Thereafter, Europe became the epicenter of the COVID-19 outbreak. Similar measures, along with the partial to complete lockdown, were imposed across the world through Brazil and Mexico were the exceptions to implement those measures soon, as they were skeptical about the danger posed by COVID-19. However, by the last week of March 2020, the governments of these two Latin American countries also decided to go for a lockdown (Londono et al. 2020).

On the other hand, India recorded its first COVID-19 positive case on January 30, 2020. As the counting soured to 500 confirmed cases, India, without any delay, announced the complete lockdown for three weeks starting from March 25, 2020. The total shutdown implementation restricted the movement of 1.3 billion populations, making it the world's largest lockdown ever imposed (Subramani and Roman 2020). Various restrictions were put into effect, such as; total restriction on domestic and international air travel and a complete restriction on public transport (trains and buses). All the shops and markets, except pharmacies and essential-commodity shops, remained shut. The lockdown was subsequently extended in different phases, and after May 17 various services partially reopened (News World-24, The Economic Times, May 1, 2020). The lockdown consisted of four phases, lasting for 68 days from March 25, 2020, to May 31, 2020. The restrictions were subsequently lifted, as the lockdown progressed. Currently, India is only second to the USA (13,759,500 with 9,571,559 confirmed COVID-19 cases, along with 139,188 mortalities (WHO 2020).

The lockdown was placed to control the rapid infection rate and develop adequate health infrastructure to cope with the current pandemic. Contrary to the aim of imposing this severe measure, the lockdown restricted the majority of economic activities, which had led to numerous regional repercussions around the world (Berman and Ebisu 2020). In addition to the health crisis, the pandemic led to poverty and socio-economic inequality, warned the UN (UNDP 2020). Even the International Monitory Fund regarded the worldwide lockdown as the worst economic slump since the great depression. The global economy will decline to a 3% growth rate as the pandemic will severely impact the global economy. (UNDP 2020) was evident when the United Kingdom economy was reported to shrink by 20.4% during March 2020 (Partington, June 2020).

The International Labour Organization had predicted in April 2020 that 30 million jobs were lost till March 2020 across the world due to lockdown 10 (Financial Times, April 2020). In India, the unemployment rate was already 6.7% in March 2020 because the GDP rate fell from 8.2% during Q1 of 2018 to 3.1% in Q1 of 2020 (Buccholz, Sept 2020). Later, due to nationwide lockdown unemployment rate increased unprecedentedly to 23% on April 19, 2020. According to the Centre for Monitoring Indian Economic estimates, 140 million people lost their jobs (Vyas 2020). Besides, cumulatively, the Indian economy had to bear the monitory loss of USD 4.5 billion each day during the first phase of lockdown. The lockdown implementation left millions of laborers stranded, mainly in the metro cities due to the shutdown of public transportation (Subramani and Roman 2020). Implementation of Unlock phases from June 2020 led to partial recovery with the national unemployment rate reversed to March 2020 level (Sharma 2020).

The lockdown period led to the colossal damage to the world's economy. On the positive note, a novel term has been coined, the 'Anthropause' for the Covid-19 induced lockdown, because it provided a substantial break in anthropogenic activities around the world, especially traveling (Rutz et al. 2020), which led to the break on the anthropogenic exploitation of the natural environment. Significant positive impacts of lockdown were evident with the substantial improvement in the air quality across the world. Drastic reduction in greenhouse gas levels was observed in Europe and major Chinese cities (Monserrate et al. 2020; European Environmental Agency 2020). A major slump in air pollution was observed in cities located in a temperate climate, and industrialized cities in Europe (He et al. 2020). Barcelona had observed a steep reduction in NO_2 and $PM_{2.5}$ of 31% and 51%, respectively (Tobías et al. 2020). Total lockdown restriction during phase-1 and phase-2 led to a reduction in 88 Indian cities (Sharma et al. 2020). Gujarat state-observed massive reduction in air pollution during lockdown (Selvam et al. 2020), where the highly polluted industrial cities of Vapi and Ankleshwar reported a steep decline in SO_2 and NO_2 of as much as 80% and 91%, respectively, in comparison to 2019 days (Nigam et al. 2020).

The present study was undertaken to determine the multi-temporal trends of SO_2, NO_x, and PM_{10} during the 37 days of Pre-lockdown (Feb 16, 2020 to March

24, 2020), 68 days long lockdown period (March 25, 2020, to May 31, 2020) and 90 days of unlocking period (June 1 to August 31, 2020) when the restriction of movement of people and services were lifted in a phased manner to get the economy back to the business as usual. We also draw a comparison with a similar period of 2019. The need to determine air pollutants' trends in Guwahati city was evident because it is the highest polluted city located in Northeast India. Moreover, it is regarded as one of the highest black carbon-emitting cities in the world, which is even responsible for increasing the mean temperature of Guwahati city by 2 °C (Duarah 2012). The lockdown phase allowed determining pollution trends in the absence of anthropogenic activities, which are the major pollution source in the study area.

1.1 Study Area

Guwahati, 26°10′20″N 91°44′45″E, is the largest city located in northeast India (Fig. 1) with a total area of 262 sq. km. The topographic elevation vary from 50 to 680 m above Mean Sea Level from the Brahmaputra river's southern bank to the foothills of the Shillong plateau (Assam Online Portal 2015). Guwahati experiences a humid subtropical climate (Koppen climate classification 'Aw') (Peel et al. 2007). The air temperature ranges from 32.2 °C in June to 23.6 °C in January. It receives substantial precipitation during the southwest monsoon period (May–September), with the mean annual precipitation around 1751.8 mm (IMD 2020). Demography wise, it is one of the fastest developing city in India, with a population of 957,352 in 2011, which was projected to increase to 1,116,267 in 2020. It is known as the gateway to North-East India (Bhushan 2005) due to its strategic location. In recent times, due to rapid industrialization, it has become the most important economic hub of North East India.

The city's urban morphology was traditionally confined to the Brahmaputra river's southern bank, which has now been extended in the form of corridors radiating towards the east, west, and south. The eastern corridor is supporting rapid residential development. At the same time, the west hosts 'Gauhati University'. The Gopinath Bordoloi International airport and the southern corridor, Guwahati-Shillong road (GS road), is the busiest of all, which houses major commercial centers and dense residential areas (Assam Online Portal 2015). The GS road also connects Assam, Dispur, with Guwahati's capital city.

1.2 Methodology

The Government of India had launched the National Air Quality Index (NAQI) in 2016 to monitor and improve the declining air quality across the country. The Central Pollution Control Board (CPCB) carries out 24-hour monitoring of air

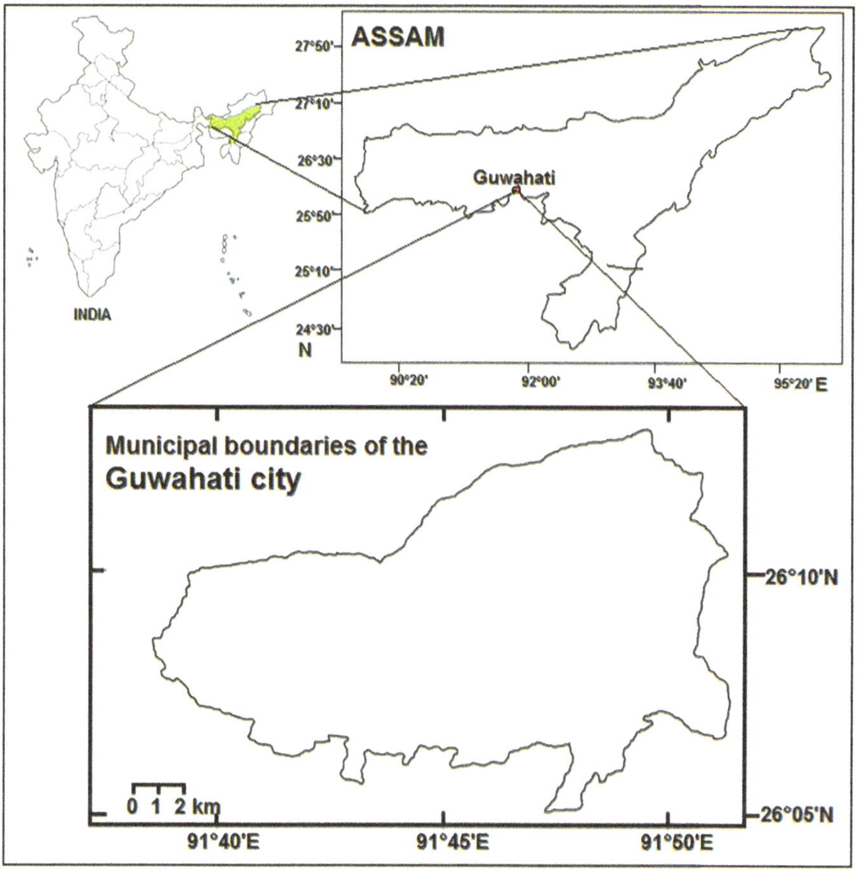

Fig. 1 Location map of study area

pollution at each monitoring station and publishes a daily NAQI bulletin. The AQI includes a 24-hour concentration of eight different major pollutants (Table 1). The objective of NAQI is to provide a numeric value, along with qualitative and color-coding categories of each pollutant, which could be understood even by the layman. In the daily NAQI bulletin, the highest single pollutant value, most commonly, the particulate matter value, is considered the AQI value.

This study has adopted the methodology given by Nigam et al. (2020). The total duration of 394 days under the scope of the present study was further divided into eight different phases (197 days of each 2020 and 2019; see Table 1). The division was based upon the phases implemented during the lockdown and unlocking period by the Government of India to contain the spread of COVID-19. The 24-hour mean concentration values ($\mu g/m^3$) of the three major air pollutants (PM_{10}, NO_x, and SO_2) for the duration of 394 days (Feb 16, 2020, to Aug 31, 2020, and a similar period of 2019) were acquired from the CPCB and Assam Pollution Control Board (APCB)

Table 1 Different phases of Covid-19 induced lockdown and unlocking in India, along with the activities allowed during each phase

S No.	Phase	Date	Activities allowed
1	Pre-lockdown	Feb 16–March 24, 2020	Business as usual.
2	Lockdown Phase-1	March 25–April 14, 2020 (21 Days)	Total shutdown except for essential services.
3	Lockdown Phase-2	April 15–May 3, 2020 (19 Days)	Similar to Lockdown phase-1.
4	Lockdown Phase-3	May 4–17, 2020 (14 Days)	Cities are divided into Red, Orange, and Green zones with the reopening of some activities in the green zone.
5	Lockdown Phase-4	May 18–31, 2020 (14 Days)	Lockdown extended in metropolitan cities. Red zone further divided into Containment zone and Buffer zone. Relaxation in the green and orange zone.
6	Unlock 1.0	June 1–June 30, 2020 (30 Days)	Reopening in some states in a phased manner having low Covid-19 caseload.
7	Unlock 2.0	July 1–31, 2020 (31 Days)	Lockdown only in the containment zones. Night curfew 10 PM-5 AM everywhere.
8	Unlock 3.0	August 1–31, 2020 (31 Days)	Night curfew uplifted, lockdown observed only in Maharashtra and Tamil Nadu.

AQI database. We also used 18-year mean rainfall data of Guwahati city from the Indian Meteorological Department.

Descriptive statistics (Mean and Standard deviation) and inferential (linear correlation) statistical techniques were used to compute the multi-temporal trend values.

The mean was computed to find out the average of the air pollutants during each phase,

$$Mean = \frac{Xi}{N} \tag{1}$$

where X_i is the sum of the daily pollutant value in the phase and N is the total number of days in the phase.

The standard deviation was computed to account for the range of fluctuation from the mean during each phase,

$$\text{Standard deviation} = \sqrt{\frac{\sum (Xi - \mu)^2}{N}}, \tag{2}$$

where X_i is the daily pollutant value in the phase, μ is the mean pollutant value of the phase and N is the total number of days in the phase.

Moreover, the difference between 2020 and 2019 pollutant values were computed using:

$$\text{Mean percentage difference} = \frac{(\textit{New Value} - \textit{Old Value})}{\textit{Old Value}}\ 100, \tag{3}$$

where, *New value* is the mean phase value for 2020 and *Old Value* is the mean phase value for 2019. We also constructed the correlation matrix plot to determine the correlation between the pollutants value and the mean daily rainfall for 2019 and 2020.

2 Results

The phase-wise distribution of the air pollutants for 2019 and 2020 is shown in Fig. 2. The standard deviation observed in pollutant levels during each phase in 2019 and 2020 is presented in Table 2. The mean difference in (%) and standard deviation (SD) difference in (%) using 2020 and 2019 in the same phase for Guwahati is shown in Table 3.

The PM_{10} during the pre-lockdown (169.22 μg/m^3) period was nearly similar to the 2019 similar period (170.28 μg/m^3) (Fig. 2). During phase-1, the mean concentration was higher (168.7 μg/m^3) than the previous year (112.32). Similarly, for phase-2 PM_{10} was (40.55 μg/m^3) compared to (19.34 μg/m^3) during a similar period in 2019. Since phase-3, the mean concentration of PM_{10} (45.69 μg/m^3) was significantly lower than that in 2019 (123.57 μg/m^3). During phase-4 it came down by

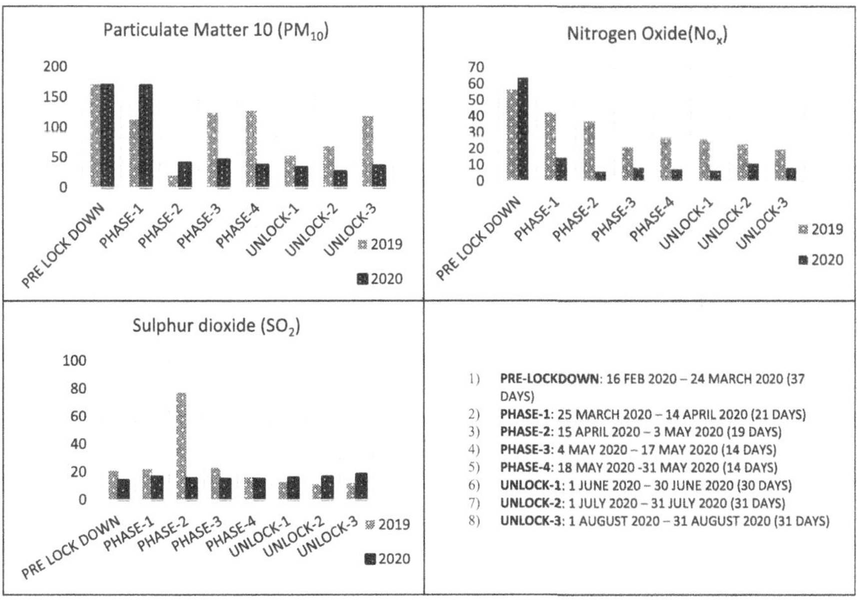

Fig. 2 Mean concentration of pollutants during each phase in 2019 and 2020

Table 2 Breakpoints for AQI scale 0–500 (unit: µg/m^3) (due to unavailability of data for all the 6 air pollutants, only 3 pollutants are considered in the present study). Source: Central Pollution Control Board

AQI Category	AQI	PM10	PM2.5	NO_2	O3	CO	SO_2
Good	0–50	0–50	0–30	0–40	0–50	0–1.0	0–40
Satisfactory	51–100	51–100	31–60	41–80	51–100	1.1–2.0	41–80
Moderately polluted	101–200	101–250	61–90	81–180	101–168	2.1–10	81–380
Poor	201–300	251–350	91–120	181–280	169–208	10–17	381–800
Very poor	301–400	351–430	121–250	281–400	209–748*	17–34	801–1600
Severe	401–500	430 +	250+	400+	748+*	34+	1600+

Table 3 Standard deviation observed in pollutant levels during each phase in 2019 and 2020

	PM_{10} (µg/m^3)		NO_x (µg/m^3)		SO_2 (µg/m^3)	
Phase	2019	2020	2019	2020	2019	2020
Pre lock down	64.36	57.58	33.56	37.6	3.89	0.488
Phase-1	63.25	42.57	25.44	8.63	4.75	1.65
Phase-2	44.08	20.65	28.04	2.95	3.33	1.84
Phase-3	253.07	17.69	10.73	4.16	8.75	0.34
Phase-4	191.49	21.5	12.86	4.998	0.29	0.24
Unlock-1	23.2	13.19	11.6	3.29	2.7	0.23
Unlock-2	66.26	9.54	9.97	15.86	1.83	0.86
Unlock-3	180.23	10.15	6.7	4.18	0.34	0.44

more than 250% to 36.92 µg/m^3, when compared to that in 2019 (126.72 µg/m^3). The unlock-1 phase coincided with the onset of the southwest rainfall, so the concentration decreased by 57% from 52.29 to 33.31 µg/m^3. Unlock-2 recorded the lowest PM_{10} concentration among all phases (26.21 µg/m^3) and (67.83 µg/m^3) in 2019. During the last phase, it increased to 35.23 µg/m^3, when compared to 117.51 µg/m^3 in 2019.

NO_2 was higher (63.54 µg/m^3) during 2020 pre-lockdown when compared to the same period in 2019 (56.17 µg/m^3). NO_2 was significantly lower during 2020 (14.25, 5.82, 8.0, 7.28, 6.3, 10.56 and 7.99 µg/m^3) from phase-1 to unlock-3, respectively, in comparison to (41.98, 36.97, 20.99, 27.01, 25.38, 22.91 and 19.58 µg/m^3) for the same period in 2019.

SO_2 concentration was 13.77 µg/m^3 during 2020 pre-lockdown when compared to the same period in 2019 (20.7 µg/m^3). A similar trend for SO_2 during 2020 was observed during all remaining phases (16.47, 15.15, 14.7, 14.86, 15, 71, 16.30, and 18.12 µg/m^3), respectively. It is noted that during 2019 the concentration was higher at 21.73, 76.85, 22.83, 16.01, 12.51, 10.94, and 11.65 µg/m^3, respectively.

A scatter plot matrix (Fig. 3), a grid (or matrix) of scatter plots, is a graphical equivalent of the correlation matrix, which is used to assess air pollutant data and to visualize the bivariate relationships between combinations of variables of pre-COVID-19 (2019) and COVID-19 (2020) for Guwahati city. Each scatter plot in the matrix visualizes the relationship between a pair of variables, allowing many relationships between several pairs of all air pollutant variables to be explored at once. Figure 3 clearly shows the variable distribution of the pollutant variables from pre-COVID-19 (2019) to the COVID-19 lockdown period (2020) in all combinations of scatter plots. From the matrix plot, a positive correlation between PM_{10} and NO_2, SO_2 is found. No correlation is observed between the air pollutants and rainfall (Table 4).

The percentage difference in PM_{10} concentration was −0.62% during the pre-lockdown. It was 50.19 and 109.6 during phase-1 and phase-2. Since the phase-3, the mean concentration dropped significantly by −63.02%, phase-4 (−70.01%). During the unlock phase, the difference narrowed down to (−36.29%), but later it again increased (−61.35%) during unlock-2 and (−70.01%) unlock-3. Rainfall played a prominent role in leading to the higher difference between 2019 and 2020, along with decreased concentrations during lockdown periods. While the difference in SD continuously increased from pre-lockdown to phase-3 (−10.53% to −93%) and then decreased in phase-4 and phase-5 by −88.77% and −43.14%, respectively.

NO_x was found to be the most reduced pollutant among the three species in 2010 when compared to that in the 2019 period. Although the concentration was higher in 2020 (12.03%), however, since phase-1 it showed continuous reduction till unlock-3. The NO_x difference during pre-lockdown decreased to (−66.07% in phase-1) and (−89.47% in phase-2). It was −61.60% in phase-3 and showed an increasing tendency in phase-4 (−61.13%) and unlock-1 (−71.63%), before it decreased to −53.90% and −59.19% during the unlock-2 and unlock-3, respectively.

SO_2 is mainly produced by vehicular emission, evident from the observed trends in % difference between 2019 and 2020. The highest difference was observed in phase-1 (−80.28%), while in phase-1 (−33.47%) and phase-3 (−61.23%). With the increase in vehicular traffic during phase-4, the difference was reduced to −7.19% subsequently, which increased during the three unlock periods (25.57%, 49.00%, 55.53%). The SD difference of SO_2 during pre-lockdown was −87.45% and with the advent of lockdown, it decreased by −65.26% and −44.74% in phase-2. The maximum difference was observed in phase-3 (−96.11%), which rapidly reduced by −17.24% and then abruptly increased by −91.48%; this anomaly needs to be understood. Later in unlock-2, it decreased by −53% before increasing by 29.41% (Table 5).

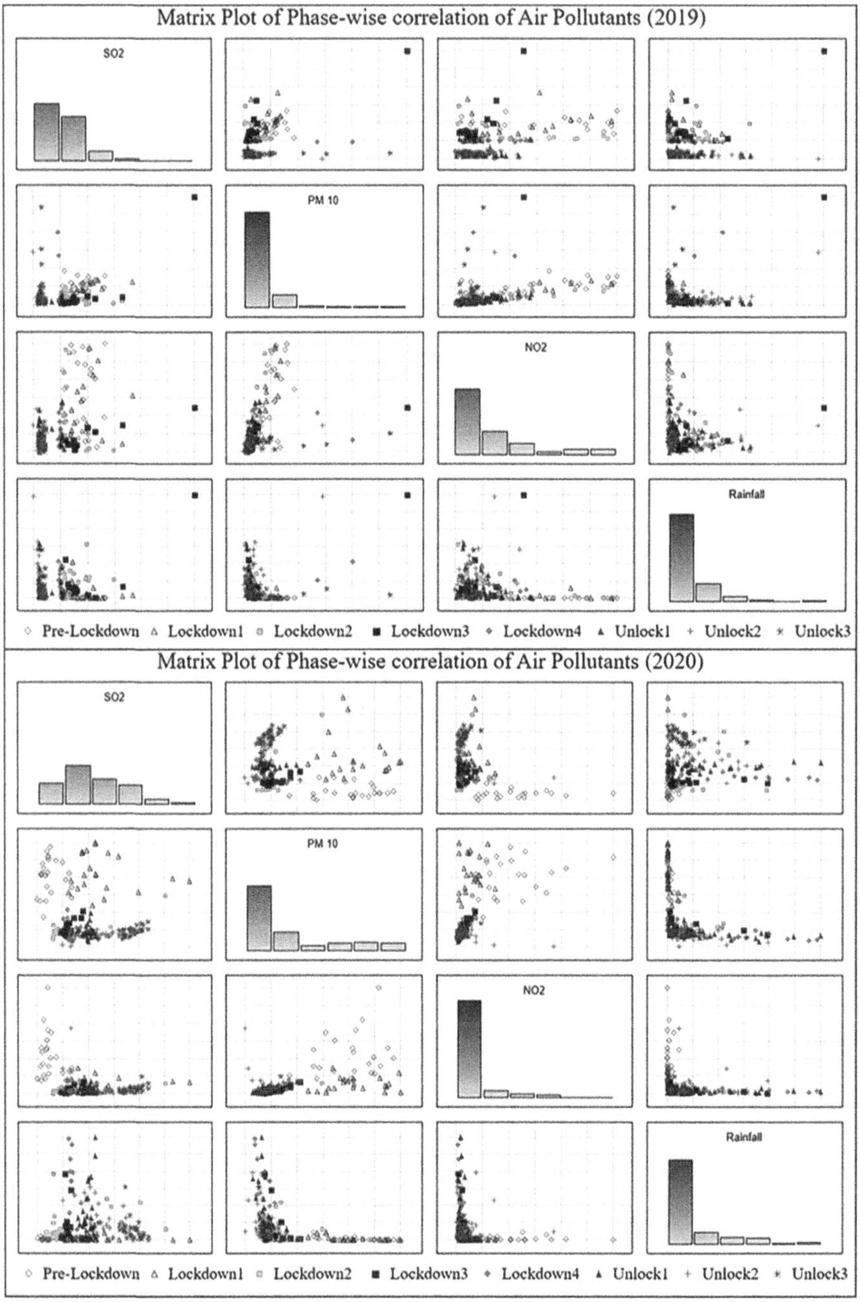

Fig. 3 Matrix plot for phase-wise distribution of air pollutants

Table 4 Mean difference in (%) and standard deviation difference in (%) observed during 2020 in comparison to 2019 similar phases in Guwahati

S. No	Phase	PM_{10} Mean difference (in %)	PM_{10} Standard deviation (in %)	NO_x Mean difference (in %)	NO_x Standard deviation (in %)	SO_2 Mean difference (in %)	SO_2 Standard deviation (in %)
1	Pre-lockdown	−0.62	−10.53	+13.12	+12.03	−33.47	−87.45
2	Phase-1	+50.19	−32.69	−66.05	−66.07	−24.20	−65.26
3	Phase-2	+109.6	−53.15	−84.25	−89.47	−80.28	−44.74
4	Phase-3	−63.02	−93.00	−61.60	−61.23	−35.61	−96.11
5	Phase-4	−70.86	−88.77	−73.05	−61.13	−7.19	−17.24
6	Unlock-1	−36.29	−43.14	−75.17	−71.63	+25.57	−91.48
7	Unlock-2	−61.35	−85.60	−53.90	+59.07	+49.00	−53.00
8	Unlock-3	−70.01	−94.36	−59.19	−37.61	+55.53	+29.41

3 Discussion

It is evident from the recent studies that reductions in most of the pollutants were observed all over India during the lockdown period. A study across 12 cities located in different spatial segments Indo-Gangetic Plain (IGP) showed a substantial decrease (35%) of $PM_{2.5}$ concentrations across the cities located in IGP after the implementation of lockdown (Das et al. 2020). In Saurashtra and South Gujarat regions, reductions up to 30–84% in NO_2 concentration were observed, while O_3 increased by 16–48% due to reduction in NO_2. The average decrease in AQI values by 58% was mainly observed in industrial cities such as Ahmedabad, Gandhinagar, Jamnagar, and Rajkot (Selvam et al. 2020). The atmospheric pollution level (NO_2, $PM_{2.5}$, and PM_{10}) in Ahmedabad city also showed a significant improvement during the study period, implying a positive response of COVID-19 imposed a lockdown on the environment (Aman et al. 2020).

In Delhi, the pollutant level came down to its five-year low during the first week of lockdown phase-1, where $PM_{2.5}$ concentration dropped to 42 μg/m^3—a scenario observed in March 2016 (The Hindu, June 30 2020). AQI reduction in Delhi was 49% compared to the previous year; thus, an improvement of about 60% was observed in the industrial and transport hub (Mahato et al. 2020). During the lockdown period, reduction in $PM_{2.5}$, among all other pollutants, was maximum in Gaya, Kanpur, Nagpur, and Kolkata (Sharma et al. 2020). The study of variation in ambient air quality during COVID-19 lockdown in Chandigarh carried out by Suman Mor et al. (2020) showed significant reductions in all air pollutants during the first and second phases of the lockdown.

The concentration of PM_{10}, $PM_{2.5}$, NO_2, and SO_2 reduced by 55%, 49%, 60%, and 19% for Delhi, and 44%, 37%, 78% and 39% for Mumbai, respectively, during the post-lockdown phase leading to a significant improvement in air quality (Kumari and Toshniwal 2020). The reduction in mean concentration from the pre-lockdown

Table 5 Correlation Matrix for PM_{10}, NO_x, SO_2 and average daily rainfall

	2020					2019				
Lockdown	Variable	PM_{10}	SO_2	NO_x	Rainfall	Variable	PM_{10}	SO_2	NO_x	Rainfall
Pre- lock down	PM_{10}	1				PM_{10}	1			
	So_2	0.16	1			SO_2	0.43	1		
	NO_x	0.45	0.13	1		NO_x	0.7	0.27	1	
	Rainfall	0.13	0.39	−0.29	1	Rainfall	−0.02	0.22	−0.19	1
Phase-1	PM_{10}	1				PM_{10}	1			
	SO_2	−0.1	1			So_2	0.53	1		
	NO_x	0.09	0.03	1		NO_x	0.81	0.23	1	
	Rainfall	−0.1	−0.19	−0.25	1	Rainfall	−0.25	0	−0.24	1
Phase-2	PM_{10}	1				PM_{10}	1			
	SO_2	0.66	1			SO_2	−0.63	1		
	NO_x	0.79	0.51	1		NO_x	0.88	−0.46	1	
	Rainfall	−0.37	−0.17	−0.15	1	Rainfall	−0.11	−0.14	−0.22	1
Phase-3	PM_{10}	1				PM_{10}	1			
	SO_2	0.81	1			SO_2	0.92	1		
	NO_x	0.93	0.8	1		NO_x	0.79	0.89	1	
	Rainfall	−0.26	−0.43	−0.24	1	Rainfall	0.11	0.15	0.12	1
Phase-4	PM_{10}	1				PM_{10}	1			
	SO_2	−0.31	1			SO_2	−0.57	1		
	NO_x	0.83	−0.3	1		NO_x	0.13	0.33	1	
	Rainfall	−0.06	0.55	−0.27	1	Rainfall	0.11	0.11	0.56	1
Unlock-1	PM_{10}	1				PM_{10}	1			
	SO_2	0.05	1			SO_2	0.03	1		
	NO_x	0.66	0.16	1		NO_x	0.8	0.02	1	
	Rainfall	−0.16	0.41	0.12	1	Rainfall	−0.17	−0.38	−0.24	1

Unlock-2	PM_{10}	1				PM_{10}	1			
	SO_2	0.61	1			SO_2	−0.47	1		
	NO_x	−0.46	−0.02	1		NO_x	0.24	−0.29	1	
	Rainfall	0.06	−0.02	−0.35	1	Rainfall	0.16	0.03	0.09	1
Unlock-3	PM_{10}	1				Pm_{10}	1			
	SO_2	0.69	1			So_2	−0.47	1		
	NO_x	0.72	0.6	1		NO_x	0.06	0.12	1	
	Rainfall	−0.02	−0.38	−0.01	1	Rainfall	0.36	−0.33	0.1	1
Total	PM_{10}	1				Pm_{10}	1			
	SO_2	−0.24	1			So_2	0.27	1		
	NO_x	0.55	−0.38	1		NO_x	0.43	0.22	1	
	Rainfall	−0.43	−0.03	−0.26	1	Rainfall	−0.04	−0.01	−0.26	1

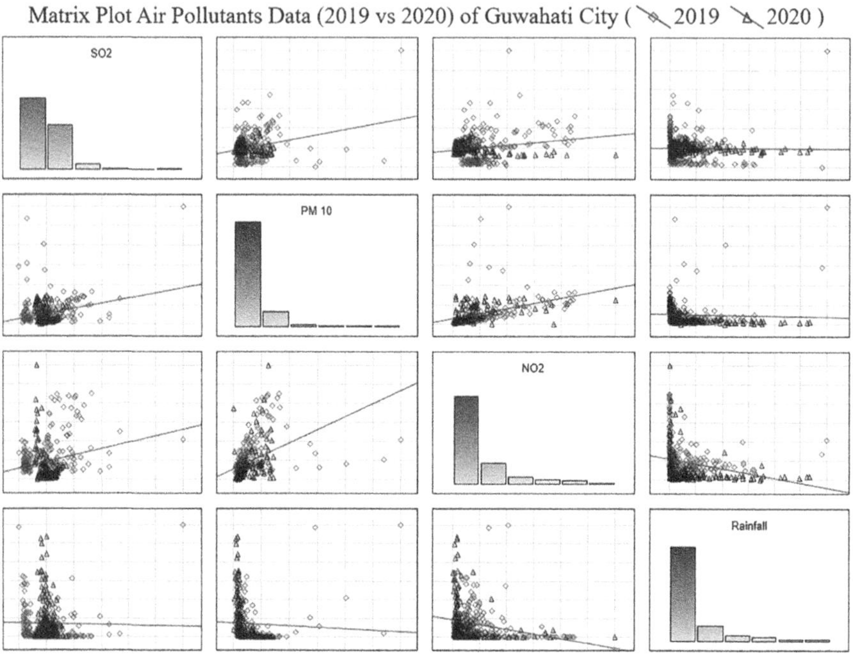

Fig. 4 Matrix Plot air pollutant data showing the linear relationship for 2019 versus 2020

phase to during lockdown of the primary air pollutants is observed in Kolkatta City (Sarkar et al. 2020). Garg et al. (2021), in a similar study on 16 cities designated as hotspot region covering almost two-thirds of India, also reported a significant reduction in the observed (mean) levels of PM_{10}, $PM_{2.5}$, and NO_2 concentration during the lockdown period from March 25 to April 25.

The current study assessed the impact of COVID-19 induced lockdown and unlocking period on the concentration of PM_{10}, NO_x, and SO_2 compared to the similar period of 2019 for Guwahati city, India. The trend analysis of air pollutants during eight different phases of lockdown and unlock (Fig. 3) has determined that high density of vehicular traffic, diesel railway engine depot at Guwahati, their continuous operation, and industrial estates around the city were the major source of high pollution in the city. The evidence of their significant share can be verified from the current study results. The COVID-19 induced lockdown restricted the transportation and industrial sectors, whose shutdown due to COVID-19 led to a higher reduction of pollutant concentrations than in 2019 similar periods (Fig. 4). PM_{10} is mainly produced by suspended dust particles triggered by vehicular movement, NO_x by incomplete combustion of fossil fuels, while SO_2 is produced by combustion of non-renewable energy in the transportation and industrial units.

The rainfall is a significant determinant that helps to lower the pollutant levels which occurred negligibly during the study period. In the absence of an adequate amount of rainfall, the difference in the trends between 2019 and 2020 is a result of the continuous operation of select industrial establishments during the lockdown (Table 2).

During phase-3, with restrictions eased, the economic activities were restarted gradually, which led to the increasing trend in the pollutants. During phase-4 (lockdown) and phase-5 (unlock-1), PM_{10} mean concentrations exhibited an increasing trend due to increased vehicular movement and industrial operations.

Concurrently, it is evident from the negative correlation between AQI values and COVID-19 lockdown days (Fig. 4) that most air pollutants decreased since March 25, 2020, as the lockdown period got extended in a phased manner. The most obvious reason was the shutting down of the industrial and transport sector since the lockdown phase-1 started, and as the days progressed, pollutants were subsequently flushed out (CPCB 2020).

However, different trends of some pollutants (minimal increment during later phases) are probably due to two primary causes. First, the diversity in the industrial setup and decline in the number of on-road vehicles (Chen et al. 2020), as Guwahati city hosts various manufacturing plants. This difference indirectly influences the emission of different pollutants (as per the raw material used for the processing) and the finished products' transportation to the market. The air quality was substantially improved during the lockdown phases. Moreover, the range of daily fluctuation of pollutants had narrowed down when compared to the similar period of 2019. However, as the unlock period started, the south-west monsoon, which plays a vital role in maintaining the local weather of Brahmaputra plains, also arrived simultaneously. The monsoon single-handedly helped to stretch the lower air pollutants trend observed since the lockdown at least until unlock-3.

The reduction in pollutants was started as early as since the lockdown phase-1; however, the present study suggested that the highest benefits were observed during the lockdown phase-2, which implies that actual benefits of total lockdown were experienced after two to three weeks since the inception of lockdown. Although it was generally observed that the lockdown and unlock period progression caused pollutant levels to dip uniformly, however, some anomalies observed in the result such as; the higher mean level of SO_2 since lockdown phase-4 till unlock-3, in comparison to the similar period of 2019 could be explained with the fact that, Guwahati is the most important commercial center of north-eastern India; thus, it plays a prominent role in goods and services transportation to other remote parts of north-east. As the restriction on transportation got relaxed during lockdown phase-4, a considerable amount of goods consignments were transported from Guwahati to other regional cities, even more than the business-as-usual days, which led to an increase in SO_2 generated from vehicular emission.

4 Conclusion

The present study assessed the trends of PM_{10}, SO_2, and NO_x during the pre-lockdown, lockdown, and post-lockdown periods and a similar period of 2019 for Guwahati city situated in northeast India. The strategic location of Guwahati in north-eastern India naturally provides it the command over the other sister cities in the region; however, on the downside, its location makes Guwahati highly prone to ever-increasing environmental degradation and severely polluted atmosphere. The meta-analysis of continuous data was performed using descriptive statistics on air pollutant data was acquired from the Central Pollution Control Board and Assam Pollution Control Board. The result revealed that as much as 63.02%, 80.28%, and 84.25% reduction in the mean concentration of PM_{10}, SO_2, and NO_x pollutants, respectively, and the fluctuation from the mean values was also drastically reduced by 93% for PM_{10}, 96.11% for SO_2 and 89.47% for NO_x in comparison to a similar period of 2019. The significance of the study is to show how the different sets of restrictions (in phases) imposed upon a city can significantly reduce pollutant concentrations.

The modified form of the present lockdown measures can be a game-changer if it also takes care of the economy and the environment. The mild form of lockdown can be imposed on certain parts of the city on specific days and diversion of traffic in that specific area. This kind of city planning would provide a trough (a gap) in uniformly high pollutant concentration levels. A sudden trough on certain days would provide relief from continuously high pollution levels. A mild form of lockdown could be the need of the hour if the economy and environment want to go hand in hand.

Acknowledgments The first author (Ritwik Nigam) acknowledges the financial support provided by the University Grant Commission (UGC), Govt. of India, New Delhi, to conduct this research. The authors also thank the University Administration for their support.

References

Aman MA, Salman MS, Ali PY (2020) COVID-19 and its impact on environment: improved pollution levels during the lockdown period – a case from Ahmedabad, India. Remote Sens Appl Soc Environ 20:100382. https://doi.org/10.1016/j.rsase.2020.100382

Assam Online Portal (2015) Capital of Assam. Assam Online Portal, Introduction. Retrieved from https://web.archive.org/web/20150807175846; http://online.assam.gov.in/web/guest/capitalofassam

Berman JD, Ebisu K (2020) Changes in US air pollution during the COVID-19 pandemic. Sci Total Environ 739:139864, ISSN 0048-9697. https://doi.org/10.1016/j.scitotenv.2020.139864

Bhushan C (2005) Assam: its heritage and culture, P182. Kalpaz Publications. ISBN 978-8178353524

Buchholz K (2020) India records second quarter of GDP losses. Statista, topic, BRIC countries. Retrieved from https://www.statista.com/chart/18245/india-quarterly-gdp-growth/

Chen LWA, Chien LC, Li Y, Lin G (2020) Nonuniform impacts of COVID-19 lockdown on air quality over the United States. Sci Total Environ. https://doi.org/10.1016/j.scitotenv.2020.141105

CPCB (Central Pollution Control Board) (Ed.). (2020) National air quality index. Retrieved from https://app.cpcbccr.com/ccr

Das M, Das A, Mandal A (2020) Examining the impact of lockdown (due to COVID-19) on domestic violence (DV): an evidences from India. Asian J Psychiatr 102335. https://doi.org/10.1016/j.ajp.2020.102335

Duarah CK (2012) Guwahati amongst most polluted cities in the world. Assam Times. Retrieved from https://www.assamtimes.org/node/6662

ESA (European Environmental Agency) (2020) Air pollution goes down as Europe takes hard measures to combat. Accessed 4 Oct 2020. Retrieved from https://www.eea.europa.eu/highlights/air-pollution-goes-down-as

Financial Times (2020) Loss of working hours to equal 195m full-time jobs, UN agency warns. Finalcial Times, Content. Retrieved from https://www.ft.com/content/d78b8183-ade7-49c2-a8b5-c40fb031b801

Garg A, Kumar A, Gupta NC (2021) Impact of lockdown on ambient air quality in COVID-19 affected hotspot cities of India: need to readdress air pollution mitigation policies. Environ Claims J 33(1):65–76. https://doi.org/10.1080/10406026.2020.1822615

He G, Pan Y, Tanaka T (2020) The short-term impacts of COVID-19 lockdown on urban air pollution in China. Nat Sustain 3:1005–1011. https://doi.org/10.1038/s41893-020-0581-y

Indian Meteorological Department (IMD) (2020) Guwahati climatological table 1981–2010. Retrieved from https://city.imd.gov.in/citywx/extreme/FEB/guwahati2.html

Kumari P, Toshniwal D (2020) Impact of lockdown measures during COVID-19 on air quality– A case study of India. Int J Environ Health Res 32:1–8. https://doi.org/10.1080/09603123.2020.1778646

Londono E, Andreoni M, Casado L, Ahmad A, McDonald B (2020) As Latin America shuts down to fight virus, Brazil and Mexico are holdouts. The New York Times, world, Americas Retrieved from https://www.nytimes.com/2020/03/25/world/americas/coronavirus-brasil-mexico.html

Mahato S, Pal S, Ghosh KG (2020) Effect of lockdown amid COVID-19 pandemic on air quality of the megacity Delhi, India. Sci Total Environ 730:815. https://doi.org/10.1016/j.scitotenv.2020.139086

Monserrate MA, Ruano MA, Alcalde LS (2020) Indirect effects of COVID-19 on the environment. Sci Total Environ 728(2020):138813. https://doi.org/10.1016/j.scitotenv.2020.138813

Mor S, Kumar S, Singh T, Dogra S, Pandey V, Ravindra K (2020) Impact of COVID-19 lockdown on air quality in Chandigarh, India: understanding the emission sources during controlled anthropogenic activities. Chemosphere 263:127978

Nigam R, Pandya K, Luis A, Sengupta R, Kotha M (2020) Positive effects of COVID-19 lockdown on air quality of industrial cities - Ankleshwar and Vapi, Gujarat, India. Manuscript submitted for publication

Partington R (2020) The Guardian. UK GDP falls by record 20.4% in April as lockdown paralyses economy. Retrieved from https://www.theguardian.com/business/2020/jun/12/britains-gdp-falls-204-in-april-as-economy-is-paralysed-by-lockdown

Peel, M. C., Finlayson, B. L., and McMahon, T. A. (2007). Updated world map of the Köppen–Geiger climate classification. Hydrol Earth Syst Sci 11 (5): 1633–1644 https://doi.org/10.5194/hess-11-1633-2007. ISSN 1027-5606

Rutz C, Loretto MC, Bates AE, Davidson SC, Duarte CM, Jetz W, Johnson M, Kato A, Kays R, Mueller T, Primack RB (2020) COVID-19 lockdown allows researchers to quantify the effects of human activity on wildlife. Nat Ecol Evol 4(9):1156–1159. https://doi.org/10.1038/s41559-020-1237-z.. ISSN 2397-334X

Sarkar M, Das A, Mukhopadhyay S (2020) Assessing the immediate impact of COVID-19 lockdown on the air quality of Kolkata and Howrah, West Bengal, India. Environ Dev Sustain 23(6):8613–8642. https://doi.org/10.1007/s10668-020-00985-7

Selvam S, Muthukumar P, Venkatramanan S, Roy PD, Manikanda Bharath K, Jesuraja K (2020) SARS-CoV-2 pandemic lockdown: effects on air quality in the industrialized Gujarat state of India. Sci Total Environ 737:140391

Sharma Y (2020) Unemployment rate falls to pre-lockdown level: CMIE. The Economic Times. Indiatimes. Retrieved from https://economictimes.indiatimes.com/news/economy/indicators/unemployment-rate-falls-to-pre-lockdown-level-cmie/articleshow/76528571.cms

Sharma S, Zhang M, Gao A, Zhang H, Kota SH (2020) Effect of restricted emissions during COVID-19 on air quality in India. Sci Total Environ 728:138878. https://doi.org/10.1016/j.scitotenv.2020.138878

Subramani MV, Roman J (2020) The coronavirus response in India- World's largest lockdown. Am J Med Sci 360(6):P742–P748. Retrieved from https://www.amjmedsci.org/article/S0002-9629(20)30351-7/fulltext#seccesectitle0006

The Hindu (2020) Coronavirus lockdown lifts Delhi's march air quality to 5-year high. Accessed 24 Dec 2020. Retrieved from https://www.thehindu.com/news/cities/Delhi/coronavirus-lockdown-lifts-delhis-march-air-quality-to-5-year-high/article31252221.ece

Tobías C, Carnerero C, Reche J, Massagué M, Minguillón MC, Alastuey A, Quero X (2020) Changes in air quality during the lockdown in Barcelona (Spain) one month into the SARS-CoV-2 epidemic. Sci Total Environ 726(2020):1–4. https://doi.org/10.1016/j.scitotenv.2020.138540

UNDP (United Nations Development Programme) (2020) Socio-Economic impact of COVID-19. Accessed 8 July 2020. Retrieved from https://www.undp.org/content/undp/en/home/coronavirus/socio-economic-impact-of-covid-19.html

Vyas M (2020) Unemployment rate touched 26%. Centre for Monitoring Indian Economy (CMIE). Retrieved from https://www.cmie.com/kommon/bin/sr.php?kall=warticle & dt=2020-04-21% 2010:40:01 & msec=873

World Health Organization (WHO) (2020) WHO Coronavirus Disease (COVID-19) Dashboard. Retrieved from https://covid19.who.int/

Impact of Lockdown on Air Quality in Megacities of India During COVID-19 Pandemic

Pallavi Pradeep Khobragade and Ajay Vikram Ahirwar

Abstract In the month of March 2020, the World Health Organization (WHO) declared COVID-19 as a pandemic. Amid the COVID-19 pandemic, Government of India has also initiated preventive measures by imposing nationwide lockdown initially for three weeks from 24th March 2020 to 14th April 2020 and extended up to 3rd May 2020. The aim of this study is to assess the impacts on air quality ($PM_{2.5}$, PM_{10}, NO_2, NH_3, SO_2, CO and O_3) during different phases of lockdown in different cities (Delhi, Mumbai, Bengaluru and Nagpur) of India. As a result, significant reduction in aerosol concentrations were observed; 40%, 53%, 45% and 34% in $PM_{2.5}$ and 37%, 53%, 35% and 20% in PM_{10} at Delhi, Mumbai, Bengaluru and Nagpur respectively. Other air quality parameters such as NO_2, NH_3, SO_2 and CO showed declining trends whereas increasing trend for O_3 was observed. The present air quality data were also compared with previous year (2019) data and found significant reduction in the current year (2020). The overall results demonstrate that air quality is significantly improved due to nationwide lockdown. From this study, it can be observed that ambient clean air can be achieved by reducing people's mobility in urban areas. This could help in reducing concentrations of air pollutants in cities. This study helps to visualize the pollutants response due to reduction in anthropogenic emissions in megacities which can help in making atmospheric governance policies.

Keywords Corona virus · SARS-CoV-2 · COVID-19 · $PM_{2.5}$ · PM_{10} · Air pollution

P. P. Khobragade (✉)
Rungta College of Engineering and Technology, Raipur, India

A. V. Ahirwar
National Institute of Technology, Raipur, Chhattisgarh, India

B. Phartiyal et al. (eds.), *Climate Change and Environmental Impacts: Past, Present and Future Perspective*, Society of Earth Scientists Series,
https://doi.org/10.1007/978-3-031-13119-6_24

1 Introduction

COVID-19 is spread by the pathogen Severe Acute Respiratory Syndrome coronavirus 2 (SARS-CoV-2) and caused severe pneumonia outbreaks globally consecutively outspread over the entire world affecting 196 countries by March 25th 2020 raising extreme attention. (Ahmadi et al. 2020; Contini and Costabile 2020; Ficetola and Rubolini 2021). Covid-19 outbursts have become one of the most challenging problems for the Government (Ficetola and Rubolini 2021). Maximum countries in the world have declared some sort of lockdown to reduce human interaction by following social distancing ceasing the spread of this novel coronavirus thus reducing the effects of COVID-19 (He et al. 2020). The world is trying to fight with this novel coronavirus in every best possible way and control rapid spread of this SARS-CoV-2 in humans (Anjum 2020). The lockdown resulted positively by bringing unintentional social benefits in terms of air quality improvement in India by dropping the pollutant concentrations (Gautam 2020; He et al. 2020). The first affected person by SARS-CoV-2 in India is observed in Kerala in late January 2020 (Gautam 2020; Gautam and Hens 2020). Due to rapid spread of the virus, Kerala was the top most state reporting maximum number of novel corona virus affected positive cases. To get the situation of spreading this pandemic under control, the prime minister of India "Narendra Modi" declared complete lockdown sessions to decrease the effects and transmissions of the novel coronavirus and practice social distancing (Gautam 2020). It was decided that social distancing and home isolation measures will help to reduce the negative impacts of COVID-19 (Han et al. 2020). Aerosols are one of the ways for rapid spread of this virus and are recognized as one of the important pollutants by national and international agencies, associated with mortality and morbidity (Dutheil et al. 2020; Gautam 2020).

During lockdown phase, the main contributing sectors to air pollution were brought into halt. On March 22, 2020, the Lt. Governor of Delhi has notified the "Delhi Epidemic Diseases, COVID-19 regulations, 2020" ordering all the hospitals to have a flu corners testing all suspected cases, checking the travel history of the suspects and have a right to impose home or institution quarantine (Gautam and Hens 2020). During 2016, 4.2 million premature deaths were reported due to ambient air pollutants (WHO 2018). The air pollution index during the lockdown phase considerably prevents premature deaths due to improved air quality. The recent data over USA and Europe after the lockdown enforcement confirms the reduction in air quality benefiting environment and living beings (Suhaimi et al. 2020).

The associated viruses need specific environment for their growth and survival (Rios and Gianmoena 2020). Climatic variables were considered as the finest drivers of worldwide variation of confirmed Covid-19 cases growth rates (Ficetola and Rubolini 2021) and have diverse influences on epidemiology of various infectious diseases (Rios and Gianmoena 2020). COVID-19 may also spread through exhaled breath droplets known as microdroplets (Riediker and Morawska 2020). The air

Fig. 1 Geographical map of study locations

pollution weakens the immune system and makes humans susceptible to various respiratory virus infections (Han et al. 2020).

In this study, we carried out a severe analysis into the issue and estimate how lockdown affected air quality in terms of $PM_{2.5}$ and PM_{10} at various locations in Indian megacities including Delhi, Mumbai, and Nagpur (Fig. 1). The air quality data from different sites in the study areas are collected covering the entire region. Also, statistical analysis is executed to evaluate the relationship between temperature and humidity in the regions affecting transmissibility of COVID-19. The decrease in

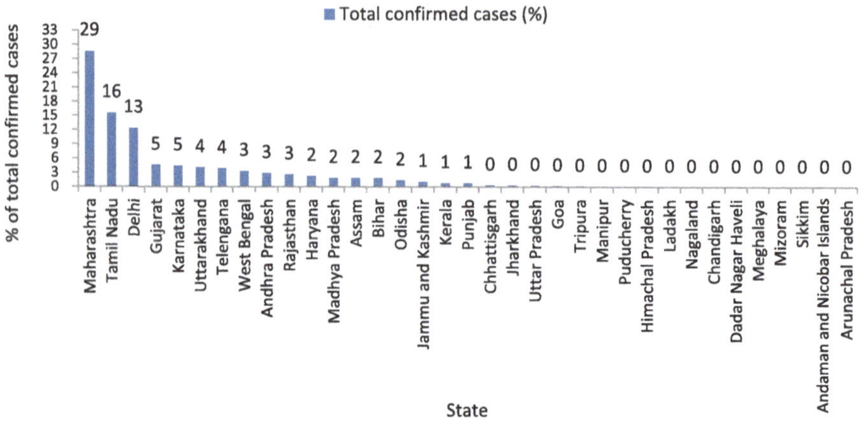

Fig. 2 Total active cases in India (%). Source: Ministry of Health and Family Welfare

particulate concentrations was observed immediately after a week due to reduced anthropogenic activities at all the locations.

Figure 2 shows that the highest number of cases has been reported in Maharashtra (29%), Tamil Nadu (16%), Delhi (13%) and Gujarat (5%) and it is increasing with time across the country. Few states in India including Arunachal Pradesh, Andaman and Nicobar Islands, Sikkim, Mizoram, Meghalaya, Dadar Nagar Haveli, etc. are still in a better position with minimum number of patients.

1.1 Site Description

Delhi, the national capital of India, is located between latitudes 28°24′17″ and 28°53′00″ N and longitudes 76°50′24″ and 77°20′37″E at 216 meters above the mean sea level (MSL). The city covers an area of approximately 1500 square kilometres. The sites selected for particulate samples data collection are spread over the entire city. Mumbai, the capital city of Maharashtra is located between latitudes 19°′4′33.924″N and 72°52′38.7336″E is the second most populous city after Delhi. Delhi and Mumbai are termed as epicentres for corona virus in India; also Maharashtra has maximum number of COVID-19 cases in which Mumbai holds first position in positive corona patients (Jain and Sharma 2020). In the south, is located at latitude 12° 58′ 20.7912′N and longitude 77° 34′ 50.3148″ E in Karnataka being another megacity is selected for analysis. Another city in Maharashtra state, Nagpur located at latitude 21° 8′ 47.8788′N and longitude 79° 5′ 19.8960″ E is selected to check the similarity in the increasing cases whether the number of positive cases in Mumbai and Nagpur are correlated. The details of the sampling stations are given in Table 1.

Table 1 Details of stations from selected cities

Sr. No	City	Number of stations considered	Name of stations
1	Delhi	10	Anand Vihar-DPCC, Ashok Vihar-DPCC, Bawana-DPC, Dwarka-Sector 8-DPCC, IGI Airport (T3)-IMD, Mundka-DPCC, North Campus, DTU-CPCB, Rohini DPCC, Sirifort-CPCB
2	Mumbai	8	Bandra, Borivali East, Chhatrapati Shivaji Int. Airport, Colaba, Kurla, Powai, Sion, Vile Parle west
3	Bengaluru	10	Bapuji Nagar, BTM Layout, BWSSB, City Railway Station, Hebbal, Hombegowda Nagar, Jayanagar 5th Block, Peenya, Sanegurava Hills, Silk Board Ballygunge
4	Nagpur	1	Opp GPO Civil Lines-MPCB

1.2 Methodology

Ambient particulate pollutant data ($PM_{2.5}$ and PM_{10}) is collected from Delhi, Mumbai, and Nagpur prior a month and during lockdown and compared it with same interval in 2019 to show fluctuations in air quality parameters. Air Quality Data i.e., 24 hours average concentrations were obtained from continuous ambient air quality monitoring 24 hours concentrations of $PM_{2.5}$, PM_{10}, NO_2, SO_2, CO, NH_3 and O_3 was collected from Central Pollution Control Board, Ministry of Environment, Forests and Climate Change, New Delhi for the period of 24th February 2020 to 12th June 2020 and in 2019 for the same time frame. The stations having more than 75% data were considered for data collection from each station in every selected city. Spatiotemporal analysis is done to see the correlation between increasing number of corona cases and meteorological parameters at all the locations. Meteorological data on temperature and relative humidity for the study locations have been collected from worldweatheronline.com. Correlation between meteorological parameters and fluctuations of confirmed corona positive cases is analysed to understand the relationship between them.

2 Results and Discussions

2.1 Effect on Air Quality

The lockdown resulted in appreciable drop in particulate matter concentrations at all the major locations in the cities. A voluntary public curfew was observed on 22nd March 2020 in India few days after which noticeable downfall are observed in $PM_{2.5}$ and PM_{10} concentrations. All air quality control units positioned in the city were selected and used to collect data for 60 days (from February 24th 2020 to 12th June

2020). $PM_{2.5}$ and PM_{10} was found to be higher in all the selected study areas before the lockdown period as per the standards set by NAAQS and WHO. $PM_{2.5}$ and PM_{10} concentrations were within NAAQS limit immediately after the lockdown period due to low pollutant concentrations. The particulate pollutant concentration showed tremendous downfall after the public curfew immediately followed by lockdown due to stopping anthropogenic activities including industrial activities and vehicular movements. The data concerning assessment of sources of $PM_{2.5}$ and PM_{10} in the cities were obtained from local study report prepared by Automatic Research Association of India (ARAI) and The Energy and Resources Institute (TERI) (https://www.teriin.org/sites/default/files/2018-08/Report_SA_AQM-Delhi-NCR_0.pdf).

Although highly variable, in Delhi, the particulate concentration for $PM_{2.5}$ at Anand Vihar, Ashok Vihar, Dwarka and Sirifort after the lockdown period is found to be comparable (mean: 90.10–100.91 μg/m^3). IGI and North Campus showed minimum $PM_{2.5}$ concentrations (54.29–65.82 μg/m^3) indicating the influence of lower anthropogenic activities. Bawana, Mundka, DTU and Rohini have comparable $PM_{2.5}$ concentrations (110.91–132.43 μg/m^3) in the city. Particulate concentrations for PM_{10} at Bawana, Dwarka, Mundka, DTU and Rohini exhibit comparable concentrations (mean: 101.54–127.60 μg/m^3), AnandVihar, Ashok Vihar and Sirifort showed comparable PM_{10} concentrations (85.64–92.61 μg/m^3) and IGI and North Campus were found to have minimum PM_{10} concentrations (78.83–85.99 μg/m^3). Within few weeks, the AQI and particulate concentrations dropped down up to 25% after the lockdown (He et al. 2020). The city observed around 50% reduction in air pollution levels (Gautam and Hens 2020). In Mumbai, after the lockdown, Colaba, Kurla, Powai, Sion and Vile Parle had $PM_{2.5}$ concentration within the NAAQS limit (22.77–33.2 μg/m^3), Borivali and Chhatrapati Shivaji Terminal (CSTM) were found to have comparable $PM_{2.5}$ concentrations (42.68–43.68 μg/m^3) slightly crossing the NAAQS limit and Bandra (50 μg/m^3) station had highest $PM_{2.5}$ concentration in Mumbai. PM_{10} concentration at CSMT were found within the NAAQS standards (38.69 μg/m^3), Colaba, Powai and Vile Parle had comparable PM_{10} concentrations (40.08–59.37 μg/m^3), Bandra, Borivali and Sion had comparable concentration level (60.8–65.85 μg/m^3) and Kurla had maximum PM_{10} concentration (72.67 μg/m^3) in Mumbai. The $PM_{2.5}$ concentrations at Bapuji, BWSSB, Hebbal, Hombegowda, Jayanagar and Silk board were within NAAQS standards in. BTM exceeded the limit (54 μg/m^3) and Peenya (58.16 μg/m^3) had Maximum $PM_{2.5}$ concentration in. PM_{10} concentrations at maximum stations were within the set limits at Bapuji, BTM, Hebbal, Jayanagar and Silk board (26.25–46.82 μg/m^3), Sanegurava Halli crossed the limit (61.95 μg/m^3) with maximum concentration at City Railway station (82.38 μg/m^3). The $PM_{2.5}$ (37.5 μg/m^3) and PM_{10} (46.78 μg/m^3) concentration were within the NAAQS standards at Civil Lines in Nagpur city. The northern India has recorded 20 year low air pollution during the lockdown phase and the data shows 30% drop in air pollution over the Northeast U.S. (NASA 2020).

In Delhi, reductions in 2020 in comparison to 2019 by (60%) NO_2 concentration (110 ± 25–44 ± 23 μg/m^3), (31%) NH_3 concentration (11 ± 3–8 ± 3 μg/m^3), (36%)

SO_2 concentration (25 ± 10–16 ± 7 μg/m^3), (2%) CO concentration (80 ± 22–79 ± 25 μg/m^3) and an increase in (68%) O_3 concentration (23 ± 18–71 ± 62 μg/m^3) is observed.

Increment in concentrations of (63%) NO_2 (21 ± 9–34 ± 44 μg/m^3) and (51%) SO_2 concentration (22 ± 10–33 ± 20 μg/m^3) with decrease in (36%) CO concentration (89 ± 9–57 ± 25 μg/m^3) and (19%) O_3 concentration (29 ± 24–24 ± 20 μg/m^3) in Mumbai is observed from 2019 to 2020.

In, reductions in 2020 in comparison to 2019 by (70%) NO_2 concentration (70 ± 16–21 ± 15 μg/m^3), (31%) NH_3 concentration (4 ± 1–3 ± 1 μg/m^3) and (40%) O_3 concentration (60 ± 43–36 ± 12 μg/m^3) and increase in (31%) SO_2 concentration (9 ± 5–11 ± 3 μg/m^3) and (7%) CO concentration (26 ± 8–28 ± 10 μg/m^3) is noticed.

The COVID-19 lockdown in Nagpur resulted in reductions in 2020 in comparison to 2019 by (36%) NO_2 concentration (52 ± 16–33 ± 18 μg/m^3), (47%) NH_3 concentration (13 ± 7–7 ± 5 μg/m^3) and (63%) O_3 concentration (140 ± 46–52 ± 16 μg/m^3) although increase (79%) SO_2 concentration (20 ± 11–4 ± 4 μg/m^3), (32%) CO concentration (31 ± 11–21 ± 8 μg/m^3) is noticed which may be due to stationary source emissions like industrial sectors by fossil fuel burning (Broomandi et al. 2020).

Significant reduction in concentrations of $PM_{2.5}$, PM_{10} and NO_2 were observed at Delhi and Nagpur during 2020 lockdown except Mumbai. Delhi witnessed increased concentration of O_3 during the lockdown period in pandemic, whereas, Mumbai and Nagpur experienced lowered concentration as compared to year 2019. $PM_{2.5}$ and NO_2 increased up to 31% and 63% in Mumbai which may be due to emergency movement of vehicles. SO_2 increased at Mumbai and up to 41% and reduced up to 58% in Delhi and Nagpur. Average CO reduction was up to 16% at all the sites. The variation in NO_2 concentration is due to variation of city traffic (Wang et al. 2020a, b). The decrease in NO_2 also showed a reduction in $PM_{2.5}$ concentration indicating that vehicular emissions are an important contributor to $PM_{2.5}$ concentrations. Decrease in NO_2 and $PM_{2.5}$ is more in comparison to SO_2 and CO, this indicates that emissions from stationary sources didn't decrease as much as traffic. Industrial activities increased in Mumbai whereas reduced at Delhi and Nagpur (Chu et al. 2021).

2.2 *Comparison Between Pre and Post COVID-19*

The lockdown resulted in appreciable drop in $PM_{2.5}$ and PM_{10} concentrations at all the major locations in the city. The day after the "Janta Curfew", a significant drop down has been noticed at all the sites indicating that anthropogenic activities contribute majorly to the particulate sources. The counter COVID actions minimised the overall $PM_{2.5}$ concentrations during the study period up to 37% and PM_{10} concentrations up to 40%. The shutting down of redundant commercial activities reduces the particulate pollution with greater effects in cities with a larger economy,

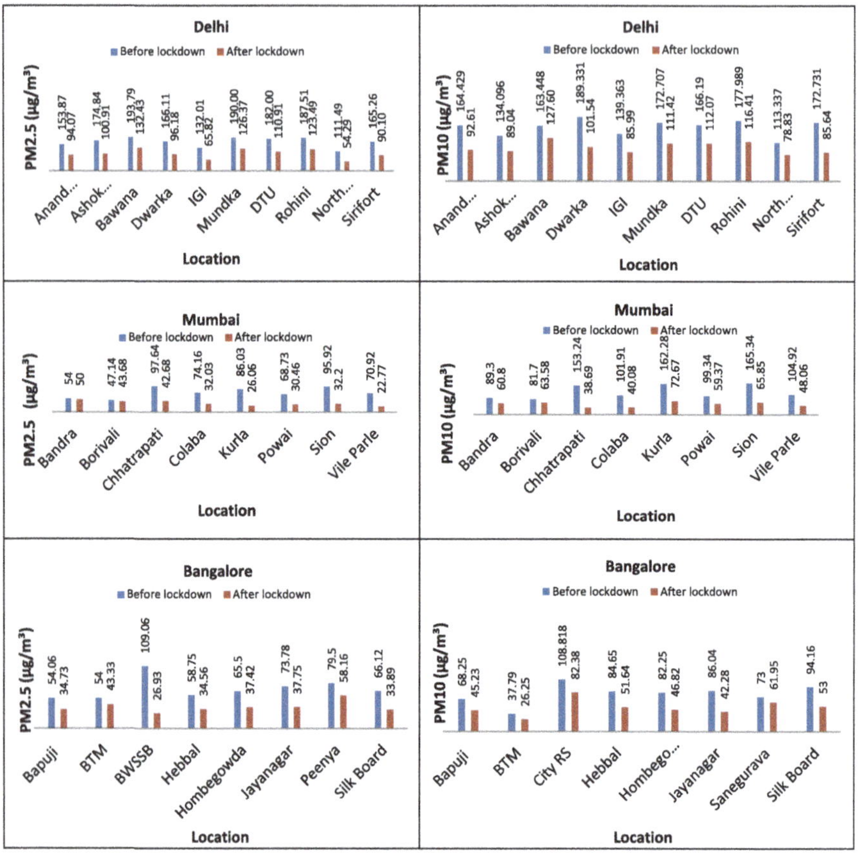

Fig. 3 Change in air pollutant concentrations at the study locations before and after lockdown

Table 2 Variation in average pollutant concentrations (µg/m^3) during 2019 and 2020

	Nagpur		Delhi				Mumbai	
	$PM_{2.5}$	PM_{10}	$PM_{2.5}$	PM_{10}	$PM_{2.5}$	PM_{10}	$PM_{2.5}$	PM_{10}
2020	41.96	51.34	117.94	121.15	46.95	59.37	22.08	35.44
2019	79.28	86.59	174.11	208.88	79.13	83.63	30.23	70.83
Percentage reduction in 2020	47	41	32	42	41	29	27	50

heavy vehicular pollution and higher industrial activities (He et al. 2020). The particulate matter level during the lockdown phase remained high, $PM_{2.5}$ concentrations was three times and two times higher than NAAQS standards respectively. People living in contaminated cities are at more risk from COVID-19 (Moftakhar et al. 2020).

Figure 3 shows the concentrations of $PM_{2.5}$ and PM_{10} at Delhi, Mumbai, and Nagpur before and after lockdown. The temporal investigation showed that there is a significant decrease in $PM_{2.5}$ and PM_{10} concentration at the study locations after the lockdown. The extent of decline in the pollutant concentration at the study locations is different. $PM_{2.5}$ showed an average change in concentration of 42%, 36%, 40% and 29% at Delhi, Mumbai, and Nagpur respectively whereas PM_{10} showed an average change in concentration of 37%, 57%, 36% and 30% at Delhi, Mumbai, and Nagpur respectively. Comparison was done in between the pollutant concentration data of 2019 and 2020 during the same time period. A significant decrease in the pollutant concentration data has been observed in 2020 as compared to 2019. Table 2 shows the decrease in the pollutant concentration during 2019 and 2020 during the same time period.

2.3 *Pre and Post-lockdown Variation in Gaseous Pollutants*

Delhi showed notable reductions from pre-lockdown period to post-lockdown in (53%) NO_2 (70 ± 17–33 ± 15 μg/m^3) and (24%) CO (93 ± 13–71 ± 26 μg/m^3) concentrations while increase in (13%) NH_3 (7 ± 2–8 ± 4 μg/m^3), (44%) SO_2 (10 ± 3–18 ± 7 μg/m^3) and (71%) O_3 (27 ± 17–93 ± 65 μg/m^3) concentrations. In Mumbai, decrease in (87%) NO_2 (85 ± 45–11 ± 10 μg/m^3), (53%) CO (93 ± 22–44 ± 9 μg/m^3) and (59%) O_3 (37 ± 24–15 ± 11 μg/m^3) concentrations is observed while increase in (47%) SO_2 (20 ± 3–38 ± 22 μg/m^3) concentration is noticed. Showed reductions in (41%) NO_2 (32 ± 3–19 ± 16 μg/m^3), (8%) SO_2 (12 ± 2–11 ± 3 μg/m^3) and (7%) CO (29 ± 10–27 ± 10 μg/m^3) and increase in (33%) NH_3 (2 ± 1–3 ± 1 μg/m^3) and (38%) O_3 (21 ± 11–34 ± 11 μg/m^3) concentrations. In Nagpur, decrease in (58%) NO_2 (65 ± 12–27 ± 11 μg/m^3), (70%) SO_2 (10 ± 8–3 ± 2 μg/m^3) and (20%) CO (25 ± 9–20 ± 8 μg/m^3) concentrations and increase in (14%) NH_3 (6 ± 2–7 ± 5 μg/m^3) and (9%) O_3 (48 ± 25–53 ± 13 μg/m^3) concentrations. Despite the decreases in particulate concentrations, increase in SO_2, NO_2, NH_3, CO and O_3 concentrations is observed which may be due to stationary sources, household cooking fuel and emergency movement of vehicles. In a similar study at Almati, non-traffic source contribution has found to be increased during the lockdown (Broomandi et al. 2020; Kerimray et al. 2020) which may be reason for enhanced concentrations of NH_3, SO_2 during the lockdown. Increased NH_3 concentrations in Delhi, and Nagpur may be due to agricultural activities and biomass burning (Bray et al. 2021), fossil fuel combustion (Wu et al. 2020). Increased SO_2 concentrations during the lockdown phase was observed at Delhi and Mumbai which again may be due to biomass burning and industrial emissions (Bari et al. 2020). However decrease in SO_2 concentrations is observed at and Nagpur during the lockdown period.

Figure 4 shows the air quality index (AQI) during 2019 and 2020 for $PM_{2.5}$ and PM_{10}. The AQI of $PM_{2.5}$ was moderate for 18% of the days, unhealthy for sensitive groups for 55% of the days, unhealthy for 26% of the days and very unhealthy

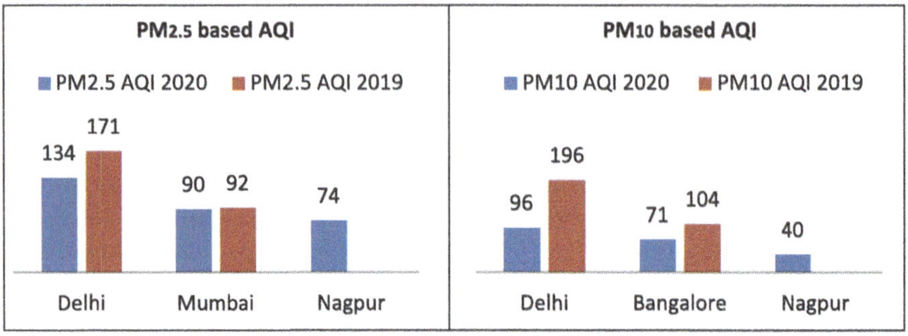

Fig. 4 Comparison between $PM_{2.5}$ based AQI and PM_{10} based AQI at study locations during 2020 and 2019

for 2% of the days and AQI for PM_{10} is good for about 6% of the days, moderate for 55% of the days, unhealthy for sensitive groups for 30% of the days, unhealthy for 8% of the days and very unhealthy for 2% of the days in Delhi during 2020. The AQI across Delhi is under "satisfactory" (AQI: 51–100) category after COVID-19 (http://safar.tropmet.res.in/). The lockdown measures in Delhi resulted in a dropdown in $PM_{2.5}$ up to 30% (Anjum 2020). The AQI for $PM_{2.5}$ in Mumbai was good for 7% of the days, moderate for 58% of the days, unhealthy for sensitive groups for 26% of the days and unhealthy for 8% of the days. The AQI across Mumbai is under "good" (AQI: 0–50) category after COVID-19 ((http://safar.tropmet.res.in/). The AQI for PM_{10} in was good for 10% of the days, moderate for 77% of the days, unhealthy for sensitive groups for 13% of the days during the study period in 2020. The AQI for $PM_{2.5}$ in Nagpur was good for 6% of the days, moderate for 87% of the days, unhealthy for sensitive groups for 8% of the days and AQI for PM_{10} was good for 89% of the days, moderate for 10% of the days and unhealthy for 1% of the days during the study period in 2020. The data for 2019 for Nagpur is not available. A significant positive association between $PM_{2.5}$, PM_{10} and confirmed corona cases has been observed indicating that the decrease in particulate concentration during the lockdown phase helped in keeping the number of positive patients low (Zhu et al. 2020). During 2020, $PM_{2.5}$ AQI is unhealthy for sensitive group at Delhi, moderate at Mumbai and Nagpur whereas during 2019, Delhi experienced unhealthy atmosphere and Mumbai had moderate pollution. PM_{10} AQI was moderate at Delhi and Mumbai and good in Nagpur during 2020 whereas it was unhealthy at Delhi and unhealthy for sensitive groups in during 2019.

2.4 Meteorological Data Collection and Processing

Meteorological factors including temperature (T) and relative humidity (RH) may enhance air pollution transmitting infection of diseases (Jiang et al. 2020). Ambient environmental factors including temperature and relative humidity show partial correlation with airborne pathogens (Suhaimi et al. 2020). Atmospheric temperature and relative humidity are found to have direct association with the spread of this pandemic (Wang et al. 2020a, b; Ahmadi et al. 2020). Virus transmission is affected by many factors including climatic variables like temperature, humidity and precipitation (Ahmadi et al. 2020; Méndez-Arriaga 2020; Menebo 2020). This makes it essential to understand relationship between climatic conditions in a geographical area and transmission of COVID-19 pandemic (Ahmadi et al. 2020). The data is collected from 24th February 2020 to 12th June 2020. The meteorological data was collected after every 3 hours in a day and daily average is calculated. An inconsistent lagged effect of temperature and humidity is observed on COVID-19 growth rate (Carleton et al. 2020). A strong relationship between local existing climate and COVID-19 increasing rates has been observed signifying the possibility of seasonal difference in the spatial design of outbreaks (Ficetola and Rubolini 2021). Ultraviolet radiation indicates a strong effect on COVID-19 growth rate with 1 kJ m^{-2} increase in hourly UV decreasing confirmed growth rate by 0.09 points causing a delay in the spread (Carleton et al. 2020).

Local temperature and humidity has found to influence directly transmitted diseases by affecting the survival and transmission of the virus outbreak (Ficetola and Rubolini 2021; Wang et al. 2020a, b). It is expected that the downfall in the air pollutant concentrations are not correlated with weather conditions (He et al. 2020). In Nagpur, no specific correlation between T, RH and number of corona cases is observed. In Mumbai and, T is found to be negatively associated with cases whereas RH is found to have positive correlation (0.6) with number of cases (Suhaimi et al. 2020). In Delhi, RH is negatively associated with number of cases whereas temperature is found to exhibit positive correlation (0.7) with number of cases. The duration of the epidemic is insufficient to study the impact of climatic factors on the transmission of SARS-Cov-2 virus (Wang et al. 2020a, b). It can be concluded that the total seasonal effects of climatic variables are indeterminate due to uncertainty in the effects of temperature and humidity.

3 Conclusion

Using a suitable and comprehensive dataset of the study areas, the positive effects of city lockdown on air quality has been analysed which partly offset the costs of the COVID-19 pandemic. The temporal investigation showed that there is a significant decrease in $PM_{2.5}$ and PM_{10} concentration at the study locations after the lockdown. The extent of decline in the pollutant concentration at the study locations is different. $PM_{2.5}$ showed an average reduction in concentration of 40%, 53%, 45% and 34% at Delhi, Mumbai and Nagpur respectively, whereas PM_{10} showed an average reduction in concentration of 37%, 53%, 35% and 30% at Delhi, Mumbai and Nagpur respectively. The result shows that the pollutants are significantly decreased, while the average level of O_3 has been increased in Delhi and slightly decreased at Mumbai and Nagpur during 2020 in comparison with 2019. Decrease in NO_2 and CO at Delhi and Mumbai (Vehicular activities) and increase in SO_2 (industrial, however in depth investigation needed) Meanwhile, around 39% and 43% was the average reduction of PM_{10} and $PM_{2.5}$ respectively, during lockdown compared with previous year owing to restricted vehicles movement, biomass burning events and construction activities. We find that the radical precautionary measures against the spread of the epidemic have a substantial impact on air quality. The remarkable enhancement in air quality might lead to considerable health benefits. The activities identified provoking the pollutants to cross the air quality standards must be controlled after the pandemic is over benefiting human health and society. Overall, the study exhibits the association between air pollution, meteorology and receptiveness to COVID-19 cases in study areas and finds that the lockdown indeed improved the air quality. The study shows that lockdown resulted in controlling pollutants for a time being, due to reduction in anthropogenic activities. This indicates that government should implement effective control policies after the lockdown to reduce pollutant concentrations in the cities.

References

Ahmadi M, Sharifi A, Dorosti S, Ghoushchi SJ, Ghanbari N (2020) Investigation of effective climatology parameters on COVID-19 outbreak in Iran. Sci Total Environ 10:729

Anjum NA (2020) Good in the worst: COVID-19 restrictions and ease in global air pollution. Preprint, 2020040069

Bari MA, Kindzierski WB, Roy P (2020) Identification of ambient SO_2 sources in industrial areas in the lower Athabasca oil sands region of Alberta, Canada. Atmos Environ 231:117505

Bray CD, Battye WH, Aneja VP, Schlesinger WH (2021) Global emissions of NH3, NOx, and N2O from biomass burning and the impact of climate change. J Air Waste Manag Assoc 71(1): 102–114

Broomandi P, Karaca F, Nikfal A, Jahanbakhshi A, Tamjidi M, Kim JR (2020) Impact of COVID-19 event on the air quality in Iran. Aerosol Air Qual Res 20(8):1793–1804

Carleton T, Cornetet J, Huybers P, Meng K, Proctor J. (2020) Ultraviolet radiation decreases COVID-19 growth rates: global causal estimates and seasonal implications. SSRN Electronic Journal 3588601

Chu B, Zhang S, Liu J, Ma Q, He H (2021) Significant concurrent decrease in PM2. 5 and NO2 concentrations in China during COVID-19 epidemic. J Environ Sci 99:346–353

Contini D, Costabile F (2020) Does air pollution influence COVID-19 outbreaks? Atmos 11(4):377

Dutheil F, Baker JS, Navel V (2020) COVID-19 as a factor influencing air pollution? Environ Pollut 263:114466. https://doi.org/10.1016/j.envpol.2020.114466

Ficetola GF, Rubolini D (2021) Containment measures limit environmental effects on COVID-19 early outbreak dynamics. Sci Total Environ 761:144432

Gautam S (2020) The influence of COVID-19 on air quality in India: a boon or inutile. Bull Environ Contam Toxicol 104(6):724–726

Gautam S, Hens L (2020) SARS-CoV-2 pandemic in India: what might we expect? Environ Dev Sustain 22:3867–3869

Han Y, Lam JC, Li VO, Guo P, Zhang Q, Wang A, Crowcroft J, Wang S, Fu J, Gilani Z, Downey J (2020) The effects of outdoor air pollution concentrations and lockdowns on Covid-19 infections in Wuhan and other provincial capitals in China. Preprints:2020030364

He G, Pan Y, Tanaka T (2020) COVID-19, city lockdowns, and air pollution: evidence from China. medRxiv

Jain S, Sharma T (2020) Social and travel lockdown impact considering coronavirus disease (COVID-19) on air quality in megacities of India: present benefits, future challenges and way forward. Aerosol Air Qual Res 20(6):1222–1236

Jiang Y, Wu XJ, Guan YJ (2020) Effect of ambient air pollutants and meteorological variables on COVID-19 incidence. Infect Control Hosp Epidemiol 41(9):1011–1015

Kerimray A, Baimatova N, Ibragimova OP, Bukenov B, Kenessov B, Plotitsyn P, Karaca F (2020) Assessing air quality changes in large cities during COVID-19 lockdowns: the impacts of traffic-free urban conditions in Almaty, Kazakhstan. Sci Total Environ 15(730):139179

Méndez-Arriaga F (2020) The temperature and regional climate effects on communitarian COVID-19 contagion in Mexico throughout phase 1. Sci Total Environ 15(735):139560

Menebo MM (2020) Temperature and precipitation associate with Covid-19 new daily cases: a correlation study between weather and Covid-19 pandemic in Oslo, Norway. Sci Total Environ 1(737):139659

Moftakhar L, Mozhgan SEIF, Safe MS (2020) Exponentially increasing trend of infected patients with COVID-19 in Iran: a comparison of neural network and ARIMA forecasting models. Iran J Public Health 49(Suppl 1):92

NASA (2020). https://earthobservatory.nasa.gov/images. Accessed 22 July 2020

Riediker M, Morawska L (2020) Low exhaled breath droplet formation may explain why children are poor SARS-CoV-2 transmitters. Aerosol Air Qual Res 20(7):1513–1515

Rios V, Gianmoena L (2020) Is there a link between temperatures and COVID-19 contagions? Evidence from Italy. medRxiv

Suhaimi NF, Jalaludin J, Latif MT (2020) Demystifying a possible relationship between COVID-19, air quality and meteorological factors: evidence from Kuala Lumpur, Malaysia. Aerosol Air Qual Res 20(7):1520–1529

Wang P, Chen K, Zhu S, Wang P, Zhang H (2020a) Severe air pollution events not avoided by reduced anthropogenic activities during COVID-19 outbreak. Resour Conserv Recycl 158: 104814

Wang J, Tang K, Feng K, Li X, Lv W, Chen K, Wang F (2020b) High temperature and high humidity reduce the transmission of COVID-19. arXiv preprint arXiv: 2003.05003

WHO Report AAP_BoD_results_May 2018. https://www.who.int/airpollution/Ambient:2018

Wu C, Wang G, Li J, Li J, Cao C, Ge S, Xie Y, Chen J, Liu S, Du W, Zhao Z (2020) Non-agricultural sources dominate the atmospheric NH3 in Xi'an, a megacity in the semi-arid region of China. Sci Total Environ 20(722):137756

Zhu Y, Xie J, Huang F, Cao L (2020) Association between short-term exposure to air pollution and COVID-19 infection: evidence from China. Sci Total Environ 20(727):138704

Understanding Urban Floods as Extreme Events and Disaster Management: A Case Study of Bengaluru

Sonal Bindal, Sandipan Samanta, and Anil Kumar Gupta

Abstract Flooding is by far the most natural hazards, affecting the greatest number of people on the planet. Severe floods in urban areas have become more common in recent years as a result of uncontrolled urbanisation and climate change, and this trend is expected to continue in the coming years. Climate change is constantly changing the weather pattern. There has been rise in annual average temperature, which has affected the variation of sea-land surface temperature. Thus, leading to change in the monsoon pattern over Indian ocean and Indian Sub-continent. Now-a-days, it is usually experienced with increase in number of summer days and lesser winter days. The monsoon system of the Indian Subcontinent is greatly affected by Indian Ocean and due to changing pattern, shorter rainfall period having higher intensity rainfall for very short period of time leading to occurrence of extreme rainfall events over isolated areas. This leads to the condition of sudden disaster like situation in urban agglomeration where due to poor water drainage structure the situation has been occurring more frequently. The results identify 18% of total area is liable to urban floods out of which 10% area indicates high risk, 50% area shows medium risk and remaining 34% area falls under low risk of flooding. Thus, making an alarm to an urban man-made disaster due to heavy rainfall in urban areas. Due to human involvement in the environment, there have been a dominance over the natural components leading to change in rainfall pattern.

Keywords Climate change · Extreme rainfall · Urban floods · Population · Land use/land cover patterns

---

S. Bindal (✉) · A. K. Gupta
National Institute of Disaster Management, Ministry of Home Affairs, Govt of India, New Delhi, India

S. Samanta
Department of Water Resources, TERI School of Advanced Studies, New Delhi, India

B. Phartiyal et al. (eds.), *Climate Change and Environmental Impacts: Past, Present and Future Perspective*, Society of Earth Scientists Series,
https://doi.org/10.1007/978-3-031-13119-6_25

1 Introduction

Over the last 40 years, flood is the most natural hazards, taking account for 43% of all disasters and distressing the world's largest population (EM-DAT 2015; Centre for Research on the Epidemiology of Disasters 2015). In recent years, the severity of flood events in urban areas has increased significantly. The most populous cities in India, including Mumbai, Bengaluru, Chennai, and Hyderabad, have suffered significant economic losses, as well as severe communication, power, and transportation failures, epidemics, and, in some cases, the loss of human lives, as a result of urban floods (Khaladkar et al. 2009; Mukherjee et al. 2018). This is due to change in patterns of cloud formations which has heavy impact on the changing rainfall intensity, duration and frequency of the rainfall pattern in future (Trenberth 2011; Pour et al. 2020).

Flooding in urban areas is also caused by unplanned growth in low-lying areas, solid waste accumulation blocking drainage routes, and altered rainfall patterns due to climate change (Pour et al. 2020). Un-managed urbanisation is pushing land use into low-lying areas and near floodplains. This cumulative settlement is to blame for altered flow paths and the sealing of natural surfaces, both of which result in significantly increased runoff volume. The majority of drainage lines in India were constructed in the past to handle rainfall intensities ranging from 12 to 20 mm (Ali et al. 2014). These drains are easily overwhelmed when heavy rains fall and as a result, the overflowing storm water spreads into neighbouring areas, resulting in flood-like conditions (Rajeevan et al. 2008). Floods of this type range from minor incidents to major events that inundate cities for several days, causing widespread societal consequences such as temporary habitation relocation, infrastructure damage, and water quality deterioration (Gupta and Nair 2011).

One of the main consequences of extreme events is urban flooding and water accumulation in cities, urban and semi-urban areas, not only on the Indian sub-continent but also globally (Goswami et al. 2006). High-intensity rainfall that occurs for a short period of time is a growing concern in cities, as it has a direct impact on the urban social structure as well as the local environment (Mukherjee et al. 2018). This type of phenomenon has an impact on the entire hydrological characteristics of the catchment. This has an impact on the elements present at the surface as well as the biotic components above and below the ground surface. This type of situation cannot be described as an extreme event in the twenty-first century because the frequency with which it is likely to occur has increased, putting coastal areas at risk. Furthermore, climate change, as well as rising sea levels and temperatures, have exacerbated the problem by causing glaciers to melt and river landscapes to change (IPCC 2007). Over the next century, a global average sea level rise of 9–88 cm is predicted. As a result, the seas and oceans are experiencing hydrological and climatological changes. Furthermore, they are a major driving force in the local cyclonic activities, monsoon pattern, and wind-pattern (Kripalani and Kumar 2004).

The oceans and seas are clearly affected by climate change and its consequences, resulting in the occurrence of extreme events such as high intensity rainfall and shorter wet days (Ali et al. 2014). In addition, rapid urbanisation and industrialization have resulted in a drastic change in local atmospheric conditions. Many of these are at risk due to lack of planning and ecological factor development, putting major cities like Mumbai, London, Chicago, Sydney, Kolkata, and Karachi at risk (Gupta and Nair 2011; Ranger et al. 2011). As urbanisation encroaches more wet lands, open spaces, tanks, and wooded areas are converted to urban and suburban areas, the surface area available for water infiltration into soils decreases. A few examples of areas that cause less soil infiltration are: residential/commercial/industrial/bituminous roads/parking lot Land paving has increased impervious surface, reducing water percolation (Chithra et al. 2015). As a result of high intensity rainfall, these drainage systems are currently insufficient, causing flooding.

Recent studies highlight the increase in extreme rainfall events due to changing climate. All this are evident as untimely rainfall, delayed monsoon, large variation in the rainfall both spatially and temporally, heavy downpour, unseasonal rains, cloudburst etc. It has made us visualize that we are at an alarming stage (Trenberth 2011; Bhandiwad 2015). The Kedarnath flash floods, Chennai floods, season waterlogging and submergence due to rain in Delhi, Kolkata, Mumbai in July 2005, Chennai in October 2005 and again in December 2005, and Bangalore in October 2005 caused heavy economic damages, loss of life, disruption of transport, etc. (Guhathakurta et al. 2011; Ramachandra et al. 2017).

As the subcontinent is likely to experience high climatic variability in the future, Indian cities will experience more flooding from high intensity and longer duration rainfall events. The primary goal of the current research is to better understand how urban floods caused by extreme rainfall events in the urban city of Bengaluru affect the frequency and magnitude of floods that occur in the city. The study evaluates various hydrological characteristics associated with precipitation and runoff factor, as well as the region's vulnerability and risk factor, which could affect millions of people in the near future. Furthermore, the study attempted to identify and understand the gaps and connections that exist in the management of urban floods (Rafiq et al. 2016).

2 Materials and Methods

2.1 Study Area

Bengaluru is the third most populous city in Southern India and the fifth most populous city in India, and it is growing at a rapid pace. The city has a total population of 10 million people, with an additional 8.5 million people living in the metropolitan area. The city is located 900 meters above sea level on the Deccan plateau. It is located in the south-eastern corner of Karnataka State, between the latitudinal parallels of 12° 39′ N and 13° 18′ N and the longitudinal meridians of 77°

22′ E and 77° 52′ E, at an elevation of about 900 meters and covering an area of about 2191 square kilometers (Bengaluru rural and urban districts). Apart from being the state's political capital, the city is also a major commercial centre for some of the state's major industrial establishments.

2.2 Climate

The state capital has a four-season climate that is suitable for living. The tropical moderate climate has a dry summer season from March to June, followed by the south-west monsoon season from June to October, which is generally driven by winds from the Indian Ocean. It also receives a fair amount of rainfall from the retreating monsoon winds in the north-east until November. Then there's the mild winter, with temperatures ranging from 30 degrees in January to as low as 15–17 degrees Celsius. The average annual rainfall is 856.9 mm, with the S-W Monsoon accounting for more than 60% of the total and the retreating monsoon accounting for the remainder (Kumar et al. 2016).

2.3 Hydrology

The city's potable water supply is provided by the Krishna river, which is located downstream. The Arkavati river is about 50 kilometers from the city, and its tributaries are used to discharge sewage from the city. The waste is disposed of in a tributary of the Vishabhavati river, which carries the waste-water load. Freshwater is obtained from either rainfall or underground sources, which are typically aquifers. Underground cracks, fishers, and rock joints are known to be confined sources of water that are extremely difficult to extract. Deep bore-wells and dug-wells are used to extract water from unconfined aquifers at depths of up to 100 metres below the earth's surface.

2.4 Land Use Land Cover Change/Pattern

Intensive encroachment for the rising city, illegal occupying of wetlands, filling of water-bodies and construction of residential blocks, breaching of natural and ancient water supply channels and lakes has led to choking of storm water in the city as the complete hydrology of the city was good as it has a good natural drainage technique because natural gradient from the plateau and percolation. Storm water drainage systems are in disrepair, causing drains and sewage to overflow and clog. Water drainage efficiency has deteriorated over time as a result of population growth.

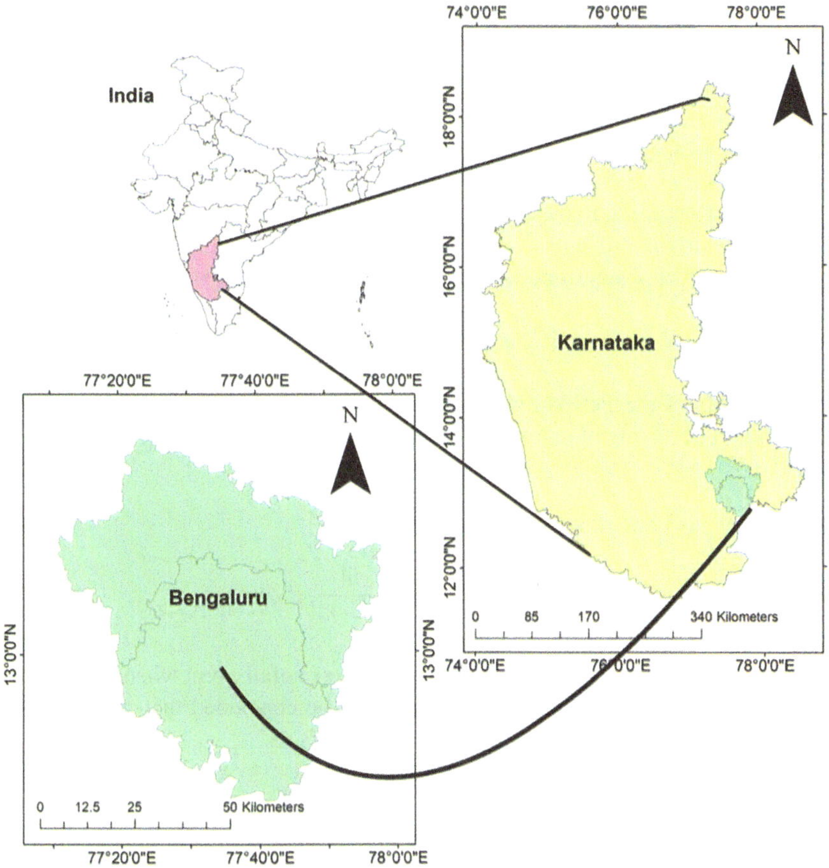

Fig. 1 Location map of the study area

In addition, precipitation and waste water runoff into common lakes, water tanks and channels pollutes the surface water rather than restoring it. So much precipitation, even on unseasonal days, causing metropolitan flooding. Foaming has also occurred in the past due to the mixing of poisons in surface water lakes. As a result, the hydrologically rich zone is transformed into a terrible limited condition, affecting people's lives (Fig. 1).

The city is losing its green cover and the water bodies, all due to the anthropogenic activities one of them is dumping of waste-residential as well as commercial and industrial. The solid waste which are non-bio-degradable in nature is dumped in water channels leading to choking of drains which in turn increase the waste load and some of the waste gets decomposed over the bed of channels and water bodies thus lowering the water volume in the cross-section of water bodies leading to additional rain leads to urban flooding.

2.5 Methodology

Bengaluru's maximum one-day rainfall of 180 mm in 1997 has a return period of 50 years over the same area and date. This predicts the likelihood of extreme rainfall. The frequency analysis of point-rainfall is commonly used to obtain such data. Annual data series are used to study the probability of an event occurring, which is applicable to random hydrological processes (Goyal 2016). Therefore, the probability of occurrence of any event for any kind of random variable is given by recurrence interval or return period of that event:

$$\mathrm{T} = 1/\mathrm{P} \tag{1}$$

Where T is the Time Period and P is the probability of occurrence. Now, if the probability of occurring of extreme rainfall event is P then non-occurrence is known as "q".

$$\mathrm{q} = (1 - \mathrm{P}) \tag{2}$$

$$P_{r,n} = C_r P_r q q^{n-r} = \frac{n!}{(n-r)!} P^r q^{n-r} \tag{3}$$

Where $P_{r,\,n}$ = probability of occurrence of extreme rainfall event where P occurring times r in n successive years. Hence, it can also be concluded that probability of occurring an event at least once in n successive years.

$$\mathrm{P}_{1=} 1 - \mathrm{q}^{\mathrm{n}} = 1 - (1 - \mathrm{P})^{\mathrm{n}} \tag{4}$$

2.6 Hydro-meterological Assessment

Rainfall is the main cause of Bangalore floods. Because rainfall intensity, duration, and frequency determine the extent and severity of flooding, rainfall analysis is critical for flood mapping. Rainfall time series analysis can be done in a variety of ways. The Indian Meteorological Department (IMD) provides daily station rainfall data (1981–2020) for this study.

Strange monsoon table is used where cumulative way to find the runoff volume generated from the daily rainfall, if considered to be daily rainfall occurring on a particular area and whole volume is considered to be the run-off volume. Considering an Empirical yet widely used formula for obtaining the runoff volume from the daily rainfall data by using three following methods. Using the Inglis and Dsouza formula as between annual runoff R in cm and P as annual precipitation in cm.

$$\text{For Ghats Region of Western India}: R = 0.85P - 30.5 \quad (5)$$

$$\text{For Deccan Plateau}: R = 1/254\, P\, (P - 17.8) \quad (6)$$

Another theory based empirical formula related to flood area-peak relationship where the area is known and on basis of it that we can get the peak of flood.

$$Q_{p=}\ 124A/(A + 10.4)^{1/2} \quad (7)$$

Where Q is the discharge generated from the A in km^2 of a particular catchment.

2.7 SCS-CN Curve Number Method

In this type generally the S parameter is used for the soil-moisture-vegetation relationship for a particular catchment.

$$S = 25{,}400/CN - 254 = 254\ (100/CN - 1) \quad (8)$$

The constant 254 is used to express S in mm. The curve number CN is related to S as

$$CN = 25{,}400/S + 254 \quad (9)$$

This formula is very widely used and it is prevalent because of the antecedent moisture condition in the various types of soil groups. Condition-I when soil is dry, but not to the wilting point. Condition-II Average condition and Condition-III Sufficient rainfall has occurred in past 5 days.

For AM- III condition

$$CN_{III} = CN_{II}/(0.427 + 0.00573\ CN_{II}) \quad (10)$$

$$Q = (P - 0.2)^2/(P + 0.8S) \quad (11)$$

For $P > 0.2S$, Where, Q is the discharge and P is the precipitation and Snyder's Number derived from above equation.

3 Result and Discussion

Rainfall patterns for Bengaluru over 40 years, from 1981 to 2020, show that annual average rainfall has varied from 875.43 mm, 1094.38 mm, 897.30 mm and 846.05 mm for the years 1981–1990, 1991–2000, 2001–2010 and 2011–2020. The results of this study reveals that the low and highest annual rainfall is also

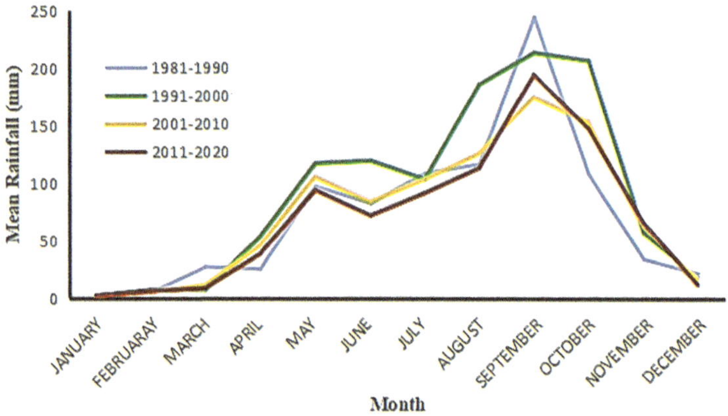

Fig. 2 Shift in the Mean Rainfall for Bengaluru in last four decades

decreasing from 1981 to 2020. Although, there is a decrease in number of rainy days, but there is an increase in the overall intensity of rainfall and shifting of rainfall periods. The Fig. 2 depicts that there is a increase in the average rainfall in the month of October. This is also seen as a direct impact of climate change (Guhathakurta et al. 2011). As a result, it is prudent to understand rainfall and its changing patterns, to be prepared to deal with its consequences, and to adapt our lives in response to it.

Bengaluru is known as India's Silicon Valley, but rapid urbanisation and modernization demands are destroying wetlands, lakes, and water bodies, affecting the local hydrological characteristics of the watershed. The results show that pre-monsoon rains used to peak in Bangalore in May, then again in September and October. Unlike the pre-monsoon season, the monsoon season has longer periods of recurrent rain. In the past, water had more time to percolate into the natural ground. However, due to the roads and buildings that cover much of our city, surface penetration has decreased (Singh et al. 2018). Storm water runoff now accounts for the majority of rainwater loss. Extreme weather events have increased, resulting in urban flooding.

Table 1 indicates that the retreating monsoon accounts more rainfall than the south-west monsoon. Over the years there is increase in the rainfall concentration over the Bengaluru capital as southern part of India is getting more rainfall than the northern region. Even the mean annual rainfall of a half-decade is 855.75 mm. This is due to various anthropogenic activities the situation of climate change and other factors has been increased leading to have extreme cases of rainfall events in India (Bhandiwad 2015).

The mass influx of the population from various parts of India has resulted in increase in the last five decades and there has been a 637% increase in urban areas in Greater Bengaluru area from 1973 to 2009 and it is still growing. The rise in built up area from 16% in 2000 to around 24 in 2009 and almost 30% today has seen a corresponding decrease in wet lands, breaching of lakes and tanks and decrease in green cover. The mean of 50 years rainfall data from year (1963 to 2013) has been

Table 1 Extreme rainfall events recorded (1963–2013)

S. No	Bengaluru	Date	Year	Rainfall (mm)	S. No	Bengaluru	Date	Year	Rainfall (mm)
1	North	11th Jan	1969	13.5	13	South	22nd Feb	1986	42.5
2	North	17th Feb	2013	153	14	South	16th March	2008	94.5
3	North	19th April	2001	108	15	South	17th April	1971	64.4
4	North	26th May	2002	95	16	South	26th May	2002	116.9
5	North	1st June	2009	84.5	17	South	28th June	2013	103
6	North	18th July	1998	123.5	18	South	22nd July	1979	67.5
7	North	4th August	1998	92	19	South	12th August	1996	103
8	North	30th September	1963	91.9	20	South	14th September	2007	108.4
9	North	23rd October	2005	120.7	21	South	1st October	1997	178.9
10	North	16th November	1991	105.6	22	South	15th November	1991	103.6
11	North	5th December	1993	66.4	23	South	5th December	1993	71.3
12	South	11th Jan	1967	24.2	24	South	15th December	1995	81.3

used in order to find the various hydrological analysis of the Bengaluru city which resulted in the changing condition of the area and occurrences of extreme rainfall events. The annual mean rainfall of the Bengaluru urban and rural is 897.3033 mm calculated. The mean rainfall of monthly data as normal rainfall is 855.75 mm. This clearly shows that the area is affected equally by S-W monsoon as well as N-E retreating monsoon. From the study we can say how maximum of the extreme rainfall events are usually occurred during the retreating monsoon as from the table inferred that October being the wettest month in overall 50 years.

The results have found the probability of occurrence of an extreme rainfall event in next 30 years was found to be 0.333. Recently, extreme rainfall event occurred on a single day in October 2015, with precipitation more than 120 mm/day in isolated pockets of Bengaluru. Additionally, risk was calculated for getting an extreme rainfall event where the probability of occurrence of the event may be at least once in 30 years is 0.4545. Hence, it clearly shows that with increase in the value of precipitation of annual mean with lesser duration, leads to increase in the chance of occurrence of extreme weather event.

By using the Eqs. (5 and 6) given by Inglis and Dsouza, annual rainfall was calculated as 254.1 mm over the area of Bengaluru city. It is an empirical formula generally used for the Western Ghats region of South India. Again, by using the Inglis formula in the flood peak-area relationship, runoff value obtained to be $3277m^3/s$ as the flood peak if it rains for 24 hrs.

Further, by using the infiltration as well as soil-vegetation- moisture condition through Eqs. (10 and 11). This has helped in understanding the red soil topography having values from silty to clayey texture of the soil. By considering the AMC type 3 condition, rainfall has occurred almost every day for the month of July. Also, as per keeping the soil to be mixture of clayey slit of red type soil Group-B soil type is used. Now, the value of infiltration capacity of the soil for the month of July having an annual mean rainfall of 126.46 mm, the value comes out to be $0.886m^3/s$. The CN number is 90 and 98 for 60% group B and 92, 98 for 40% group C. In that case 96.917 is the weighted AMC_{III} condition.

The calculation of probabilities of rainfall using different equations helped in understanding the various scenarios leading to increase in the number of extreme weather events causing urban floods. The study aids in defining the influence of anthropogenic activities has resulted in changing hydrological characteristics of a certain area which may be a Silicon Valley of India.

4 Summary and Conclusion

Urban flood is significantly different from rural floods and modeling of such events is a challenge to the researchers due to its unpredictable nature. A vast number of numerical models have evolved over the past few years capable of flood mapping; most of them are commercial, rigorous and need extensive datasets to generate precise results. The booming population due to migration to the city for education and job has increased the demand for residential, commercial and industrial elements. The disturbances in the hydrological cycles, demand and supply for domestic and industrial needs, all leads to shift in the localized climate of Bengaluru city. More than 79% of the total water-bodies have been vanished, more specifically got sacrificed due to the rapid changing demographic scenario of the city.

This research presents a simple sophisticated approach to analyze extreme rainfall events based on past critical events and synthetic hyetographs for the Bengaluru city. Google Earth projection of risk maps visualizes the areas under high risk of flooding, namely Gali Anjaneya Swamy Temple junction, Bannerghatta Road, Jayadeva flyover, Nayandahalli Junction, Silk Board junction, Magrath-Brigade Road junction, Wilson Garden Quarters in South Bengluru, HRBR Layout and Veerannapalya in eastern Bengaluru, Rajajinagar Industrial Area in West Bengaluru, LBS Nagar in Yelahanka zone and Sarjapur Road, Varthur, Panathur, Belagere and Hoodi in the Mahadevapura zone. Hence, high priority should be assigned to these zones during flooding circumstances.

The model produces relative flooding based on underlying terrain quality and input parameters. The critical events can be effectively regenerated using developed 2D flood model for various critical storm conditions. Thus, it proves to be a reliable approach in urban environment with limited data. Thus, these models can be developed for and applied to any region with minimum input data requirement as rainfall event, digital elevation model (DEM) and study area catchment properties. The past rainfall events can be simulated using developed 2D model to visualize the flow pattern, identify the low-lying areas susceptible to flooding and generate flood inundation maps. The flood inundation maps so developed can be used for developing flood risk maps, setup early warning system to alert population before extreme events and develop combat strategies. Furthermore, the model can be used to simulate real-time flood event to map flooded area and accordingly decide combat strategy and adopt relief measures.

Further, ancient water tanks and the natural water tanks are considered to be the potential sources of drinking water as well as they are natural way of rejuvenation of water and soil. The water bodies act as infiltrators and at the same time helps in maintain the local temperature and the water balance in the ecosystem. The nearby by soil gets hydrated which leads to recharging of underground aquifers which are the sources of drinking water. Ancient structures include water tanks, jalashays, water supply pipes, irrigation reservoirs, and channels (Shivanna and Vyshnavi 2019). After the monsoon season, the tanks or water bodies can store water for up to six months. Every city is located near a river, but it is worth noting that there is no major river running through the city. The restoration and resilience of these ancient strutures must be a part of the city's disaster resilience plan.

Occurrence of such events cannot be controlled but the risk and damage can effectively be reduced with the help of modern flood modeling techniques. The areal extent and depth of flood inundation reflect the most important parameters, particularly in mapping of flood hazards (Rangari et al. 2019). Since decades there is a water dispute of Krishna river of Karnataka state with the state of Tamil Nadu due to water sharing of the state. This is due to the reason that the Karnataka being at the upstream takes water from the river and more energy is required to lift the water to 900 m to the Bengaluru city to get potable water. This has occurred due to the fact that water bodies inside the city has been destructed by encroachment, dumping of pollution and debris of construction materials and tremendous concretization has led to the effect of water being flooded or doesn't gets an ecological path to go inside the ground or flows to the downstream.

NDWI studies by Ministry of Jal Shakti states that by 2025, grey will dominate 98.5% of the city means there is so much of urbanization that only 1.5% will remain as green cover in the city. The probability of return period for an extreme rainfall event occur once in 30 years comes out to be nearly half the probability. On top of that if reducing frequency and getting more rainfall in shorter period then the probability will increase thus the extreme event with dilapidated condition of the sewerage and poor hydrological condition will further aggravate the problem leading to loss of lives, property and damage to environment as well.

The climate change is a global situation but if locally act, there can be a tremendous change in the hydrological and climatological characteristics of the state (Bhandiwad 2015). The area gets rain from the winds of Arabian Sea but while retreating the rainfall duration is less but the intensity is more. This has resulted to change in the vegetation and biodiversity of the nearby districts of Hosur, Mysuru and other places. This is surely an alarm ringing as extensive destruction of ecology and the water resources has led to destruction of mankind. It is very much necessary to adopt various techniques to combat the sudden disasters which are occurring in the area.

Acknowledgement The authors acknowledge the support and motivation from Maj. Manoj Kumar Bindal, Executive director, NIDM.

References

Ali H, Mishra V, Pai DS (2014) Observed and projected urban extreme rainfall events in India. J Geophys Res Atmos 119(22):12–621

Bhandiwad R (2015) Variability of rainfall in Bangalore city - mapping of hourly variation of heavy rainfall event. Am Int J Res Sci Technol Eng Math (AIJRSTEM) 15–160:124–139

Centre for Research on the Epidemiology of Disasters UNISDR (2015) The human cost of weather related disasters 1995–2015. https://www.unisdr.org/we/inform/publications/46796 Accessed Dec 2018

Chithra SV, Nair MH, Amarnath A, Anjana NS (2015) Impacts of impervious surfaces on the environment. Int J Eng Sci Invention 4(5):27–31

EM-DAT, CRED (2015) Centre for Research on the Epidemiology of Disasters. The International Disaster Database. Available online www.emdat.be. Accessed 6 Feb 2019

Goswami BN, Venugopal V, Sengupta D, Madhusoodanan MS, Xavier PK (2006) Increasing trend of extreme rain events over India in a warming environment. Science 314(5804):1442–1445

Goyal MK (2016) Engineering hydrology. PHI Learning Pvt. Ltd., Delhi

Guhathakurta P, Sreejith OP, Menon PA (2011) Impact of climate change on extreme rainfall events and flood risk in India. J Earth Syst Sci 120(3):359–373

Gupta AK, Nair SS (2011) Urban floods in Bangalore and Chennai: risk management challenges and lessons for sustainable urban ecology. Curr Sci 100:1638–1645

IPCC (2007) Impacts, adaptation and vulnerability. Contribution of Working Group II to the fourth assessment report of the Intergovernmental Panel on Climate Change, 976

Khaladkar RM, Mahajan PN, Kulkarni JR (2009) Alarming rise in the number and intensity of extreme point rainfall events over the Indian region under climate change scenario. Indian Institute of Tropical Meteorology. Research Report no. RR-123

Kripalani RH, Kumar P (2004) Northeast monsoon rainfall variability over south peninsular India vis-à-vis the Indian Ocean dipole mode. Int J Climatol 24(10):1267–1282

Kumar P, Geneletti D, Nagendra H (2016) Spatial assessment of climate change vulnerability at city scale: a study in Bangalore, India. Land Use Policy 58:514–532

Mukherjee S, Aadhar S, Stone D, Mishra V (2018) Increase in extreme precipitation events under anthropogenic warming in India. Weather Clim Extremes 20:45–53

Pour SH, Abd Wahab AK, Shahid S, Dewan A (2020) Low impact development techniques to mitigate the impacts of climate-change-induced urban floods: current trends, issues and challenges. Sustain Cities Soc 20:102373

Rafiq F, Ahmed S, Ahmad S, Khan AA (2016) Urban floods in India. Int J Res Sci Eng Res 7(1): 721–734

Rajeevan M, Bhate J, Jaswal AK (2008) Analysis of variability and trends of extreme rainfall events over India using 104 years of gridded daily rainfall data. Geophys Res Lett 35(18):L18707

Ramachandra TV, Vinay S., Bharath HA (2017) Frequent floods in Bangalore: causes and remedial measures (vol. 123). ENVIS Technical Report No. ETR123

Rangari VA, Umamahesh NV, Bhatt CM (2019) Assessment of inundation risk in urban floods using HEC RAS 2D. Model Earth Syst Environ 5(4):1839–1851

Ranger N, Hallegatte S, Bhattacharya S, Bachu M, Priya S, Dhore K, Rafique F, Mathur P, Naville N, Henriet F, Herweijer C (2011) An assessment of the potential impact of climate change on flood risk in Mumbai. Clim Chang 104(1):139–167

Shivanna S, Vyshnavi DV (2019) Geomorphological study of Yelahanka watershed, Bangalore rural district, Karntaka, India. Int J Adv Sci Res Manag 4(8):73–77

Singh P, Sinha VSP, Vijhani A, Pahuja N (2018) Vulnerability assessment of urban road network from urban flood. Int J Disaster Risk Reduct 28:237–250

Trenberth KE (2011) Changes in precipitation with climate change. Clim Res 47(1–2):123–138

Engendered Climate Risk Analysis: A Precursor to Gender Equality and Empowerment

Savita Aggarwal, Vandana Sharma, Smita Chakravarty, and Jagriti Kher

Abstract Rural and urban poor women are much more vulnerable to climatic extremes and disasters, as they have very limited resources and have to undertake multiple tasks of the household, child care as well production to sustain their families. As natural resources become scarce due to climatic changes coupled with demographic factors, the existing gender inequalities may increase worsening the situation for women. The Conference of Parties of the UNFCCC have emphasized time and again the need for gender balance and gender equity in climate action as well as negotiations. The LIMA work program on gender and consequently the Gender Action Plan have highlighted one of the priority areas, as the need for metrics to create evidence of differential vulnerability and risk to climate change.

Though several climate and gender related indices have been developed from time to time, they fail to capture the climate related dimension which make women differentially vulnerable to climate change. In the present study, a new index called Gender based Climate Risk Index (GCRI) has been used to capture the vulnerability and resultant risk faced by both men and women at the sub-national (state) level in India. A combination of the IPCC's risk assessment framework and Caroline Moser's Gender Analysis Framework has been used to arrive at the index. GCRI has been validated and tested for reliability keeping the availability of data at the required spatial scales. The study has shown that despite facing the same climate related hazards and exposure, women faced greater climatic risk as compared to men, due to their greater sensitivity and low levels of adaptive capacity. Mapping the index at the sub-national scale has also shown different states need to examine their causes of vulnerability and incorporate targeted action in their climate adaptation plans for improving the lives of women and empowering them. It is important that climate related policy making is made more gender sensitive and gender transformative to enable women to lead climate resilient lives.

Keywords Climate change · Gender · GCRI · Risk · Vulnerability

S. Aggarwal (✉) · V. Sharma · S. Chakravarty · J. Kher
Department of Development Communication and Extension, Institute of Home Economics, University of Delhi, Delhi, India

B. Phartiyal et al. (eds.), *Climate Change and Environmental Impacts: Past, Present and Future Perspective*, Society of Earth Scientists Series,
https://doi.org/10.1007/978-3-031-13119-6_26

1 Introduction

Climate change continues to impact populations around the globe by extreme events and disasters. The brunt of its impacts is borne much more by the vulnerable sections of the population comprising of rural and urban poor and much more so women, who have to undertake multiple tasks of the household as well of production to sustain their families. Other reasons for women's greater vulnerability are low education, lack of information, training and institutional support and poor participation in decision-making at various levels (Goh 2012). As natural resources become scarce due to demographic factors coupled with climatic changes, women are the most affected since they are the prime managers of natural resources (MOSPI 1998–99; RFSTE 2005; UNICEF/WHO 2010, 2012; UN Women 2010; Sonalde et. al. 2010; Koolwal and van de Walle 2011; Sorenson et al. 2011, Karim et al. 2013; Arora 2014; Asaba et al. 2014; NSSO 2014; Graham et al. 2016).

Women's inequitable access and control of resources also makes them more vulnerable to climate change and extremes (Arora-Johnson 2011; Resurreccion 2011; FAO 2011; Nellemann et al. 2011; Ravera et al. 2016). Due to time poverty for undertaking economic activities as well as leadership roles, women continue to remain disadvantaged leading to high gender inequalities in different domains thus failing to achieve the Sustainable Development Goals of gender equality and empowerment. The poorer performance of women is apparent by the fact that the global average of HDI value for women (0.705) is almost 6% lower than that of men (0.749) showing poorer achievement of females as compared to males in different domains (UNDP 2018). Among the developing regions, South Asia has the widest gender gap of 16.3% in levels of human development (UNDP 2018), reflecting the gender-based inequalities. India's performance in development and gender-based indices is much lower as it ranks 131 out of 189 countries in HDI (UNDP 2020), 123 out of 160 countries in Gender Inequality Index (UNDP 2020) and 112 out of 153 countries in Gender Gap Index (WEF 2020).

It is reported that climate change will make the situation even worse by increasing the existing gender inequalities thereby contributing to greater climate related vulnerability of women (IPCC 2014; Sultana 2018). Studies have shown strong links between climate related disasters and more deaths, higher injuries and fatalities of women (Peterson 2007; Cotthem 2009; Neumayer and Plumper 2007; WEDO 2008; Haigh and Vallely 2010; Bradshaw and Fordham 2013). It is important to note that patriarchy reinforces gender imbalance by preserving certain benefits pertaining to nutrition, health and power in favour of men. Higher deaths among women during the time of disasters are also linked to patriarchy induced vulnerability of women (Arora-Johnson 2011; Onwutuebe Chidiebere 2019).

Thus, climate change multiplies the existing gender inequalities and increases the problem of marginalization faced by women. It is well accepted globally that vulnerable communities especially women and female children capacity to cope and adapt in the face of natural disasters, climate change and other extreme weather events. Despite working very closely with environment, women face greater

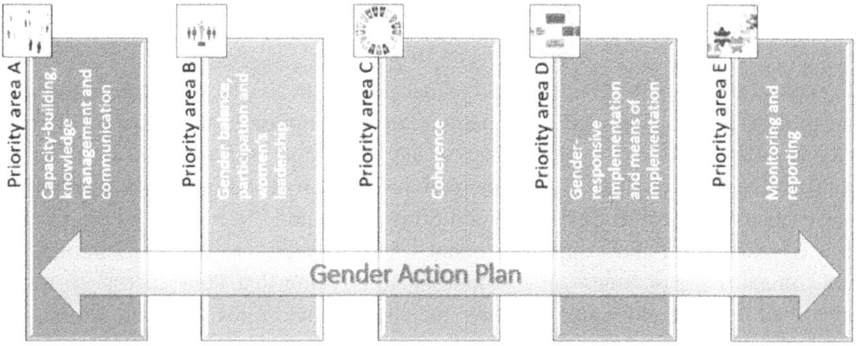

Fig. 1 Five priority areas of the Gender Action Plan (Source: UNFCCC, Gender Action Plan https://unfccc.int/topics/gender/workstreams/the-gender-action-plan)

challenges due to changing climate and climatic extremes. The Conference of Parties of the UNFCCC have emphasized time and again the need for gender balance and gender equity in climate action as well as negotiations. The LIMA work program on gender and consequently the Gender Action Plan (Fig. 1) have highlighted one of the priority areas as the need for metrics to create evidence of differential vulnerability and risk to climate change.

Several indices have been developed from time to time to quantify vulnerability/ climate related vulnerability of populations and regions such as Water Poverty Index, Climate Vulnerability Index (CVI), (Sullivan 2002; Sullivan and Meigh 2005), Climate vulnerability Index for Water at the Household Level (CVI-WH) (Kher and Aggarwal 2015), Global Climate Change Risk Index (German Watch 2013). Some gender specific indices have also been developed over time such as Gender Inequality Index, Gender Vulnerability Index (GVI) and Women Resilience Index (WRI). However, all these indices fail to capture the specific dimensions which make women differentially vulnerable to climate change.

There is thus a need for a more direct and holistic index which can assess the risk and vulnerabilities faced by women due to climatic variability, shocks and disasters. Further, since vulnerability varies across temporal and spatial scales, identifying and quantifying vulnerability at the sub-national as well as smaller spatial scales is extremely essential to take targeted and pro-active measures to reduce the risk and enhance the adaptive capacity and resilience of populations especially women. In the present study, a new tool called Gender based Climate Risk Index (GCRI) has been developed to quantify the differential climate related vulnerability and risk faced by men and women. The tool (GCRI) is in line with the UNFCCC agenda of bolstering the role of women in climate action and decisions through the gender action plan.

2 Methodology

For conceptualizing the gender-based climate risk index, it was important to consider the multifarious roles played by women in family, community and society. Different Gender Analysis Frameworks were considered for the purpose of examining the roles and needs of women. The framework proposed by Caroline Moser comprising of Practical-Gender-Needs (PGNs) and Strategic-Gender-Needs (SGNs) of women was considered the most appropriate and was integrated into the IPCC framework for climate risk assessment. It is important to mention that PGNs comprise needs identified by men and women in their socially accepted roles in society (arranging freshwater, fuel, food, fodder in rural and peri-urban areas). On the other hand, SGNs refer to the needs of education, income earning, legal status and rights, ownership and control over productive resources, participation in decision making (Moser 1993).

A holistic 'Gender based Climate Risk Index' henceforth referred to as GCRI, was therefore, developed based on the IPCC AR 5 Risk Assessment Framework. Figure 2, depicts how 'Risk' to climate change and extremes is an outcome of interplay between 'Hazards', 'Exposure' and 'Vulnerability'. The latter (vulnerability) further constitutes 'Sensitivity' and 'Lack of Adaptive Capacity' as its sub-components. The sub-components related to PGNs and SGNs were integrated into the vulnerability component of climate related risk.

The GCRI had 4 components namely Hazards, Exposure, Sensitivity and Lack of adaptive Capacity. Each of the four components had several sub-components and indicators selected after an extensive review of literature and consultation with experts. Overall GCRI values were based on a total of 54 indicators. The validity of GCRI was established by an expert consultation workshop with specialists from the fields of climate change, gender studies, health and nutrition, energy, water sectors from national and international organizations. The reliability of the index

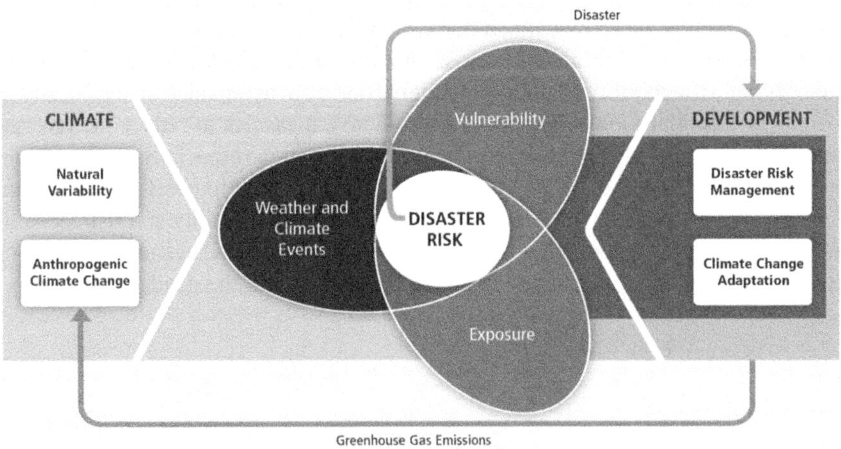

Fig. 2 The interrelation between various components of climate related risks (Source: IPCC 2014)

was established by using Cronbach alpha test for internal consistency among the indicators of sub-components.

The index was calculated using the methodology of UNDP Human Development Index (2009). GCRI was computed as follows:

$$I_{(GCRI)} = I_{H\ (1-H)} + I_{E(1-E)} + I_{V(1-V)}$$

$$\frac{I_V = I_{S(1-S)} + I_{LAC(1-LAC)}}{2}$$

*where H is hazards, E is exposure, V is vulnerability, S is sensitivity, LAC is lack of adaptive capacity.

Secondary data was collected from a variety of sources such as India Meteorological Department, Census of India, Ministry of Water Resources, Central Ground Water Board, Central Water Commission, Ministry of Women and Child Development, Ministry of Environment Forests and Climate Change.

In order to permit comparison between males and females, the indicators were normalized using maximum and minimum values from the combined data of males and females in different states in India. The Indian states were classified into six equidistant categories namely no risk, low risk, moderate risk, high risk, very high and finally extreme risk based on their GCRI score. Values closer to zero indicated least vulnerability/risk whereas, values closer to one depicted high risk.

3 Results

The sub-index values for all the four components of GCRI have shown significant variation across states. It is important to mention that the index values for hazards and exposure cannot be segregated by gender and have therefore been considered the same for males and females. In order to quantify the of Hazard index, as many as 11 climatic parameters were used. The results have shown that the index values ranged from 0.11 in Sikkim (least susceptible) to 0.67 in Rajasthan (most susceptible). As many as 12 states, most of which were large and highly populated were either moderately or highly susceptible to hazards. Interestingly these states occupied almost 64% of the India's geographical area and were home to almost three-fourth of the total population of India. India on the whole was moderately susceptible to hazards with an index value of 0.45.

In terms of exposure index, which was captured by 7 indicators, 12 states experienced moderate or high exposure comprising of almost 60% of the geographical area and home to 74% of the total population of India. These states had a large number of families living below poverty line, a high rate of forest degradation and considerable number of landless households. India, on the whole was moderately susceptible to exposure with an index value of 0.39.

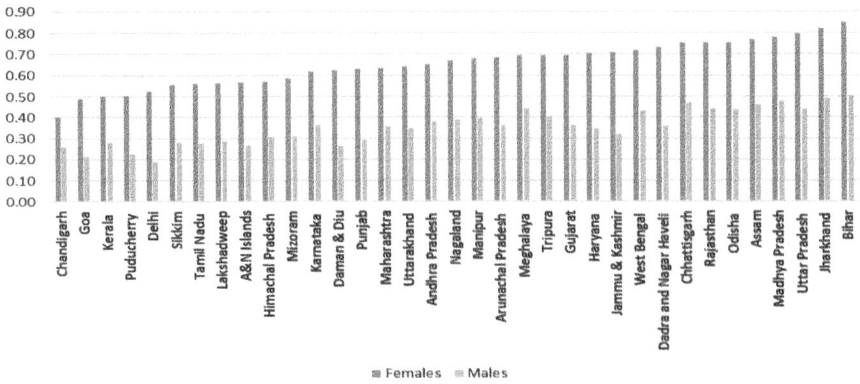

Fig. 3 Comparison of vulnerability index of females and males across states and UTs in India

In terms of the vulnerability component, the male-female differentials in various indicators were examined and quantified. There were large scale gender-based differentials across states in vulnerability to climate change and extremes. The vulnerability index values for females ranged from 0.40 in Chandigarh to as high as 0.85 in Bihar (Fig. 3). In as many as 31 states, women constituting 97% of the country's female population faced either high, very high or extreme vulnerability to climate change and extremes. On the contrary, males showed much lower levels of vulnerability ranging from 0.19 in Delhi to 0.50 in Bihar. Males experienced low vulnerability in 14 states but faced moderate levels of vulnerability in 21 states. Weighted average of vulnerability for India as a whole stood at 0.72 for females and 0.42 for males. Such huge disparities in the vulnerability of males and females can be attributed to the high sensitivity and low levels of lack of adaptive capacity of women, because of their poorer health and reproductive health, low participation in income earning and decision-making roles as well as other socio-economic, cultural barriers and patriarchal mindsets which grant women much lower privileges and resources.

In terms of GCRI, not only a huge variation was found across states in India but also there were gender differentials in risk to climate change and extremes. The GCRI values of females ranged from 0.25 (low risk) in Goa to as high as 0.62 (high risk) in Bihar (Fig. 4). By comparison, males faced much lower level of risk as the GCRI values ranged from as low as 0.16 in Goa to 0.51 in Bihar (Fig. 5). A weighted average of GCRI scores reflect a ten percent point difference between the climate-based risk faced by males and females. Most of the highly populous states such as West Bengal, Maharashtra, Odisha, Rajasthan, Andhra Pradesh, Madhya Pradesh, Uttar Pradesh and Bihar, comprising of 55% of geographical area of the country and home to 64% of the total female population, were in the category of high risk for females. These states performed poorly in all the four components of GCRI, especially in sensitivity and lack of adaptive capacity which put together comprise the vulnerability component. In the remaining states, women faced moderate levels of risk. By comparison, only 9% of the male population of India, comprising of

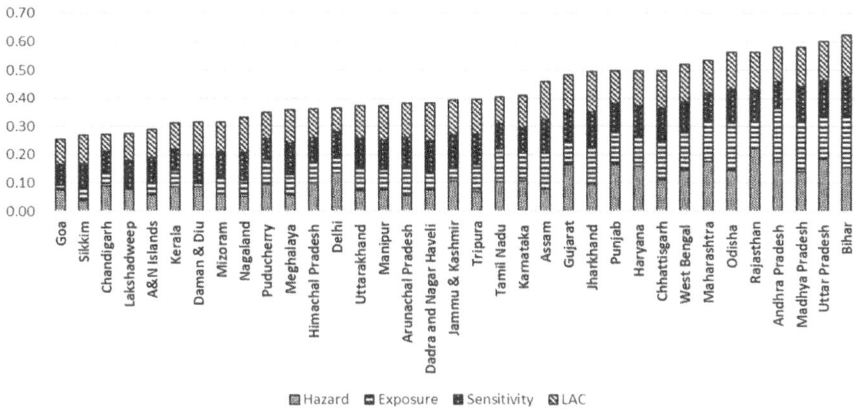

Fig. 4 Variation in GCRI across states and UTs in India—Females

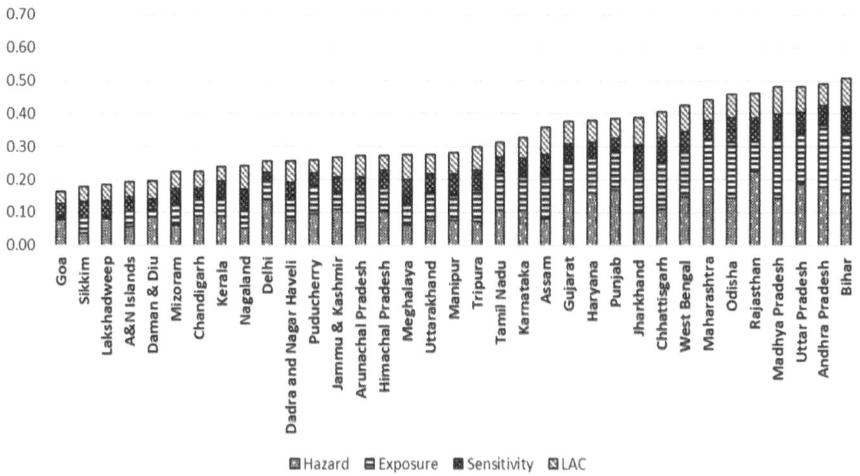

Fig. 5 Variation in GCRI across states and UTs in India—Males

almost 3% of geographical area of India faced high risk to climatic extremes and 72% were at moderate risk. The remaining male population faced no or low risk. It is important to mention that in none of the states, females were in the no risk zone, indicating wide gender-based disparity in the level of climate related risk faced by them.

4 Conclusion

The results have shown large gender-based differentials among the states in terms of risk to climate change despite males and females being exposed to the same climate related hazards and exposure. This was due to much higher vulnerability of women reflected by their greater sensitivity and limited adaptive capacity. Higher vulnerability of women is because of their poor health and reproductive health status, inadequate provision of household facilities including sanitation and very limited access to productive opportunities and resources as well as restricted participation in community roles. Women were also limited in their adaptive capacity because of lower literacy levels, poor income and extremely low participation in decision making at all levels in these states. By contrast, men faced much lower vulnerability as compared to the female counterparts because of lower sensitivity and greater adaptive capacity by virtue of better performance in all the above-mentioned domains. Thus, the developed index has provided a strong evidence of the much greater sensitivity and poorer adaptive capacity of women resulting in greater vulnerability to climate change as compared to men. Mapping the index at the sub-national scale has also shown that different states need to examine their causes of vulnerability and incorporate targeted action in their climate adaptation plans in order to improve the lives of women and empower them.

Considering that gender-based distribution of work continues till date in most developing countries, it is extremely important to bring a change in the mindset of people to bring about more balance in domestic and economic participation of women. At the same time, it is necessary to meet the Practical Gender Needs of women. In turn, these will prevent time and energy poverty in women and lead to fulfilling their *Strategic-Gender-Needs* of undertaking economic activities, education, skill development and participation in leadership and community activities. This will help in reducing their sensitivity to climate related factors and enhance their human capacity to adapt to climate change and extremes. It is extremely important to invest in women's education (for increased awareness and knowledge), access to medical facilities and sanitation (better nutritional status and reproductive health), economic participation (for income generation and consequent participation in adaptation and mitigation efforts and adoption of sustainable adaptation strategies) as well as participation in community, political and administrative domains (to enhance gender sensitive and transformative decision making at all levels). Consequently, the reduced sensitivity and enhanced adaptive capacity will have the potential to enable women to lead climate resilient lives.

References

Arora D (2014) Gender differences in time poverty in rural Mozambique. Working paper, no: 2014–05, Department of Economics, University of Utah

Arora-Johnson S (2011) Virtue and vulnerability. Discourses on women, gender and climate change. Glob Environ Chang 21:744–751

Asaba RB, Fagan Honor G, Kabonesa C, Mugumaya F (2014) Women and access to water in rural Uganda: a review. J Gend Water 3(1):1–8

Bradshaw S, Fordham M (2013) Women, girls and disasters a review for DFID. https://assets.publishing.service.gov.uk/government/uploads/system/uploads/attachment_data/file/844489/withdrawn-women-girls-disasters.pdf

Cotthem WV (2009) Women and climate change: issues of impact, equity and adaptation from http://desertification.wordpress.com/2009/01/18/women-and-climate-change Accessed 20 Nov 2020

Desai SB et al (2010) Human development in India. Challenges for a society in Transition. Oxford University

FAO (2011) Assessing forest degradation towards the development of globally applicable guidelines. Forest resources assessment working paper 177. Food and Agriculture Organization of the United Nations, Rome, Italy

German Watch (2013) Harmeling, S., & Eckstein, D. Global climate risk index, 2013. Who suffers most from extreme weather events? Weather related loss events in 2011 and 1992 to 2011. German Watch, Bonn

Goh AHX (2012) A literature review of the gender-differentiated impacts of climate change on women's and men's assets and well-being in developing countries. CAPRi working paper no. 106. International Food Policy Research Institute, Washington, DC

Graham JP, Hirai M, Kim SS (2016) An analysis of water collection labor among women and children in 24 sub-Saharan African Countries. PLoS One 11(6):0155981. https://doi.org/10.1371/journal.pone.0155981

Haigh C, Vallely B (2010) Gender and the climate change agenda: the impacts of climate change on women and public policy. Women's Environmental Network (WEN), London

Intergovernmental Panel on Climate change, IPCC (2014) Climate change 2014. Impacts, adaptation and vulnerability Part A: global and sectoral aspects. Working Group II contribution of the fifth assessment report of the Intergovernmental panel on climate change. Cambridge, New York

Karim RKM, Emmelin M, Resurreccion B., Wamala S (2013) Water, women and marital violence in a Bangladesh village. Global water Forum. Retrieved from http://www.globalwaterforum.org/2013/03/25/water-women-and-marital-violence-in-a-bangladesh-village/?pdf=6365

Kher J, Aggarwal S, Punhani G (2015) Vulnerability of poor urban women to climate-linked water insecurities at the household level: a case study of slums in Delhi. Indian J Gend Stud 22(1):15–40. https://doi.org/10.1177/0971521514556943. ISSN: 0971-5215

Koolwal G., van de Walle D (2011) Better access to water raises welfare, but not women's off-farm work. Poverty Reduction and Economic Management Network (PREM), The World Bank

Ministry of Statistics and Programme Implementation (MoSPI) (1998–99) Time use survey: summary of findings. Retrieved from http://www.mospi.gov.in/mospi_time_use_survey.htm

Moser CON (1993) Gender planning and development Theory, practice and training. Routledge, London. ISBN 0-203-41194-3

Nellemann C, Verma R, Hislop L (2011) 'Women at the frontline of climate change: gender risk and hopes', A Rapid Response Assessment, United Nations Environment Programme, GRID-Arendal

Neumayer E, Plumper T (2007) The gendered nature of natural disasters: the impact of catastrophic events on the gender gap in life expectancy, 1981–2002. Ann Assoc Am Geogr 97(3):551–566

NSSO (2014). Retrieved from https://www.hindustantimes.com/india/rural-women-go-the-extra-mile-in-walk-for-water/story-KGcr9rkU62Qiw66cSifnVN.html

Onwutuebe Chidiebere J (2019) Patriarchy and women vulnerability to adverse climate change in Nigeria. SAGE Open 9(1) http://mospi.nic.in/Mospi_New/site/inner.aspx?status=3 & menu_id=31

Peterson K (2007) Reaching out to women when disaster strikes. Soroptimist White Paper, Philadelphia
Ravera F, Martin-Lopez B, Pascual U, Drucker A (2016) The diversity of gendered adaptation strategies to climate change of Indian farmers: a feminist intersectional approach. Ambio 45 (Suppl. 3):S335–S351. https://doi.org/10.1007/s13280-016-0833-2
Resurreccion BP (2011) The gender and climate debate: more of the same or new pathways of thinking and doing. Asia security initiative policy paper 10. RSIS Centre for Non-Traditional Security (NTS) Studies, Singapore
RFSTE/Navdanya. 2005. Women and water: a report by Research Foundation for Science, Technology and Ecology for National Commission for Women, New Delhi
Sonalde BD, Dubey A, Joshi BL, Sen M, Sharif A, Vanneman R (2010) Human development in India. Challenges for a society in Transition. Oxford University Press
Sorenson BS, Morssink C, Campos PA (2011) Safe access to safe water in low-income countries: water fetching in current times. Soc Sci Med 72:1522–1526
Sullivan CA (2002) Calculating a Water Poverty Index. World Dev 30:1195–1210
Sullivan C, Meigh J (2005) Targeting attention on local vulnerabilities using an integrated index approach: the example of climate vulnerability index. Water Sci Technol 51(5):69–78
Sultana F (2018) Gender and water in a changing climate: challenges and opportunities. In: Frohlich C, Gioli G, Cremades R, Myrttinen H (eds) Water security across the gender divide. Springer, Cham, pp 17–33
UNFCCC Gender action plan. https://unfccc.int/topics/gender/workstreams/the-gender-action-plan
United Nations (2010) World's women 2010: trends and statistics. Economic and Social affairs. UN
United Nations Development Programme (2009) Human development report 2009: overcoming barriers: human mobility and development. UNDP, New York
United Nations Development Programme (2018) Human development indices & indicators. Statistical update. UNDP, Oxford University Press, New York
United Nations Development Programme (2020) Human development report 2020. The next frontier human development and the Anthropocene. UNDP, Oxford University Press, New York
WHO/UNICEF Joint Monitoring Programme (JMP) (2010) Progress on sanitation and drinking water: 2010 update. New York, USA
WHO/UNICEF Joint monitoring programme for water supply and sanitation. (2012) Progress on drinking water and sanitation 2012 update. New York, USA
Women's Environment and Development Organization (2008) Gender, climate change and human security lessons from Bangladesh, Ghana and Senegal (ELIAMEP). WEDO
World Economic Forum (2020) Global gender gap report 2020. World Economic Forum

Covid Lockdown Improves the Health of River Yamuna: A Pilot Study

Divya Ghildyal, M. B. Santhosh Kumar, and K. P. Singh

Abstract With government announcing a sudden lockdown due to the spread of COVID-19 pandemic in the mid of March 2020, all major human activities came to a standstill. The present study analyses the impact of this lockdown on the health of river Yamuna. Eighteen sample stations were chosen for analysis, through which Yamuna river flows, two at Noida, three at Vrindawan, four at Mathura, three at Agra, one each at Firozabad and Etawah and four at Allahabad, Prayagraj. Four water quality parameters DO (Dissolved Oxygen), BOD (Biological Oxygen Demand), TC (Total Coliforms), FC (Feacal Coliforms) were studied at these sampling stations. Period of study was three months prior to lockdown (January, February, March) and four months post-lockdown (April, May, June, July) data was then analysed pre- and post- lockdown and comparative studies were done. This mere comparison showed a marked increase in river Yamuna's health. DO the biggest indicator of rivers health increase at all stations except at four stations of Prayagraj. Coliform's which pile up because of human activity or tourist inflow specially at tourist spots, were found to decrease to a remarkable extent, letting the river breathe. If such parameters could be maintained, then the high costs that go into river cleaning projects can be saved and save the river from drying.

Keywords Yamuna River · COVID-19 · Mass lockdown · DO · BOD · Coliforms

---

D. Ghildyal (✉) · K. P. Singh
Department of Physics, J. S. S. Academy of Technical Education, Noida, Uttar Pradesh, India

M. B. Santhosh Kumar
Department of Civil Engineering, J. S. S. Academy of Technical Education, Noida, Uttar Pradesh, India

B. Phartiyal et al. (eds.), *Climate Change and Environmental Impacts: Past, Present and Future Perspective*, Society of Earth Scientists Series,
https://doi.org/10.1007/978-3-031-13119-6_27

1 Introduction

River Yamuna, flowing in north India is the longest and the second largest tributary of river Ganga. It originates from the Yamunotri Glacier at a height of approximately 6387 meters on the south western slopes of Banderpooch peaks in the uppermost region of the Lower Himalayas in Uttarakhand, (Misra 2010), after which it travels through various parts of Uttarakhand, Himachal Pradesh, Haryana, Uttar Pradesh, and Delhi and ultimately merges with the Ganges at a sacred spot known as Triveni Sangam at Prayagraj. In its long journey, the Yamuna flows through seven states where industrial units discharge their effluents, mostly untreated, into it. Between Haryana's Panipat and Delhi alone, over 300 units of industrial discharge are released into river Yamuna (Datta 1992) making it the country's most polluted river. The river picks up approximately 80% of its pollutants from Delhi, Agra and Mathura the NCR (National Capitol Region) (Misra 2010). Due to high density here population and rapid industrialization, river Yamuna has dumpage of untreated or partially treated wastes (Gopal & Chauhan 2007).

Major industries are discharging their waste treated and untreated into Yamuna river including pulp and paper, sugar, textiles, leather, chemical, pharmaceuticals, oil refineries, thermal power plant etc. (Sharma and Bhadauriya 2019; Kaur et al. 2021; Sharma 2019). The Central Pollution Control Board (CPCB), Central Water Commission (CWC), Delhi Pollution Control Committee (DPCC), State Pollution Control Board (SPCB) regularly monitors the river at 19 locations. Literature studies have revealed about the health of river Yamuna stating that its water is unfit for agricultural purposes (Sharma 2015; Rai 2015) domestic use and daily use (Sharma 2015; Suruchi et al. 2015). In a recent study it has been mentioned that the occurrence of antibiotic resistance in the river caused due to sewage discharge (Lamba et al. (2020). Critical pathogens listed by WHO (World Health Organization) have also been found in the river water indicating possible association of resistant bacteria with fecal matter, which is the cause of many diseases (Lamba et al. 2020).

The major sources contributing to the pollution of Yamuna are untreated sewage, industrial effluents (Malik et al. 2014). Dumping of garbage and dead bodies, immersion of idols and pollution due to in-stream uses of water. The treated, untreated or partially treated sewage from these STPs (Sewage Treatment Plants) is discharged directly or through a carrier drain into the river (Bhardwaj et al. 2017).

Due to unavoidable reasons such as power failures, mechanical problems or maintenance issues, these STPs are unable to operate continuously. This poses a major threat to water quality, as the collected sewage is discharged into the river at a few locations without any treatment. Previous reports have shown that Yamuna river water fails to meet the Water Quality (WQ) (Sharma and Arun 2011; Sharma 2009; Singh et al. 2018; Sharma and Kansal 2011) standards prescribed by CPCB. Very few stations meet the criteria of outdoor bathing as far as BOD and coliform content is concerned. The Water Quality Parameters monitored before lockdown show how Yamuna water is a big threat to mankind causing water borne diseases, disrupting the aquatic ecosystem. Hence, it is mandatory that steps be taken to reduce the biological and metallurgical effluent load deposited into the river. With lockdown being declared due to COVID-19 pandemic human activities came to a standstill and the river finally managed to breathe after a long time.

Major activities such as tourism, fairs, bathing and cloth washing near the *ghats* were curtailed during the lockdown period, showing an improvement in rivers health. Our findings also draw further attention to what will happen after the lockdown period. The rivers are surrounded by densely populated areas. Although the socio-economic effect of COVID-19 spread will remain for some period, the pollutants will eventually start returning to the river from industrial activities. We tested this hypothesis using datasets (DO, BOD, Total Coliform, Fecal Coliforms) due to outbreak during February–March 2020, which may be attributed to the mass lockdowns. Our findings suggest that there is ample scope for restoring the global environment from the ill-effects of anthropogenic activities through temporary shutdown measures. In the name of economic growth, most rivers and streams have been turned into sewer canals and are getting difficult to be treated (Kumar et al. 2015; Reddy 2004; Rani et al. 2013; Muthaiyah 2020; Parween 2017).

The present paper is an attempt to evaluate the impact of the lockdown due to the COVID-19 spread on water quality. This paper attempts to understand the effect of lockdown due to COVID-19 on major water quality parameters DO, BOD, Total Coliforms, Feacal Coliforms. With tourism coming to a standstill, coliforms were found to decrease to a remarkable level, infact Prayagraj reported a good decline in variation in coliform content. (Sharifi and Khan 2020).

2 Study Area

The present study was performed to assess the changes in four water quality parameters for DO, BOD, total coliform, fecal coliform three months prior lockdown (January, February, March) and four months after (April, May, June, July). River Yamuna flows through major tourist towns in Uttar Pradesh, Agra, Mathura, Delhi, Prayagraj, Allahabad. Major waste dumping in river is from the NCR (National Capital Region), for the present study, total eighteen sampling sites (S1–S18) were selected, CPCB (Centre for Pollution Control board) have permanent monitoring stations at these sites, here they monitor Water Quality parameters regularly. The sampling sites and study area are mentioned in Table 1. S1 and S2 sites are located at Noida, Okhla Barrage and Village Gharbara, as such these are not tourist sites, nor any major industry here, but dumping of sewage waste takes place at these sites. S4 and S5 sites are located at Vrindawan a major tourist spot as per Hindu culture, sites (S6–S9) located at different spots at Mathura, birthplace of Lord Krishna, once again a major tourist spot for people around the world. Sites (S10–S12) are located at Agra, it is 206 kilometers south of the national capital New Delhi (Fig. 1, shows sampling sites).

Table 1 List of sampling sites along with their geospatial locations

S. no.	Site no	Site location	Latitude	Longitude
1	S1	U/S Okhla Barrage, Noida	28.5458 N	77.3147E
2	S2	D/S Village Gharbara/Tilwara, Noida	28.4081 N	77.4969E
3	S3	U/S Vrindavan	27.9144 N	77.5178E
4	S4	Kesi Ghat Vrindavan	27.5861 N	77.7008E
5	S5	D/s Vrindavan	27.5825 N	77.5178E
6	S6	U/s Mathura	27.5144 N	77.6808E
7	S7	Shahpur, Mathura	27.4947 N	77.6977E
8	S8	Vishram Ghat, Mathura	27.3686 N	77.7475E
9	S9	D/S Mathura	27.3058 N	77.8016E
10	S10	U/s Kailashghat, Agra	27.2361 N	78.2361E
11	S11	U/s Waterworks, Agra	27.0233 N	78.2508E
12	S12	D/s Tajmahal, Agra	27.1758 N	78.0841E
13	S13	u/s Firozabad	27.1247 N	78.3691E
14	S14	D/s Etawah	26.7505 N	78.0180E
15	S15	U/s Water Intake, Allahabad	25.4139 N	81.8681E
16	S16	D/S Balua Ghat Praygraj	25.4312 N	81.8680E
17	S17	D/s Chhachhar nala, Praygraj	25.4227 N	81.8433E
18	S18	D/s Emergency Outfall, Prayagraj	25.4252 N	81.8563E

Taj Mahal is the most visited tourist spot in India, attracting nearly 6.9 million visitors in 2018–19 (Sharma 2019). Pouring more load in river Yamuna because of massive tourists arrival. Site S13 is located at Firozabad is a city near Agra in the state of Uttar Pradesh in India. It is the centre of India's glassmaking industry, responsible for dumpage of industrial waste in the river. Site S14 is located at Etawah is a city on the banks of Yamuna River in the state of Western Uttar Pradesh in India. The city lies 300 km southeast of the national capital New Delhi, and 230 km northwest of the state capital Lucknow. Etawah is about 120 km east of Agra and is about 165 km west of Kanpur. Sites (S15–S18) are located at Allahabad officially known as Prayagraj, is a metropolis in the Indian state of Uttar Pradesh; Allahabad lies close to Triveni Sangam, the "three-river confluence" of the Ganges, Yamuna and Sarasvati rivers. It plays a central role in Hindu scriptures. A place of religious importance and the site for historic Kumbh Mela held every 12 years; over the years it has also been the site of immersion of ashes for Hindus (Martin et al. 2016; Singh 2019; Rout 2017). It attracts tourist around the year causing major pollution to the river. These sites under study count for highly polluted stretches of Yamuna. At few sites data could not be collected due to lockdown restriction.

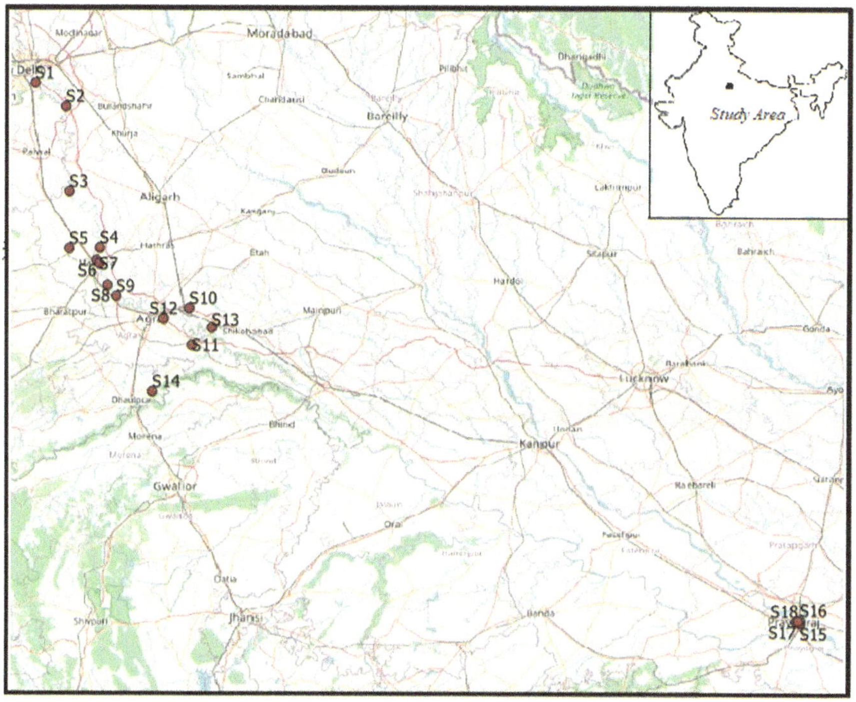

Fig. 1 Study area showing sampling sites

3 Methodology

Water samples were collected from all the selected eighteen sites (S1–S18) on monthly basis from January 2020 to July 2020. A total of four parameters DO (Dissolved Oxygen), BOD (Bio Oxygen Demand), Total Coliforms (TC) and Fecal Coliforms (FC) were analysed using standard methods as prescribed by (APHA 1997). Detail analysis can be found in CPCB manual.

3.1 Statistical Analysis

All values analyzed are presented in Tables 2, 3, 4 and 5 along with increase or decrease after lockdown compared to values before lockdown. Table 6 gives a summary of the analysis stating the count of sites where water has been found fit either for drinking or outdoor bathing after lockdown, in short the effects of lockdown have been analyzed for water quality of river Yamuna. All data analysis was done using Excel 2007 (Microsoft Office 2007, Microsoft Corp., USA).

Table 2 Water quality of River Yamuna in UP sites (1–5)

		Water quality of River Yamuna in UP								
Site					Year 2020					
1	**Month**	**January**	**Feb.**	**March**	**Before lockdown**	**April**	**May**	**June**	**July**	**After lockdown**
	DO (mg/l)	1.5	0.8	1	1.1	7.4	7.4	3.5	4.5	5.7
	BOD (mg/l)	30	24	22.5	25.5	12	12	18	15	14.25
2	**Month**	**January**	**Feb**	**March**	**Before lockdown**	**April**	**May**	**June**	**July**	**After lockdown**
	DO (mg/l)	0	0	0	0	2.8	2.8	1.8	0	1.85
	BOD (mg/l)	51	42	45	46	27	27	30	36	30
3	**Month**	**January**	**Feb**	**March**	**Before lockdown**	**April**	**May**	**June**	**July**	**After lockdown**
	DO (mg/l)	4.8	4.2	5.3	4.76666667	7.2	7.1	7.4	6.5	7.05
	BOD (mg/l)	8	9	10	9	6.6	6.6	7.8	8.2	7.3
	Total Coliform (MPN/100 ml)	_	62,000	70,000	66,000	49,000	46,000	68,000	72,000	58,750
	Fecal Coliform (MPN/100 ml)	_	49,000	63,000	56,000	0	33,000	46,000	49,000	32,000
4	**Month**	**January**	**Feb**	**March**	**Before lockdown**	**April**	**May**	**June**	**July**	**After lockdown**
	DO (mg/l)	4.4	4	5.2	4.53333333	7.3	6.9	7.2	6.1	6.875
	BOD (mg/l)	9	9.2	10.4	9.53333333	7.2	7.4	8	8.8	7.85
	Total Coliform (MPN/100 ml)	75,000	80,000	88,000	81,000	49,000	50,000	75,000	78,000	63,000
	Fecal Coliform (MPN/100 ml)	_	52,000	65,000	58,500	31,000	34,000	50,000	50,000	41,250
5	**Month**	**January**	**Feb**	**March**	**Before lockdown**	**April**	**May**	**June**	**July**	**After lockdown**
	DO (mg/l)	4.4	4	5.2	4.53333333	7.3	6.9	7.2	6.1	6.875
	BOD (mg/l)	9	9.2	10.4	9.53333333	7.4	7.4	8	8.8	7.9
	Total Coliform (MPN/100 ml)	75,000	62,000	88,000	75,000	54,000	50,000	75,000	78,000	64,250
	Fecal Coliform (MPN/100 ml)	58,000	52,000	65,000	58333.3333		34,000	50,000	50,000	44666.667

Table 3 Water quality of River Yamuna in UP sites (6–10)

		Water quality of River Yamuna in UP								
Site					Year 2020					
6	**Month**	**January**	**Feb**	**March**	**Before lockdown**	**April**	**May**	**June**	**July**	**After lockdown**
	DO (mg/l)	4.2	7.1	4.9	5.4	7	6.8	7.4	6.3	6.875
	BOD (mg/l)	9	7.6	11	9.2	8	7.8	8.4	9	8.3
	Total Coliform (MPN/100 ml)	98,000	80,000	92,000	90,000	70,000	67,000	80,000	85,000	75,500
	Fecal Coliform (MPN/100 ml)	75,000	72,000	78,000	75,000		46,000	54,000	54,000	51333.333
7	**Month**	**January**	**Feb**	**March**	**Before lockdown**	**April**	**May**	**June**	**July**	**After lockdown**
	DO (mg/l)	4.8	5.4	4.8	5	6.8	6.6	6.1	6.7	6.55
	BOD (mg/l)	8	7.6	10	8.53333333	8	9	9.8	9	8.95
	Total Coliform (MPN/100 ml)	110,000	98,000	110,000	106,000	80,000	78,000	85,000	88,000	82,750
	Fecal Coliform (MPN/100 ml)	80,000	72,000	85,000	79,000	54,000	63,000	54,000	65,000	59,000
8	**Month**	**January**	**Feb**	**March**	**Before lockdown**	**April**	**May**	**June**	**July**	**After lockdown**
	DO (mg/l)	4.1	6.9	4.6	5.2	7.4	6.5	7.2	6.5	6.9
	BOD (mg/l)	10	8	12	10	9.2	8.2	9.6	9.6	9.15
	Total Coliform (MPN/100 ml)	110,000	88,000	80,000	92666.6667	78,000	92,000	92,000	98,000	90,000
	Fecal Coliform (MPN/100 ml)	92,000	75,000	80,000	82333.3333	49,000	54,000	63,000	98,000	66,000
9	**Month**	**January**	**Feb**	**March**	**Before lockdown**	**April**	**May**	**June**	**July**	**After lockdown**
	DO (mg/l)	4	6.8	4.6	5.13333333	6.8	6.6	7	6.2	6.65
	BOD (mg/l)	10	8	11.6	9.86666667	9	8.4	9.2	9.8	9.1
	Total Coliform (MPN/100 ml)	140,000	110,000	130,000	126666.667	78,000	85,000	98,000	110,000	92,750
	Fecal Coliform (MPN/100 ml)	98,000	85,000	88,000	90333.3333		49,000	72,000	75,000	65333.333
10	**Month**	**January**	**Feb**	**March**	**Before lockdown**	**April**	**May**	**June**	**July**	**After lockdown**
	DO (mg/l)	4.7	4.7	5.4	4.93333333	6.3	7.8	7.1	7.2	7.1
	BOD (mg/l)	11.6	11.6	13.2	12.1333333	9.6	8.4	9.2	11.6	9.7
	Total Coliform (MPN/100 ml)	38,000	33,000	52,000	41,000	33,000	33,000	34,000	32,000	33,000
	Fecal Coliform (MPN/100 ml)	22,000	18,000	21,000	20333.3333	13,000	13,000	16,000	11,000	13,250

Table 4 Water Quality of River Yamuna in UP sites (11–14)

		Water quality of River Yamuna in UP								
Site					Year 2020					
11	**Month**	**January**	**Feb**	**March**	**Before lockdown**	**April**	**May**	**June**	**July**	**After lockdown**
	DO(mg/l)	4.1	4.3	5.2	4.53333333	5.7	7.3	6.7	6.9	6.65
	BOD(mg/l)	12.4	14.4	13.6	13.4666667	10.4	9.2	10.8	12.8	10.8
	Total Coliform(MPN/100 ml)	52,000	48,000	63,000	54333.3333	49,000	44,000	45,000	42,000	45,000
	Fecal Coliform(MPN/100 ml)	30,000	21,000	22,000	24333.3333	17,000	18,000	17,000	15,000	16,750
12	**Month**	**January**	**Feb**	**March**	**Before lockdown**	**April**	**May**	**June**	**July**	**After lockdown**
	DO(mg/l)	3.9	4	5.1	4.33333333	5.4	7.1	5.6	6.3	6.1
	BOD(mg/l)	14.8	16.4	14.4	15.2	11.2	10	11.6	14.8	11.9
	Total Coliform(MPN/100 ml)	110,000	110,000	110,000	110,000	89,000	79,000	78,000	70,000	79,000
	Fecal Coliform(MPN/100 ml)	35,000	44,000	46,000	41666.6667	52,000	46,000	43,000	28,000	42,250
13	**Month**	**January**	**Feb**	**March**	**Before lockdown**	**April**	**May**	**June**	**July**	**After lockdown**
	DO(mg/l)	_	_	5.3	5.3	7	6.6	5.6	6	6.3
	BOD(mg/l)	16.4	13.2	13.6	14.4	11.2	12.8	12.8	15.6	13.1
14	**Month**	**January**	**Feb**	**March**	**Before lockdown**	**April**	**May**	**June**	**July**	**After lockdown**
	DO(mg/l)	_	_	5.2	5.2	6.9	6.6	5.5	5.9	6.225
	BOD(mg/l)	15.2	19	16	16.7333333	11.6	13.2	15.6	16.4	14.2

Table 5 Water quality of River Yamuna in UP sites (15–18)

		Water quality of River Yamuna in UP								
Site					Year 2020					
15	**Month**	**January**	**Feb**	**March**	**Before lockdown**	**April**	**May**	**June**	**July**	**After lockdown**
	DO(mg/l)	10.6	9.3	8.5	9.46666667	9.3	8.4	8.9	7.6	8.55
	BOD(mg/l)	2	2.2	2	2.06666667	1.7	1.8	1.9	2	1.85
	Total Coliform(MPN/100 ml)	3400	2700	2500	2866.66667	1700	1400	1500	1700	1575
	Fecal Coliform(MPN/100 ml)	1300	780	910	996.666667	400	450	400	600	462.5
16	**Month**	**January**	**Feb**	**March**	**Before lockdown**	**April**	**May**	**June**	**July**	**After lockdown**
	DO(mg/l)	10.8	11.5	8.1	10.1333333	9.3	9.1	9	7.3	8.675
	BOD(mg/l)	2.1	2.3	2.4	2.26666667	2	2	1.8	2	1.95
	Total Coliform (MPN/100 ml)	3900	3300	3200	3466.66667	2100	1700	1700	1700	1800
	Fecal Coliform (MPN/100 ml)	1700	1700	1300	1566.66667	310	600	550	550	502.5
17	**Month**	**January**	**Feb**	**March**	**Before lockdown**	**April**	**May**	**June**	**July**	**After lockdown**
	DO (mg/l)	10.3	9.1	8	9.13333333	9.2	8.1	8.3	7.4	8.25
	BOD (mg/l)	2.2	2.6	2.3	2.36666667	2	2.1	2.2	2.3	2.15
	Total Coliform (MPN/100 ml)	4300	3400	2700	3466.66667	2100	2100	2000	2100	2075
	Fecal Coliform (MPN/100 ml)	2100	1700	1300	1700	310	680	610	610	552.5
18	**Month**	**January**	**Feb**	**March**	**Before lockdown**	**April**	**May**	**June**	**July**	**After lockdown**
	DO (mg/l)	10.1	9	7.8	8.96666667	9	7.9	8.2	7.3	8.1
	BOD (mg/l)	2.1	2.5	2.4	2.33333333	2	2.1	2	2.2	2.075
	Total Coliform (MPN/100 ml)	3900	3500	2600	3333.33333	2000	2000	1700	2000	1925
	Fecal Coliform (MPN/100 ml)	1700	1700	1100	1500	550	580	610	680	605

Table 6 Number of sites with improvement in water quality

Parameters	Fit for outdoor bathing (BL)	Fit for outdoor bathing (AL)	Fit for Drinking (BL)	Fit for Drinking (AL)
DO	10	17	4	16
BOD	4	4	0	3
FC	1	4	0	0

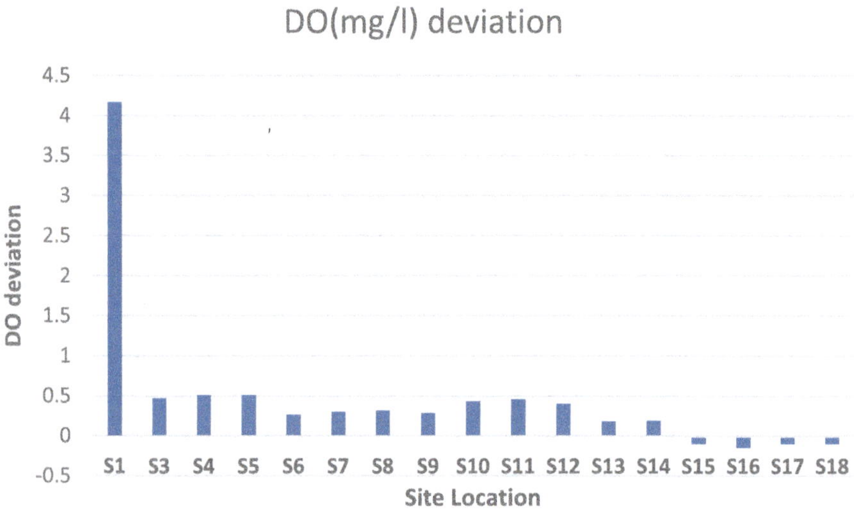

Fig. 2 Deviation of DO before and after lockdown

4 Results and Discussion

Dissolved Oxygen (DO) is an essential parameter in water quality assessment and reflects the physical and biological process prevailing in the water which indicates the degree of pollution in water bodies. During the study, the DO varied with a minimum value being recorded at Okhla Barrage, Noida 1.1 mg/l to a maximum value of 10.13 mg/l at Balua Ghat, Prayagraj, before lockdown, and a marked increase in values was observed after the lockdown values showed a major increase of 80.7% at Okhla Barrage, Noida creating history in itself. No study has reported such an abnormal value in DO due to any experiment or Action Plan which this lockdown did. With improved DO aquatic life got an opportunity to breathe. Figure 2 shows the deviation of Dissolved Oxygen before and after lockdown. While at other stations a good increase of approximately 30% to 35% was found at Vrindawan, and Agra, this could have happened because of the sudden dip in tourism at Agra and Vrindawan, due to sudden announcement of lockdown, while sampling station at Firozabad showed the least increase of approximately 16%.

While DO at four stations sampled at Prayagraj, Allahabad found a minor dip in value of approximately 9% to 10%. Highest decrease was at Balua Ghat, Prayagraj approximately 16%. Before lockdown 10 sites were found fit for outdoor bathing while after lockdown marked improvement in data was observed and 17 sites were found fit to meet criteria of outdoor bathing as prescribed by CPCB for DO. Table 6. Only four sites managed to meet the criteria of drinking water for DO before lockdown. But a major increase was found after lockdown with 16 sites meeting the criteria for drinking water as prescribed by CPCB for DO. No such study till date has reported such a massive improvement, showing the failure of big projects for cleaning the river.

Biochemical oxygen demand (BOD) is the amount of oxygen used by microorganism to decompose the organic matter (Jouanneau et al. 2014). It is an empirical standarised laboratory test which measures oxygen requirement for aerobic oxidation of decomposable organic matter and certain inorganic materials in water, polluted waters and wastewater under controlled conditions of temperature and incubation period. (Montgomery 1967).

The quantity of oxygen required for above oxidation is a measure of the test. BOD was sampled at 18 stations out of which 17 stations showed a decline only one station at Shahpur Mathura showed an increase of 4.88%. Maximum values recorded before lockdown were at Village Gharbara, Noida 46 mg/l followed by 25.5 mg/l at Okhla Barrage, Noida. Such values show that river cannot be considered a river at these places it seems to be more of a Nala at Noida. Vrindawan and Mathura recorded values in the range of 9 to 10 mg/l. While Agra, Firozabad and Etawah recorded values in the range of 12 to 16 mg/l. while Prayagraj recorded values of 2.26 mg/l. Once again the results showed the positive impact of COVID lockdown as BOD decline shows an increase in DO (The two bear an inverse relation). Figure 3 shows deviation of BOD before and after lockdown. Sampling station Okhla Barrage Noida, showed the highest decrease of 78.95%. D/S Village Gharbara/Tilwara, Noida showed a decrease of 53.3%. A good decrease of approximately 30% was found at three sampling stations in Agra. Four stations of Prayagraj, Allahabad showed a decrease of10–16%. Below 10% decrease was found at Vishram Ghat, Mathura, D/S Mathura (Fig. 3).

Total Coliforms Coliform bacteria include all aerobic and facultative anaerobic gram negative, non-sore forming, rod-shaped bacteria that produce gas upon fermentation in prescribed culture media within 48 hr. at 35 °C. (Kocatepe et al. 2016) Presence of these bacteria presents significant health hazards. Though CPCB has prescribed a limit for presence of these coliforms between 500 to 2500 MPN/100 ml (Table 2) for bathing purpose and 50 or less MPN/100 ml for drinking purpose (Table 1) none of the sampling stations managed to meet the criteria for drinking water before and after lockdown. But four sampling stations at Prayagraj (Allahabad; S15–S18) managed to meet the criteria of water being fit for outdoor bathing before and after lockdown with coliform content declining massively after lockdown to 68% as shown in Fig. 4. Analysis for Total Coliform showed very high count in fact very hazardous for meeting any criteria before lockdown. Such high values show water being highly unfit for any purpose. The highest value was observed at

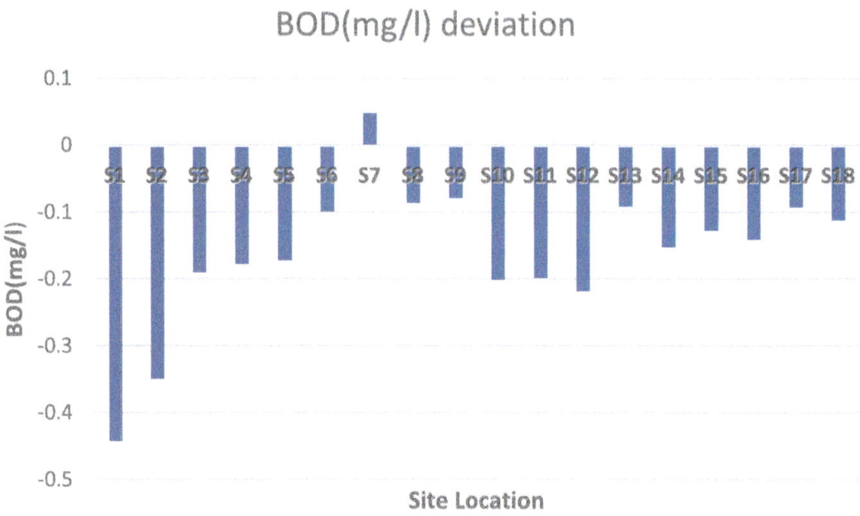

Fig. 3 Deviation of BOD before and after lockdown

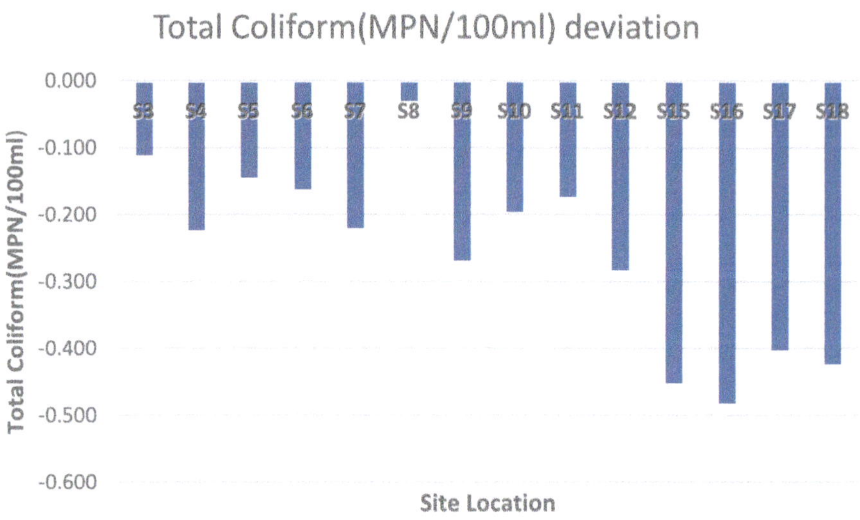

Fig. 4 Deviation of total Coliforms before and after lockdown

Mathura, 126,666 MPN/100 ml before lockdown, which declined to 92750MPN/100 ml after lockdown. This could be due to stoppage of sewage discharge, industrial discharge and subdued tourist activities during lockdown.

Highest decrease was found at Balua Ghat, Prayagraj approximately 93%. Seems COVID lockdown measures were followed here with full strictness (Fig. 4). Also Prayagraj being a major tourist attraction because of this river found a dip in tourists visiting the ghats and polluting the same, people using less of the river side to

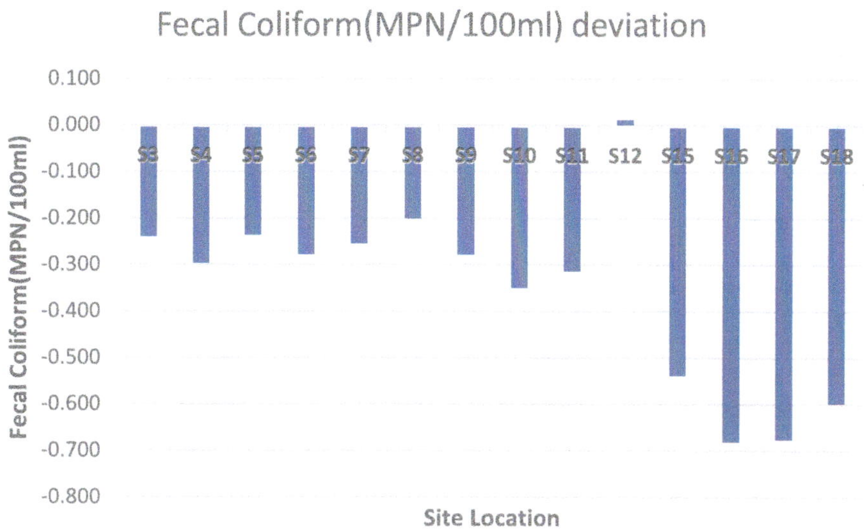

Fig. 5 Variation of Fecal Coliform before and after lockdown

discharge wastes are the major reasons for this result. No study in literature has ever recorded such results, a 93% decline in Coliform content after lockdown should be an eye opener for government to impose partial lockdowns in future for survival of river. Followed by the Water Intake station, Allahabad showing a decline of 82%. While Emergency outfall station Prayagraj showed a decline of 73.16%, followed by Chhachhar Nala, Prayagraj of 67%. Sampling stations of Vrindawan and Mathura also showed decline of 12 to 28%. The lowest decline was observed at Vishram Ghat, Mathura of 2.96%.

Fecal coliform: commonly used as an indicator of water contamination and possible presence of pathogenic microbes. Fecal coliforms particularly originate from the intestinal tract of warm-blooded animals and are mostly associated with sewage (fecal matter). Most of the drains receive household waste, agricultural wastes, hospital wastes but the substantial part of it is sewage. The abundance of fecal coliform varied in the samples collected from the sites. Lowest values were recorded at Emergency outfall Prayagraj. In fact, water was found meeting quality norms for outdoor bathing and decline was highest at Praygraj approximately 68% after lockdown. Before lockdown highest count was obtained from Mathura 90333MPN/100 ml before lockdown this value showed a decline of 38,27% after lockdown. Stations at Mathura showed very high values before lockdown showing the poor maintenance of sewage system at the place. Fourteen stations were sampled for fecal coliform all showed a decline except Taj Mahal station at Agra which showed a minor increase of 1.38% in fecal coliform content after lockdown. Highest decline was observed at Prayagraj, 70%. A good decline compared to usual days shown in Fig. 5 depicting variation of fecal coliform before and after lockdown. Followed by Chhachharnala, Prayagraj showing a decline of 67.5%, followed by

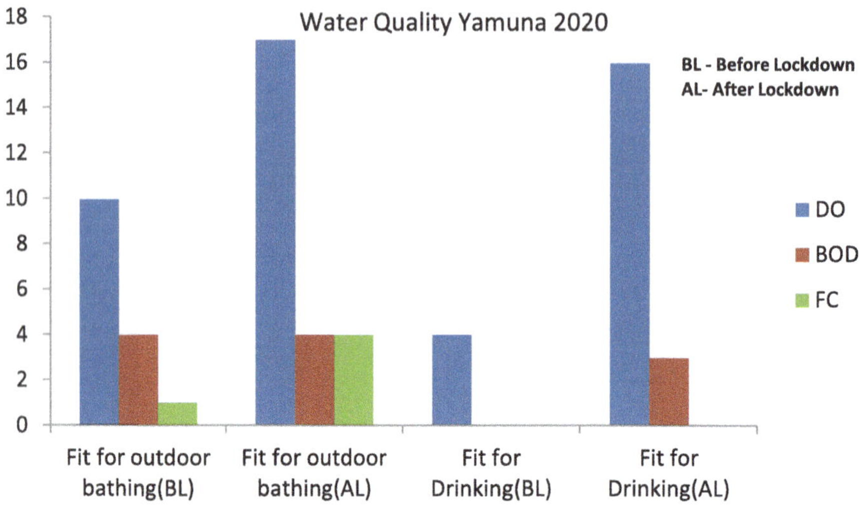

Fig. 6 Summary of water quality improvement after lockdown

station Emergency Outfall, Prayagraj showing a decline of 60%. Two other sampling stations of Agra showed a decline of 45% and above. While Mathura and Vrindawan stations showed a decline in between 25 to 40%.

No literature survey to date has reported such excellent improvement in Water Quality Parameters due to any type of experimental or river project for cleaning the river, which this lockdown showed. Results of coliform also conclude that it is high time the government takes strict measures to stop the discharge of sewage waste in rivers. All parameters showed improvement as shown in Fig. 6 after lockdown even concluding that water is fit for drinking purpose as far as DO and BOD are concerned (Table 6).

The overall summary of water quality of Yamuna River could be classified as excellent after lockdown for DO and Coliforms. The results clearly indicate that rivers overall pollution has decreased to alarming levels after lockdown. Hence, we conclude that the river falls in good condition in terms of DO, BOD, and coliforms after lockdown.

5 Conclusion and Recommendations

With all major polluting industries being closed, the lockdown significantly reduced the toxic load in the river. A massive dip in the number of visitors at ghats in Mathura, Vrindawan and Prayagraj also helped improve the river water quality.

Although the partial lockdown has contributed to a positive impact on water quality, our findings suggest that there is ample scope for restoring the global environment from the ill-effects of anthropogenic activities through temporary shutdown measures. This would ultimately help in improving the quality of river water and result in its healthy use. The outcome of the present study needs careful monitoring of tourists at Prayagraj, Agra, major tourist attractions, dumpage of industrial and sewage waste in NCR for the river to survive. This study highlights the fact, that the cleanliness of rivers lies in the hands of human beings; no doubt several issues with lockdown will come up, especially religious, but, before the river vanishes and water becomes a scarcity it is high time strict measures are taken.

5.1 Data Availability Statement

Major datasets presented in this study can be found online on UPPCB and CPCB sites.

References

American Public Health Association (1997) Standard method for the examination of waste water, 19th edn. American Public Health Association, Washington, DC

Bhardwaj R, Gupta A, Garg KJ (2017) Evaluation of heavy metal contamination using environ metrics and indexing approach for River Yamuna, Delhi stretch, India. Water Res 31:52–66. https://doi.org/10.1016/j.wsj.2017.02.002

Datta C (1992) Yamuna River turned sewer. Econ Polit Wkly 27:2633–2636

Gopal B, Chauhan M (2007) River Yamuna from source to Delhi: human impacts and approaches to conservation. In: Martin P, Gopal B, Southey C (eds) Restoring River Yamuna: concepts, strategies and socio-economic considerations. National Institute of Ecology, New Delhi

Jouanneau S, Recoules L, Durand MJ, Boukabache A, Picot V, Primault Y, Lakel A, Sengelin M, Barillon B, Thouand G (2014) Methods for assessing biochemical oxygen demand (BOD): A review. Water Res 49:62–82. https://doi.org/10.1016/j.watres.2013.10.066

Kaur L, Rishi MS, Arora NK (2021) Deciphering pollution vulnerability zones of River Yamuna in relation to existing land use land cover in Panipat, Haryana, India. Environ Monit Assess 193: Article 120. https://doi.org/10.1007/s10661-020-08832-y

Kocatepe T, Taskaya G, Turan H (2016) Microbiological investigation of wild, cultivated mussels (Mytilus galloprovincialis L. 1819) and stuffed mussels in Sinop–Turkey. Ukr Food J 5:299–305. https://doi.org/10.24263/2304-974X-2016-5-2-9

Kumar P, Kaushal RK, Nigam AK (2015) Assessment and management of Ganga river water quality using multivariate statistical techniques in India. Asian J Water Environ Pollut 12:61–69. https://doi.org/10.3233/AJW-150018

Lamba M, Sreekrishnan TR, Shaikh Z (2020) Sewage mediated transfer of antibiotic resistance to River Yamuna in Delhi, India. J Environ Chem Eng 8:102088. https://doi.org/10.1016/j.jece.2017.12.041

Malik D, Singh S, Thakur J, Singh RK, Kaur A, Nijhawan S (2014) Heavy metal pollution of the Yamuna River: an introspection. Int J Curr Microbiol Appl Sci 3:856–863

Martin NH, Trmčić A, Hsieh T-H, Boor KJ, Wiedmann M (2016) The evolving role of coliforms as indicators of unhygienic processing conditions in dairy foods. Front Microbiol 7:1549. https://doi.org/10.3389/fmicb.2016.01549

Misra AK (2010) A river about to die: Yamuna. J Water Resour Prot 2:489–500. https://doi.org/10.4236/jwarp.2010.25056

Montgomery AC (1967) The determination of biochemical oxygen demand by respirometric methods. Water Res 1:631–662

Muthaiyah NP (2020) Rejuvenating Yamuna River by wastewater treatment and management. Int J Energy Environ Eng 5:14–29. https://doi.org/10.11648/j.ijees.20200501.13

Parween M (2017) Waste water management and water quality of river Yamuna in the megacity of Delhi. Int J Environ Sci Technol 14(1):1280–1288. https://doi.org/10.1007/s13762-017-1280-8

Rai OP (2015) Pollution and conservation of river Yamuna. Chhattisgarh Law J 2:150–152

Rani M, Akolkar P, Bhamrah HS (2013) Water quality assessment of River Yamuna from origin to confluence to River Ganga, with respect to biological water quality and primary water quality criteria. J Entomol Zool Stud 1:1–6

Reddy SK (2004) Water pollution and the law. Ind Judicial Rev 1:190–194

Rout C (2017) Assessment of water quality: A case study of river Yamuna. Int J Earth Sci Eng 10: 398–403

Sharifi A, Khan R (2020) The COVID-19 pandemic: impacts on cities and major lessons for urban planning, design, and management. Sci Total Environ 749:1–14. https://doi.org/10.1016/j.scitotenv.2020.142391

Sharma PM (2009) Water quality profile of Yamuna River, India. Hydro Nepal J Water Ener Environ 3:1–7

Sharma K (2015) Pollution study of river Yamuna: the Delhi story. Int J Sci Res 6:1718–1622

Sharma A (2019) Tourists up at Taj Mahal and Red Fort but Qutub Minar loses its No. 2 Spot. The Economic Times, 10 July 2019

Sharma D, Arun K (2011) Water quality analysis of River Yamuna using water quality index in the national capital territory, India (2000–2009). Appl Water Sci 1(3):147–157

Sharma P, Bhadauriya M (2019) Yamuna River water pollution: A review. Int J Eng Sci Comput 9(5):21906–21910

Sharma D, Kansal A (2011) Water quality analysis of river Yamuna using water quality index in the national capital territory, India (2000-2009). Appl Water Sci 1:147–157

Singh R (2019) A comparative statistical study on water pollution between the Ganga and the Yamuna Rivers. Int J Eng Res Technol 8:899–902

Singh B, Khan S, Shahjahan (2018) Assessment of water quality on the Yamuna river using principle component analysis: A case study. Int J Res Anal Rev 5:951–954

Suruchi, Ratnani S, Gurjar S, Manish (2015) Study of pollution level in Yamuna River due to untreated Delhi drains. Int J Sci Res 4:587–588. https://doi.org/10.21474/IJAR01/3131

The manufacturer's authorised representative in the EU is Springer Nature Customer Service Centre GmbH, Europaplatz 3, 69115 Heidelberg, Germany. If you have any concerns regarding our products, please contact ProductSafety@springernature.com

Printed and bound by CPI Group (UK) Ltd, Croydon, CR0 4YY
30/09/2026
02225244-0016